Microbial Allies

Olubukola Oluranti Babalola
Ayansina Segun Ayangbenro
Editors

Microbial Allies

Understanding Plant-Microbe Interactions in Sustainable Agriculture

Springer

Editors
Olubukola Oluranti Babalola
Food Security and Safety Focus Area
Faculty of Natural and Agricultural Sciences
North-West University
Mmabatho, South Africa

Department of Life Sciences
Imperial College London
Ascot, Berkshire, UK

Ayansina Segun Ayangbenro
Food Security and Safety Focus Area
Faculty of Natural and Agricultural Sciences
North-West University
Mmabatho, South Africa

ISBN 978-3-031-90529-2 ISBN 978-3-031-90530-8 (eBook)
https://doi.org/10.1007/978-3-031-90530-8

This Springer imprint is published by the registered company Springer Nature Switzerland AG
The registered company address is: Gewerbestrasse 11, 6330 Cham, Switzerland

Preface

Crop production faces challenges from soil nutrient depletion, soil degradation, and climate change. Conventional farming practices, heavily reliant on chemical inputs, have impaired soil, water, and ecosystems, putting food production at the precipice. This book, "Microbial Allies: Understanding Plant-Microbe Interactions in Sustainable Agriculture," seeks to shed light on the transformative potential of beneficial microbes in environmental sustainability and agricultural productivity.

A predicted global population of nearly 10 billion by 2050 means global demand for food and energy is expected to grow exponentially. Yet, the planet's ability to meet this demand, arable land, freshwater, and biodiversity, is directly threatened by climate change. Climate change exacerbates these problems adding frequency of drought and floods and higher temperatures that compromise crop production across the globe. Accelerating the microbial transformation is therefore imperative not just to combat climate change but also to ensure humanity can feed itself.

In the face of these challenges, current practices that are dependent on chemical inputs are unsustainable. The long-term use of synthetic fertilizers and pesticides have led to water pollution, soil degradation, loss of biodiversity, and the emergence of resistant pests and pathogens. Furthermore, these practices increase greenhouse gas emissions, exacerbating climate change problem. Thus, a paradigm shift is required to fulfil the increasing demand for food required by a growing global population. The functional potential of the plant-associated microorganisms and its innovation should be used in crop production.

Microorganisms have played crucial roles in sustainable agriculture by influencing the overall health of the soils, supporting plant growth, and contributing to stress management. This book sheds light on the intricate potential of microbes in sustainable crop production. In this way, by examining these interactions in detail, it is possible to uncover the possibilities for improving the yields of agricultural crops. It initiates with a discussion on how plants interact with pathogens. Therefore, there is a need to develop mechanisms that would enhance the sustainable management of plant diseases. These fundamentals create the foundation upon which the reader can understand how and in what ways those microbes can be beneficial in maintaining the health of plants.

The book equally explores how microorganisms facilitate nutrient cycling, reduce biotic stresses, control hormones, and encourage plant growth. These organisms can be leveraged to improve efficiency of nitrogen and phosphorus nutrient utilizing microbes, decrease the use of synthetic chemicals, and increase plants' and crop resistance to biotic and abiotic stress.

This book has been organized into four main sections, starting from soil microbial diversity and its role in agricultural ecosystems (Part I), to advancing plant growth and stress management (Part II), plant protection and pathogen management (Part III), and innovations in microbial applications and technologies (Part IV). This book is a call to action for researchers, agriculturists, politicians, and other stakeholders to take a revolutionary step toward a more sustainable and regenerative approach to agriculture. It is not just a compilation of scientific facts. It promotes a paradigm shift in order to address the global concerns of environmental sustainability and food security by utilizing microbial allies.

Mmabatho, South Africa Olubukola Oluranti Babalola
Mmabatho, South Africa Ayansina Segun Ayangbenro

Contents

Editors and Contributors

About the Editors

Olubukola Oluranti Babalola (Pr.Sci.Nat, MASSAF, FASLP, FTWAS, FAS, FAAS, FISC), a Provost Visiting Professor at Imperial College London, the Vice President of both the Organization for Women in Science for the Developing World (OWSD) and The World Academy of Sciences (TWAS). She is a National Research Foundation (NRF)-rated B2 internationally acclaimed scientist with over 21 years of research experience focusing on rhizosphere metagenomics, and holds an MBA in General Management. Without reservation, the former Research Director of the Food Security & Safety Focus Area at North-West University (NWU), South Africa, is still leading as the Head and Principal Investigator of a Microbial Ecology Research laboratory. Olubukola is a member of the editorial board for *Microbiology Spectrum*, *Applied and Environmental Microbiology* (American Society for Microbiology, USA), *BMC Microbiology* (Elsevier), Transgenic Research (Springer), and *Biochemistry and Biophysics Reports* (Elsevier). She is an International Advisor to F1000Research Plant Science Gateway, Nano-Horizon and Heliyon (Cell Press). Olubukola has received numerous awards, including being a finalist for the GenderInSite 2020 and a Clarivate Highly Cited Research top 1% honoree in the field for 2022 and 2023. Recently, she received the 2024 South African Women in Science Awards (SAWiSA). Olubukola joined NWU in 2009; at NWU, Olubukola is passionate about capacity building and has graduated 37 doctoral fellows, 25 master's, and numerous Honors students. Several of her PhD students are postdocs or staff in America, Australia, Belgium, Canada, Italy, Sweden, and the UK and scattered across Africa. Prof. Babalola is a prolific author with over 500 publications. She is an alumnus of the American Association for the Advancement of Science (AAAS)-TWAS Science Diplomacy program, having gained valuable insight into policymaking, diplomacy, and diplomatic engagement, which has impacted the world. She is taking action using the gender lens and forging collaborations with diplomats. She is a fellow of the Nigerian Academy of

Science (NAS), The World Academy of Sciences (TWAS), and the Academy of Science of South Africa (ASSAF) and an honorary fellow of the International Science Council (ISC). Her wealth of international experience spans the Americas, Asia, Europe, and Oceania, encompassing 64 countries. She has international collaborations globally, research grants, and many awards. Olubukola has an H-index of 79, with over 30,000 citations.

Ayansina Segun Ayangbenro is an extraordinary Senior Lecturer at North-West University, South Africa, and a Y2-rated researcher recognized by the National Research Foundation (NRF). His academic journey is marked by significant achievements, including being a recipient of TWAS/NRF African Renaissance Doctoral Scholarship NRF Postdoctoral Fellowship Award, and Académie de recherche et d'enseignement supérieur Coopération au développement (ARES) Postdoctoral Research Fellowship Award. His research interest lies at the intersection of microbiology, environmental science, and sustainable agricultural practices, addressing critical environmental challenges such as pollution, biodiversity preservation, and climate change mitigation in crop production. Through his expertise, he explores microbial solutions for environmental sustainability, particularly in soil health management and bioremediation strategies. Ayansina has authored and co-authored many peer-reviewed journal articles published in reputable national and international journals.

Contributors

Abisola E. Ajakaye Department of Public Health, University of Suffolk, Ipswich, UK

Pharmacy Department, Camp David Pharmacy and Stores, Abeokuta, Nigeria

Akinlolu Olalekan Akanmu Food Security and Safety Focus Area, Faculty of Natural and Agricultural Sciences, North-West University, Mmabatho, South Africa

Sulistya Ika Akbari Department of Biology, Faculty of Mathematics and Natural Sciences, Universitas Padjadjaran, Jatinangor, West Java, Indonesia

Saheed A. Akinola Food Security and Safety Focus Area, Faculty of Natural and Agricultural Sciences, North-West University, Mmabatho, South Africa

Department of Microbiology and Parasitology, School of Medicine and Pharmacy, College of Medicine and Health Sciences, University of Rwanda, Butare, Rwanda

Ashim Das Astapati Department of Botany, Gurucharan College, Silchar, Assam, India

Theresa Abimbola Awotundun Department of Microbiology and Biotechnology, Faculty of Natural Sciences, Ajayi Crowther University, Oyo, Oyo State, Nigeria

Ayansina Segun Ayangbenro Food Security and Safety Focus Area, Faculty of Natural and Agricultural Sciences, North-West University, Mmabatho, South Africa

School of Agricultural Sciences, Faculty of Agriculture and Natural Sciences, University of Mpumalanga, Mbombela, 1200, South Africa

Modupe S. Ayilara Food Security and Safety Focus Area, Faculty of Natural and Agricultural Sciences, North-West University, Mmabatho, South Africa

Department of Biotechnology, Environmental Pollution Science and Technology Limited, Ido-Ijesha, Nigeria

Mohammed Baala Phytopathology Unit, Department of Plant Protection, Ecole National of Agriculture Meknes, Meknes, Morocco

Olubukola Oluranti Babalola Food Security and Safety Focus Area, Faculty of Natural and Agricultural Sciences, North-West University, Mmabatho, South Africa

Department of Life Sciences, Silwood Park Campus, Imperial College London, Buckhurst road, Ascot, Berkshire, SL5 7PY, United Kingdom

Kolawole Banwo Department of Microbiology, University of Ibadan, Ibadan, Nigeria

Zineb Belabess Plant Protection Laboratory, Regional Center of Agricultural Research of Meknes, National Institute of Agricultural Research, Meknes, Morocco

Oumaima Benaissa Phytopathology Unit, Department of Plant Protection, Ecole National of Agriculture Meknes, Meknes, Morocco

Debasish Bhattacharjee Department of Zoology, Gurucharan College, Silchar, Assam, India

Cristina Cruz Department of Plant Biology, Faculdade de Ciências, Universidade de Lisboa, Lisbon, Portugal

Homero Reyes de la Cruz Instituto de Investigaciones Químico-Biológicas, Universidad Michoacana de San Nicolás de Hidalgo, Morelia, Michoacán, Mexico

Sangeeta Dey Department of Ecology and Environment Science, Gurucharan College, Silchar, Assam, India

Febri Doni Department of Biology, Faculty of Mathematics and Natural Sciences, Universitas Padjadjaran, Jatinangor, West Java, Indonesia

Department of Global Development, Cornell University, Ithaca, USA

Olabode O. Efunwoye Department of General Studies, Federal College of Agriculture, Akure, Ondo State, Nigeria

Moussa El Jarroudi Water, Environment and Development Unit, SPHERES Research Unit, Department of Environmental Sciences and Management, University of Liège, Arlon, Belgium

Ben Jesuorsemwen Enagbonma Food Security and Safety Focus Area, Faculty of Natural and Agricultural Sciences, North-West University, Mmabatho, South Africa

Rachid Ezzouggari Phytopathology Unit, Department of Plant Protection, Ecole National of Agriculture Meknes, Meknes, Morocco

Ayomide Emmanuel Fadiji Food Security and Safety Focus Area, Faculty of Natural and Agricultural Sciences, North-West University, Mmabatho, South Africa

Hawkesbury Institute for the Environment, Western Sydney University, Penrith, NSW, Australia

Camilla Fagorzi Department of Biology, University of Florence, Florence, Italy

Abdelaaziz Farhaoui Phytopathology Unit, Department of Plant Protection, Ecole National of Agriculture Meknes, Meknes, Morocco

Racheal Oluwayemisi Fashogbon Department of Microbiology and Biotechnology, Faculty of Natural Sciences, Ajayi Crowther University, Oyo, Oyo State, Nigeria

Samuel A. Fasusi Department of Crop Science and Biotechnology, College of Bioresource Science, Dankook University, Cheonan, South Korea

Alessandra Pereira Fávero Embrapa Pecuária Sudeste, São Carlos, Brazil

Ana Carolina Vasconcelos Fernandes Laboratory of Microbiology and Biomolecules, Department of Morphology and Pathology, Federal University of São Carlos, São Carlos, Brazil

Biotechnology Graduate Program, Federal University of São Carlos, São Carlos, Brazil

Reza Ghasemi-Fasaei Department of Soil Science, School of Agriculture, Shiraz University, Shiraz, Iran

Youness Ighissane Phytopathology Unit, Department of Plant Protection, Ecole National of Agriculture Meknes, Meknes, Morocco

Emmanuel Edoghogho Imade Food Security and Safety Focus Area, Faculty of Natural and Agricultural Sciences, North-West University, Mmabatho, South Africa

Mohammed Khadiri Phytopathology Unit, Department of Plant Protection, Ecole National of Agriculture Meknes, Meknes, Morocco

Laboratoire de Biotechnologies Microbiennes et Protection des Végétaux, Faculté des Sciences, Université Ibn Zhor, Agadir, Morocco

Mohamed Kouighat Research Unit of Plant Breeding and Plant Genetic Resources Conservation, Regional Agricultural Research Center of Meknes, National Institute of Agricultural Research, Rabat, Morocco

Paulo Teixeira Lacava Laboratory of Microbiology and Biomolecules, Department of Morphology and Pathology, Federal University of São Carlos, São Carlos, Brazil

Biotechnology Graduate Program, Federal University of São Carlos, São Carlos, Brazil

Rachid Lahlali Phytopathology Unit, Department of Plant Protection, Ecole National of Agriculture Meknes, Meknes, Morocco

Adedayo Ayodeji Lanrewaju Department of Biotechnology and Food Science, Faculty of Applied Science, Durban University of Technology, Durban, South Africa

Jesús Salvador López-Bucio Catedrático CONACYT-Instituto de Investigaciones Químico-Biológicas, Universidad Michoacana de San Nicolás de Hidalgo, Morelia, Michoacán, Mexico

José López-Bucio Instituto de Investigaciones Químico-Biológicas, Universidad Michoacana de San Nicolás de Hidalgo, Morelia, Michoacán, Mexico

José López-Hernández Instituto de Investigaciones Químico-Biológicas, Universidad Michoacana de San Nicolás de Hidalgo, Morelia, Michoacán, Mexico

Jorge Martín-García Recognised Research Group AGROBIOTECH, UIC-370 (JCyL), Department of Plant Production and Forest Resources, Higher Technical School of Agricultural Engineering of Palencia, University Institute for Research in Sustainable Forest Management (iuFOR), University of Valladolid, Palencia, Spain

Alessio Mengoni Department of Biology, University of Florence, Florence, Italy

Mia Miranti Department of Biology, Faculty of Mathematics and Natural Sciences, Universitas Padjadjaran, Jatinangor, West Java, Indonesia

Muhamad Shakirin Mispan Institute of Biological Sciences, Faculty of Science, Universiti Malaya, Kuala Lumpur, Malaysia

David Mxolisi Modise Faculty of Natural and Agricultural Sciences, North West University, Potchefstroom, South Africa

Ava Mohrazi Department of Soil Science, School of Agriculture, Shiraz University, Shiraz, Iran

Soumitra Nath Bioinformatics Centre, Gurucharan College, Silchar, Assam, India

Department of Biotechnology, Gurucharan College, Silchar, Assam, India

Afolake Atinuke Olanbiwoninu Department of Microbiology and Biotechnology, Faculty of Natural Sciences, Ajayi Crowther University, Oyo, Oyo State, Nigeria

Omowunmi R. Oluwole Department of General Studies, Federal College of Agriculture, Akure, Ondo State, Nigeria

Iacopo Passeri Department of Biology, University of Florence, Florence, Italy

Isadora Camargo Pedrino Laboratory of Microbiology and Biomolecules, Department of Morphology and Pathology, Federal University of São Carlos, São Carlos, Brazil

Biotechnology Graduate Program, Federal University of São Carlos, São Carlos, Brazil

Bukola Margaret Popoola Department of Microbiology and Biotechnology, Faculty of Natural Sciences, Ajayi Crowther University, Oyo, Oyo State, Nigeria

Jorge Poveda Recognised Research Group AGROBIOTECH, UIC-370 (JCyL), Department of Plant Production and Forest Resources, Higher Technical School of Agricultural Engineering of Palencia, University Institute for Research in Sustainable Forest Management (iuFOR), University of Valladolid, Palencia, Spain

Dedat Prismantoro Department of Biology, Faculty of Mathematics and Natural Sciences, Universitas Padjadjaran, Jatinangor, West Java, Indonesia

Shiying Qu Recognised Research Group AGROBIOTECH, UIC-370 (JCyL), Department of Plant Production and Forest Resources, Higher Technical School of Agricultural Engineering of Palencia, University Institute for Research in Sustainable Forest Management (iuFOR), University of Valladolid, Palencia, Spain

Mohammed Radi Phytopathology Unit, Department of Plant Protection, Ecole National of Agriculture Meknes, Meknes, Morocco

Everlon Cid Rigobelo Agricultural and Livestock Microbiology Postgraduate Program, School of Agricultural and Veterinarian Sciences, São Paulo State University (UNESP), Jaboticabal, SP, Brazil

Nia Rossiana Department of Biology, Faculty of Mathematics and Natural Sciences, Universitas Padjadjaran, Jatinangor, West Java, Indonesia

Oyindamola John Samson Department of Microbiology and Biotechnology, Faculty of Natural Sciences, Ajayi Crowther University, Oyo, Oyo State, Nigeria

Óscar Santamaría Recognised Research Group AGROBIOTECH, UIC-370 (JCyL), Department of Plant Production and Forest Resources, Higher Technical School of Agricultural Engineering of Palencia, University Institute for Research in Sustainable Forest Management (iuFOR), University of Valladolid, Palencia, Spain

Dhrubjyoti Sharma Bioinformatics Centre, Gurucharan College, Silchar, Assam, India

Deepti Mala Singha Department of Ecology and Environment Science, Gurucharan College, Silchar, Assam, India

Rizky Riscahya Pratama Syamsuri Doctorate Program in Biotechnology, Graduate School, Universitas Padjadjaran, Bandung, West Java, Indonesia

Mohammed Taoussi Phytopathology Unit, Department of Plant Protection, Ecole National of Agriculture Meknes, Meknes, Morocco

Francesca Vaccaro Department of Biology, University of Florence, Florence, Italy

List of Figures

List of Tables

Part I
Soil Microbial Diversity and Its Role in Agricultural Ecosystems

Chapter 1
Microbial Diversity in Agricultural Soils: Implications for Soil Functionality and Productivity

Ayansina Segun Ayangbenro, Akinlolu Olalekan Akanmu, Ayomide Emmanuel Fadiji, and Olubukola Oluranti Babalola

Abstract Soil is a dynamic and complex biological system. Mineral mining, continuous cropping, soil erosion and unsuitable land-use practices have mostly caused the loss of fertile soil. Because these soils are intensively used for crop production, their productivity and functionality are severely impaired. Soil degradation is a continuous process, and the consequences of climate change exacerbate it. Climate change and environmental pressures are particularly harmful to soils and crops. Soil microbial diversity and ecosystem services offered by soils are under threat in many agro-ecosystems due to various natural and human factors. The loss of biodiversity has become a worldwide issue as evidence mounts that it will negatively impact ecosystem services. Modern agronomic methods that disrupt soil trophic networks and reduce long-term soil productivity are common in agricultural soils. The use of

A. S. Ayangbenro
Food Security and Safety Focus Area, Faculty of Natural and Agricultural Sciences, North-West University, Mmabatho, South Africa

School of Agricultural Sciences, Faculty of Agriculture and Natural Sciences , University of Mpumalanga, Mbombela 1200, South Africa

A. O. Akanmu
Food Security and Safety Focus Area, Faculty of Natural and Agricultural Sciences, North-West University, Mmabatho, South Africa

A. E. Fadiji
Food Security and Safety Focus Area, Faculty of Natural and Agricultural Sciences, North-West University, Mmabatho, South Africa

Hawkesbury Institute for the Environment, Western Sydney University, Richmond, NSW, Australia

O. O. Babalola (✉)
Food Security and Safety Focus Area, Faculty of Natural and Agricultural Sciences, North-West University, Mmabatho, South Africa

Department of Life Sciences, Silwood Park Campus, Imperial College London, Buckhurst road, Ascot, Berkshire SL5 7PY, United Kingdom
e-mail: olubukola.babalola@nwu.ac.za

O. O. Babalola, A. S. Ayangbenro (eds.), *Microbial Allies*,
https://doi.org/10.1007/978-3-031-90530-8_1

toxic agrochemicals, mono-cropping, extensive tillage, and other agricultural management practices affects soil microbial community structure and soil sustainability. As a result, changes occur in the soil microbial community structure, affecting the physical, chemical, and biological characteristics of the soil. In this chapter, we discuss the importance of soil microbial diversity and its role in soil quality and fertility, the contribution of soil microbes to soil structure, nutrient acquisition and circulation, and pollutant degradation.

Keywords Climate change · Ecosystem function · Food production · Land degradation · Loss of biodiversity · Structural diversity

1.1 Introduction

Two of the most severe global issues affecting terrestrial ecosystems are land degradation and biodiversity loss (Akhtar-Schuster et al. 2017). Approximately 23% of the world's terrestrial area is currently degraded, with 5–10 million extra hectares degraded each year, and about 1.5 billion people worldwide are significantly impacted by land degradation (Barbier and Hochard 2016; Muñoz-Rojas 2018). To maintain ecosystem function and productivity, limit climate change, safeguard biodiversity, and ensure food production and resource provision, initiatives to reduce degradation and restore degraded areas are critically needed (Muñoz-Rojas 2018). Most of the functions of soil ecosystem are difficult to examine directly; therefore, they are inferred from measurable qualities, such as soil quality indicators, which include various biological, chemical, and physical characteristics.

The maintenance of different ecosystem processes relies heavily on complex and diverse soil microbial populations (Han et al. 2022). Soil microorganisms constitute a large portion of the earth's biodiversity and play an important role in biogeochemical cycles and ecosystem function. Soil microbiome plays a critical role in nutrient cycling, litter decomposition, pollutant degradation, climate regulation, and other ecosystem functions and services (Delgado-Baquerizo et al. 2016). These functions are extremely important to both farmers and society. Thus, determining whether these soil ecosystem services are being hampered is critical for sustainable agriculture. Table 1.1 summarizes the functions of soil microorganisms in maintaining soil health.

Recent studies have revealed that the intensity of agricultural management practices has increased and can have a significant positive or negative impact on soil functionality (Ding and Wang 2021; Guo et al. 2021) and overall diversity of soil microorganisms (Wagg et al. 2014). Increasing agricultural practices, such as tillage, irrigation, fertilization, pesticides, herbicides, and livestock density can have severe environmental consequences (Tilman et al. 2002), such as depletion of soil fertility and reduced soil microbial diversity. Intensification can also reduce soil

Table 1.1 Impact of soil microorganisms on ecosystem processes

Ecosystem service	Impact of soil microbes	Selected references
Nutrient cycling and plant growth	Enhance organic matter decomposition; facilitate nutrient uptake by plants; protects against pathogens; influences the composition of soil and plant diversity	Van Der Heijden et al. (2008), Martin and Van Der Heijden (2024) and Singh Rawat et al. (2023)
Soil structure formation and aggregation	Production of polymers of carbon that enhance soil aggregates and contribute substantially to the promotion of soil structure	Ayangbenro and Babalola (2021) and Jansson and Hofmockel (2020)
Water uptake and reduction in nutrient leaching losses	Reduction in nutrient leaching; enhance water uptake and infiltration	Cavagnaro et al. (2015)
Climate change mitigation	Promotes CO_2 storage; enhance carbon stocks; mitigate emission of greenhouse gases	Jansson and Hofmockel (2020)
Bioremediation	Degradation of a variety of pollutants through metabolic activities	Ayangbenro (2017), Ayangbenro and Babalola (2021) and Ndeddy Aka and Babalola (2016)
Biodiversity conservation and stress resilience	Conservation of biodiversity and ecosystem stability; confers resistance and resilience against disturbance and stress	Brussaard et al. (2007) and Singh Rawat et al. (2023)
Soil fertility	Enhance nutrient uptake by plants; facilitate plant growth; improve soil structure resulting in enhanced fertility.	Brussaard et al. (2007)

resilience to extreme weather events due to climate change, pests, biological invasion, and pollution (Montanarella et al. 2015). Studies have shown that management practices in many agro-ecosystems have a significant impact on soil microbial diversity, resulting in variations in individual species abundance. For instance, to maintain sustainable soil health and crop productivity, zero-tillage practices are preferred over conventional tillage. Compared with conventional tillage without mulching, zero-tillage systems with crop residue mulch have higher microbial activity in the surface layer of soils (Ngosong et al. 2010).

Reduced biodiversity in soils has increased, raising growing concerns that it could jeopardize various ecological services, including plant nutrient absorption and resource cycling between aboveground and belowground populations (Lal et al. 2017). Thus, to develop sustainable agriculture, an understanding of the cascading linkages between agricultural practices and soil microorganisms, multifunctionality, and their interrelationships is required. This may provide fresh insights into the mechanisms and processes at work and can help alleviate the harmful effects of biodiversity loss.

However, due to ongoing climate change, soil multifunctionality is in jeopardy. Climate change causes not only a long-term increase in temperature but is also linked to an increase in the frequency of extreme weather events, such as prolonged droughts and significant flooding. Increased land-use intensities, mining, and pollution pose additional challenges (Schloter et al. 2018). Because of the persistent threat of soil degradation due to anthropogenic and climatic factors, robust soil conservation measures are required.

According to Lodge et al. (1996), microorganisms can indirectly affect ecosystem processes by altering the diversity of other organisms. Microorganisms play crucial and often excellent roles in ecosystem functioning and maintaining healthy biospheres; their diversity is important for soil functioning and, as a result, ecosystem goods and services related to soils. Thus, soil microbial diversity must be preserved to protect these functions. Molecular technologies are now being used to overcome previous methodological obstacles in characterizing agricultural soil microbial diversity. Thus, significant progress is being made in opening the "black box" of microbial diversity of soil (Allison and Martiny 2008), especially in determining the normal operating ranges of soil microbial diversity under various soil, climatic, land-use, and agricultural practices scenarios (Montanarella et al. 2015). This is critical for understanding broader soil processes and predicting the expected effects of land-use practices or climate change on soil microbial diversity and soil ecosystem services (Montanarella et al. 2015). Additionally, it is important to link microbial diversity to specific soil functions. This will contribute to a better understanding of the critical roles that soil microbes play in mediating soil services.

It is therefore important to quantify and comprehend the effects of various soil management practices on microbial populations, including functional groups. This review provides an overview of the importance of microbial diversity in agro-ecosystems, their role in soil quality and fertility, the contributions of soil microbes to nutrient acquisition and circulation, and threats to microbial soil diversity in agro-ecosystems.

1.2 Diversity of Microbes in Agricultural Soils and Their Impact on Soil Functionality

All biological variations, from genes to species, to communities and ecosystems, are considered part of biodiversity. Microbial diversity refers to the diversity and complexity of microbes at various levels of biological structure (Torsvik and Øvreås 2002). This includes taxon richness and evenness, functional groups in communities, and genetic variability within taxon (species). It is also defined as the distribution and amount of information directly applicable to the total genetic complexity or variation in a group (Torsvik and Øvreås 2002).

Soil ecologists have been forerunners in many ways because of the vast diversity of soils. Much earlier than aboveground ecologists, these researchers explored soil biodiversity from a functional standpoint. However, challenges remain because soils are a highly varied ecosystem, and because many species are small, interactions and processes occur at scales that seem incomprehensible to humans (Turbé et al. 2010). Agricultural practices that encourage the accumulation and retention of soil organic matter boost bacterial diversity. Both species extinction and changes in the relative abundance of present species can affect ecosystem functioning and, consequently, ecosystem services. Under diverse management approaches, Li et al. (2019) found that the structural community of the *Camellia oleifera* forest bacterial species richness and the fungal community considerably varied due to farming activities, although the fungal species richness remained constant. The quantity of diverse beneficial and harmful microorganisms in soil, as well as changes in the microbial community structure, affect the stability and productivity of agroecosystems under different management practices (Francioli et al. 2016).

Tillage has been identified as the most important driver of agricultural soil microbial communities and bacterial diversity in particular (Dong et al. 2017). It has a harmful impact on soil bacterial diversity, but not on fungal and functional diversity (Gupta et al. 2022). Wang et al. (2016) conducted a 5-year tillage treatment study that altered soil moisture, texture, nutrient levels, and bacterial community structure in dryland in China. Tillage has a major impact on soil moisture, organic carbon, and physical qualities. Tillage practices disrupt soil aggregates, compact soil, and reduce gaps between soil particles, causing changes in water and gas transport into the soil and thereby affecting soil function as a habitat for microbes and microbial functional diversity (Holland 2004). Because these variables have a large impact on soil enzyme activity, the type of tillage method directly affects soil enzyme activity (Gupta et al. 2022). Conventional tillage and increased rates of chemical inputs have been intensively exploited to increase productivity to meet the expanding food demands of an ever-increasing human population (Gupta et al. 2022). As a result, it has caused severe land degradation and put the concept of sustainable agriculture in jeopardy.

Dong et al. (2017) reported that agricultural soil with no tillage practices resulted in a considerable increase in the alpha diversity of bacteria, as seen in unique and higher operational taxonomic units (OTUs), evenness, richness, and Shannon index, as well as a lower Simpson index compared to conventional tillage soil. This indicates that zero-tillage practices in winter wheat soil have greater species richness and diversity than soil undergoing conventional tillage practices. Acidobacteria, Actinobacteria, Bacteroidetes, Chloroflexi, and Proteobacteria are the major phyla that dominate both conventional and no-tillage agricultural soils with different relative abundances (Dong et al. 2017; Gupta et al. 2022).

Afforestation is another management practice that affects microbial community structure and diversity. Grassland afforestation is a common land-use conversion that occurs worldwide (Ding and Wang 2021). Grassland afforestation has recently exploded as a means of sequestering carbon dioxide, preserving soil and water resources, and obtaining lumber and other wood products. The initial homeostasis of soil abiotic environment may be disrupted by grassland afforestation, which may alter soil functions (Deng et al. 2016; Kang et al. 2018).

Ding and Wang (2021) observed a decline in the species richness and diversity of soil bacteria and fungi in afforested soil compared to grassland soil. For soil bacteria, there were significant decreases in the observed OTUs and diversity indices.

Acidobacteria, Actinobacteria, Bacteroidetes, Chloroflexi, Cyanobacteria, Proteobacteria, Nitrospirae, and Verrucomicrobia are some of the bacterial phyla that make up the bacterial kingdom. There were no significant differences in the abundances of individual phyla between grassland and afforested soils. Ascomycota, Basidiomycota, Chytridiomycota, Glomeromycota, Rozellomycota, and Zygomycota are phyla found in soil fungal communities (Ding and Wang 2021). The authors also reported an increase in the functional profiles of bacterial communities in afforested soil compared with grassland soil. Sulfur oxidation, fermentation, anoxygenic photoautotrophs, aerobic chemoheterotrophic, nitrate denitrification, nitrite respiration, denitrification, nitrous oxide denitrification, and nitrogen respiration were some of the functions that were higher in afforested soils than in grassland soils, whereas plant pathogens were lower. According to the functional profiles of soil fungi, pathotrophs, saprotrophs, and symbiotrophs dominated the trophic modes. In afforested soils, symbiotrophic were more abundant and pathotrophs were less abundant in grassland soils (Ding and Wang 2021).

In comparison with conventional farming, organic farming increases microbial diversity (Fadiji et al. 2020). Organic farming holds great potential for addressing the negative effects of agricultural intensification, such as biodiversity loss, even though the dangers and advantages of organic agriculture remain debatable (De Graaff et al. 2019). In a meta-analysis assessment of the effects of agricultural intensification on microbial diversity, De Graaff et al. (2019) observed that nitrogen fertilization increased fungal and bacterial diversity by 13% and 3%, respectively, whereas arbuscular mycorrhizal fungi were negatively impacted. The addition of nitrogen as an organic fertilizer or as a combination of synthetic and organic fertilizer drives the significant increase in bacterial diversity. The study reported that the length of time that N fertilizer was applied had a substantial influence on the diversity of bacteria, with short-term (<5 years) studies lacking effect and long-term (≥5 years) N addition resulting in enhanced bacterial diversity. In addition, with N fertilizer application, the functional diversity of diverse microbial groups increased, and the degree of beneficial responses was identical regardless of the quantity, type, or duration of treatment (De Graaff et al. 2019).

Nitrogen fertilizer directly affects bacterial diversity by altering soil chemical composition. Urea and ammonium fertilizers, when applied at high rates, can suppress soil microbes because of their increased ionic strength and ammonia toxicity (Omar and Ismail 1999). The processes underlying the enhanced saprophytic variety caused by N fertilizer applications are unknown. However, similar to the bacterial community's responses, the number of species with a more copiotrophic lifestyle, such as yeasts, which have a stronger genetic aptitude for inorganic N uptake than other fungi groups, appears to have increased in the fungal community (Treseder and Lennon 2015; De Graaff et al. 2019).

It can be difficult to separate the effects of organic farming on soil microbial diversity because the particular management practices associated with conventional and organic farming vary substantially between tests (De Graaff et al. 2019). Generally, differences occur in one or more of the following parameters between conventional and organic farming systems: tilling intensity, herbicide use, fertilization practices, and crop species or genotype. The type and number of factors that differentiate these systems are likely to have a significant impact on soil microbial diversity (De Graaff et al. 2019). In addition, microbial diversity also varies in space and time, thus, the effects of these practices on microbial diversity may be hidden (Decaëns 2010). Soil microbial diversity is higher in the rhizosphere than in bulk soil (Lupwayi et al. 2001), and management effects differ in the rhizosphere (Higo et al. 2015). Comparing bulk soil to aggregates, measurement of diversity increases detection (Lupwayi et al. 2001). Depending on plant developmental states or changing abiotic variables, soil biodiversity responses to agricultural management approaches alter during the growing season (Higo et al. 2015). As a result, understanding biodiversity variability over space and time is critical.

1.3 Current Threats to Soil Biodiversity and the Ecosystem's Regenerative Potential

Soil is a site of infinite reactions that governs an array of services to humanity and the environment, as it performs various tasks and services that support life on Earth, including biogeochemical cycling, biomass production, water flow regulation, pollution control, and climate management (Daily et al. 1997). Soil biota is responsible for a range of important ecosystem functions, such as mineralization, decomposition, and nutrient cycling. In addition, it coordinates other ecosystem services such as disease suppression and pollutant degradation through water infiltration, bioremediation, soil formation, and climate regulation (Carrascosa et al. 2014). The important abiotic variables affecting soil biodiversity include climate, precipitation, temperature, soil structure and texture, salt, and pH. Climate influences the physiology of soil organisms, with higher temperatures and soil moisture levels boosting their activity and development (Furtak and Galazka 2019). Although soil biodiversity is important for controlling processes in terrestrial ecosystems that supply essential services, the relationship between soil diversity and ecosystem functioning is neither fully understood nor effectively harnessed for human gain.

One of the main factors driving interest in soil diversity and agricultural sustainability is the claim by the insurance hypothesis that high biodiversity protects the malfunctioning of ecosystems when stressed or disturbed (Wagg et al. 2018; Siebert et al. 2019). Although no clear links exist between soil biodiversity and ecosystem functioning, this attribute provides insights into the importance of soil biodiversity in coping with stress and resistance to disturbances. The resistance and resilience of soil biodiversity for resource sustainability tailored toward the effective use of natural resources such as nutrients and water is another factor of interest, especially for its promising agricultural and conservation values (El Mujtar et al. 2019; Huang et al. 2020). Furthermore, the management and enhancement of the values of soil biodiversity have been scarcely explored, which motivates further investigation into their sustainability. Sustainable soil management is defined as the regulation, provisioning, supporting, and maintaining the cultural services provided by soils without significantly impairing the soil functions that enable those services or biodiversity (Wolff and Kaphengst 2017).

The reduction in forms of life existing in soils, both in terms of number and variety, is often referred to as a decline in soil biodiversity. In contrast, soil biodiversity loss can substantially affect the capacity of soil to function, adapt to disturbances, and recover from disturbances (Turbé et al. 2010). This tiered complexity is, however, accountable for the difficulty in estimating the impacts of threats on soil biodiversity. The diversity of soil organisms varies from microorganisms to macro- and mega-fauna. In contrast, every threat likely affects a particular entity of soil-dwelling organisms in a distinct manner, and the contributing roles and functions of each soil biota component could be affected by pressure (Almond et al. 2020). There is no available framework to assess the intensity of a threat, which predisposes soil biodiversity to the actual risk of each threat. Since soils are influenced by a variety of elements, including biotic and abiotic, the primary factors that pose enormous threats to soil diversity have therefore been categorized into; the loss of organic matter, erosion, salinization, and soil pollution, which includes industrial and nuclear pollution (radioactivity), physical deterioration (compaction), floods, soil sealing, landslides, habitat fragmentation, and climate change (global warming) (Cachada et al. 2018). Although the loss of soil biodiversity has not been specifically identified as a concern for soil, it is implicitly accepted because it is influenced by the impacts of the stated soil problems. However, some of the threats to soil emanating from anthropogenic activities that have been observed to have detrimental consequences on soil biodiversity. These include land-use change, intensive exploitation by humans, and decreases in soil organic matter; the use of genetically modified organisms, the introduction and diffusion of invasive species, and environmental pollution by microplastics have been identified as emerging threats (Tibbett et al. 2020).

Human activities directly affect soil biota, which depends on the biotic and abiotic environment. Since soil is the bedrock of civilization, changes in soil biota are connected to the rapid increase in the human population, with a rising proportion of land surface being sealed to build cities and other infrastructure, every year an amount of open soil with a surface gets sealed (Al-Kaisi and Lowery 2017). Apart from the land space required for increasing infrastructure, land is also used for mining and agriculture. It has been estimated that by 2050, more than one billion hectares of land could be converted to agriculturally managed land for food production alone (Benke and Tomkins 2017). However, the use of heavy machinery, irrigation, and chemical inputs in intensified agriculture results in habitat modifications that not only alter soil properties and structures but also reduce soil biodiversity (Techen et al. 2020). The effects of which are evident in the disruption of soil microbial diversity and many soil fauna because of the persisting effects of chemicals, such as pesticides, such as neonicotinoid, and herbicides like glyphosate, which can linger in soils for years and influence non-target species, are all potential threats to soil biodiversity. This effect has been associated with decreases in aboveground insects and soil invertebrates such as earthworms (Saha et al. 2020). Heavy metal pollution due to agricultural, mining, and smelting activities can destroy soil microbial taxa, and change the composition of soil (Stefanowicz et al. 2020).

1.4 Regenerative Potential of the Ecosystem

The protection of biodiversity, functions, and services provided by forest ecosystems in human-modified tropical landscapes depends on the maintenance of old growth forest remnants, forest regeneration in abandoned agricultural areas, and restoration of degraded lands (Arroyo-Rodríguez et al. 2020). Thus, natural regeneration of ecosystems has multiple benefits for both nature and humans, and this plays a major role in national restoration plans (Chazdon and Uriarte 2016). Therefore, protecting or conserving remnant ecosystems is one of the first points of call for ecosystem regeneration, which can be achieved through nature reserves, restoration of degraded ecosystems, provision of connections between remnant habitats to reduce fragmentation, and management of urban vegetation and/or structures to increase biodiversity (Aguilera et al. 2020).

Certain conditions have been found essential for the successful regeneration of degrading ecosystems, and one of the foremost in this regard is the restoration of forests (Graham et al. 2017). This is because biodiversity protection, carbon sequestration, soil conservation, and water conservation are all significant ecosystem services provided by forest ecosystems (Ramachandra and Bharath 2020). The impact of plantation management strategies on ecosystem service varies; the primary focus being on increasing ecosystem services. Hence, to maintain ecosystem functions and values for the future, forest restoration initiatives learn from the past and productively navigate restoration plans to improve regeneration systems and techniques, integrate ecological knowledge, overcome biotic and abiotic disturbances, and adapt for future forest landscapes.

Regenerative agriculture is other means of achieving food productivity with minimal disturbance to the ecosystem. This system has been offered as a potential alternative method of food production with lower or even positive environmental and/or social implications (Benke and Tomkins 2017). Regenerative agriculture has shown potential for sustainable food production and is a suitable strategy for mitigating the impact of climate change because the system sustains the health of soil by restoring its carbon content, and in turn improves its productivity, a situation that has been described as the opposite of conventional agriculture (Gosnell et al. 2019). Moreover, regenerative cropping has been estimated to reduce or sequester 14.5–22 gigatons of CO_2 annually (Newton et al. 2020). However, the primary goal of regenerative agriculture is to improve soil health and restore degraded soils, which in turn improves vegetation, water quality, and land productivity.

1.5 Contribution of Soil Microbial Diversity to the Rejuvenation of Soil and Crop Health

Soil health is frequently employed in the context of agricultural sustainability (Kibblewhite et al. 2008; Frąc et al. 2018), particularly in the perspective of soil as a dynamic living organism that functions holistically rather than as an inert substrate (Cardoso et al. 2013; Frąc et al. 2018). As a result, we shall be concentrating on the health of the soil rather than the quality. Soil health is often referred to as the soil's ability to preserve and sustain biological production, environmental quality and support animal, crop health and human (Doran and Jones 2018). In recent years, the cultivation of soil microbial biodiversity as a way to improve the quality of soil and boost the productivity of agroecosystems has been presented as a promising and innovative initiative for plant productivity (Bagyaraj and Ashwin 2017). Implementing such approach may provide an alternative to the current misuse of agrochemicals by allowing complex adjustments to plant production. Soil microbial biodiversity helps plants develop by decomposing organic waste and supplying nutrients. Their biocontrol functions in plant defence against pathogens critically affect soil health (Frąc et al. 2015). The use of soil microbial biodiversity as a quality indicator should not be restricted to diversity indices alone; a structural examination of the microbial population is also required to establish its major functions in enhancing crop health and soil quality.

The application of various types of organic manure has a considerable impact on soil health, both directly (by changes in soil microbial communities) and indirectly (through changes in physicochemical features). All agricultural systems require good soil management, and reducing soil deterioration is a top goal to sustain future output. However, it is through soil microbial biodiversity that this impact can be realized. Aside from their recognized impacts on soil-borne diseases (Abawi and Widmer 2000; Frąc et al. 2018), all management practices, such as the application of composts, tillage regimes, and rotational and cover crops, may also have significant impacts on other groups of soil microbes, particularly on beneficial fungi populations.

In agro- and horticultural soils, arbuscular mycorrhizal fungi (AMF) constitute the most significant class of helpful microbes (Smith and Read 2010). In various tests, significant improvements in crop yield were observed after AMF inoculation (Bagyaraj and Ashwin 2017). Improvement of plant establishment and root system, nutrient cycling stimulation, increased tolerance of plants to stress, improvement of soil structure, enhancement of uptake of low mobility ions, and increase in plant community diversity are some of the major effects of AMF symbiosis (Azcón-Aguilar and Barea 1997). Some antagonistic fungi that inhibit fungal infections, such as Trichoderma spp. and Glomus spp., can help control agricultural plant diseases (Dawidziuk et al. 2016). *Trichoderma* species (*T. viride*, *T. asperellum*, *T. atroviride*, *T. virens*, and *T. harzianum*) are commonly used antagonistic agents and biostimulants for horticultural crops (López-Bucio et al. 2015). Microbial inoculation of consortia of AMF with plant growth-promoting rhizobacteria (PGPR) and others, such as microbes involved in N fixation and P solubilization, has various favorable impacts on crop and soil health (Bagyaraj and Ashwin 2017). PGPR and AMF have synergistic and beneficial effects on the general development of horticultural crops alongside soil microbial activity and diversity (De Coninck et al. 2015).

1.6 Future Perspectives: Harnessing the Potentials of Microbial Diversity in Agriculture

Agroforestry and conventional smallholder agricultural land-use has long been recognized as contributing to biodiversity conservation (Garrity 2004), particularly when land-use preserves sufficient tree cover, which offers complementary habitats, resources, and connections for local biota. Hence, agriculture serves as a diverse interface between microorganisms and plants that is essential to ecosystem stability and plant growth (Hakim et al. 2021). The variety of unicellular organisms, including archaea, bacteria, fungi, and protists, constitute microbial diversity; they thrive all across the biosphere, establishing the bounds of life and fostering the survival and evolution of other living things (Kulasooriya 2019). Meanwhile, microbes differ in their appearance, physiology, and cellular metabolism, as well as in their ecological ranges and activities, expression, genetic structure, and evolution.

The diversity of microorganisms now existing on Earth is well recognized and assumed to be immense, although the exact amount of microbial diversity has not been quantified (Vitorino and Bessa 2018). Although the quantitative and qualitative makeup of microbial communities differs, the proportion of community members that are affected by metabolic, physiological, and physicochemical changes in the environment generated by organisms is prone to change (Dasgupta and Brahmaprakash 2021). Thus, important niches for microbial diversity occur in highly specific associations, and beneficial microorganisms appear to provide an alternative opportunity to overcome various adverse effects that affect ecosystems, either through natural or human activities, such as the excessive use of synthetic pesticides. Furthermore, the involvement of microbes in crucial activities, such as soil structure formation, organic matter breakdown, toxins elimination, and carbon, phosphorus, nitrogen, and sulfur cycling, indicates that microorganisms in soil are critical to the preservation of soil function in both natural and managed agricultural soils (Yadav et al. 2021). Additionally, microorganisms play an important role in the suppression of soil-borne plant diseases, plant development, and vegetation changes (Vida et al. 2020).

A close relationship between soil microbes and plants has been established in several studies, and plant-associated microbiomes have been found to be diverse and vast, and have been employed in the agricultural sector to increase plant adaptability and yield. Despite technical improvements in the study of soil microorganisms, soil microbial communities and their assembly methods have not been fully explored (Dubey et al. 2020). Recently, investigations into high-throughput omics approaches such as genomics, transcriptomics, proteomics, enzymology, and metabolomics have been conducted to quantify soil microbial abundance, diversity, activity, and functional potential (Park et al. 2020). The introduction of next-generation sequencing techniques such as complete genome shotgun sequencing, and single-molecule long-read sequencing, in particular, has enabled the identification of soil microorganism communities, whereas the other omics approaches mentioned previously have enabled the investigation of the biological functions of these communities (Park and Kim 2016; Tedersoo et al. 2021). In addition, the numerous omics technologies that have emerged globally and are introducing key novelties to many sectors of the agricultural sciences have similarly made bioinformatics assume a critical role in these scientific domains, as it provides crucial help in characterizing the many facets of biological functions. Thus, the current turn of events has afforded microbial ecology researchers to examine genetic components from environmental samples on a massive scale and make necessary deductions that will advance the course of ecosystem functioning and agricultural revolution.

1.7 Conclusion

Global food demand is becoming more diverse and increasing, placing increasing pressure on agricultural productivity and posing a challenge to modern agriculture. Hence, agricultural intensification is becoming unavoidable to feed the growing population, particularly in developing nations where food shortages are more prevalent. Addressing this increasing challenge will involve improving crop production through sustainable use of available resources and leveraging innovative technologies to enhance efficiency and minimize waste. Beneficial changes in soil microbial populations through appropriate soil management can enhance plant productivity by altering water dynamics and enzyme activities. Effective management practices in agriculture that increase the rates of soil organic matter formation and stabilization can also significantly improve the microbial status of soil. Therefore, to maintain healthy soil conditions and ensure food security in a changing climate, it is essential to adopt sustainable technologies, such as the inoculation of beneficial microbes and their persistence in soil. Soil microbiota play a significant role in preserving soil health, fertility, and ecosystem function. The beneficial roles of microbes include their ability to decompose, recycle nutrients, and degrade pollutants and other elements to support plant development. However, intensive farming practices such as frequent tilling, monoculture cropping, and indiscriminate agrochemical use can negatively affect soil microbial communities and their diversity. The changing climate further exerts increasing environmental pressure, resulting in adverse effects on soil microbial ecosystems and the services provided by the microbes. To mitigate these threats and promote sustainable agriculture, appropriate soil conservation measures and sustainable practices must be implemented to improve soil microbes. This entails the use of minimum tillage systems, crop rotation, the smart use of fertilizers, and other agro inputs. Molecular methods have revealed the composition of soil microbial communities and have made it possible to determine the effects of different practices and changes on the associated microbes. Thus, it is essential to maintain and promote soil microbial diversity to safeguard the essential functions and promote ecosystem resilience.

References

Abawi G, Widmer T (2000) Impact of soil health management practices on soilborne pathogens, nematodes and root diseases of vegetable crops. Appl Soil Ecol 15:37–47

Aguilera MA, Tapia J, Gallardo C, Núñez P, Varas-Belemmi K (2020) Loss of coastal ecosystem spatial connectivity and services by urbanization: natural-to-urban integration for bay management. J Environ Manag 276:111297

Akhtar-Schuster M, Stringer LC, Erlewein A, Metternicht G, Minelli S, Safriel U, Sommer S (2017) Unpacking the concept of land degradation neutrality and addressing its operation through the Rio Conventions. J Environ Manag 195:4–15

Al-Kaisi MM, Lowery B (2017) Soil health and intensification of agroecosystems. Academic

Allison SD, Martiny JB (2008) Resistance, resilience, and redundancy in microbial communities. Proc Natl Acad Sci 105:11512–11519

Almond RE, Grooten M, Peterson T (2020) Living planet report 2020-bending the curve of biodiversity loss. World Wildlife Fund

Arroyo-Rodríguez V, Fahrig L, Tabarelli M, Watling JI, Tischendorf L, Benchimol M, Cazetta E, Faria D, Leal IR, Melo FP (2020) Designing optimal human-modified landscapes for forest biodiversity conservation. Ecol Lett 23:1404–1420

Ayangbenro A (2017) Biodegradation of natural bitumen by *Providencia stuartii* isolated from heavy oil contaminated soil. Global NEST J 19:353–358

Ayangbenro AS, Babalola OO (2021) Reclamation of arid and semi-arid soils: the role of plant growth-promoting archaea and bacteria. Curr Plant Biol 25:100173

Azcón-Aguilar C, Barea J (1997) Applying mycorrhiza biotechnology to horticulture: significance and potentials. Sci Hortic 68:1–24

Bagyaraj D, Ashwin R (2017) Soil biodiversity: role in sustainable horticulture. Biodivers Hortic Crops 5:1–18

Barbier EB, Hochard JP (2016) Does land degradation increase poverty in developing countries? PLoS One 11:e0152973

Benke K, Tomkins B (2017) Future food-production systems: vertical farming and controlled-environment agriculture. Sustain Sci Pract Policy 13:13–26

Brussaard L, De Ruiter PC, Brown GG (2007) Soil biodiversity for agricultural sustainability. Agric Ecosyst Environ 121:233–244

Cachada A, Rocha-Santos T, Duarte AC (2018) Soil and pollution: an introduction to the main issues. In: Soil pollution. Elsevier, pp 1–28

Cardoso EJBN, Vasconcellos RLF, Bini D, Miyauchi MYH, Santos CAD, Alves PRL, Paula AMD, Nakatani AS, Pereira JDM, Nogueira MA (2013) Soil health: looking for suitable indicators. What should be considered to assess the effects of use and management on soil health? Sci Agric 70:274–289

Carrascosa M, Sánchez-Moreno S, Alonso-Prados JL (2014) Relationships between nematode diversity, plant biomass, nutrient cycling and soil suppressiveness in fumigated soils. Eur J Soil Biol 62:49–59

Cavagnaro TR, Bender SF, Asghari HR, Van Der Heijden MG (2015) The role of arbuscular mycorrhizas in reducing soil nutrient loss. Trends Plant Sci 20:283–290
Chazdon RL, Uriarte M (2016) Natural regeneration in the context of large-scale forest and landscape restoration in the tropics. Biotropica 48:709–715
Daily GC, Matson PA, Vitousek PM (1997) Ecosystem services supplied by soil. In: Nature's services: societal dependence on natural ecosystems. Island Press, Washington, DC, pp 113–132
Dasgupta D, Brahmaprakash G (2021) Soil microbes are shaped by soil physico-chemical properties: a brief review of existing literature. Int J Plant Soil Sci 33(1):59–71
Dawidziuk A, Popiel D, Kaczmarek J, Strakowska J, Jedryczka M (2016) Optimal Trichoderma strains for control of stem canker of brassicas: molecular basis of biocontrol properties and azole resistance. BioControl 61:755–768
De Coninck B, Timmermans P, Vos C, Cammue BP, Kazan K (2015) What lies beneath: belowground defense strategies in plants. Trends Plant Sci 20:91–101
De Graaff M-A, Hornslein N, Throop HL, Kardol P, Van Diepen LTA (2019) Effects of agricultural intensification on soil biodiversity and implications for ecosystem functioning: a meta-analysis. In: Advances in agronomy. Elsevier, pp 1–44
Decaëns T (2010) Macroecological patterns in soil communities. Glob Ecol Biogeogr 19:287–302
Delgado-Baquerizo M, Maestre FT, Reich PB, Jeffries TC, Gaitan JJ, Encinar D, Berdugo M, Campbell CD, Singh BK (2016) Microbial diversity drives multifunctionality in terrestrial ecosystems. Nat Commun 7:10541
Deng Q, Cheng X, Hui D, Zhang Q, Li M, Zhang Q (2016) Soil microbial community and its interaction with soil carbon and nitrogen dynamics following afforestation in central China. Sci Total Environ 541:230–237
Ding L, Wang P (2021) Afforestation suppresses soil nitrogen availability and soil multifunctionality on a subtropical grassland. Sci Total Environ 761:143663
Dong W, Liu E, Yan C, Tian J, Zhang H, Zhang Y (2017) Impact of no tillage vs. conventional tillage on the soil bacterial community structure in a winter wheat cropping succession in northern China. Eur J Soil Biol 80:35–42
Doran JW, Jones AJ (2018) Soil quality and health: indicators of sustainability. In: Doran JW, Jones AJ (eds) Methods for assessing soil quality, SSSA special publication number 49. Soil Science Society of America, Madison, pp XI–XIV
Dubey RK, Tripathi V, Prabha R, Chaurasia R, Singh DP, Rao CS, El-Keblawy A, Abhilash PC (2020) Unravelling the soil microbiome: perspectives for environmental sustainability. Springer
El Mujtar V, Muñoz N, Mc Cormick BP, Pulleman M, Tittonell P (2019) Role and management of soil biodiversity for food security and nutrition; where do we stand? Glob Food Sec 20:132–144
Fadiji AE, Ayangbenro AS, Babalola OO (2020) Organic farming enhances the diversity and community structure of endophytic archaea and fungi in maize plant: a shotgun approach. J Soil Sci Plant Nutr 20:2587–2599
Frąc M, Jezierska-Tys S, Yaguchi T (2015) Occurrence, detection, and molecular and metabolic characterization of heat-resistant fungi in soils and plants and their risk to human health. Adv Agron 132:161–204
Frąc M, Hannula SE, Bełka M, Jędryczka M (2018) Fungal biodiversity and their role in soil health. Front Microbiol 9:707
Francioli D, Schulz E, Lentendu G, Wubet T, Buscot F, Reitz T (2016) Mineral vs. organic amendments: microbial community structure, activity and abundance of agriculturally relevant microbes are driven by long-term fertilization strategies. Front Microbiol 7:1446
Furtak K, Galazka A (2019) Edaphic factors and their influence on the microbiological biodiversity of the soil environment/Czynniki Edaficzne I Ich Wplyw Na Bioroznorodnosc Mikrobiologiczna Srodowiska Glebowego. Adv Microbiol 58:375–385
Garrity DP (2004) Agroforestry and the achievement of the millennium development goals. Agrofor Syst 61:5–17

Gosnell H, Gill N, Voyer M (2019) Transformational adaptation on the farm: processes of change and persistence in transitions to 'climate-smart' regenerative agriculture. Glob Environ Chang 59:101965

Graham LL, Giesen W, Page SE (2017) A common-sense approach to tropical peat swamp forest restoration in Southeast Asia. Restor Ecol 25:312–321

Guo Y, Luo H, Wang L, Xu M, Wan Y, Chou M, Shi P, Wei G (2021) Multifunctionality and microbial communities in agricultural soils regulate the dynamics of a soil-borne pathogen. Plant Soil 461:309–322

Gupta A, Singh UB, Sahu PK, Paul S, Kumar A, Malviya D, Singh S, Kuppusamy P, Singh P, Paul D, Rai JP, Singh HV, Manna MC, Crusberg TC, Kumar A, Saxena AK (2022) Linking soil microbial diversity to modern agriculture practices: a review. Int J Environ Res Public Health 19:3141

Hakim S, Naqqash T, Nawaz MS, Laraib I, Siddique MJ, Zia R, Mirza MS, Imran A (2021) Rhizosphere engineering with plant growth-promoting microorganisms for agriculture and ecological sustainability. Front Sustain Food Syst 5:16

Han Z, Xu P, Li Z, Lin H, Zhu C, Wang J, Zou J (2022) Microbial diversity and the abundance of keystone species drive the response of soil multifunctionality to organic substitution and biochar amendment in a tea plantation. GCB Bioenergy 14(4):481–495

Higo M, Isobe K, Yamaguchi M, Torigoe Y (2015) Impact of a soil sampling strategy on the spatial distribution and diversity of arbuscular mycorrhizal communities at a small scale in two winter cover crop rotational systems. Ann Microbiol 65:985–993

Holland JM (2004) The environmental consequences of adopting conservation tillage in Europe: reviewing the evidence. Agric Ecosyst Environ 103:1–25

Huang L, Bai J, Wen X, Zhang G, Zhang C, Cui B, Liu X (2020) Microbial resistance and resilience in response to environmental changes under the higher intensity of human activities than global average level. Glob Chang Biol 26:2377–2389

Jansson JK, Hofmockel KS (2020) Soil microbiomes and climate change. Nat Rev Microbiol 18:35–46

Kang H, Gao H, Yu W, Yi Y, Wang Y, Ning M (2018) Changes in soil microbial community structure and function after afforestation depend on species and age: case study in a subtropical alluvial island. Sci Total Environ 625:1423–1432

Kibblewhite M, Ritz K, Swift M (2008) Soil health in agricultural systems. Philos Trans R Soc B Biol Sci 363:685–701

Kulasooriya S (2019) The amazing world of microorganisms. Ceylon J Sci 48:303–310

Lal R, Mohtar RH, Assi AT, Ray R, Baybil H, Jahn M (2017) Soil as a basic nexus tool: soils at the center of the food–energy–water nexus. Curr Sustain Renew Energy Rep 4:117–129

Li J, Wu Z, Yuan J (2019) Impact of agro-farming activities on microbial diversity of acidic red soils in a Camellia Oleifera Forest. Rev Bras Ciênc Solo 43. https://doi.org/10.1590/18069657rbcs20190044

Lodge DJ, Hawksworth DL, Ritchie BJ (1996) Microbial diversity and tropical forest functioning. In: Biodiversity and ecosystem processes in tropical forests. Springer, pp 69–100

López-Bucio J, Pelagio-Flores R, Herrera-Estrella A (2015) Trichoderma as biostimulant: exploiting the multilevel properties of a plant beneficial fungus. Sci Hortic 196:109–123

Lupwayi N, Monreal M, Clayton G, Grant C, Johnston A, Rice W (2001) Soil microbial biomass and diversity respond to tillage and Sulphur fertilizers. Can J Soil Sci 81:577–589

Martin FM, Van Der Heijden MGA (2024) The mycorrhizal symbiosis: research frontiers in genomics, ecology, and agricultural application. New Phytol 242:1486–1506

Montanarella L, Badraoui M, Chude V, Costa I, Mamo T, Yemefack M, Aulang M, Yagi K, Hong SY, Vijarnsorn P (2015) Status of the world's soil resources: main report. Embrapa Solos-Livro científico (ALICE)

Muñoz-Rojas M (2018) Soil quality indicators: critical tools in ecosystem restoration. Curr Opin Environ Sci Health 5:47–52

Ndeddy Aka RJ, Babalola OO (2016) Effect of bacterial inoculation of strains of *Pseudomonas aeruginosa*, *Alcaligenes feacalis* and *Bacillus subtilis* on germination, growth and heavy metal (Cd, Cr, and Ni) uptake of *Brassica juncea*. Int J Phytoremediation 18:200–209

Newton P, Civita N, Frankel-Goldwater L, Bartel K, Johns C (2020) What is regenerative agriculture? A review of scholar and practitioner definitions based on processes and outcomes. Front Sustain Food Syst 4:194

Ngosong C, Jarosch M, Raupp J, Neumann E, Ruess L (2010) The impact of farming practice on soil microorganisms and arbuscular mycorrhizal fungi: crop type versus long-term mineral and organic fertilization. Appl Soil Ecol 46:134–142

Omar S, Ismail M (1999) Microbial populations, ammonification and nitrification in soil treated with urea and inorganic salts. Folia Microbiol 44:205–212

Park ST, Kim J (2016) Trends in next-generation sequencing and a new era for whole genome sequencing. Int Neurourol J 20:S76

Park J, Jeon HW, Jung H, Lee H-H, Kim J, Park AR, Kim N, Han G, Kim J-C, Seo Y-S (2020) Comparative transcriptome analysis of pine trees treated with resistance-inducing substances against the nematode Bursaphelenchus xylophilus. Genes 11:1000

Ramachandra T, Bharath S (2020) Carbon sequestration potential of the forest ecosystems in the Western Ghats, a global biodiversity hotspot. Nat Resour Res 29:2753–2771

Saha L, Kishor V, Bauddh K (2020) Impacts of synthetic pesticides on soil health and non-targeted flora and fauna. In: Ecological and practical applications for sustainable agriculture. Springer, pp 65–88

Schloter M, Nannipieri P, Sørensen SJ, Van Elsas JD (2018) Microbial indicators for soil quality. Biol Fertil Soils 54:1–10

Siebert J, Thakur MP, Reitz T, Schädler M, Schulz E, Yin R, Weigelt A, Eisenhauer N (2019) Extensive grassland-use sustains high levels of soil biological activity, but does not alleviate detrimental climate change effects. In: Advances in ecological research. Elsevier, pp 25–58

Singh Rawat V, Kaur J, Bhagwat S, Arora Pandit M, Dogra Rawat C (2023) Deploying microbes as drivers and indicators in ecological restoration. Restor Ecol 31:e13688

Smith SE, Read DJ (2010) Mycorrhizal symbiosis, 3rd edn. Academic, London

Stefanowicz AM, Kapusta P, Zubek S, Stanek M, Woch MW (2020) Soil organic matter prevails over heavy metal pollution and vegetation as a factor shaping soil microbial communities at historical Zn–Pb mining sites. Chemosphere 240:124922

Techen A-K, Helming K, Brueggemann N, Veldkamp E, Reinhold-Hurek B, Lorenz M, Bartke S, Heinrich U, Amelung W, Augustin K (2020) Soil research challenges in response to emerging agricultural soil management practices. Adv Agron 161:179–240

Tedersoo L, Albertsen M, Anslan S, Callahan B (2021) Perspectives and benefits of high-throughput long-read sequencing in microbial ecology. Appl Environ Microbiol 87:e00626–e00621

Tibbett M, Fraser TD, Duddigan S (2020) Identifying potential threats to soil biodiversity. PeerJ 8:e9271

Tilman D, Cassman KG, Matson PA, Naylor R, Polasky S (2002) Agricultural sustainability and intensive production practices. Nature 418:671–677

Torsvik V, Øvreås L (2002) Microbial diversity and function in soil: from genes to ecosystems. Curr Opin Microbiol 5:240–245

Treseder KK, Lennon JT (2015) Fungal traits that drive ecosystem dynamics on land. Microbiol Mol Biol Rev 79:243–262

Turbé A, De Toni A, Benito P, Lavelle P, Lavelle P, Camacho NR, Van Der Putten WH, Labouze E, Mudgal S (2010) Soil biodiversity: functions, threats and tools for policy makers. Bio Intelligence Service, IRD, and NIOO, Report for European Commission (DG Environment)

Van Der Heijden MGA, Bardgett RD, Van Straalen NM (2008) The unseen majority: soil microbes as drivers of plant diversity and productivity in terrestrial ecosystems. Ecol Lett 11:296–310
Vida C, De Vicente A, Cazorla FM (2020) The role of organic amendments to soil for crop protection: induction of suppression of soilborne pathogens. Ann Appl Biol 176:1–15
Vitorino LC, Bessa LA (2018) Microbial diversity: the gap between the estimated and the known. Diversity 10:46
Wagg C, Bender SF, Widmer F, Van Der Heijden MGA (2014) Soil biodiversity and soil community composition determine ecosystem multifunctionality. Proc Natl Acad Sci 111:5266–5270
Wagg C, Dudenhöffer JH, Widmer F, Van Der Heijden MG (2018) Linking diversity, synchrony and stability in soil microbial communities. Funct Ecol 32:1280–1292
Wang Z, Liu L, Chen Q, Wen X, Liao Y (2016) Conservation tillage increases soil bacterial diversity in the dryland of northern China. Agron Sustain Dev 36:28
Wolff F, Kaphengst T (2017) The UN Convention on biological diversity and soils: status and future options. In: International yearbook of soil law and policy 2016. Springer, pp 129–148
Yadav AN, Kour D, Kaur T, Devi R, Yadav A, Dikilitas M, Abdel-Azeem AM, Ahluwalia AS, Saxena AK (2021) Biodiversity, and biotechnological contribution of beneficial soil microbiomes for nutrient cycling, plant growth improvement and nutrient uptake. Biocatal Agric Biotechnol 33:102009

Chapter 2
Microbial Interactions in Soil Ecosystems: Facilitating Plant Growth, Nutrient Cycling, and Environmental Dynamics

Dhrubjyoti Sharma, Ashim Das Astapati, Sangeeta Dey, Debasish Bhattacharjee, Deepti Mala Singha, and Soumitra Nath

Abstract Soil organisms play a pivotal role in shaping the environment through their contributions to soil formation, nutrient cycling, stimulation of plant growth, and the detoxification of both organic and inorganic pollutants. Within soil food webs, soil biota are essential for processes such as mineralization, decomposition of organic matter, nitrification, ammonification, and nutrient release. Furthermore, the microbial populations in soil ecosystems are instrumental in managing and sustaining tropical grasslands worldwide. Specific microbial species inhabiting the rhizosphere, the region surrounding plant roots, have a profound impact on ecosystem development, plant vitality, and even weather monitoring. These rhizospheric microbes rely on carbon sources found in root exudates and plant litter. Additionally, certain autotrophic organisms fulfill their nitrogen requirements through nitrate salts or ammonium as their preferred nitrogen sources. Soil microbial communities also play a critical role in regulating plant growth by supplying nutrients and participating in the mineralization of organic matter in the surface soil. Interactions between populations of soil microbes can be mutually beneficial or detrimental to one of the participants. These interactions involve the exchange of genetic and molecular information, and various mechanisms, such as quorum sensing systems, biofilm formation, production of secondary metabolites, and transduction signalling, are implicated. The review comprehensively examines a wide array of soil

D. Sharma
Bioinformatics Centre, Gurucharan College, Silchar, Assam, India

A. D. Astapati
Department of Botany, Gurucharan College, Silchar, Assam, India

S. Dey · D. M. Singha
Department of Ecology and Environment Science, Gurucharan College, Silchar, Assam, India

D. Bhattacharjee
Department of Zoology, Gurucharan College, Silchar, Assam, India

S. Nath (✉)
Bioinformatics Centre, Gurucharan College, Silchar, Assam, India

Department of Biotechnology, Gurucharan College, Silchar, Assam, India

O. O. Babalola, A. S. Ayangbenro (eds.), *Microbial Allies*,
https://doi.org/10.1007/978-3-031-90530-8_2

microbes, delving into their diverse roles and intricate interactions that collectively enhance soil functionality.

Keywords Rhizosphere · Nutrient cycling · Soil microorganism · Mineralization · Quorum · Secondary metabolites

2.1 Introduction

The topmost layer of the earth's surface, i.e., the soil, is a mixture of air, water, living organisms and minerals. Soil is one of the most dynamic and significant natural resources on our planet because of the unique ways these four components interact. Soils perform many vital processes, such as recycling nutrients, filtering water, and exchanging gases with the atmosphere (Raj et al. 2019). Water and air in the soil are in a dynamic equilibrium inside the soil pores, which plays an important role in promoting plant growth and improving soil activities. There are also many different species of bacteria, protozoa, algae, viruses, and fungi found in soil (Pepper and Gentry 2015). Biodiversity and biomass of soil microorganisms are crucial to ecosystem processes and maintaining healthy soils and healthy soil structures. The microbes assist the plants in growing and also help improve the ecosystem's functioning. Additionally, they decompose dead organic materials and release nutrients, aiding in the increase of organic content in the soil that supports the healthy growth of plants. Microbes can also be used to bioremediate environmental pollutants like xenobiotics, hydrocarbons, and heavy metals (Arunraja et al. 2023; Ghosh et al. 2021; Nath et al. 2018).

The microbial community is cosmopolitan, but those specifically associated with plants thrive in the rhizosphere. This area, in close proximity to the roots of plants, is where soil microbes depend directly upon the characteristics of the soil for their growth, activity, and nutrition. Plant roots exude a variety of substances, such as sugars, amino acids, and organic acids, into the soil, which are collectively known as root exudates (Ahlawat et al. 2024). These exudates serve as a primary energy source for soil microbes, facilitating various soil functions and significantly altering the rhizosphere's environment relative to the bulk soil environment. The release of exudates by plant roots not only promotes microbial activity but also enhances nutrient cycling, soil structure, and plant health. Soil microbes just like the higher plants depend directly upon the characteristics of soil for their growth, activity and nutrition. Soil ecosystems are affected by many factors, such as pH, temperature, moisture, and texture (Brangarí et al. 2020; Msimbira and Smith 2020). The activity of the organisms can change as a result of variations in any one or more of these factors, which in turn can impact the soil fertility level. Interactions with other microorganisms or hosts primarily mediate colonization and establishment of microorganisms in different environments. This interaction involves all ecological components, which include genetic exchange, physiochemical alterations, metabolite conversion and exchange, signalling, and chemotaxis, which further leads to the selection of the species genotype (Braga et al. 2016). Depending on the relationship between the two species, the interaction may be positive, negative, or neutral.

A variety of soil processes rely on microbial enzymes, including nutrient cycling, microbe transformation, mineralization of organic matter, and pollution degradation (Martin and Schwab 2012, 2013; Rao et al. 2014). The enzymes act on specific substrates, catalyze the decomposition of plant residues and release nutrients for soil and plant growth. The review discusses various types of soil microbes, their roles, and their interactions to improve soil functionality.

2.2 Plant Root-Microbe Interactions

The rhizosphere is a term derived from the Greek word rhiza, meaning root, that was first coined by the German agronomist and plant physiologist Hiltner in 1904 (Hartmann et al. 2008). It is the soil region that immediately attaches to the plant root system (Fig. 2.1a). This region contains unaccountable numbers of microorganisms such as bacteria, actinomycetes and fungi. The high microbial load in the rhizospheric region is due to some chemical compounds called "exudates" that make the rhizosphere environment quite different from the normal soil due to its unique characteristics establishing a confluence of plants and soil with the microorganisms, nutrients and water (Fig. 2.1c). These "exudates" are essentially a mixture

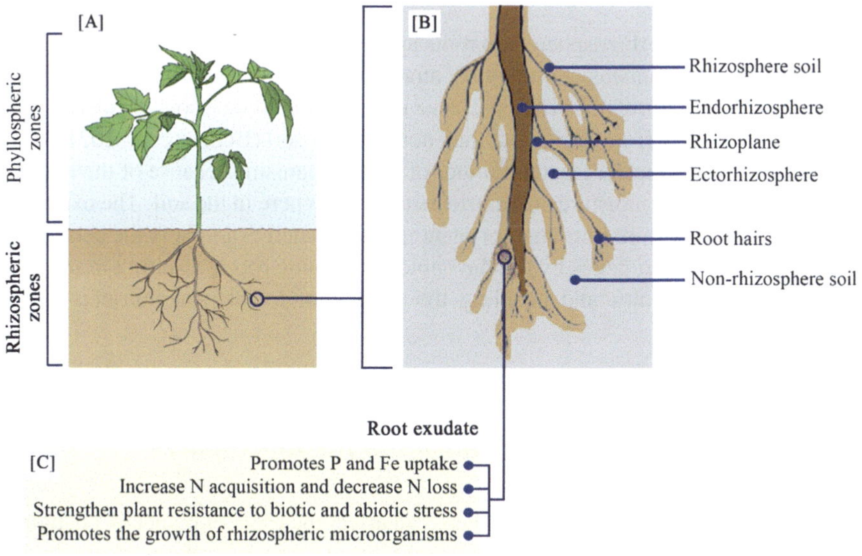

Fig. 2.1 (**a**) Two different regions of a plant, i.e., phyllospheric zone and rhizospheric zone. (**b**) Rhizospheric zone are made up of different components. (**c**) The rhizoplane region of plant roots release various carbon compounds, called exudates, which perform a variety of functions

of amino compounds, organic acids, nucleotides, flavones, enzymes, growth hormones, and carbohydrates (Rasmann and Turlings 2016). The rhizospheric region is classified into endorhizosphere, rhizoplane and ectorhizosphere (Fig. 1b). In endorhizosphere, microorganisms and cations occupy the free space between the endodermis and the cortex. The rhizoplane is the place where plant exudates are released, and it is also the site for water and nutrient uptake. The outermost zone, known as ectorhizosphere, extends from the rhizoplane into the soil (Katiyar 2022).

2.2.1 Functions of Roots in the Rhizosphere

Plant root exudates consist of carbohydrates, organic acids, amino acids, mucilage and proteins. They stimulate the biological functions and physical interactions happening between the soil microbes and plant roots (Wang et al. 2021). The exudates alter the rhizosphere's biochemical and physical characteristics, which aids in root development and plant survival. Secretions of exudates from the plant roots can be (i) Low molecular-weight substances, namely phenolics, carbohydrates, and organic acids, and (ii) High-molecular-weight exudates like proteins and mucilage (polysaccharides).

2.2.2 Functions of Root Exudates

Exudates defend the rhizosphere and roots against harmful pathogenic microorganisms. The microorganisms continuously attack the root cells and survive by secreting certain defence proteins. They preserve the soil around the roots of rhizosphere soils moist and save from drying out than normal bulk soil (Bodner et al. 2021). The rhizosphere soil attracts a greater number of microorganisms because of the release of root exudates and moisture and nutrients than elsewhere in the soil. The exudates further absorb and store nutrients for plant use. One such example is the activation of *Rhizobium meliloti* genes by the flavonoids in legume roots, resulting in the formation of root nodules and enabling the plant root to take up nutrients (Feki et al. 2020).

2.2.3 Influence of Microorganisms in the Rhizosphere

Microorganisms found in the rhizosphere can affect the roots of plants by changing the movement of carbon compounds or nutrients starting from the roots to shoots (Pett-Ridge et al. 2021). While many soil microorganisms do not interact with plant roots due to the release of specific antimicrobial root exudates, other factors also

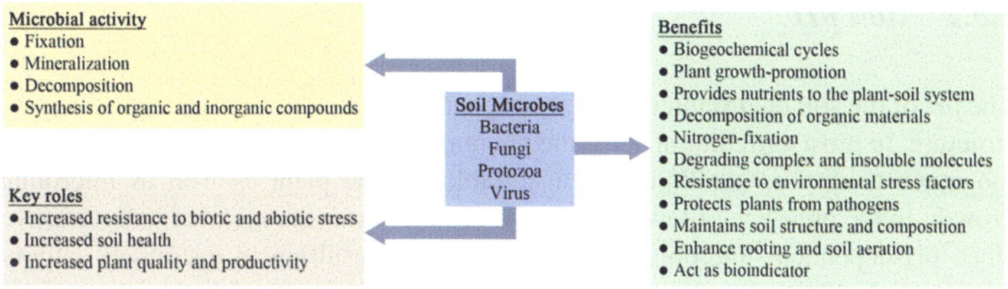

Fig. 2.2 Beneficial role of microbes in agricultural soils

contribute to this non-interaction. These factors include the physical and chemical properties of the soil, competition among microbial species, and the presence of plant immune responses (Chepsergon and Moleleki 2023). However, there are some microorganisms that do interact with the plant root system. However, there are some microorganisms which interact with the plant root system. These interaction can be of various types, such as symbiotic (beneficial for plant growth), pathogenic (invading and killing plant roots), saprophytic (living on dead roots and plants). The beneficial interactions are legume nodulation, mycorrhizae, and the production of some antimicrobial compounds that inhibit pathogenic growth (Phour et al. 2020). A schematic diagram representing the beneficial role of microorganisms in the rhizospheric region of agricultural soil is shown in Fig. 2.2.

2.3 Factors Affecting Microbial Diversity in the Soil

The composition of microbial communities in agricultural soils is affected by several factors. A combination of physical, chemical, and biological factors influences microbial community composition, soil type, and soil texture.

2.3.1 Soil Structure

An ideal soil structure is granular or crumbly, which allows air and water to pass through a wide range of pore sizes. These soil aggregates are held together by clay-humus complexes combined with calcium, which helps bind the aggregates together (Naresh et al. 2018). The spaces between the aggregates allow plant roots to extend down and soil microbes to move through them. In addition to being stable, an ideal soil structure is also resistant to erosion. Soil stability is enhanced by glutinous by-products of soil bacteria and hair-like threads of actinomycetes and fungi mycelium (Musbau et al. 2021).

2.3.2 Soil pH

The pH of the soil environment directly impacts microbial diversity and community structure. In most agricultural soils, the pH ranges between 4 and 8.5, and soils with too much acidity or alkalinity are unsuitable for the plant as well as microbial growth. In the presence of low pH, most microorganisms are metabolically inactive since pH-dependent microbial enzymes denature, resulting in decreased activity (Wang et al. 2019). In addition, low pH affects nutrition intake, cytoplasmic pH homeostasis and DNA stability (Gashu et al. 2021). A variety of factors may contribute to soil acidification, including nutrient cycling, acidic rainfall, organic matter decomposition, plant growth, fertilizer application and weathering. On the other hand, the natural eroding of carbonate-containing compounds and silicates as well as over-liming and mining bauxite residues, causes soil alkalinity (Khalidy et al. 2021). Additionally, droughts can also contribute to soil alkalinity by limiting water flow to the upper soil profile resulting in salt accumulation (Msimbira and Smith 2020). In acidic soils, *Acidobacteria* and *Alphaproteobacteria* dominate, whereas *Actinobacteria* become abundant as the soil becomes more alkaline (Shi et al. 2021). In general, fungal species can grow over a pH range of 5–9 without experiencing significant growth inhibition. However, some fungal species can indeed thrive in more acidic environments, capable of growing at pH levels below 5 (Taylor and Bhatnagar 2024). This allows certain fungi to colonize and persist in highly acidic habitats. In order to identify microorganisms based on their pH requirements, they are often divided into three groups: acidophiles (acid-loving), neutrophiles (grow at neutral pH), and alkalophiles (grow at alkaline pH).

2.3.3 Temperature

The temperature of soil surface usually ranges between 0 and 60 °C, but may fluctuate daily, weekly, or seasonally. Furthermore, insolation angle, slope, shading, soil colour, depth and moisture affect soil temperature. The majority of microbes are able to thrive at temperatures around 40 °C (Fan et al. 2021). Based on their temperature growth response, microorganisms are classified into three groups: thermophilic (warm-lovers), mesophilic (temperate-lovers) and cryophilic or psychrophilic (cold-lovers). In most microorganisms, high temperatures cause irreversible denaturation of proteins, which interrupts metabolism (Fu et al. 2020). In addition to affecting microbial cell membrane permeability, high temperatures affect the integrity of cell walls. While, at cold temperature, the membranes become stiff and brittle (Siliakus et al. 2017). Therefore, microbial communities prefer mild temperatures and mesophilic conditions for their growth and development.

2.3.4 Rainfall

Rainfall and precipitation events severely affect soil microbial growth and community composition. An erratic precipitation event may cause flooding, which can overwhelm pores and dramatically increase moisture levels. Microbial respiratory responses may be re-adapted because of altered precipitation regimes. However, prolonged flooding can deplete soil resources, disrupt soil chemistry, impact microbial communities and restrict their metabolism (Ye et al. 2020). A cycle of drying and rewetting can stress soil bacteria and inhibit the growth of fungi (Li et al. 2023). Excessive content of soil water can lead to a reduction in the diffusion of O_2 since diffusion of O_2 in water is significantly lower than that of air, which reduces the activity of aerobic bacteria but may boost the activity of anaerobes.

2.3.5 Moisture

Moisture regulates respiration in soil and therefore serves as an important factor for determining soil dynamics. An insufficient amount of moisture limits organic matter decomposition, microbial activity, and the mineralization of nutrients (Wang et al. 2016). On the other hand, the physiological stress imposed on soil microorganisms by increased soil moisture may increase nitrogen demand, which can result in a reduction in carbon use efficiency (CUE) and an increase in carbon losses through respiration (Iven et al. 2023). Additionally, increasing soil moisture content decreases redox potential and enhances oxygen consumption. Microorganisms may adapt to decreasing soil moisture by using strategies such as osmotic acclimation, resource changes, extracellular polymer production, and reducing metabolism (Brangarí et al. 2020).

2.3.6 Soil Compaction

Compression of soil reduces porosity and air permeability, restricting air and oxygen availability, and alters the microbial population and function of the soil (Beylich et al. 2010). The compaction of soil increases bulk density and changes pore size distributions. Consequently, soil health and the microhabitats of microorganisms are substantially impacted, affecting soil functions and ecosystem health.

2.3.7 Clay Mineral

In most soils, clay is the dominant mineral fraction influencing the growth of microorganisms. Clay minerals act as growth inhibitors or promoters for microorganisms by giving them food and a place to live. Additionally, it influences microbial activity by modifying the chemical and physical features of the microbial habitat. As a cyclical source of inorganic compounds, clays provide microorganisms with K^+ and NH^-, which are essential for growth and development (Fomina and Skorochod 2020). Clay minerals are also associated with other microbes like fungi and actinomycetes, but bacteria occupy a bigger niche. Furthermore, the formation, degradation, transformation, dissolution, redox, and adsorption of clay minerals are mostly carried out by bacteria (Perelomov 2023).

2.3.8 *Soil Biofilm*

A soil biofilm is inhabited by bacteria, fungi and other microorganisms, which colonise mineral surfaces, pores and plant roots in most cases. In developing a biofilm matrix, extracellular polymeric substances (EPS) mediate surface interactions and maintain a biofilm's cohesiveness (Costa et al. 2018). A diverse and metabolically active microbial community is good for soil fertility and ecosystem health development due to soil biofilm formation (Wu et al. 2019). In some cases, different microbial species may share a common micro-niche. Their coexistence benefits the microorganism's metabolic activities and contributes to a healthy ecosystem and plant development (Ge et al. 2023). Also, biofilm-covered soil responds quickly to nutrient amendment and speeds up biogeochemical processes in the soil.

2.3.9 *Pore Size*

The pores in soil aggregates facilitate the entry of microorganisms and their colonization. The pore size also influences water and air diffusion through aggregates. In the classification proposed by Zhao et al. (2017), pores are categorized based on their equivalent diameter into four classes: micropores (<30 μm), mesopores (30–75 μm), mesopores (75–100 μm), and macropores (>100 μm). In aggregates, microorganisms typically occupy 0.2–0.4% of the pore space; however, they are more frequently found on the exterior (Wu et al. 2023). There is the possibility that microbes can be trapped within aggregates and remain there until they are dispersed.

2.3.10 *Reduction-Oxidation Potential*

The reduction-oxidation reaction (or redox reaction) are common in anoxic environments, and microbes react to it in a variety of ways. Living organisms consist primarily of six elements, oxygen being a powerful oxidant and hydrogen being a powerful reductant. The remaining four elements with the largest redox number amplitudes are carbon, nitrogen, phosphorus, and sulphur (Dayo-Olagbende et al.

2019). Redox potential is essential in determining substrate availability in the soil environment and regulating microbial community richness and biodiversity. Aerobic bacteria such as *Actinomyces* sp. or *Azotobacter* sp. require high redox potential values for their growth (Tokarz and Urban 2015). Certain anaerobic bacteria survive best in greatly reduced conditions. Fungi may thrive in oxidised and moderately reduced soils, but their survival may suffer in highly reduced soils (Gadd 2007).

2.3.11 Tillage and Management Effects

Soil microorganisms are significantly influenced by tillage practices, which alter the physicochemical characteristics of the soil. Changes in microbial populations can affect soil fertility, nutrient availability, and other ecological processes (Astapati and Nath 2023; Philippot et al. 2023). Tillage with a mouldboard plough causes mechanical soil inversion, mixing plant residues into the tilled layer and limiting weed growth (Oduma et al. 2020). However, intensive ploughing can damage soil structure over time and accelerate the breakdown of organic matter (Rahman et al. 2020). During tillage, soil nutrients are periodically inverted and evenly distributed across the tilled layer, leading to homogenized soils with reduced microbial biomass and diversity. Conversely, conservation tillage methods, including no-tillage, promote a shift from bacteria-dominated to fungi-dominated communities due to less frequent soil disturbance. These methods also support free-living diazotrophs and enhance rhizosphere nitrogen fixation (Acosta-Martinez et al. 2007) Soil management practices further influence microbial diversity and distribution. The plough layer, directly affected by tillage, shows increased microbial activity at all depths post-tillage due to aggregate disturbance, improved aeration, and the incorporation of organic materials (Chellappa et al. 2021). No-tillage techniques, often employed to reduce soil erosion, involve minimal soil disturbance during planting operations. This approach results in stratified microbial activity and populations near the soil surface (Carretta et al. 2021).

2.3.12 Fertilizer and Other Chemical Applications

In agricultural soil, fertilizer application greatly affects microbial populations and diversity, besides altering soil pH drastically. Fertilizers contain high chemical concentrations, which cause salt stress and chemical toxicity (Pahalvi et al. 2021). The microbial populations, especially fungi, are more adversely affected by fungicides and fumigants because they are much more specific to the microorganisms. While the text mentions that herbicides and foliar insecticides are not usually applied at harmful concentrations, there have been reports indicating instances where these chemicals are applied at high concentrations. Such applications, as well as runoff, can potentially impact soil microorganisms adversely. For instance, Xiao et al.

(2023) observed that elevated levels of herbicide runoff can lead to significant shifts in soil microbial communities and reduce microbial diversity. Similarly, high concentrations of foliar insecticides have been documented to cause soil contamination, affecting both microbial population and activity (Naha et al. 2020; Yu et al. 2023).

2.3.13 *Nutrient Availability and Cycling*

Microbial metabolic activity in the soil is closely tied to the balance of nutrient stoichiometry. Microorganisms play a crucial role in nutrient cycling, which is essential for making nutrients available to plants (Iqbal et al. 2023). They utilize various processes such as oxidation, ammonification, and nitrogen fixation to convert nitrogen into forms that plants can use. Additionally, microorganisms decompose organic matter, releasing essential inorganic nutrients like nitrate, sulphate, and phosphate into the soil through processes like nitrification, sulphur oxidation, and phosphate mineralization (Basu et al. 2021). These activities ensure the cycling of important inorganic nutrients, maintaining a healthy and functional soil system. However, soil disturbances such as land use change or soil cultivation can disrupt microbial populations and negatively impact the soil nutrient cycle (Józefowska et al. 2020).

2.3.14 *Soil Salinity*

High salt concentrations reduce microbe activity and biomass, resulting in altered microbial communities. The reduction in microbial biomass by salinity is primarily caused by osmotic stress, which results in cell drying and lysis (Haj-Amor et al. 2022). In addition, microbes have difficulties removing water from soil with low osmotic potential. However, some microbes adapt to low osmotic potential by accumulating osmolytes, requiring a large amount of energy for synthesis, resulting in diminished activity and growth (Meena et al. 2022). Microbes that can tolerate a high salt concentration accumulate proline, glycine and betaine as organic osmolytes, while potassium cations make up a large amount of inorganic osmolytes (Tahjib-Ul-Arif et al. 2021). A study on soil enzymes conducted by Yan et al. (2015) reported that salinity strongly inhibits urease, alkaline phosphatase, and beta-glucosidase activity, whereas dehydrogenase and catalase are less affected.

2.4 Role and Diversity of Microbes in Soil Ecosystem

The soil ecosystem is a thriving habitat for an extensive and diverse community of microorganisms that significantly influence soil fertility and enhance crop production. Soil microorganisms are important in nutrient cycling, litter decomposition, and crop yield (Fig. 2.3). The physical characteristics of soil, such as bulk density, exhibit variations from the surface to the subsurface layers of the soil profile. These variations contribute to significant differences in the population of microorganisms with depth (Fig. 2.4). In agricultural soil, bacteria regulate various processes, including plant growth hormone, pathogen inhibition, soil composition improvement, and microbial leaching. The majority of indigenous bacteria are involved in the restoration or cleaning up of contaminated environments (Arunraja et al. 2023). According to Ingham (2009), soil bacteria are divided into four main functional groups: pathogens, lithotrophs, decomposers and mutualists. Fungi, another pivotal component of soil microbial communities, contribute significantly to the degradation and stabilization of organic materials. Their extensive mycelial networks break down complex organic compounds, releasing vital nutrients into the soil. Noteworthy examples include fungi like *Streptomyces* sp., which plays a crucial role in complex material decomposition, humus formation, and nutrient recycling (Buzón-Durán et al. 2020). Additionally, fungi establish mutualistic associations with various organisms, such as mycorrhizal relationships with plants. Ectomycorrhizae, involving fungi like *Amanita muscaria*, form external sheaths around plant roots, enhancing nutrient uptake and promoting plant growth (Dreischhoff et al. 2020).

Algae, as simple autotrophic organisms, are dynamic contributors to soil development. They are found in various soil types, and reported to enhance soil properties such as texture, aeration, and carbon content (Ibraheem 2007). Their production of hormones, vitamins, amino acids, and organic acids promotes plant growth, making them valuable bio-fertilizers (Choudhary et al. 2024). Another group of soil microorganism is protozoa, which represent a vital component of the soil microbiota. Ranging from flagellates to ciliates and amoebae, these organisms contribute to nutrient turnover by consuming bacteria, fungi, and organic materials. Examples such as amoebae, exhibiting phagotrophic feeding strategies, play a crucial role in maintaining microbial balance in soil aggregates and influencing nutrient availability for plants (Murase and Asiloglu 2023). Soil viruses, although often overlooked, also play a crucial role in microbial community dynamics, organic matter decomposition, and nutrient cycling (Kuzyakov and Mason-Jones 2018). The diversity and abundance of soil viruses, particularly in temperate agricultural soils, contribute to gene transfer and microbial mortality. Understanding the impact of soil pH, organic matter content, and moisture on viral abundance and diversity is crucial for comprehending their role in soil ecosystems (Bhatti et al. 2017).

2.4.1 Bacteria

Bacteria are ubiquitous, free-living organisms found in almost all habitats. There are an estimated 60,000 different types of bacteria, each with its own set of functions (Etesami et al. 2017). Free-living soil bacteria frequently colonize plant roots, supplying organic nitrogen and other vital nutrients to the plant-soil system. These plant growth-promoting rhizobacteria (PGPR) help to reduce the use of nitrogen

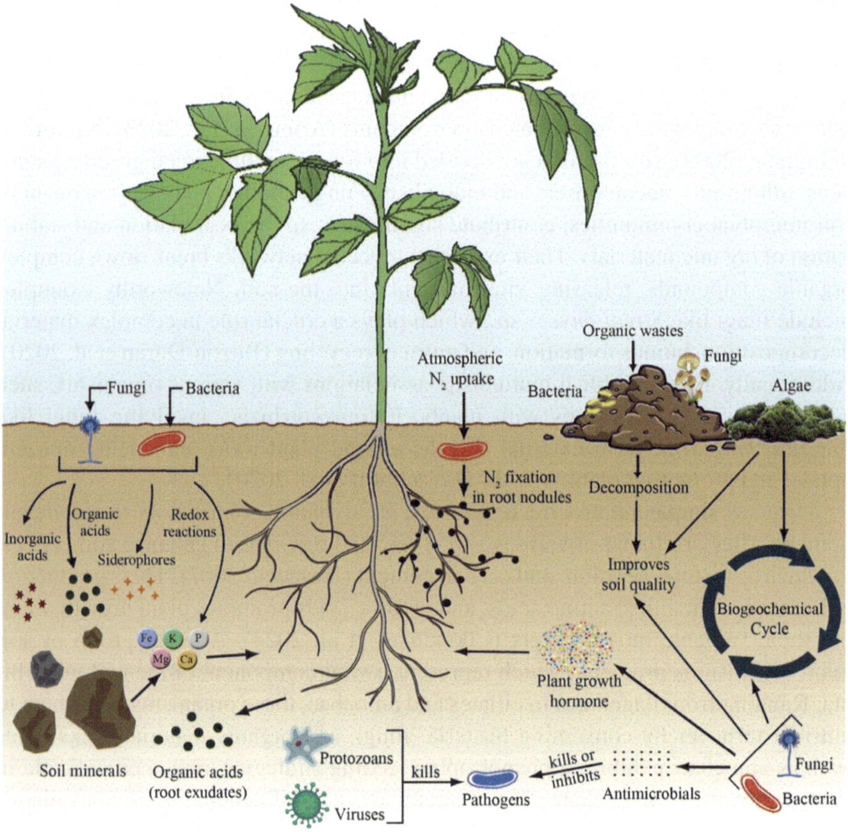

Fig. 2.3 Interrelation of soil, microbes and plants in the ecosystem

and phosphorus fertilizers. Soil bacteria also play an important role in biogeochemical cycles, and their interaction in the rhizosphere improves plant health and soil fertility. Bacteria are particularly critical for decomposing organic materials in the early phases of decomposition when moisture levels are high. They break down organic materials and release simple molecules into the environment that other living things can use. Some noteworthy examples of decomposing bacteria are *Xanthomonas oryzae*, *Bacillus subtilis*, *Streptomyces* sp., *Acidothermus* sp., *Pseudomonas* sp., etc.

Nitrogen is a critical plant nutrient that is frequently used to fertilize crops and increase yields. The root nodules of leguminous plants are symbiotically associated with nitrogen-fixing bacteria, which fix atmospheric nitrogen and produce ammonia that the plants can use. *Azospirillum* sp., *Azotobacter* sp., *Herbaspirillum* sp., *Bacillus* sp., *Burkholderia* sp., *Pseudomonas* sp., *Rhizobium* sp., and

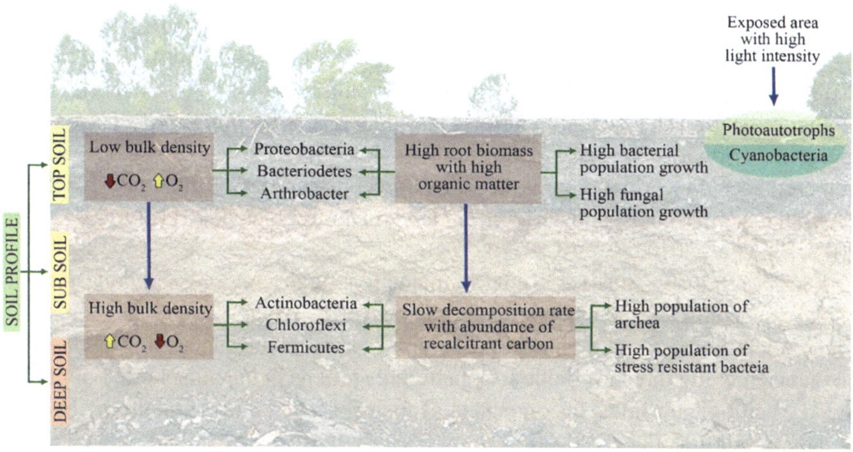

Fig. 2.4 Existence of microbial categories in the soil profile corresponding to the vertical soil depths

Gluconacetobacter sp. are some examples of nitrogen-fixing bacteria (Pedraza 2008). Bacteria use various processes to produce energy for growth and other activities, and based on that, they are broadly categorized as aerobic, anaerobic, and facultative. Aerobic bacteria require molecular oxygen as a terminal electron acceptor and cannot grow in its absence. Some important aerobic bacteria in soil are *Coccidioides* sp., *Streptomyces sp.*, etc. In the case of anaerobic bacteria, they do not need oxygen for their growth and may find oxygen toxic. They depend on other substances as electron acceptors. *Bacteroides* sp., *Fusobacterium* sp. and *Prevotella* sp., are some important examples of anaerobic bacteria. Facultative microbes are more versatile that can engage in aerobic respiration to provide energy but can switch to anaerobic respiration depending on the oxygen levels and fermentable materials in the surrounding environment (Hentges 1996). Actinobacteria is a group of gram-positive bacteria that shares both the characteristics of fungi and bacteria. Actinobacteria such as *Nocardioform* sp., *Micromonospora* sp., *Streptomyces* sp., etc. are widely spread in the soil environment where they play an important role in complex material decomposition, humus formation, secondary metabolite production and nutrient recycling in the soil (Anandan et al. 2016).

2.4.2 *Fungi*

Fungi of the plant kingdom are a significant group in their contribution to biodiversity. In the ecosystem, fungi play a functional role, particularly in carbon cycling, nutrient uptake and the manifestation of plant diseases. Fungi are distributed worldwide, but the maximum diversity of fungal species occurs in warm, humid tropical and terrestrial soils. Fungi in soil share a significant proportion of the microbial population and exhibit their vital role as decomposers, parasites and symbionts (Peay et al. 2008), affecting soil fertility and plant productivity (Maron et al. 2011). There are limitations in the traditional culture techniques for the isolation and identification of species; therefore, less than 2% of the fungal species are enumerated worldwide. Contrary to the old methodology, DNA sequencing has recently gathered acceptance in studying the fungal diversity of various habitats, viz. forest ecosystems, grasslands, and agricultural lands (Dumbrell et al. 2011; Orgiazzi et al. 2012; Shi et al. 2014).

Fungi even thrive in adverse environments like tundra regions and undergo beneficial association with autotrophic algae to form lichen (Biradar et al. 2023). Metabolically, fungi can disintegrate complex organic substances and release the elements, making them available for plants and animals in the biosphere. Fungi play the key ecological role as decomposers of the ecosystem and maintain the balance of nature. Fungi act as a recycling agent because they make available the elements of life as nitrogen and phosphorus, into the atmosphere and soil. As a result of the metabolic activity, the organic compounds locked in the rotten woods and other decaying matter are transformed into simpler elements, including trace elements essential for growth and development. The mode of nutrition and the production of exoenzymes digest the nutrients, thereby degrading complex and insoluble molecules. Fungal enzymes either act upon the substrate or remain bound outside the cell wall, breaking the large molecules into smaller ones (Kumar and Shah 2021). A humid environment is congenial for active growth as the transportation of smaller molecules and enzymes into the cell occurs in the presence of water. One of the basic modes of nutrition in fungi is saprophytic, which helps replenish the ecosystem and thereby maintains sustainability.

Fungi form mutualistic associations with a variety of organisms like cyanobacteria, algae, plants, and animals. To justify the phenomenon of mutualism, the prominent example is the association of fungi with innumerable angiospermic species forming mycorrhiza. In a mycorrhizal association, the fungal mycelia extensively spread their hyphal network covering a large surface area of the soil, thereby mobilizing minerals and water for uptake by the roots of the plants. The higher plants, in return, provide the finished product of photosynthesis to fungi for effective metabolic activities. Mycorrhizae are of various types. Ectomycorrhizae, meaning "external" mycorrhiza, where fungi envelop the roots in a sheath called a mantle as well as a 'Hartig net' of hyphae that penetrate the cells of roots (Baptista et al. 2011). In ectomycorrhizal association, the fungal partner belongs to the class Ascomycota, Basidiomycota or Zygomycota. In another type, fungi belonging to Glomeromycetes form vesicular–arbuscular connections with arbuscular mycorrhiza, also called endomycorrhizae (Lara-Capistrán et al. 2021). In this type, arbuscules formed by fungi penetrate root cells and represent the region of metabolic activity between the fungus and the host plant. Morphologically, arbuscules have a shrubby appearance. The epiphytic orchids also have mycorrhizal associations. They produce small seeds that will germinate only with the association of a mycorrhizal partner (a Basidiomycetes fungi). In the nutrient-poor seeds, the fungal partner promotes germination and growth of the orchid by providing necessary micro and macro nutrients. In a positive interaction, hyphae of *Amanita muscaria* (fly amanita) infect *Pinus radiata* (Monterey pine) roots, inducing the production of many tiny, branched rootlets (Bigg 2000). The hyphal network covers the small roots with a rigid mantle. Other examples related to fungus–plant mutualism include endophytes, i.e., fungi that live inside tissue without causing harm to the host plant. Endophytic fungi release toxins that benefit the higher plants in their growth and development. The endophytes drive away herbivores, confer resistance to environmental stresses such as drought and heavy metals in soil, and protect against other infectious organisms (Basit et al. 2021).

2.4.3 Virus

In soil ecosystems, viruses play an important role in microbial community dynamics, microbial metabolism, organic matter decomposition, and nutrient cycling (Emerson 2019). The abundance and diversity of soil viruses, especially in temperate agricultural soils, are high, usually around 10^9 per gram of soil dry weight (Liang et al. 2019). A positive correlation exists between virus production and virus-host contact frequency in the environment. Minerals and organic matter, which are important in viral attachment and detachment, may dictate the speeds at which viruses contact the host through the soil environment as a heterogeneous porous media. Furthermore, previous studies have shown that soil pH, organic matter content, and moisture can also affect the viral abundance, diversity, and reproduction strategies (Finke et al. 2017; Narr et al. 2017).

Soil viruses also play a significant role in infecting host cells through lysis and lysogeny. In the process of viral lysis, dissolved organic carbon and other nutrients are returned to the dissolved pool (Sullivan et al. 2017). The release of nutrients through viral transmission greatly affects soil respiration, fertility, and nutrient cycling. Lysogeny facilitates bacterial niche expansion and increases bacterial dispersal. It has been shown that lysogenic conversion by temperate phages can alter rhizobial phenotypes in various ways, affecting root nodule colonization, N-fixation efficiency, and crop productivity (Hyde et al. 2024). Bacteriophages in soils may have the direct capacity to regulate the population size, structure and activity of the bacterial communities (Schoelz and Stewart 2018).

2.4.4 Protozoa

Protozoa are unicellular, heterotrophic, eukaryotic creatures that consume bacteria, fungi, and organic materials. They are ubiquitous and reside in soil ecosystems that are devoid of higher species due to harsh environmental circumstances. According to their mode of eating and movement, protozoa are classified as ciliates, amoebae, and flagellates. The effectiveness of soil formation and soil fertility are both impacted by variations in the abundance and richness of protozoa. Certain variables, such as absorbent space structure, a permanent water source, and interactions with other organisms, are required for protozoan existence and action (Esteban and Fenchel 2020). These factors serve as the foundation for two critical adaptive responses: small body size and the ability to form protective quiescent cysts. Protozoa release excess nitrogen when they consume bacteria, which plants and other food web members may use.

The free-living protozoa of soils belong to two phyla, the Sarcomastigophora and the Ciliophora. The oldest group with a polyphyletic origin is the flagellates, named for their one or more flagella or whip-like propulsive organs. The majority of them have a modest and basic cell structure, and their body length ranges from 4 to 15 mm. Flagella are used for anchoring and swimming. Moreover, it can also be used to create a feeding current that draws bacterial cells, which are then carried to the base of the flagellum and consumed. They contribute significantly to nutrient turnover, with bacteria serving as their primary prey (Rønn et al. 2012). Another active and diverse group of soil protozoa is naked amoebae, which are ravenous and are most commonly found in agricultural, grassland, and forest soils (Chandarana and Amaresan 2022). They have a body resembling a jelly lump, measuring between 15 and 100 mm in width but only 1 mm in thickness when adhered to a surface. In general, this group of organisms has morphological features known as pseudopodia of various shapes. The mechanism of nutrition for amoeba is by engulfing, also known as phagotrophic, and as part of its feeding strategy, it can consume fungus, algae, bacteria, and other organisms, including tiny particle organic matter accounting for the bulk of the ingested material (Clarholm et al. 2007). They are able to enter incredibly small openings or cavities in soil aggregates and eat microorganisms that are normally thought to be unreachable to predators. On the other hand, the testate amoebae are a dynamic polyphyletic group that produces a shell-like covering. They are mostly found in wetlands, freshwater and terrestrial habitats, although they are common in damp, acidic soils and peats with high levels of organic matter and low nutrient yields. They are correlated with mosses in a dusty environment, like those that grow on rocks and tree bark. In moist environments, ciliates are the largest protozoans, which move *via* hair-like cilia. When ciliates penetrate soil cavities and pores, they become exceedingly active and thus start consuming food sources of bacteria (Foissner 1987). Similar to other protozoa, they have encysted forms, which appear when food is available, and the environment is conducive for growth and reproduction.

Protozoa stimulate plant growth in a number of ways. They can do so directly by increasing the availability of nitrogen in the soil and evaporating water, or they can perform it indirectly by preferentially feeding on soil microorganisms and changing the amounts of plant hormones in the rhizosphere. Protozoa stimulate the growth of bacterial populations by consuming bacteria, thereby enhancing decomposition rates and soil aggregation (Shah et al. 2021). They also aid in disease suppression by competing with or feeding on pathogens, as they are key food sources for other soil organisms. They contribute significantly to mineralizing nutrients by making them accessible to plants and other soil organisms. Protozoa use more nitrogen than they require because they feed on bacteria. It often occurs near a plant's root system, boosting root system development and nutrient intake. In the ammonification process, most of the ammonium is taken up by bacteria and other organisms, with very little being utilized by the plant. Phosphate is the least mobile nutrient in the soil, and continual micro-disturbance by protozoa is especially crucial for microbial phosphate availability. Protozoa also reduce phosphate leaching significantly.

2.5 Microbial Interactions in the Soil

Soil microorganisms in plant roots in the rhizosphere interact with soil elements and other microbial communities (Barea et al. 2005). These contacts between cells can be beneficial to one or more organisms or lead to competition between the cells that harm one or more species in the environment (Fig. 2.5). The interactions between microbial communities contribute towards improved plant growth and overall agro-ecosystem sustainability. Microbial interaction in the rhizospheric region falls into intraspecific and interspecific interactions.

2.5.1 Intraspecific Interactions

Intraspecific interactions occur between individuals of particular microbial populations. It is further classified into positive and negative interactions.

2.5.1.1 Positive Interaction

Positive interactions are friendly relationships between species that improve at least one species' development, reproduction, and survival without having an adverse effect on the other species (Tarnita 2017). This cooperative interaction is often observed when microorganisms use enzymes to break down non-soluble substrates like cellulose and lignin (Berg et al. 2020). Additionally, through genetic exchange between members of a population through conjugation, transformation or transduction, positive interaction aids in developing tolerance to various abiotic pressures (Aminov 2011).

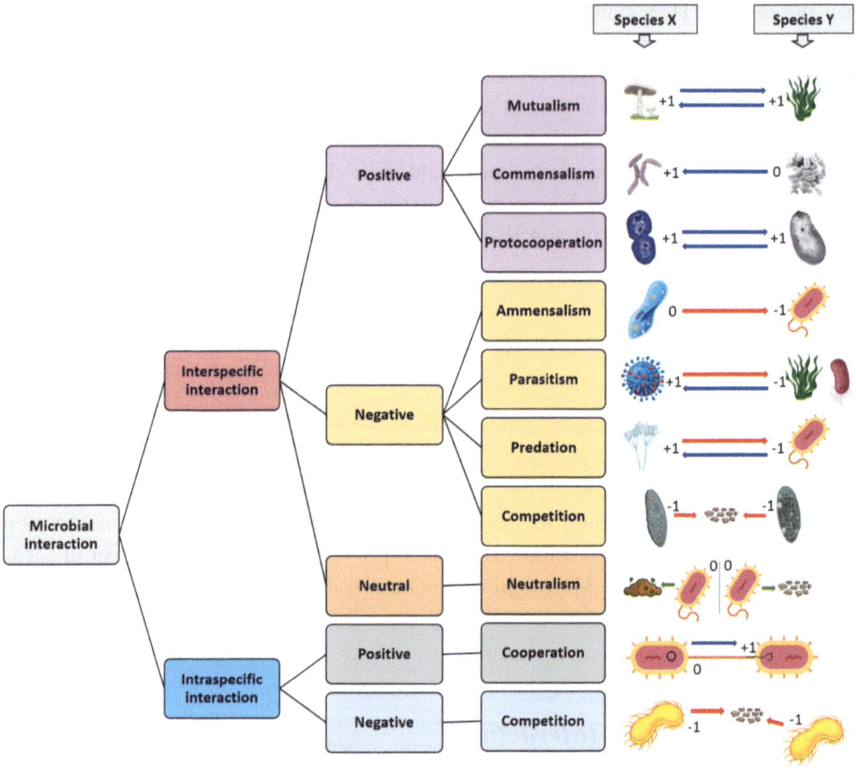

Fig. 2.5 Classification of biotic interaction in the rhizospheric region. The participating species are denoted as X and Y. The interactions between the species are denoted as +1 (positive effect), −1 (negative effect) and 0 (no effect)

2.5.1.2 Negative Interaction

Intraspecific competition arises when individuals within the same population interact negatively (i.e., competition within a population). Such interactions are often observed in microbial species in natural habitats where the nutrient availability is low and substrate requirements are similar. Intraspecific interaction also ensures that individuals in a population are distributed equally in an ecosystem (Atlas 1993). It is thus considered one of the most important characteristics in shaping soil microbiota density and size.

2.5.2 *Interspecific Interactions*

Interspecific interactions play a significant role in population dynamics. This interaction takes place between the members of different species. Positive interactions facilitate by increasing the growth rate, whereas negative interactions reduce the growth rate. Most microbial interactions are represented by assuming that they occur between two distinct populations, X and Y (Table 1) (Martin and Schwab 2012, 2013). Interspecific interactions is categorised into neutral, positive and negative interactions (Table 2.1).

2.5.2.1 Neutral Interaction (Neutralism)

Neutralism is an unaltered interaction of different microorganisms that exist in the same environment. Such correlations tend to be sporadic because environmental factors fluctuate constantly. A neutral interaction occurs when two species differ in cultural traits and have dissimilar nutritional requirements (Selim and Zayed 2017; Weiner et al. 2012).

2.5.2.2 Positive Interaction

Positive interactions among individuals, whether of the same or different species, give rise to cooperative relationships that mutually benefit the parties involved. Organisms from separate populations can establish stable, temporary, and necessary relationships through this interaction. The three most prevalent positive interactions are mutualism, protocooperation, and commensalism. Mutualism entails a specific interaction between two populations with an obligatory association, resulting in

Table 2.1 Types of interaction between microorganisms and host (Martin and Schwab 2012, 2013)

Types	Characteristics	Species X	Species Y
Neutralism	No interaction	0	0
Mutualism	Interaction needed to survive in a habitat, and specific species are required	+1	+1
Commensalism	One is benefited & the other is not harmed or helped	+1	0
Protocooperation	Interaction is necessary for survival in a habitat, but a specific type of species is not necessary	+1	+1
Ammensalism	Product of one impact other	0	−1
Parasitism	Host is usually compromised	+1	−1
Predation	Host is usually compromised	+1	−1
Competition	Organisms in the environment attempt to acquire limited nutrients	−1	−1

The participating species are denoted as X and Y. The interactions between the species are denoted as +1 (positive effect), −1 (negative effect) and 0 (no effect)

mutual benefits for both partners (Koskella and Bergelson 2020). In this association, the mutualist and the host are metabolically dependent on each other. During cohabitation, individual symbionts exhibit different physical tolerances and metabolic rates. Lichen serves as a classic example of mutualism, involving specific ascomycetes (mycobiont) and particular genera of either cyanobacteria or green algae (phycobiont) (Sanders and Masumoto 2021). The carbohydrates produced by the photosynthesis of cyanobacteria or green algae benefit the fungus, which, in turn, obtains nutrients through fungal hyphae projections, performing respiration using the O_2 produced during phycobiont photophosphorylation. The fungus provides a stable substrate, creating an environment for cyanobacteria or green algae to flourish.

Protocooperation, also known as non-obligatory mutualism, also represents a positive type of population interaction. It involves a temporary association where two interacting species benefit each other. In this interaction, although both species can survive independently, they interact favorably with one another. Unlike mutualism, it does not depend on the association for the survival of any one species. An example of protocooperation is the association between *Desulfovibrio* sp. and *Chromatium* sp. in the sulfur and carbon cycles (Sarsan et al. 2021). Studies had reported that interaction between cellulolytic bacteria, such as *Cellulomonas* sp. and N_2-fixing bacteria also exhibits protocooperation (Harindintwali et al. 2020). Another type of positive interaction is commensalism, in which members of different species engage with each other, with one species deriving benefits while the other is neither harmed nor benefited. The two populations coexist without physical contact, with each deriving benefits without affecting the other. Such interactions are often observed in soil bacteria, where waste material from one species serves as a substrate for another (Rodríguez-Verdugo 2021). Bacteria like *Nitrosomonas* sp. form nitrite using ammonium as a substrate during the nitrification process. Additionally, the bacteria *Nitrobacter* associated with *Nitrosomonas* sp. uses nitrite as a substrate to obtain energy for growth and further oxidizes it to nitrate (Baskaran et al. 2020).

2.5.2.3 Negative Interaction

Interspecific interactions also exert a negative effect on participating species or a group of populations. In such negative interactions, different populations compete with each other for food or prey on members of the other population. Negative interactions among populations include predation, parasitism, amensalism, and competition.

In predation, one organism, also called a predator, engulfs or attacks other organisms or prey, usually resulting in the death of the prey. One such interaction is protozoan bacteria in soil, where protozoans can feed on different bacterial populations and thus help to keep the bacteria count in the soil at an ideal level (Rossine et al. 2022). Parasitism is another type of association in which one population (i.e., the parasite) benefits by receiving nutrients from another (the host), causing harm to the host. The relationship between a host and a parasite is characterized by a considerable amount of time spent, either physically or metabolically. Some parasites, known as endoparasites, reside inside the host cell, whereas others, known as ectoparasites, reside outside the host cell (Görres and Amador 2021). Viruses are obligate intracellular parasites that exhibit great host specificity and act upon bacteria, fungi, algae, protozoa, etc. (Hurst 2011). On the other hand, bacteria such as *Bdellovibrio* act as ectoparasites to many gram-negative bacteria (Ebana et al. 2019).

Amensalism, also known as antagonism, is a negative type of interaction where one species causes harm to another organism without any benefits to itself. As a result of such interactions, organisms that produce inhibitory substances remain unaffected, or they may gain a competitive advantage and survive in the habitat, while organisms that do not produce inhibitory substances are inhibited. This chemical inhibition is known as antibiosis. An example of antibiosis is the interaction between *Penicillium* and bacteria (Boddy 2016). The mold *Penicillium* creates the secretion known as penicillin, which is extremely toxic to bacteria. In another type of interspecific interactions interaction, two microbial populations are negatively impacted by competition in terms of their growth and survival. This type of interaction occurs when two populations utilize the same resources, such as the same nutrition or space. It leads to a decrease in the growth rate of both populations. The microbial population competes for growth-limiting resources such as nitrogen, carbon, vitamins, growth factors, phosphorus, etc. This type of interaction is observed between *Paramecium caudatum* and *Paramecium aurelia*, where both species of *Paramecium* feed on the same bacteria population (DeLong et al. 2023).

2.6 Soil Enzymes

Soil enzymes play a crucial role in the biochemical processes involved in recycling organic matter within soil systems. The activities of these enzymes are closely tied to soil physical properties, soil organic matter (SOM), and microbial activity or biomass. Soil enzymes, acting as essential catalysts, significantly influence the transformation of energy, environmental quality, and agricultural productivity (Rao et al. 2014). Hydrolase and glucosidase, among various enzyme types, are particularly responsible for breaking down organic matter in the soil. Additionally, enzymes like amylase, urease, phosphatase, and sulfatases contribute to nutrient mineralization. Phosphatase activity facilitates the interrelationship between enzyme activity and nutrient availability or crop production (Neemisha and Sharma 2022).

Microbial communities and enzymes play vital roles in carbon (C), nitrogen (N), phosphorus (P), and sulfur (S) cycling in ecosystems. It is evident that soil microorganisms and enzyme activities regulate nutrient cycling in ecosystems. Various enzymes, such as amidases, arylsulfatase, cellulases, dehydrogenase, glucosidases, laccase, phosphatases, and urease, are responsible for the biochemical reactions in these cycles (Neemisha and Sharma 2022; Piotrowska-Długosz 2020). Soil enzymes are considered biological indicators, aiding in the understanding of soil

Table 2.2 Some important soil enzymes and their function in nutrient cycling

Enzyme	Reaction catalyzed	Product	Source	Functions
α-Amylase	Starch hydrolysis	Glucose	Plants and animals	C-cycling
β-Amylase	Starch hydrolysis	Maltose	Mainly plants	C-cycling
Dehydrogenase	Oxidation	NAD to NADP	Microorganisms	C-cycling, microbial oxidative activity
Endo-1, 4-β glucanase	Cellulose Endo-hydrolysis	Oligosaccharides	Microorganisms and protozoa	C-cycling
Phenol oxidase	Lignin hydrolysis	C compounds	Plants and microorganisms	C-cycling
Urease	Urea hydrolysis	Ammonia (NH_3) and CO_2	Plants and microorganisms	N-cycling
Alkaline phosphatase	Hydrolysis of esters and anhydrides of phosphoric acid	Phosphate (PO_4)	Mainly Bacteria	P-cycling
Arylsulfatase	Hydrolysis of sulfate esters	Sulfate (SO_4^{-2})	Microorganisms, plants, and animals	S-cycling
Protease	N mineralization	Plant available N	Microorganisms and plants	N-cycling
Chitinase	Degradation and hydrolysis of chitin	Carbohydrates and inorganic nitrogen	Microorganisms and plants	C- and N-cycling
β-glucosidase	Cellulose hydrolysis	Glucose	Bacteria, fungi, plants, and animals	C-cycling

physicochemical properties. Factors such as enzyme and substrate concentration, temperature, pH, cofactors, inhibitors, and ionic environment control enzyme activities (Nannipieri et al. 2018). Furthermore, enzyme activities in soil are influenced by cropping history, soil amendments, vegetation, agricultural chemicals, industrial pollutants, etc. (Burns et al. 2013; Zuccarini et al. 2023). A list of some essential soil enzymes contributing to soil stability is summarised in Table 2.2.

2.7 Conclusion

Microbial activity is critical to soil productivity and sustainability because it promotes a variety of fundamental soil properties. Typically, microbes are found in the rhizosphere of the soil and these rhizosphere zones are near the plant roots. They ensure that the relationship between plant microorganisms remains stable. There are many factors that have an impact on these microbial communities, like pH, temperature, moisture, rainfall, etc. There are various types of microbes present in soil which interact with each other. They have a wide application in soil ecosystems, such as biomass decomposition, mineralization, soil formation, nutrient cycling, etc. Soil microbes are important for the better functioning of the soil as well as the environment. With climate change and the rise of population, the microbial population and soil components can be affected. Thus, fundamental and strategic researches related to the conservation and mechanism of microbial interaction need to be done in order to improve specific soil functions and interactions.

Availability of Data and Materials Not applicable.

Consent for Publication Not applicable.

Competing Interests The authors declare that they have no competing interests.

References

Acosta-Martinez V, Mikha MM, Vigil MF (2007) Microbial communities and enzyme activities in soils under alternative crop rotations compared to wheat–fallow for the Central Great Plains. Appl Soil Ecol 37(1–2):41–52

Ahlawat OP, Yadav D, Walia N, Kashyap PL, Sharma P, Tiwari R (2024) Root exudates and their significance in abiotic stress amelioration in plants: a review. J Plant Growth Regul 43(6):1–26

Aminov RI (2011) Horizontal gene exchange in environmental microbiota. Front Microbiol 2:158

Anandan R, Dharumadurai D, Manogaran GP (2016) An introduction to actinobacteria. In: Actinobacteria-basics and biotechnological applications. IntechOpen

Arunraja D, Romauld SI, Devi PB, Thiruvengadam S, Kumar V (2023) Genetically engineered microbes for bioremediation and phytoremediation of contaminated environment. In: Metagenomics to bioremediation. Elsevier, pp 709–721

Astapati AD, Nath S (2023) The complex interplay between plant-microbe and virus interactions in sustainable agriculture: harnessing phytomicrobiomes for enhanced soil health, designer plants, resource use efficiency, and food security. Crop Design 2, 100028(1)

Atlas R (1993) Interactions among microbial populations. In: Microbial ecology: fundamentals and applications. Pearson Education India

Baptista P, Tavares RM, Lino-Neto T (2011) Signaling in ectomycorrhizal symbiosis establishment. In: Diversity and biotechnology of ectomycorrhizae. Springer, pp 157–175

Barea J-M, Pozo MJ, Azcon R, Azcon-Aguilar C (2005) Microbial co-operation in the rhizosphere. J Exp Bot 56(417):1761–1778

Basit A, Shah ST, Ullah I, Ullah I, Mohamed HI (2021) Microbial bioactive compounds produced by endophytes (bacteria and fungi) and their uses in plant health. In: Plant growth-promoting microbes for sustainable biotic and abiotic stress management. Springer, pp 285–318

Baskaran V, Patil PK, Antony ML, Avunje S, Nagaraju VT, Ghate SD, Nathamuni S, Dineshkumar N, Alavandi SV, Vijayan KK (2020) Microbial community profiling of ammonia and nitrite oxidizing bacterial enrichments from brackishwater ecosystems for mitigating nitrogen species. Sci Rep 10(1):5201

Basu S, Kumar G, Chhabra S, Prasad R (2021) Role of soil microbes in biogeochemical cycle for enhancing soil fertility. In: New and future developments in microbial biotechnology and bioengineering. Elsevier, pp 149–157

Berg B, McClaugherty C, Berg B, McClaugherty C (2020) Decomposer organisms. In: Plant litter: decomposition, humus formation, carbon sequestration. Springer, pp 45–65

Beylich A, Oberholzer H-R, Schrader S, Höper H, Wilke B-M (2010) Evaluation of soil compaction effects on soil biota and soil biological processes in soils. Soil Tillage Res 109(2):133–143

Bhatti AA, Haq S, Bhat RA (2017) Actinomycetes benefaction role in soil and plant health. Microb Pathog 111:458–467

Bigg WL (2000) Fungal succession and diversity in ectomycorrhizal associations: a case study approach. In: Proceedings of the California Forest Soils Council Conference on forest soils biology and forest management: February 23–24, 1996, Sacramento, California. US Department of Agriculture, Forest Service, Pacific Southwest Research Station, p 43

Biradar S, Ramya R, Sankaranarayanan A (2023) Mycobionts interactions in lichen. In: Microbial symbionts. Elsevier, pp 215–233

Boddy L (2016) Interactions between fungi and other microbes. In: The fungi. Elsevier, pp 337–360

Bodner G, Mentler A, Keiblinger K (2021) Plant roots for sustainable soil structure management in cropping systems. In: The root systems in sustainable agricultural intensification. Wiley, pp 45–90

Braga RM, Dourado MN, Araújo WL (2016) Microbial interactions: ecology in a molecular perspective. Braz J Microbiol 47(Suppl 1):86–98

Brangarí AC, Manzoni S, Rousk J (2020) A soil microbial model to analyze decoupled microbial growth and respiration during soil drying and rewetting. Soil Biol Biochem 148:107871

Burns RG, DeForest JL, Marxsen J, Sinsabaugh RL, Stromberger ME, Wallenstein MD, Weintraub MN, Zoppini A (2013) Soil enzymes in a changing environment: current knowledge and future directions. Soil Biol Biochem 58:216–234

Buzón-Durán L, Pérez-Lebeña E, Martín-Gil J, Sánchez-Báscones M, Martín-Ramos P (2020) Applications of Streptomyces spp. enhanced compost in sustainable agriculture. In: Biology of composts. Springer, pp 257–291

Carretta L, Tarolli P, Cardinali A, Nasta P, Romano N, Masin R (2021) Evaluation of runoff and soil erosion under conventional tillage and no-till management: a case study in northeast Italy. Catena 197:104972

Chandarana KA, Amaresan N (2022) Soil protists: an untapped microbial resource of agriculture and environmental importance. Pedosphere 32(1):184–197

Chellappa J, Sagar KL, Sekaran U, Kumar S, Sharma P (2021) Soil organic carbon, aggregate stability and biochemical activity under tilled and no-tilled agroecosystems. J Agric Food Res 4:100139

Chepsergon J, Moleleki LN (2023) Rhizosphere bacterial interactions and impact on plant health. Curr Opin Microbiol 73:102297

Choudhary N, Tripathi A, Singh PK, Parikh HS, Tiwari A (2024) Application of algae for enhanced plant growth and food productivity. Syst Microbiol Biomanufacturing 4(2):564–574

Clarholm M, Bonkowski M, Griffiths B, Van Elsas J, Jansson J, Trevors J (2007) Protozoa and other protista in soil. In: Modern soil microbiology. Marcel Dekker, Amsterdam, pp 147–175

Costa OY, Raaijmakers JM, Kuramae EE (2018) Microbial extracellular polymeric substances: ecological function and impact on soil aggregation. Front Microbiol 9:1636

Dayo-Olagbende GO, Adejoro SA, Ewulo BS, Awodun MA (2019) Effects of oxidation-reduction potentials on soil microbes. Agricultura 16(1–2):35–42

DeLong JP, Coblentz KE, Uiterwaal SF, Akwani C, Salsbery ME (2023) Temperature and predators as interactive drivers of community properties. Ecol Evol 13(11):e10665

Dreischhoff S, Das IS, Jakobi M, Kasper K, Polle A (2020) Local responses and systemic induced resistance mediated by ectomycorrhizal fungi. Front Plant Sci 11:1908

Dumbrell AJ, Ashton PD, Aziz N, Feng G, Nelson M, Dytham C, Fitter AH, Helgason T (2011) Distinct seasonal assemblages of arbuscular mycorrhizal fungi revealed by massively parallel pyrosequencing. New Phytol 190(3):794–804

Ebana R, Edet U, Anosike I, Etok C (2019) Bdellovibrio and like organisms: the much-anticipated "magic bullet". World News Nat Sci 23:233–249

Emerson JB (2019) Soil viruses: a new hope. MSystems 4(3):e00120–e00119

Esteban GF, Fenchel TM (2020) Ecology of protozoa. In: The biology of free-living phagotrophic protists. Springer, Cham

Etesami H, Emami S, Alikhani HA (2017) Potassium solubilizing bacteria (KSB): mechanisms, promotion of plant growth, and future prospects—a review. J Soil Sci Plant Nutr 17(4):897–911
Fan Y-X, Zhang J-Z, Zhang Q, Ma X-Q, Liu Z-Y, Lu M, Qiao K, Li F-L (2021) Biofuel and chemical production from carbon one industry flux gas by acetogenic bacteria. Adv Appl Microbiol 117:1–34
Feki K, Brini F, Mrabet M, Mhadhbi H (2020) Legume symbiotic interaction from gene to whole plant. In: Sustainable agriculture reviews 45: legume agriculture and biotechnology, vol 1. Springer, pp 137–157
Finke JF, Hunt BP, Winter C, Carmack EC, Suttle CA (2017) Nutrients and other environmental factors influence virus abundances across oxic and hypoxic marine environments. Viruses 9(6):152
Foissner W (1987) Soil Protozoa: fundamental problems, ecological significance, adaptations in ciliates and tetaceans, bioindicators and guide to the literature. Prog Protistol 2:69–212
Fomina M, Skorochod I (2020) Microbial interaction with clay minerals and its environmental and biotechnological implications. Fortschr Mineral 10(10):861
Fu Y, Jia Y, Fan J, Yu C, Yu C, Shen C (2020) Induction of Escherichia coli O157: H7 into a viable but non-culturable state by high temperature and its resuscitation. Environ Microbiol Rep 12(5):568–577
Gadd GM (2007) Geomycology: biogeochemical transformations of rocks, minerals, metals and radionuclides by fungi, bioweathering and bioremediation. Mycol Res 111(1):3–49
Gashu D, Nalivata PC, Amede T, Ander E, Bailey E, Botoman L, Chagumaira C, Gameda S, Haefele S, Hailu K (2021) The nutritional quality of cereals varies geospatially in Ethiopia and Malawi. Nature 594(7861):71–76
Ge J, Li D, Ding J, Xiao X, Liang Y (2023) Microbial coexistence in the rhizosphere and the promotion of plant stress resistance: a review. Environ Res 222:115298
Ghosh S, Sharma I, Nath S, Webster TJ (2021) Bioremediation—the natural solution. In: Microbial ecology of wastewater treatment plants. Elsevier, pp 11–40
Görres JH, Amador JA (2021) The soil fauna. In: Principles and applications of soil microbiology. Elsevier, pp 191–212
Haj-Amor Z, Araya T, Kim D-G, Bouri S, Lee J, Ghiloufi W, Yang Y, Kang H, Jhariya MK, Banerjee A (2022) Soil salinity and its associated effects on soil microorganisms, greenhouse gas emissions, crop yield, biodiversity and desertification: a review. Sci Total Environ 843:156946
Harindintwali JD, Zhou J, Yu X (2020) Lignocellulosic crop residue composting by cellulolytic nitrogen-fixing bacteria: a novel tool for environmental sustainability. Sci Total Environ 715:136912
Hartmann A, Rothballer M, Schmid M (2008) Lorenz Hiltner, a pioneer in rhizosphere microbial ecology and soil bacteriology research. Plant Soil 312(1):7–14
Hentges DJ (1996) Anaerobes: general characteristics. In: Baron S (ed) Medical microbiology. University of Texas Medical Branch at Galveston, Galveston
Hurst CJ (2011) Defining the ecology of viruses. In: Studies in viral ecology: microbial and botanical host systems, vol 1. Wiley, pp 1–40
Hyde JR, Armond T, Herring JA, Hope S, Grose JH, Breakwell DP, Pickett BE (2024) Diversity and conservation of the genome architecture of phages infecting the Alphaproteobacteria. Microbiol Spectr 12(1):e02827–e02823
Ibraheem I (2007) Cyanobacteria as alternative biological conditioners for bioremediation of barren soil. Egypt J Phycol 8(1):99–117
Ingham E (2009) Soil biology primer, Chapter 4: soil fungus. In: Soil and water conservation. Soil & Water Conservation Society, Ankeny, pp 22–23
Iqbal B, Li G, Alabbosh KF, Hussain H, Khan I, Tariq M, Javed Q, Naeem M, Ahmad N (2023) Advancing environmental sustainability through microbial reprogramming in growth improvement, stress alleviation, and phytoremediation. Plant Stress 10:100283
Iven H, Walker TW, Anthony M (2023) Biotic interactions in soil are underestimated drivers of microbial carbon use efficiency. Curr Microbiol 80(1):13

Józefowska A, Loaiza-Usuga JC, Schmidt O (2020) Consequences of land-use changes for soil quality and function, with a focus on the EU and Latin America. In: Climate change and soil interactions. Elsevier, pp 207–228

Katiyar D (2022) Rhizosphere: hotspot of soil-microbes-plant interaction. Just Agric 2(12):17

Khalidy R, Haque F, Chiang YW, Santos RM (2021) Monitoring pedogenic inorganic carbon accumulation due to weathering of amended silicate minerals in agricultural soils. J Vis Exp 172:e61996

Koskella B, Bergelson J (2020) The study of host–microbiome (co) evolution across levels of selection. Philos Trans R Soc B 375(1808):20190604

Kumar V, Shah MP (2021) Role of fungi and their enzymes in degradation and decolorization of distillery effluent for environmental safety. In: Fungi bio-prospects in sustainable agriculture, environment and nano-technology. Elsevier, pp 277–300

Kuzyakov Y, Mason-Jones K (2018) Viruses in soil: nano-scale undead drivers of microbial life, biogeochemical turnover and ecosystem functions. Soil Biol Biochem 127:305–317

Lara-Capistrán L, Hernádez-Montiel LG, Reyes-Pérez JJ, Zulueta-Rodríguez R, Jazayeri SM, Villamar-Torres RO (2021) Diversity and ecology of arbuscular mycorrhization fungi. In: Innovations in biotechnology for a sustainable future. Springer, pp 185–201

Li J-T, Xu H, Hicks LC, Brangarí AC, Rousk J (2023) Comparing soil microbial responses to drying-rewetting and freezing-thawing events. Soil Biol Biochem 178:108966

Liang X, Radosevich M, Löffler F, Schaeffer SM, Zhuang J (2019) Impact of microbial iron oxide reduction on the transport of diffusible tracers and non-diffusible nanoparticles in soils. Chemosphere 220:391–402

Maron JL, Marler M, Klironomos JN, Cleveland CC (2011) Soil fungal pathogens and the relationship between plant diversity and productivity. Ecol Lett 14(1):36–41

Martin BD, Schwab E (2012) Symbiosis:"Living together" in chaos. Stud History Biol 4(4):7–25

Martin BD, Schwab E (2013) Current usage of symbiosis and associated terminology. Int J Biol 5(1):32

Meena B, Jatav H, Dudwal B, Kumawat P, Meena S, Singh V, Khan M, Sathyanarayana E (2022) Fertilizer recommendations by using different geospatial technologies in precision farming or nanotechnology. Ecosyst Serv 14:241–257

Msimbira LA, Smith DL (2020) The roles of plant growth promoting microbes in enhancing plant tolerance to acidity and alkalinity stresses. Front Sustain Food Syst 4:106

Murase J, Asiloglu R (2023) Protists: the hidden ecosystem players in a wetland rice field soil. In: Biology and fertility of soils, pp 1–15. https://doi.org/10.1007/s00374-023-01705-9

Musbau SA, Ayinde BH, Omowunmi O, Motunrayo E, Belay E, Teshome B, Sarwar M (2021) Micro and macro (organisms) and their contributions to soil fertility. Front Environ Microbiol 7:44

Naha A, Nath DC, Nath S (2020) Impact of pesticides on the growth of Coriandrum sativum. Int J Plant Physiol Biochem 12(1):17–25

Nannipieri P, Trasar-Cepeda C, Dick RP (2018) Soil enzyme activity: a brief history and biochemistry as a basis for appropriate interpretations and meta-analysis. Biol Fertil Soils 54:11–19

Naresh R, Gupta R, Dhaliwal S, Tyagi K, Tyagi S, Prasad KK, Jat L, Tyagi P (2018) Clay-humus stability of soil organic matter and microbial biomass under conservation tillage and residue management practices of rice-wheat cropping system: a review. J Pharmacogn Phytochem 7(4):3020–3043

Narr A, Nawaz A, Wick LY, Harms H, Chatzinotas A (2017) Soil viral communities vary temporally and along a land use transect as revealed by virus-like particle counting and a modified community fingerprinting approach (fRAPD). Front Microbiol 8:1975

Nath S, Deb B, Sharma I (2018) Isolation of toxic metal-tolerant bacteria from soil and examination of their bioaugmentation potentiality by pot studies in cadmium-and lead-contaminated soil. Int Microbiol 21(1):35–45

Neemisha, Sharma S (2022) Soil enzymes and their role in nutrient cycling. In: Structure and functions of pedosphere. Springer, pp 173–188

Oduma O, Okeke C, Umunna M, Ehiomogue P, Orji F (2020) Effect of tillage operation on the physical and mechanical properties of soil in south–east Nigeria. Futo J Ser (FUTOJNLS) 6(2):1–13

Orgiazzi A, Lumini E, Nilsson RH, Girlanda M, Vizzini A, Bonfante P, Bianciotto V (2012) Unravelling soil fungal communities from different Mediterranean land-use backgrounds. PLoS One 7(4):e34847

Pahalvi HN, Rafiya L, Rashid S, Nisar B, Kamili AN (2021) Chemical fertilizers and their impact on soil health. In: Microbiota and biofertilizers, vol 2: ecofriendly tools for reclamation of degraded soil environs. Springer, pp 1–20

Peay KG, Kennedy PG, Bruns TD (2008) Fungal community ecology: a hybrid beast with a molecular master. Bioscience 58(9):799–810

Pedraza RO (2008) Recent advances in nitrogen-fixing acetic acid bacteria. Int J Food Microbiol 125(1):25–35

Pepper IL, Gentry TJ (2015) Microorganisms found in the environment. In: Environmental microbiology. Elsevier, pp 9–36

Perelomov L (2023) The role of interactions between bacteria and clay minerals in pedochemical processes. Geochem Int 61(10):1026–1035

Pett-Ridge J, Shi S, Estera-Molina K, Nuccio E, Yuan M, Rijkers R, Swenson T, Zhalnina K, Northen T, Zhou J (2021) Rhizosphere carbon turnover from cradle to grave: the role of microbe–plant interactions. In: Rhizosphere biology: interactions between microbes and plants. Springer, pp 51–73

Philippot L, Chenu C, Kappler A, Rillig MC, Fierer N (2023) The interplay between microbial communities and soil properties. Nat Rev Microbiol 22:1–14

Phour M, Sehrawat A, Sindhu SS, Glick BR (2020) Interkingdom signaling in plant-rhizomicrobiome interactions for sustainable agriculture. Microbiol Res 241:126589

Piotrowska-Długosz A (2020) Significance of the enzymes associated with soil C and N transformation. In: Carbon and nitrogen cycling in soil. Springer, pp 399–437

Rahman MM, Alam MS, Kamal MZU, Rahman GM (2020) Organic sources and tillage practices for soil management. In: Resources use efficiency in agriculture. Springer, pp 283–328

Raj A, Jhariya MK, Yadav DK, Banerjee A, Meena RS (2019) Soil for sustainable environment and ecosystems management. In: Sustainable agriculture, forest and environmental management. Springer, pp 189–221

Rao M, Scelza R, Gianfreda L (2014) Soil enzymes. In: Enzymes in agricultural sciences. OMICS Group eBooks, Foster City, pp 10–43

Rasmann S, Turlings TC (2016) Root signals that mediate mutualistic interactions in the rhizosphere. Curr Opin Plant Biol 32:62–68

Rodríguez-Verdugo A (2021) Evolving interactions and emergent functions in microbial consortia. Msystems 6(4):e00774–e00721

Rønn R, Vestergård M, Ekelund F (2012) Interactions between bacteria, protozoa and nematodes in soil. Acta Protozool 51(3):223–235

Rossine FW, Vercelli GT, Tarnita CE, Gregor T (2022) Structured foraging of soil predators unveils functional responses to bacterial defenses. Proc Natl Acad Sci 119(52):e2210995119

Sanders WB, Masumoto H (2021) Lichen algae: the photosynthetic partners in lichen symbioses. Lichenologist 53(5):347–393

Sarsan S, Pandiyan A, Rodhe AV, Jagavati S (2021) Synergistic interactions among microbial communities. In: Microbes in microbial communities: ecological and applied perspectives. Springer, pp 1–37

Schoelz JE, Stewart LR (2018) The role of viruses in the phytobiome. Ann Rev Virol 5:93–111

Selim SM, Zayed MS (2017) Microbial interactions and plant growth. In: Plant-microbe interactions in agro-ecological perspectives. Springer, pp 1–15

Shah K, Tripathi S, Tiwari I, Shrestha J, Modi B, Paudel N, Das B (2021) Role of soil microbes in sustainable crop production and soil health: a review. Agric Sci Technol (1313–8820) 13(2):109–118

Shi L-L, Mortimer PE, Ferry Slik J, Zou X-M, Xu J, Feng W-T, Qiao L (2014) Variation in forest soil fungal diversity along a latitudinal gradient. Fungal Divers 64(1):305–315

Shi Y, Li Y, Yang T, Chu H (2021) Threshold effects of soil pH on microbial co-occurrence structure in acidic and alkaline arable lands. Sci Total Environ 800:149592

Siliakus MF, van der Oost J, Kengen SW (2017) Adaptations of archaeal and bacterial membranes to variations in temperature, pH and pressure. Extremophiles 21:651–670

Sullivan MB, Weitz JS, Wilhelm S (2017) Viral ecology comes of age. Environ Microbiol Rep 9(1):33–35

Tahjib-Ul-Arif M, Shorna SI, Polash MAS, Haque MR, Kibria MG, Rhaman MS, Hossain MA (2021) Role of organic solutes on mineral stress tolerance. In: Organic solutes, oxidative stress, and antioxidant enzymes under abiotic stressors. CRC Press, pp 83–104

Tarnita CE (2017) The ecology and evolution of social behavior in microbes. J Exp Biol 220(1):18–24

Taylor DL, Bhatnagar JM (2024) Fungi in soil: a rich community with diverse functions. In: Soil microbiology, ecology and biochemistry. Elsevier, pp 75–129

Tokarz E, Urban D (2015) Soil redox potential and its impact on microorganisms and plants of wetlands. J Ecol Eng 16(3):20–30

Wang Q, Zeng Z, Zhong M (2016) Soil moisture alters the response of soil organic carbon mineralization to litter addition. Ecosystems 19:450–460

Wang C-Y, Zhou X, Guo D, Zhao J-H, Yan L, Feng G-Z, Gao Q, Yu H, Zhao L-P (2019) Soil pH is the primary factor driving the distribution and function of microorganisms in farmland soils in northeastern China. Ann Microbiol 69(13):1461–1473

Wang NQ, Kong CH, Wang P, Meiners SJ (2021) Root exudate signals in plant–plant interactions. Plant Cell Environ 44(4):1044–1058

Weiner A, Schopf S, Wanner G, Probst A, Wirth R (2012) Positive, neutral and negative interactions in cocultures between Pyrococcus furiosus and different methanogenic archaea. Microbiol Insights 5:MBI. S8516

Wu Y, Cai P, Jing X, Niu X, Ji D, Ashry NM, Gao C, Huang Q (2019) Soil biofilm formation enhances microbial community diversity and metabolic activity. Environ Int 132:105116

Wu Y, Fu C, Peacock CL, Sørensen SJ, Redmile-Gordon MA, Xiao K-Q, Gao C, Liu J, Huang Q, Li Z (2023) Cooperative microbial interactions drive spatial segregation in porous environments. Nat Commun 14(1):4226

Xiao R, Kuang B, Hu Y, Wang Y, Bai J, Zhang L, Wang C, Wei Z, Zhang K, Jorquera MA (2023) Ecological risk assessment of glyphosate and its possible effect on bacterial community in surface sediments of a typical shallow Lake, northern China. Ecotoxicol Environ Saf 249:114445

Yan N, Marschner P, Cao W, Zuo C, Qin W (2015) Influence of salinity and water content on soil microorganisms. Int Soil Water Conserv Res 3(4):316–323

Ye C, Butler OM, Chen C, Liu W, Du M, Zhang Q (2020) Shifts in characteristics of the plant-soil system associated with flooding and revegetation in the riparian zone of Three Gorges Reservoir, China. Geoderma 361:114015

Yu Z, Lu T, Qian H (2023) Pesticide interference and additional effects on plant microbiomes. Sci Total Environ 888:164149

Zhao D, Xu M, Liu G, Yao X, Tuo D, Zhang R, Xiao T, Peng G (2017) Quantification of soil aggregate microstructure on abandoned cropland during vegetative succession using synchrotron radiation-based micro-computed tomography. Soil Tillage Res 165:239–246

Zuccarini P, Sardans J, Asensio L, Peñuelas J (2023) Altered activities of extracellular soil enzymes by the interacting global environmental changes. Glob Chang Biol 29(8):2067–2091

Chapter 3
Exploring Plant-Microbe Interactions in Extreme Environments: Lessons from Arid and Desert Ecosystems

Ben Jesuorsemwen Enagbonma, Emmanuel Edoghogho Imade, Olubukola Oluranti Babalola, David Mxolisi Modise, Everlon Cid Rigobelo, and Cristina Cruz

Abstract Desert and arid environments are among the harshest ecosystems on Earth, characterized by nutrient-deprived soils, extreme temperatures, and minimal rainfall. Despite these challenging conditions, these ecosystems support an impressive diversity of plant life, largely due to the intricate interactions between plants and microbes. Here, we presented insights into the dynamics of plant-microbe interactions in these extreme environments, shedding light on plant adaptation mechanisms and offering new approaches for ecosystem management and sustainable agriculture. This review is organized into four sections: (i) challenges of arid and desert environments, (ii) plant adaptations in arid and desert ecosystems, (iii) the role of microbes in plant adaptation, and (iv) lessons learned and applications. Exploring plant-microbe interactions in desert and arid environments reveals the remarkable resilience of life in extreme environments. By understanding the mechanisms through which plants and microbes coexist and cooperate under these conditions, we gain deeper insights into ecosystem dynamics and uncover new opportunities for innovation in agriculture and ecological management.

B. J. Enagbonma (✉) · E. E. Imade · O. O. Babalola
Food Security and Safety Focus Area, Faculty of Natural and Agricultural Sciences, North-West University, Mmabatho, South Africa
e-mail: 30775450@nwu.ac.za

D. M. Modise
Faculty of Natural and Agricultural Sciences, North West University, Potchefstroom, South Africa

E. C. Rigobelo
Agricultural and Livestock Microbiology Postgraduate Program, School of Agricultural and Veterinarian Sciences, São Paulo State University (UNESP), Jaboticabal, SP, Brazil

C. Cruz
Department of Plant Biology, Faculdade de Ciências, Universidade de Lisboa, Lisbon, Portugal

O. O. Babalola, A. S. Ayangbenro (eds.), *Microbial Allies*,
https://doi.org/10.1007/978-3-031-90530-8_3

Keywords Dry environment · Environmental sustainability · Food security · Plant-microbe nexus

3.1 Introduction

Arid and desert ecosystems are regions of the earth where annual rainfall is less than 250 mm, with random and irregular precipitation, and some years may experience no rainfall at all (Wang et al. 2022). Apart from oceans, arid areas are the most widespread environment on earth, occupying approximately 6×10^7 km^2 of the world's land area (Spinoni et al. 2021). Based on the Aridity Index (AI), which measures the ratio between average annual precipitation and evapotranspiration, the United Nations Environmental Programme grouped these regions into arid deserts (AI < 0.2), hyperarid deserts (AI < 0.05), and semi-arid (AI < 0.5) (Fig. 3.1) (Sharma et al. 2021). Arid deserts experience uneven and short precipitation events, averaging around 250 mm annually, while hyperarid deserts receive less than 25 mm of rainfall per year. Semi-arid regions are comparatively cooler than dry, hot deserts, experiencing long, dry summers followed by winters with occasional rainfall (D'Odorico et al. 2013; Wang et al. 2022; Abotalib et al. 2021). Survival in these environments is confronted by severe ecological circumstances including temperature fluctuations, elevated levels of solar radiation, water constraints, nutrient dearth and soil salinity, all of which severely impact plant survival and growth (Xu et al. 2021). In such extreme arid and desert circumstances, soil microbiomes must not only cope with the lack of water but also endure intense solar flux, high salinity,

Fig. 3.1 Map showing the universal distribution of the desert areas based on their Aridity index

oligotrophic conditions, prolonged desiccation, high evaporation rates, extreme temperature fluctuations, and lethal levels of UV radiation (Coclet et al. 2022).

Given that arid and deserts environments cover one-third of Earth's land surface, utilizing these vast, desiccated and secluded landscapes for cultivation could help address some of the challenges associated with achieving global food security for an ever-increasing human population (Forsyth 2016). This review, therefore, aims to unravel the dynamics of plant-microbe interactions in arid and desert environments, which not only enhances our understanding of the adaptation mechanisms of plants but also divulges novel methods for ecosystem management and sustainable agriculture. To achieve this, the review is structured into four key sections: (a) challenges of arid and desert environments, (b) plant adaptations in arid and desert ecosystems, (c) role of microbes in plant adaptation, and (d) lessons learned and applications.

3.2 Challenges of Arid and Desert Environments

Arid and desert regions are intricate ecosystems that pose numerous challenges (Table 3.1) for animal husbandry and crop cultivation (Zhang et al. 2023). For instance, water resources in desert areas are severely limited due to low precipitation and high evaporation, complicating the availability of water for domestic, industrial and agricultural use (Morante-Carballo et al. 2022). Gober (2010) reported that arid regions and their urban centers are increasingly vulnerable to water shortages as climate change threatens to reduce supply while rapid population growth intensifies demand. According to Ajjur and Al-Ghamdi (2021), water scarcity has doubled over the past three decades, as reflected in the water stress levels in North Africa and Middle Eastern deserts, which increased from 8411% in 1987 to 16,422% in 2017. Changes in precipitation patterns and prolonged droughts affect not only human populations but also agricultural irrigation, potentially causing economic losses equivalent to 14% of the region's GDP over the next 30 years (Winkler and Brooks 2020).

Extreme temperatures also pose a significant challenge to biota in the arid and desert regions, with temperatures rising above 40 °C (104 °F) or dropping far below (Greenville et al. 2012). In the Chihuahuan Desert which spans Mexico and the United States, temperatures can fluctuate dramatically within a single day. Nighttime temperatures can fall below freezing point (0 °C or 32 °F), while daytime temperatures can exceed 37 °C (100 °F) (McIntosh et al. 2019; Kidron and Gutschick 2017; Flores et al. 2017). The high temperatures and water scarcity in these regions often result in very low humidity, insufficient to generate cloud cover (Zhuo and Zhou 2022). As a result, cloudless skies allow the sun's rays to directly heat the land, leading to soil erosion (Xu and Zhang 2020). This can reduce soil fertility, increase salinization, deplete soil nutrient and limit vegetation growth (Yuan et al. 2022).

Table 3.1 Major challenges faced within some respective deserts across each continent

Continent	Desert	Challenge
Africa	Sahara	Extreme temperatures (40%), water scarcity (30%), sandstorms (30%)
	Kalahari	Extreme heat (35%), limited water sources (40%), habitat loss (25%)
Antarctica	Antarctic Desert	Extreme cold (50%), strong winds (30%), ice and snow cover (20%)
	McMurdo Dry Valleys	Harsh cold (50%), low humidity (30%), strong katabatic winds (20%)
Asia	Gobi	Harsh winters (35%), dust storms (35%), temperature extremes (30%)
	Arabian	Intense heat (40%), water scarcity (35%), sandstorms (25%)
Europe	Accona	Water scarcity (40%), soil erosion (35%), temperature extremes (25%)
	Oltenian Sahara	Desertification (40%), limited water resources (35%), soil degradation (25%)
North America	Mojave	Extreme heat (40%), flash floods (25%), limited water (35%)
	Sonoran	High temperatures (40%), scarce rainfall (35%), Biodiversity threats (25%)
South America	Atacama	Extreme aridity (45%), water scarcity (35%), harsh climate (20%)
	Patagonian Desert	Cold conditions (40%), strong winds (35%), limited vegetation (25%)
Australia	Great Victoria	Extreme temperatures (40%), scarce water (35%), remote location (25%)
	Simpson	High temperatures (40%), limited rainfall (35%), isolation (25%)

The % represents a rough estimation of the impact of each challenge within some respective world deserts. The table was extrapolated from (Laity 2009; Warner 2009; Whitford and Duval 2019; Cooke and Warren 2023; Evenari et al. 1982; Middleton 2017)

Wind is a significant abiotic feature that influences the dynamics of arid and desert ecosystems (Saif et al. 2021). In some desert, winds can reach speeds of approximately 60 miles per hour (100 km/h), and with sparse vegetation to impede them, these winds are capable of transporting dust and sand across entire continents (Al-Dousari et al. 2019). For example, windstorms in the Sahara lift large amounts of material into the air, with African dust occasionally crossing the Atlantic Ocean (Madhavan 2016).

3.3 Plant Adaptations in Arid and Desert Ecosystems

Plants in arid and desert regions exhibit an extraordinary range of adaptive strategies to survive in extreme environments characterized by high solar radiation, limited water availability, and severe temperatures (Zahedi et al. 2021). These adaptations can be behavioral, physiological, or morphological (Table 3.2), each playing a critical role in enhancing water use efficiency and maintaining essential

Table 3.2 Adaptation strategies utilized by desert plants for survival

Adaptation type	Strategy	Description	References
Morphological	Deep root systems	Roots that extend deep into the ground to access underground water sources.	Tariq et al. (2024a)
	Reduced leaf surface area	Smaller or fewer leaves to minimize water loss through transpiration.	Li et al. (2021)
	Thick cuticles	Waxy, thick outer layer on leaves and stems to reduce water evaporation.	Skrzydeł et al. (2021)
	Succulence	Storing water in fleshy leaves, stems, or roots.	Li et al. (2021)
	Leaf modifications	Leaves are modified into spines to reduce water loss and deter herbivores.	Peguero-Pina et al. (2020)
	Hairy or reflective surfaces	Leaf surfaces are covered with hairs or reflective coatings to reduce water loss and heat absorption.	Ehleringer and Mooney (1978) and Streb et al. (1997)
Physiological	CAM photosynthesis	Crassulacean Acid Metabolism, fixing CO_2 at night to reduce water loss during the day.	Mok et al. (2023)
	C4 photosynthesis	More efficient photosynthesis under high temperatures and low CO_2 conditions.	Wang (2007)
	Osmoregulation	Regulation of osmotic pressure to maintain cell turgor and water retention.	Shameem Kausar and Hameed (2023)
	Drought tolerance mechanisms	Biochemical processes to survive extended periods of drought.	Hõrak (2017)
Behavioral	Stomatal regulation	Opening stomata at night to decrease water-loss and closing them during the day.	Müller et al. (2017)
	Drought-deciduous strategy	Shedding leaves during extreme drought to conserve water.	Marchin et al. (2010)
	Leaf rolling or folding	Rolling or folding leaves to reduce surface area exposed to sunlight and transpiration.	Kusaka et al. (2005)

functions (Alkhedir and Taniguchi 2024). Morphological adaptations include features such as deep root systems to access groundwater, reduced leaf surface area to curtail water loss, and thick cuticles to prevent desiccation (Hu et al. 2021). Physiological strategies involve changes in photosynthetic pathways, such as crassulacean acid metabolism, which allows plants to fix carbon dioxide at night, thereby reducing water loss during the hot daytime (Dos Santos et al. 2022).

Behavioral adaptations in desert plants include opening stomata at night to lessen transpiration and using drought-deciduous approaches, where plants shed their leaves during periods of severe drought (Meng et al. 2024). For instance, Li et al. (2022) investigated the physiological and morphological responses of *Calligonum arborescens* Litv and *Haloxylon ammodendron* (C.A. Mey.) Bunge, two desert species from the Taklimakan Desert shelter belt, to drought stress. They reported a significant difference in water potential, stomatal conductance and morphological traits between the two species. It showed that *C. arborescens* Litv developed significantly thicker and shorter leaves and accumulated non-structural carbohydrates primarily in its roots. In contrast, *H. ammodendron* (C.A. Mey.) Bunge produced thinner, shorter leaves and accumulated non-structural carbohydrates throughout its body, with no significant change in the roots, stems, or leaves. The authors concluded that both species exhibited distinct acclimatization responses to drought, with *C. arborescens* Litv. having a higher water-use efficiency, likely due to differences in leaf water potential (Ψl).

In another study, García-López et al. (2021) reported that *Larrea tridentata,* a species found in Mojave, Sonoran and Chihuahuan deserts, produces several metabolites such as lignans and nordihydroguaiaretic acid, which serve as natural defense mechanisms against threats like insects, other predatory animals, and microorganisms including viruses, bacteria, protozoa, and fungi. Furthermore, to mitigate stress-induced oxidative damage, desert plants enhance their antioxidant defense system to scavenge reactive oxygen species. Their cells also produce and amass elevated concentrations of well-suited metabolites (Table 3.3), like soluble sugars, and some amino acids, like glycine betaine and proline, which help lower osmotic potential. Understanding these adaptations not only provides insights into plant resilience but also offers valuable knowledge for developing sustainable agricultural practices in arid regions.

3.4 Role of Microbes in Plant Adaptation

In response to the harsh conditions of arid environments, plants have evolved various strategies to cope with environmental stress, including establishing beneficial relationships with microbes (Fig. 3.2) (Soussi et al. 2016; Imade and Babalola 2021). Recent research highlights the essential role of microbes in enhancing plant adaptation to arid climates (Ayangbenro et al. 2023; Afridi et al. 2022; Tariq et al. 2024b). Microbes interact with plants through multiple mechanisms that influence plant growth, development, and stress tolerance (Enagbonma et al. 2023; Fadiji

Table 3.3 Biochemicals produced by some desert plant in mitigating stress

Desert plant	Biochemical produced	Function in stress mitigation	References
Wild prickly pear fruit	Phenolic, flavonoids, betacyanins, betaxanthins, and betalains.	Antioxidant defense	Valero-Galván et al. (2021)
Creosote bush (*Larrea tridentata*)	Phenolic compounds (e.g., nordihydroguaiaretic acid), terpenes, phenolic acids, and flavonoids	UV protection, antimicrobial activity	Arteaga et al. (2005) and Sagaste et al. (2019)
Aloe vera (*Aloe barbadensis* Miller)	Entire phenolic composites, water-soluble polymers, cell wall polysaccharides, and aloin	Moisture retention, UV protection, wound healing	Golmohammadİ (2022) and González-Delgado et al. (2023)
Date palm (*Phoenix dactylifera* L.)	Flavonoids, phenolic acids, proline	Antioxidant activity, osmotic regulation	Gibson et al. (2015)
Agave (*Agave Americana* and *Agave salmiana*)	Saponins, flavonoids	Anti-inflammatory, antioxidant activity, water retention	Morreeuw et al. (2021) and Leal-Díaz et al. (2015)
Mesquite (*Prosopis juliflora* (Sw.) DC.)	Alkaloids, phenolic acids, tannins, aldehyde, anthocyanidine	Herbivory deterrence, UV protection, antioxidant activity	Darwish et al. (2022)
Desert lavender (*Lamiaceae* family)	Essential oils (e.g., limonene, camphor)	Deterrent to herbivores, antimicrobial activity	Popa et al. (2021)
Desert Marigold (*Tagetes minuta* L.)	Alkaloids, tannins, saponins, flavonoids and terpenoids	UV protection, antioxidant activity	Sadia et al. (2015)
Ocotillo (*Fouquieria splendens*)	Phenolics, flavonoids, tannins	Water retention, deterrent to herbivores	López-Romero et al. (2022) and Monreal-García et al. (2019)
Ephedra (*Ephedra fragilis*)	Ephedrine, flavonoids, tannins	Tolerance, antioxidant activity	Guenaou et al. (2021) and Soumaya et al. (2020)

et al. 2022). These interactions can be broadly categorized into several key approaches:

Firstly, through symbiotic relationships, arbuscular mycorrhizal fungi (AMF) form associations with desert plant roots, significantly enhancing nutrient and water uptake. In this mutualistic relationship, the plant receives essential nutrients such as phosphorus, while the fungi obtain carbohydrates from the plant (Watkinson 2016). Under heat stress for example, plant roots often struggle to absorb water and nutrients; however, AMF improve tolerance through increased uptake, resulting in better growth and yield (Zhu et al. 2017). This is evidenced by higher water use efficiency, water holding capacity, and relative water content in AMF-inoculated maize (*Zea mays*) grown at 40 °C (Zhu et al. 2011). Additionally,

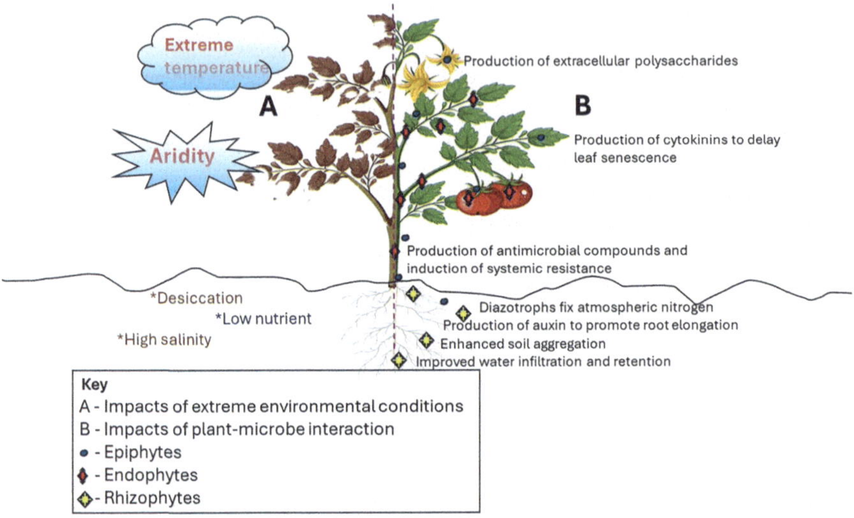

Fig. 3.2 Plant-microbe interaction mitigates extreme condition of arid region

AMF-inoculated asparagus (*Asparagus officinalis*) accumulates more macronutrients (N, P, K) and micronutrients (Ca, Mg, Fe) compared to non-AMF plants under heat stress (Yeasmin et al. 2019). This multifaceted interaction underscores the critical role of AMF in enabling plants to adapt and thrive in challenging environmental conditions.

Apart from the mycorrhizal symbiosis between plants and fungi, another crucial plant-microbe interaction involves biological nitrogen fixation (Kucho et al. 2017; Bellabarba et al. 2019). Although 80% of Earth's atmosphere consists of dinitrogen gas, plants cannot utilize it directly and rely on various bacteria to convert, or "fix," it into a usable form (Amoo et al. 2021; Vacheron et al. 2015). These bacteria form symbiotic relationships with leguminous plants, converting atmospheric nitrogen into ammonia, which plants can then use. This process significantly enhances plant growth in nitrogen-deficient soils (Sellstedt and Richau 2013; Svistoonoff et al. 2014).

In a study assessing the microbiome properties of *Prosopis cineraria* root nodules, Rashid et al. (2024) investigated microbial diversity in the roots and nodules of two *P. cineraria* populations from the Al Ain Desert, UAE. They collected samples from three mature trees in the open desert and three nodulating trees in a frequently irrigated polyhouse tunnel at a desert farm. The desert trees exhibited no detectable root nodules, whereas the farm trees had abundant nodules. Nodulation in *P. cineraria* roots was associated with a taxonomic shift in the root-associated microbiome. Nodule microbiomes from *P. cineraria* seedlings grown in a growth chamber exhibited enrichment patterns specific to nodule compartments and genotypes.

Trees in desert farms, experiencing less abiotic stress, produced abundant root nodules, similar to other Caesalpinioideae plants. Furthermore, the ultrastructure of root nodules, from the research, revealed intracellular Bacteroides in some cortical cells, with 18% and 21% Actinobacteria on the nodule surface and in the nodule endosphere, respectively. Under optimal growth chamber conditions, the healthiest nodules contained 2.6% and 0.5% Actinobacteria in the nodule surface microbiome and in the nodule endospore microbiome compartments, respectively, indicating distinct microbial communities in growth chambers versus farm environments. Some actinobacterial isolates from *P. cineraria* exhibited a transcriptionally active ACC deaminase gene, known for conferring drought tolerance (Singh et al. 2015; Nascimento et al. 2016).

OTU14 (*Actinophytocola* sp.) was found exclusively in roots but not in nodules, and *nifH* gene fragments of *Actinophytocola* sp. were detected in the root metagenomes and metatranscriptome of *Colophospermum mopane* trees (Burbano et al. 2015). Actinobacteria are more abundant in root microbiomes under conditions of nutrient limitation, salt stress, water stress, and heat stress, and they provide abiotic stress tolerance to many plants (Naylor and Coleman-Derr 2018; Hoskisson and Fernández-Martínez 2018; Xu et al. 2018). Some Actinobacteria also produce nitrogenase enzymes, facilitating associative nitrogen fixation (Van Nguyen and Pawlowski 2017).

Furthermore, arid plant also rely on the services of endophytic microbes for survival (Elsheikh et al. 2021). Endophytes, a special group of bacteria and fungi residing within plant tissues, establish a spectrum of relationships with their host plants, ranging from latent pathogens or saprotrophs to mutualistic associations (Babalola et al. 2020). These microorganisms, particularly soil and rhizospheric microbes, play crucial roles in ecosystem functioning and plant adaptation to environmental stresses (Imade and Babalola 2021). Studies have indicated that bacterial endophytes from arid plants can enhance plant growth by synthesizing indole-3-acetic acid (IAA). Different endophytic microbes produce varying levels of IAA, which differentially affects plant growth. Dobbelaere et al. (2003) found that IAA produced by various bacterial endophytes promotes plant growth by expanding root areas, thereby improving nutrient uptake from the soil. Dawwam et al. (2013) reported that bacterial endophytes synthesized IAA in quantities ranging from 0.36 to 14.77 mg/mL and from 0.6 to 10.73 mg/mL. Additionally, two endophytic fungi, *Penicillium chrysogenum* and *Alternaria alternata*, along with a non-spore-forming fungus, were isolated from the medicinal plant *Asclepias sinaica*. These fungi exhibited high IAA production, between 100 and 160 mg/mL, and significantly increased the root length of maize plants (Fouda et al. 2015).

Microbial activity also enhances soil structure by producing extracellular polysaccharides that bind soil particles together, forming stable aggregates. These aggregates improve soil porosity, water infiltration, and retention, which are crucial in arid environments. Rillig et al. (2019) demonstrated that AMF and soil bacteria significantly improved soil aggregation, reducing erosion and enhancing water retention capabilities. Recent research has provided substantial evidence of the benefits of microbial interactions in arid environments. For instance, Bashan et al. (2014) showed that inoculating plants with *Azospirillum brasilense* in arid regions improved plant growth and drought tolerance by enhancing nitrogen uptake and water use efficiency. Another study by Chitarra et al. (2016) illustrated how mycorrhizal fungi inoculation in grapevines not only improved nutrient uptake but also increased drought resilience, leading to healthier and more productive plants.

Phyllosphere microbes, which reside on plant leaf surfaces, play a critical role in plant survival in arid regions. These microbes enhance drought tolerance by producing protective extracellular polysaccharides and osmolytes, which reduce water loss and prevent leaf desiccation (Ruiz-Lozano et al. 2016). They also secrete antioxidants and phytohormones, mitigating oxidative stress and promoting growth under arid conditions (Egamberdieva et al. 2021; Raza et al. 2022). In addition, phyllosphere bacteria can inhibit pathogenic microbes through the production of antimicrobial compounds, thus enhancing plant health (Hoskisson and Fernández-Martínez 2018). By modulating the leaf microenvironments and providing protection against biotic and abiotic stressors, phyllosphere microbes significantly contribute to plant resilience in arid ecosystems. For instance, Araya et al. (2020) identified cold-tolerant, hyper-ACC-degrading bacteria from the phyllospheres of *Colobanthus quitensis* and *Deschampsia antarctica*. These microbes could be further explored as potential microbial inoculants to alleviate abiotic stresses in plants that inhabit extreme environments.

In addition, beneficial microbes can effectively outcompete pathogenic organisms for the limited nutrients and space available in soil and on plant surfaces. By colonizing these ecological niches, beneficial microbes reduce the resources available to pathogens, thereby limiting their growth and capacity to infect plants (Ab Rahman et al. 2018). This competition is particularly important in arid regions, where resources are scarce and microbial competition is intense. For instance, root-colonizing bacteria like *Pseudomonas* spp. produce siderophores—molecules that tightly bind iron, an essential nutrient—thereby reducing its availability to pathogenic microbes and inhibiting their proliferation (El-Sayed et al. 2014).

3.5 Lessons Learned and Applications

A thorough understanding of the complex interplay between plants and microorganisms in harsh environments provides significant insights into the adaptability and resilience of life (Gamalero et al. 2020). The extreme conditions of deserts and arid regions offer a natural laboratory for investigating these interactions, thereby guiding biotechnological innovations, ecological conservation, and sustainable agricultural practices (Rodríguez-Núñez et al. 2020). By synthesizing knowledge of these interactions, researchers can develop biofertilizers and biostimulants that enhance crop resilience to nutrient and drought stress. For instance, Shi et al. (2015) reported that *Glomus mosseae* and *Glomus etunicatum* (Arbuscular mycorrhizal fungi) significantly improved the growth of *Plantago minuta* and increased the uptake of nitrogen, phosphorous and potassium in both the roots and shoots of this desert plant under variable soil moisture conditions. Additionally, the colonialization of desert plant roots by mycorrhizal fungi leads to the protrusion of fungal hyphae into the surrounding soil, increasing the surface area for nutrient and water absorption, which allows plants to thrive in nutrient-poor soils.

Understanding the contributions of desert microbiome also informs conservation efforts. Maintaining microbial diversity is crucial for conserving ecosystem stability and function, as these organisms strengthen many processes essential for plant survival. Conservation strategies should aim to protect both microbial and floral diversity, ensuring the sustainability of these fragile environments. Maldonado et al. (2022), in their study titled "Partners to survive: *Hoffmannseggia doellii* root-associated microbiome at the Atacama Desert," revealed that the root-associated soils of *Hoffmannseggia doellii* exhibited a higher number of species and greater phylogenetic diversity of fungi and bacteria compared to control soils across various locations and to soil microorganisms associated with other plant species. In addition, the *H. doellii*-associated microbiome demonstrated an increased number of positive interactions and more extensive within-kingdom fungal collaborations than those observed in the control network. Their findings suggest that the microbiomes associated with *H. doellii* soils may play a key role in the plant's ability to survive in such extreme environments.

The extremophilic nature of microbes in desert and arid environments makes them valuable resources for biotechnology. Metabolites and enzymes derived from these microorganisms have potential applications across various industries, ranging from biofuels to pharmaceuticals. This is supported by research from Rateb et al. (2011), which identified a variety of metabolites, including hygromycin A, 5″-dihydrohygromycin A, deferroxamine E, and chaxalactins A–C. Notably, some of these compounds, such as chaxalactins A–C, exhibited significant activity against Gram-positive bacterial strains.

3.6 Conclusion

The review of plant-microbe interactions in desert and arid ecosystems highlights the remarkable adaptability of life in extreme conditions. These nexuses emphasize the significance of symbiosis, nutrient cycling, microbial bioactive compounds, and evolutionary adaptations in supporting plant survival in harsh environments. By applying these insights to conservation, agriculture, and biotechnology, we can develop innovative solutions to address global challenges such as environmental degradation, food security, and sustainable development.

Acknowledgements The authors thank North-West University and members of the Food Security and Safety Focus Area for providing a conducive environment for conducting this research.

Author's Contributions All authors contributed equally.

Funding B.J.E would like to thank North-West University for its postdoctoral bursary and research support. O.O.B. would like to thank the National Research Foundation, South Africa, for grants that support research in her lab.

References

Ab Rahman SFS, Singh E, Pieterse CM, Schenk PM (2018) Emerging microbial biocontrol strategies for plant pathogens. Plant Sci 267:102–111

Abotalib AZ, Heggy E, El Bastawesy M, Ismail E, Gad A, Attwa M (2021) Groundwater mounding: a diagnostic feature for mapping aquifer connectivity in hyper-arid deserts. Sci Total Environ 801:149760

Afridi MS, Javed MA, Ali S, De Medeiros FHV, Ali B, Salam A, Sumaira, Marc RA, Alkhalifah DHM, Selim S (2022) New opportunities in plant microbiome engineering for increasing agricultural sustainability under stressful conditions. Front Plant Sci 13:899464

Ajjur SB, Al-Ghamdi SG (2021) Evapotranspiration and water availability response to climate change in the Middle East and North Africa. Clim Chang 166(3):28

Al-Dousari A, Al-Nassar W, Al-Hemoud A, Alsaleh A, Ramadan A, Al-Dousari N, Ahmed M (2019) Solar and wind energy: challenges and solutions in desert regions. Energy 176:184–194

Alkhedir GM, Taniguchi T (2024) Morphological and physiological adaptation of a desert shrub, *Encelia farinosa*, under drought stress. Acta Oecol 122:103976

Amoo AE, Enagbonma BJ, Ayangbenro AS, Babalola OO (2021) Biofertilizer: an eco-friendly approach for sustainable crop production. In: Food security and safety: African perspectives. Springer, pp 647–669

Araya MA, Valenzuela T, Inostroza NG, Maruyama F, Jorquera MA, Acuña JJ (2020) Isolation and characterization of cold-tolerant hyper-ACC-degrading bacteria from the rhizosphere, endosphere, and phyllosphere of antarctic vascular plants. Microorganisms 8(11):1788

Arteaga S, Andrade-Cetto A, Cárdenas R (2005) *Larrea tridentata* (Creosote bush), an abundant plant of Mexican and US-American deserts and its metabolite nordihydroguaiaretic acid. J Ethnopharmacol 98(3):231–239

Ayangbenro AS, Odelade KA, Babalola OO (2023) Seasonal variation drives microbial diversity in cowpea rhizosphere in a semi-arid region. Int J Agric Biol 23(3):201–208

Babalola OO, Fadiji AE, Enagbonma BJ, Alori ET, Ayilara MS, Ayangbenro AS (2020) The nexus between plant and plant microbiome: revelation of the networking strategies. Front Microbiol 11:548037

Bashan Y, de-Bashan LE, Prabhu SR, Hernandez J-P (2014) Advances in plant growth-promoting bacterial inoculant technology: formulations and practical perspectives (1998–2013). Plant Soil 378(1):1–33

Bellabarba A, Fagorzi C, diCenzo GC, Pini F, Viti C, Checcucci A (2019) Deciphering the symbiotic plant microbiome: translating the most recent discoveries on rhizobia for the improvement of agricultural practices in metal-contaminated and high saline lands. Agronomy 9(9):529

Burbano CS, Grönemeyer JL, Hurek T, Reinhold-Hurek B (2015) Microbial community structure and functional diversity of nitrogen-fixing bacteria associated with *Colophospermum mopane*. FEMS Microbiol Ecol 91(4):fiv030

Chitarra W, Pagliarani C, Maserti B, Lumini E, Siciliano I, Cascone P, Schubert A, Gambino G, Balestrini R, Guerrieri E (2016) Insights on the impact of arbuscular mycorrhizal symbiosis on tomato tolerance to water stress. Plant Physiol 171(2):1009–1023

Coclet C, Cowan D, Lebre PH (2022) Survival under stress: microbial adaptation in hot desert soils. In: Ramond J-B, Cowan DA (eds) Microbiology of hot deserts. Springer, Cham, pp 293–317

Cooke RU, Warren A (2023) Geomorphology in deserts. University of California Press

D'Odorico P, Bhattachan A, Davis KF, Ravi S, Runyan CW (2013) Global desertification: drivers and feedbacks. Adv Water Resour 51:326–344

Darwish RS, Fares MAA, Hammoda HMA, Yassin AM, Ghareeb DA, Harraz FM, Shawky E (2022) Comparative metabolomics applied for valorization of Mesquite (*Prosopis juliflora* (Sw. DC.)) by-products as potential source of anti-inflammatory functional constituents. Ind Crop Prod 176:114344

Dawwam G, Elbeltagy A, Emara H, Abbas I, Hassan MM (2013) Beneficial effect of plant growth promoting bacteria isolated from the roots of potato plant. Ann Agric Sci 58(2):195–201

Dobbelaere S, Vanderleyden J, Okon Y (2003) Plant growth-promoting effects of diazotrophs in the rhizosphere. Crit Rev Plant Sci 22(2):107–149

Dos Santos TB, Ribas AF, de Souza SGH, Budzinski IGF, Domingues DS (2022) Physiological responses to drought, salinity, and heat stress in plants: a review. Stress 2(1):113–135

Egamberdieva D, Alimov J, Alaylar B, Karadayi M, Arora NK (2021) Extremophiles in saline environment: potential for sustainable agriculture. In: Microbial communities and their interactions in the extreme environment. Springer Singapore, pp 1–16

Ehleringer J, Mooney H (1978) Leaf hairs: effects on physiological activity and adaptive value to a desert shrub. Oecologia 37:183–200

El-Sayed WS, Akhkha A, El-Naggar MY, Elbadry M (2014) In vitro antagonistic activity, plant growth promoting traits and phylogenetic affiliation of rhizobacteria associated with wild plants grown in arid soil. Front Microbiol 5. https://doi.org/10.3389/fmicb.2014.00651

Elsheikh EA, El-Keblawy A, Mosa KA, Okoh AI, Saadoun I (2021) Role of endophytes and rhizosphere microbes in promoting the invasion of exotic plants in arid and semi-arid areas: a review. Sustain For 13(23):13081

Enagbonma BJ, Fadiji AE, Ayangbenro AS, Babalola OO (2023) Communication between plants and rhizosphere microbiome: exploring the root microbiome for sustainable agriculture. Microorganisms 11(8):2003

Evenari M, Shanan L, Tadmor N (1982) The Negev: the challenge of a desert. Harvard University Press

Fadiji AE, Babalola OO, Santoyo G, Perazzolli M (2022) The potential role of microbial biostimulants in the amelioration of climate change-associated abiotic stresses on crops. Front Microbiol 12:829099

Flores J, Pérez-Sánchez RM, Jurado E (2017) The combined effect of water stress and temperature on seed germination of Chihuahuan Desert species. J Arid Environ 146:95–98

Forsyth I (2016) Desert journeys: from exploration to covert operations. Geogr J 182(3):226–235

Fouda AH, Hassan SE-D, Eid AM, Ewais EE-D (2015) Biotechnological applications of fungal endophytes associated with medicinal plant Asclepias sinaica (Bioss.). Ann Agric Sci 60(1):95–104

Gamalero E, Bona E, Todeschini V, Lingua G (2020) Saline and arid soils: impact on bacteria, plants, and their interaction. Biology 9(6):116

García-López J, Contreras-Servín C, Álvarez-Fuentes G, Herrera-Medina R (2021) Creosote bush (*Larrea tridentata*) phytochemical traits and its different uses: a review. J Appl Life Sci Int 24(2):34–45

Gibson M, Ghanaem A, Brooks S-L, Ghaly DA (2015) Solvent extraction of antioxidants, phenols and flavonoids from Saudi Arabia dates. Chemistry, Environmental Science. http://hdl.handle.net/10222/64666

Gober P (2010) Desert urbanization and the challenges of water sustainability. Curr Opin Environ Sustain 2(3):144–150

Golmohammadİ F (2022) Medical plant of Aloe vera in desert regions of Iran: greenhouses, economic importance, development, extension, processing and marketing. Black Sea J Agric 5(1):1–15

González-Delgado M, Minjares-Fuentes R, Mota-Ituarte M, Pedroza-Sandoval A, Comas-Serra F, Quezada-Rivera JJ, Sáenz-Esqueda Á, Femenia A (2023) Joint water and salinity stresses increase the bioactive compounds of Aloe vera (*Aloe barbadensis* Miller) gel enhancing its related functional properties. Agric Water Manag 285:108374

Greenville AC, Wardle GM, Dickman CR (2012) Extreme climatic events drive mammal irruptions: regression analysis of 100-year trends in desert rainfall and temperature. Ecol Evol 2(11):2645–2658

Guenaou I, Nait Irahal I, Errami A, Lahlou FA, Hmimid F, Bourhim N (2021) Bioactive compounds from *Ephedra fragilis*: extraction optimization, chemical characterization, antioxidant and antiglycation activities. Molecules 26(19):5998

Hõrak H (2017) Learning from the experts: drought resistance in desert plants. New Phytol 216(1):5–7

Hoskisson PA, Fernández-Martínez LT (2018) Regulation of specialised metabolites in Actinobacteria–expanding the paradigms. Environ Microbiol Rep 10(3):231–238

Hu D, Lv G, Qie Y, Wang H, Yang F, Jiang L (2021) Response of morphological characters and photosynthetic characteristics of Haloxylon ammodendron to water and salt stress. Sustain For 13(1):388

Imade EE, Babalola OO (2021) Biotechnological utilization: the role of Zea mays rhizospheric bacteria in ecosystem sustainability. Appl Microbiol Biotechnol 105(11):4487–4500

Kidron GJ, Gutschick VP (2017) Temperature rise may explain grass depletion in the Chihuahuan Desert. Ecohydrology 10(4):e1849

Kucho K-i, Tamari D, Matsuyama S, Nabekura T, Tisa LS (2017) Nitrogen fixation mutants of the actinobacterium *Frankia casuarinae* CcI3. Microbes Environ 32(4):344–351

Kusaka M, Ohta M, Fujimura T (2005) Contribution of inorganic components to osmotic adjustment and leaf folding for drought tolerance in pearl millet. Physiol Plant 125(4):474–489

Laity JJ (2009) Deserts and desert environments, vol 3. Wiley

Leal-Díaz AM, Santos-Zea L, Martínez-Escobedo HC, Guajardo-Flores D, Gutiérrez-Uribe JA, Serna-Saldivar SO (2015) Effect of *Agave americana* and *Agave salmiana* ripeness on saponin content from aguamiel (agave sap). J Agric Food Chem 63(15):3924–3930

Li S, Wang H, Gou W, White JF, Kingsley KL, Wu G, Su P (2021) Leaf functional traits of dominant desert plants in the Hexi Corridor, Northwestern China: trade-off relationships and adversity strategies. Glob Ecol Conserv 28:e01666

Li C, Han H, Ablimiti M, Liu R, Zhang H, Fan J (2022) Morphological and physiological responses of desert plants to drought stress in a man-made landscape of the Taklimakan desert shelter belt. Ecol Indic 140:109037

López-Romero JC, Torres-Moreno H, Rodríguez-Martínez KL, Ramírez-Audelo V, Vidal-Gutiérrez M, Hernández J, Robles-Zepeda RE, Ayala-Zavala JF, González-Ríos H, Valenzuela-Melendres M, Villegas-Ochoa MA, Salazar-López NJ, González-Aguilar GA (2022) *Fouquieria splendens*: a source of phenolic compounds with antioxidant and antiproliferative potential. Eur J Integr Med 49:102084

Madhavan S (2016) Saharan dust detection using multi-sensor satellite remote sensing measurements. PhD, George Mason University, United States, Virginia

Maldonado JE, Gaete A, Mandakovic D, Aguado-Norese C, Aguilar M, Gutiérrez RA, González M (2022) Partners to survive: Hoffmannseggia doellii root-associated microbiome at the Atacama Desert. New Phytol 234(6):2126–2139

Marchin R, Zeng H, Hoffmann W (2010) Drought-deciduous behavior reduces nutrient losses from temperate deciduous trees under severe drought. Oecologia 163:845–854

McIntosh MM, Holechek JL, Spiegal SA, Cibils AF, Estell RE (2019) Long-term declining trends in Chihuahuan Desert forage production in relation to precipitation and ambient temperature. Rangel Ecol Manag 72(6):976–987

Meng D, Ma J, Min X, Zang Y, Sun W (2024) Nocturnal stomatal behaviour and its impact on water use strategies of desert herbs in the Gurbantunggut Desert, Northwest China. Sci Total Environ 929:172749

Middleton NJ (2017) Desert dust hazards: a global review. Aeolian Res 24:53–63

Mok D, Leung A, Searles P, Sage TL, Sage RF (2023) CAM photosynthesis in *Bulnesia retama* (Zygophyllaceae), a non-succulent desert shrub from South America. Ann Bot 132(4):655–670

Monreal-García HM, Almaraz-Abarca N, Ávila-Reyes JA, Torres-Ricario R, González-Elizondo MS, Herrera-Arrieta Y, Gutiérrez-Velázquez MV (2019) Phytochemical variation among populations of *Fouquieria splendens* (Fouquieriaceae). Bot Sci 97(3):398–412

Morante-Carballo F, Montalván-Burbano N, Quiñonez-Barzola X, Jaya-Montalvo M, Carrión-Mero P (2022) What do we know about water scarcity in semi-arid zones? A global analysis and research trends. Water 14(17):2685

Morreeuw ZP, Castillo-Quiroz D, Ríos-González LJ, Martínez-Rincón R, Estrada N, Melchor-Martínez EM, Iqbal HMN, Parra-Saldívar R, Reyes AG (2021) High throughput profiling of flavonoid abundance in *Agave lechuguilla* residue-valorizing under explored mexican plant. Plan Theory 10(4):695

Müller HM, Schäfer N, Bauer H, Geiger D, Lautner S, Fromm J, Riederer M, Bueno A, Nussbaumer T, Mayer K (2017) The desert plant *Phoenix dactylifera* closes stomata via nitrate-regulated SLAC 1 anion channel. New Phytol 216(1):150–162

Nascimento FX, Rossi MJ, Glick BR (2016) Role of ACC deaminase in stress control of leguminous plants. In: Plant growth promoting actinobacteria: a new avenue for enhancing the productivity and soil fertility of grain legumes. Springer, Singapore, pp 179–192

Naylor D, Coleman-Derr D (2018) Drought stress and root-associated bacterial communities. Front Plant Sci 8:303756

Peguero-Pina JJ, Vilagrosa A, Alonso-Forn D, Ferrio JP, Sancho-Knapik D, Gil-Pelegrín E (2020) Living in drylands: functional adaptations of trees and shrubs to cope with high temperatures and water scarcity. Forests 11(10):1028

Popa CL, Lupitu A, Mot MD, Copolovici L, Moisa C, Copolovici DM (2021) Chemical and biochemical characterization of essential oils and their corresponding hydrolats from six species of the Lamiaceae Family. Plan Theory 10(11):2489

Rashid A, Chaluvadi SR, Wang X, Hazzouri KM, Sudalaimuthuasari N, Rafi M, Al-Nuaimi M, Sasi S, Antepenko E, Bennetzen JL, Amiri KMA (2024) Microbiome properties in the root nodules of Prosopis cineraria, a leguminous desert tree. Microbiol Spectrum 12(6):e03617–03623

Rateb ME, Houssen WE, Harrison WT, Deng H, Okoro CK, Asenjo JA, Andrews BA, Bull AT, Goodfellow M, Ebel R (2011) Diverse metabolic profiles of a Streptomyces strain isolated from a hyper-arid environment. J Nat Prod 74(9):1965–1971

Raza A, Salehi H, Rahman MA, Zahid Z, Madadkar Haghjou M, Najafi-Kakavand S, Charagh S, Osman HS, Albaqami M, Zhuang Y (2022) Plant hormones and neurotransmitter interactions mediate antioxidant defenses under induced oxidative stress in plants. Front Plant Sci 13:961872

Rillig MC, Aguilar-Trigueros CA, Camenzind T, Cavagnaro TR, Degrune F, Hohmann P, Lammel DR, Mansour I, Roy J, van Der Heijden MG (2019) Why farmers should manage the arbuscular mycorrhizal symbiosis. New Phytologist 222(3):1171–1175

Rodríguez-Núñez K, Rodríguez-Ramos F, Leiva-Portilla D, Ibáñez C (2020) Brown biotechnology: a powerful toolbox for resolving current and future challenges in the development of arid lands. SN Appl Sci 2:1–23

Ruiz-Lozano JM, Aroca R, Zamarreño ÁM, Molina S, Andreo-Jiménez B, Porcel R, García-Mina JM, Ruyter-Spira C, López-Ráez JA (2016) Arbuscular mycorrhizal symbiosis induces strigolactone biosynthesis under drought and improves drought tolerance in lettuce and tomato. Plant Cell Environ 39(2):441–452

Sadia S, Qureshi R, Khalid S, Nayyar BG, Zhang J-t (2015) Role of secondary metabolites of wild marigold in suppression of Johnson grass and Sun spurge. Asian Pac J Trop Biomed 5(9):733–737

Sagaste CA, Montero G, Coronado MA, Ayala JR, León JÁ, García C, Rojano BA, Rosales S, Montes DG (2019) Creosote bush (*Larrea tridentata*) extract assessment as a green antioxidant for biodiesel. Molecules 24(9):1786

Saif J, Wright A, Khattak S, Elfadli K (2021) Keeping cool in the desert: using wind catchers for improved thermal comfort and indoor air quality at half the energy. Buildings 11(3):100

Sellstedt A, Richau KH (2013) Aspects of nitrogen-fixing actinobacteria, in particular free-living and symbiotic Frankia. FEMS Microbiol Lett 342(2):179–186

Shameem Kausar FA, Hameed M (2023) Osmoregulation mediated by structural and functional features in cenchrus ciliaris L. under saline conditions. Pak J Bot 55(4):1229–1243

Sharma LK, Raj A, Somawat K (2021) Spatio-temporal assessment of Environmentally Sensitive Areas (ESA) in the Thar Desert India, to combat desertification under UNCCD framework. J Arid Environ 194:104609

Shi Z, Mickan B, Feng G, Chen Y (2015) Arbuscular mycorrhizal fungi improved plant growth and nutrient acquisition of desert ephemeral Plantago minuta under variable soil water conditions. J Arid Land 7:414–420
Singh RP, Shelke GM, Kumar A, Jha PN (2015) Biochemistry and genetics of ACC deaminase: a weapon to "stress ethylene" produced in plants. Front Microbiol 6:149440
Skrzydeł J, Borowska-Wykręt D, Kwiatkowska D (2021) Structure, assembly and function of cuticle from mechanical perspective with special focus on perianth. Int J Mol Sci 22(8):4160
Soumaya B, Yosra E, Rim BM, Sarra D, Sawsen S, Sarra B, Kamel M, Wissem A-W, Isoda H, Wided M-K (2020) Preliminary phytochemical analysis, antioxidant, anti-inflammatory and anticancer activities of two Tunisian Ephedra species: *Ephedra alata* and *Ephedra fragilis*. S Afr J Bot 135:421–428
Soussi A, Ferjani R, Marasco R, Guesmi A, Cherif H, Rolli E, Mapelli F, Ouzari HI, Daffonchio D, Cherif A (2016) Plant-associated microbiomes in arid lands: diversity, ecology and biotechnological potential. Plant Soil 405:357–370
Spinoni J, Barbosa P, Cherlet M, Forzieri G, McCormick N, Naumann G, Vogt JV, Dosio A (2021) How will the progressive global increase of arid areas affect population and land-use in the 21st century? Glob Planet Chang 205:103597
Streb P, Tel-Or E, Feierabend J (1997) Light stress effects and antioxidative protection in two desert plants. Funct Ecol 11(4):416–424
Svistoonoff S, Hocher V, Gherbi H (2014) Actinorhizal root nodule symbioses: what is signalling telling on the origins of nodulation? Curr Opin Plant Biol 20:11–18
Tariq A, Graciano C, Sardans J, Zeng F, Hughes AC, Ahmed Z, Ullah A, Ali S, Gao Y, Peñuelas J (2024a) Plant root mechanisms and their effects on carbon and nutrient accumulation in desert ecosystems under changes in land use and climate. New Phytol 242(3):916–934
Tariq M, Tahreem N, Zafar M, Raza G, Shahid M, Zunair M, Iram W, Zahra ST (2024b) Occurrence of diverse plant growth promoting bacteria in soybean [*Glycine max* (L.) Merrill] root nodules and their prospective role in enhancing crop yield. Biocatal Agric Biotechnol 57(1):103072
Vacheron J, Renoud S, Muller D, Babalola O, Prigent-Combaret C (2015) Handbook for Azospirillum. Technical issues and protocols. Everything practical you always wanted to know about Azospirillum sp. but were afraid to ask. In: Cassan FD, Okon Y, Cecilia M, Creus MC (eds) Alleviation of abiotic and biotic stresses in plants by Azospirillum. Springer, Cham, pp 333–365
Valero-Galván J, González-Fernández R, Sigala-Hernández A, Núñez-Gastélum JA, Ruiz-May E, Rodrigo-García J, Larqué-Saavedra A, Martínez-Ruiz NDR (2021) Sensory attributes, physicochemical and antioxidant characteristics, and protein profile of wild prickly pear fruits (*O. macrocentra Engelm., O. phaeacantha Engelm.*, and *O. engelmannii* Salm-Dyck ex Engelmann.) and commercial prickly pear fruits (*O. ficus-indica* (L.) Mill.). Food Res Int 140:109909
Van Nguyen T, Pawlowski K (2017) Frankia and actinorhizal plants: symbiotic nitrogen fixation. In: Rhizotrophs: plant growth promotion to bioremediation. Springer, Singapore, pp 237–261
Wang R (2007) C 4 plants in the deserts of China: occurrence of C 4 photosynthesis and its morphological functional types. Photosynthetica 45:167–171
Wang X, Geng X, Liu B, Cai D, Li D, Xiao F, Zhu B, Hua T, Lu R, Liu F (2022) Desert ecosystems in China: past, present, and future. Earth Sci Rev 234:104206
Warner TT (2009) Desert meteorology. Cambridge University Press
Watkinson SC (2016) Mutualistic symbiosis between fungi and autotrophs. In: The fungi. Elsevier, pp 205–243
Whitford WG, Duval BD (2019) Ecology of desert systems. Academic
Winkler DE, Brooks E (2020) Tracing extremes across iconic desert landscapes: socio-ecological and cultural responses to climate change, water scarcity, and wildflower superblooms. Hum Ecol 48(2):211–223
Xu E, Zhang H (2020) Change pathway and intersection of rainfall, soil, and land use influencing water-related soil erosion. Ecol Indic 113:106281

Xu L, Naylor D, Dong Z, Simmons T, Pierroz G, Hixson KK, Kim Y-M, Zink EM, Engbrecht KM, Wang Y (2018) Drought delays development of the sorghum root microbiome and enriches for monoderm bacteria. Proc Natl Acad Sci 115(18):E4284–E4293

Xu H-F, Raanan H, Dai G-Z, Oren N, Berkowicz S, Murik O, Kaplan A, Qiu B-S (2021) Reading and surviving the harsh conditions in desert biological soil crust: the cyanobacterial viewpoint. FEMS Microbiol Rev 45(6):fuab036

Yeasmin R, Bonser SP, Motoki S, Nishihara E (2019) Arbuscular mycorrhiza influences growth and nutrient uptake of asparagus (*Asparagus officinalis* L.) under heat stress. HortScience 54(5):846–850

Yuan K, Xu H, Zhang G, Yan J (2022) Aridity and high salinity, rather than soil nutrients, regulate nitrogen and phosphorus stoichiometry in desert plants from the individual to the community level. Forests 13(6):890

Zahedi SM, Karimi M, Venditti A (2021) Plants adapted to arid areas: specialized metabolites. Nat Prod Res 35(19):3314–3331

Zhang Y, Tariq A, Hughes AC, Hong D, Wei F, Sun H, Sardans J, Peñuelas J, Perry G, Qiao J, Kurban A, Jia X, Raimondo D, Pan B, Yang W, Zhang D, Li W, Ahmed Z, Beierkuhnlein C, Lazkov G, Toderich K, Karryeva S, Dehkonov D, Hisoriev H, Dimeyeva L, Milko D, Soule A, Suska-Malawska M, Saparmuradov J, Bekzod A, Allin P, Dieye S, Cissse B, Whibesilassie W, Ma K (2023) Challenges and solutions to biodiversity conservation in arid lands. Sci Total Environ 857:159695

Zhu X-C, Song F-B, Liu S-Q, Liu T-D (2011) Effects of arbuscular mycorrhizal fungus on photosynthesis and water status of maize under high temperature stress. Plant Soil 346:189–199

Zhu X, Song F, Liu F (2017) Arbuscular mycorrhizal fungi and tolerance of temperature stress in plants. In: Arbuscular mycorrhizas and stress tolerance of plants. Springer, Singapore, pp 163–194

Zhuo L, Zhou L (2022) Moisture transport and water vapour budget over the Sahara Desert. Int J Climatol 42(13):6829–6843

Chapter 4
Enhancing Bioremediation Efficiency in Agricultural Soil Through Microbial Consortia Optimization

Ava Mohrazi and Reza Ghasemi-Fasaei

Abstract Unrestrained anthropogenic activities (industrialization, deforestation, urbanization and building dams) cause production of a wide range of pollutants, such as heavy metals, organic and inorganic compounds, and various agrochemicals, have severely affected the global ecosystem. Therefore, finding efficient, inexpensive, and environmentally friendly methods for reducing environmental pollution is a condition for sustainable development worldwide. Among the treatment methods, special attention has been paid to bioremediation.

Bioremediation is an eco-friendly and green remediation method that involves the use of microorganisms (bacteria, fungi, or a combination of both) or their enzymes. It can be applied in situ or ex situ. However, the disadvantages include slow processes that depend on environmental conditions and microbial species. The application of chemical and microbial treatments to enhance remediation efficiency is referred to as assisted remediation. A microbial consortium is made up of diverse microorganisms that can work together in a community. These microorganisms can use organic pollutants as food and carbon sources for growth and reproduction, transforming them into safer or non-harmful substances. To enhance the efficiency of a microbial consortium, the selection of species, composition of the consortia, and environmental manipulation should be appropriate. In order for all these factors to be optimized for the best results, each factor must be in an optimal state. Statistical methods such as the response surface method and the Taguchi approach can be used to achieve this goal.

If the right conditions and toxins are present, bioremediation can effectively clean up contaminated areas. However, it does have downsides, such as limited efficacy, site-specificity, slow development, and the need for assessment and monitoring. Despite these drawbacks, bioremediation remains a crucial tool in environmental cleanup efforts. This introductory chapter will explore the fundamentals, advantages, disadvantages, and perspectives surrounding bioremediation methods using microbial consortiums.

A. Mohrazi (✉) · R. Ghasemi-Fasaei
Department of Soil Science, School of Agriculture, Shiraz University, Shiraz, Iran

O. O. Babalola, A. S. Ayangbenro (eds.), *Microbial Allies*,
https://doi.org/10.1007/978-3-031-90530-8_4

Keywords Microbial consortium · Bioremediation · Pollution · Optimization

4.1 Introduction

Soil is a complex body of minerals and organic matter found within the outermost layer of the earth. This system is crucial to support life on Earth (Ghaneei-Bafghi et al. 2024). Soil pollution is defined as "the build-up in soils of persistent toxic compounds, chemicals, salts, radioactive materials, or disease-causing agents, which have adverse effects on plant growth and animal health". It is important to recognize that in most cases, pollution can be referred to as contamination and that pollutants in this case can be natural (Shi et al. 2023). An important point to consider is that pollution exceeds the natural background level of the substance. Therefore, in cases where substances are at or just above the background level and do not affect the soil or organisms, they are not considered pollutants. This is important to consider because in some cases, the pollutant might not have adverse effects (Yang et al. 2023). The detrimental effects on soil organisms, plant health, and impacts that extend to human activity should also be considered pollutants based on the magnitude of their impact on the environment and society. Environmental pollution (soil and water) has become a global concern because it can have irreparable effects on human health and other living beings (Wu et al. 2022). Therefore, it is necessary to understand the factors influencing the aggravation and creation of pollution. On the other hand, expanding and improving the methods of purifying the environment from pollution is a protective strategy against pollutants. The current research aims to review the biological methods of environmental remediation.

4.1.1 Causes of Soil Pollution

Soil pollution results from both natural occurrences and human activities. Utilizing compounds such as pesticides and irrigation with contaminated water are important factors in soil pollution. Some human activities, including mining activities and industrial operations, are additional factors that contribute to soil contamination. These activities make the soil unsuitable for plants and other organisms by releasing harmful compounds such as heavy metals, chemicals, and organic compounds like antibiotics and dyes (Zhou et al. 2022).

4.1.2 Effects of Soil Pollution

Soil pollution has dangerous consequences that affect the environment, human health, and the economy. It causes a reduction in soil fertility and productivity, affecting the growth of plants and the productivity of agronomy crops. In general,

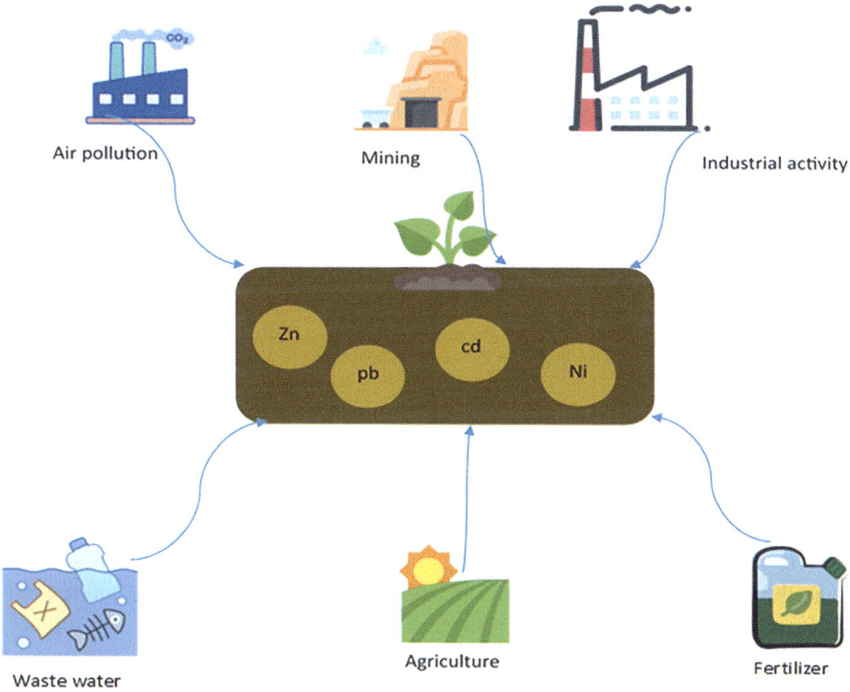

Fig. 4.1 Illustrates the various sources of heavy metal pollution

pollutants such as dyes and heavy metals can cause skin and respiratory diseases. These pollutants seriously endanger the health of humans and animals (Chen et al. 2022). The term "heavy metal" (HM) is used to describe elements with a density greater than 5 g per cubic centimeter, and pollution with heavy metals is a serious threat to human and animal life (Adnan et al. 2024). Heavy metals can enter the environment through various sources such as industrial and agricultural activities (Fig. 4.1). In fact, this pollution can cause dangerous illnesses like cancer, brain damage, and more (Angon et al. 2024). It also has a negative effect on plant health (Table 4.1).

4.1.3 Soil Remediation

Soil remediation is an important technique for removing pollutants from water and soil, restoring it to its natural state. The remediation process depends on several factors, such as the type and dose of pollutants.

Different techniques, such as physical, chemical, and biological approaches, are applied to remediate pollutants from water and soil pollution (Zeng et al. 2024).

Table 4.1 Effect of HMs in plant uptake and human health

Element	Plant effect	References	Human effect	References
Zn	Necrotic spotting, genetic-related disorders and plant growth	Yahaghi et al. (2019)	Brain damage	Adnan et al. (2024)
Cd	Reactive oxygen species production, photosynthesis, nutrient uptake	Huybrechts et al. (2021)	Effect on thyroid, bladder cancer	Adnan et al. (2024)
Pb	Seed germination, chlorophyll synthesis and plant growth	Yahaghi et al. (2019)	Effect on thyroid, brain damage	Adnan et al. (2024)
Ni	Seed germination and plant development are inhibited by inhibiting amylase and protease activity	Koza et al. (2022)	Pulmonary disease, bladder cancer	Adnan et al. (2024)

Physical methods focus on processes such as excavation, soil washing, and thermal desorption to remove pollutants from the soil. Physical remediation mainly includes soil replacement methods and thermal desorption. Soil replacement involves replacing or partially replacing contaminated soil with clean soil to dilute the pollutant concentration, increase the soil's environmental capacity, and remediate the soil (Yao et al. 2012). Soil replacement consists of three types: soil replacement, soil spading, and new soil importing. (i) Soil replacement involves removing the contaminated soil and replacing it with new soil. This method is suitable for contaminated soil with a small area. Additionally, the replaced soil should be treated properly to prevent secondary pollution. (ii) Soil spading involves deeply digging the contaminated soil to spread the pollutant into deeper sites, achieving dilution and natural degradation. (iii) New soil importing entails adding clean soil to contaminated soil, covering it on the surface or mixing it to decrease the pollutant concentration. Soil replacement can effectively isolate the soil and ecosystem, reducing its impact on the environment. However, this technology requires a lot of work and is costly. It is also suitable for soils with a small area and heavy pollution (Qian and Liu 2000; Zhang et al. 2004).

In chemical methods, chemical agents are used to convert pollutants into less harmful substances. Chemical leaching involves washing contaminated soil with fresh water, reagents, and other liquids or gases that can leach pollutants from the soil. Through ion exchange, precipitation, adsorption, and chelation, heavy metals in the soil are transferred from soil to the liquid phase and then recovered from the leachate. The leachate mainly contains mineral detergents, chelation agents, surfactants, etc. (Yao et al. 2012).

Biological methods are green methods that use microorganisms (bacteria, fungi, plants, and combinations of them) to degrade or detoxify pollutants in the soil (Liao et al. 2022; Liu et al. 2023; Kaya et al. 2024).

The choice of a good remediation technique depends on several factors, such as the type and concentration of pollutants. Among these methods, bioremediation is

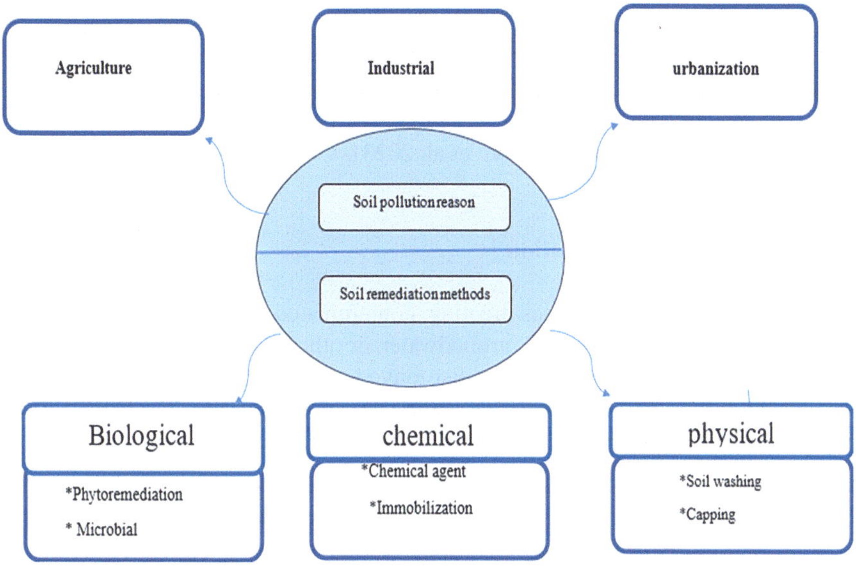

Fig. 4.2 The summary of remediation approaches

one of the most appropriate choices for cleaning soil pollution. The illustration of the remediation method is shown in Fig. 4.2.

4.2 What Is Bioremediation?

Bioremediation is a green method that utilizes living organisms (plants, bacteria, and fungi) to neutralize or remove pollutants from soil and water (Azubuike et al. 2016; Bala et al. 2022). By harnessing natural processes for the detoxification and degradation of various contaminants, bioremediation emerges as a crucial method for eliminating hazardous substances from soil and ecosystems (Bala et al. 2022).

The versatility of bioremediation strategies, encompassing both in situ and ex situ applications, highlights its broad potential in addressing environmental challenges. As a key component of environmental biotechnology, bioremediation leverages microbial metabolism and phytoremediation to transform or eliminate environmental pollutants, providing a promising alternative for ecological restoration. By utilizing microbial consortia and biodegradation processes, this technique has become a forefront solution for environmental remediation, offering practical means to reclaim polluted water and soil and safeguard human health and the environment (Tedesco et al. 2024).

4.2.1 Types of Bioremediations

In general, bioremediation is an efficient method for removing pollutants (organic and inorganic compounds) and can be categorized into two types based on the location site: in situ and ex situ (Davoodi et al. 2023).

4.2.1.1 In-Situ Bioremediation

In-situ bioremediation involves treating contaminated material directly on-site without the need to transport soil, groundwater, or other materials to a separate location for cleanup. This method is less disruptive to the environment and often more cost-effective. Techniques include bioventing, biosparging, and natural attenuation, where native microorganisms are used to degrade pollutants.

1. Natural attenuation: utilizes the capabilities of native microorganisms (bacteria and fungi) to degrade pollutants over time without human intervention (Azubuike et al. 2016; Bala et al. 2022).
2. Bioventing/Biosparging: These techniques involve the injection of air/nutrients to stimulate the growth of native microbial populations that can break down contaminants (Sravya and Sangeetha 2022; Kaur et al. 2020).

4.2.1.2 Ex-Situ Bioremediation

Ex-situ remediation involves transporting contaminated soil or water to another site for cleanup. This method is crucial when pollutants (organic and inorganic) are too concentrated or the site is too large (Azubuike et al. 2016; Bala et al. 2022).

4.2.2 Key Mechanisms of Bioremediation

Bioremediation involves phytoremediation, bioremediation, and combined remediation. This context focuses on the bioremediation method. While microorganisms may not directly degrade and destroy hazardous materials, they can influence the migration and transformation of these substances by altering their physical and chemical properties. The remediation mechanisms include extracellular complexation, precipitation, oxidation-reduction reactions, and intracellular accumulation. In simpler terms, bioremediation has two stages:

1. Microbial Activity: Relies on promoting the growth of specific microorganisms that use contaminants as a food and energy source to degrade common pollutants such as hydrocarbons, hazardous materials, and pesticides.

2. Conversion Process: Microorganisms convert harmful compounds into less harmful ones during the bioremediation process. Optimal environmental conditions enhance the efficiency of bioremediation, where suitable temperatures, nutrients, and food sources must be maintained. If natural conditions are not ideal, they can be improved by adding amendments like molasses, vegetable oil, or air to stimulate microbial growth (Hu et al. 2023; Schommer et al. 2023).

4.2.3 Bioremediation Dynamics

Bioremediation operates through a complex interaction within microbial cells, involving energy production through redox reactions essential for respiration and other biological functions necessary for cell maintenance and reproduction. A typical bioremediation system requires a delivery mechanism that provides an energy source and essential nutrients to facilitate healing processes. Additionally, the approach involves the degradation, removal, immobilization, or detoxification of various chemical wastes and hazardous materials through microbial activity (Abatenh et al. 2017; Saleem et al. 2021; Ojuederie and Babalola 2017).

4.2.4 Bioremediation Techniques

Bioremediation techniques utilize various microorganisms, including bacteria, algae, fungi, or a combination of them, to efficiently degrade complex compounds (organic and inorganic). These methods are categorized based on factors such as the site of application, type of microorganisms, type of contaminants, and specific environmental conditions (Muhammad et al. 2024).

4.2.5 Classification Based on Microorganisms and Contaminants

1. Microorganisms Used: Techniques vary depending on whether bacteria, fungi, algae, or yeast are employed, each with unique capabilities to handle different pollutants (Zhou et al. 2023).
2. Type of Contaminants: Strategies are tailored based on the pollutants present, such as hydrocarbons or heavy metals, to ensure effective degradation (Yu et al. 2023).

4.2.6 Site-Specific Techniques

Bioremediation methods are often distinguished by the location of their application—in situ or ex situ—considering factors such as cost, site characteristics, and pollutant concentration. In-situ methods treat contaminants on-site, while ex-situ techniques involve the removal of contaminated material to another location for treatment (Rendón-Castrillón et al. 2024).

4.2.7 Method of Application

Different approaches used for bioremediation implementation (Azubuike et al. 2016) include: Bioventing, Biosparging, Bioslurping, Bioaugmentation, and Biostimulation (Mohammadi et al. 2024).

4.3 The Role of Microorganisms in Remediation

Microorganisms play a vital role in remediation, utilizing their natural metabolic activities to transform pollutants into less harmful substances. These organisms possess unique enzymatic systems that enable them to degrade a wide range of environmental contaminants, such as heavy metals and pesticides (Wei et al. 2024).

4.3.1 Microbial Processes and Mechanisms

1. Degradation and Conversion: Microorganisms use pollutants as food and energy sources, breaking them down into nontoxic, less harmful compounds (Chen et al. 2024).
2. Enzymatic Actions: Microbial enzymes play a crucial role in the remediation of pollutants, enhancing microbial capacity to handle complex chemical structures, especially in the degradation of pollutants (Narayanan et al. 2023).
3. Advanced Microbial Methods: Recent innovations in microbial bioremediation include the development of bioelectrochemical systems, microbial biofilms, and glycoconjugates, improving the efficiency of this method (Ferreira et al. 2024).

Microbial consortia, often a combination of various microorganisms, ensure the degradation of complex pollutants and support ecological balance, preventing secondary pollution. Genetically engineered microorganisms have been introduced to enhance bioremediation capabilities, equipped with augmented degradative enzymes that can tackle a broader spectrum of contaminants, from heavy metals and hydrocarbons to synthetic chemicals. They also play a crucial role in the

Table 4.2 Microbial consortium efficiency on pollutant removal

Microbial consortium	Pollutant	Removal efficiency (%)	References
Proteobacteria + bacteroidota	Polycyclic aromatic hydrocarbons (PAHs)	88	Zhou et al. (2024)
(Azospirillum sp., Pseudomonas sp., Saccharomyces sp., and Rhizobium sp.).	Lead (Pb)	56	Ferreira et al.(2024)
Bacteria Pandoraea + Ascomycota + Basidiomycota	Cadmium	95	Li et al. (2023)
Sphingomonas + *Arthrobacter*	PAHs	45	Zhou et al. (2024)
Trichoderma. saturnisporum + *Trichoderma citrinoviride*	Cyanide	99	Rishi et al. (2023)
(Azospirillum sp., Pseudomonas sp., Saccharomyces sp., and Rhizobium sp.).	Cd	60	Ferreira et al.(2024)
Halomonas *pacifica* + Shewanella *alga*	Malachite green dye	77	Alghamdi et al. (2023)

immobilization and mobilization of contaminants, using mechanisms to either sequester harmful substances or convert them into mobile forms for easier extraction and treatment. This dual capability makes microorganisms indispensable agents in the ongoing efforts to cleanse and restore contaminated sites worldwide. Examples of microbial consortium efficiency in pollutant removal are provided in Table 4.2.

4.4 Advantages of Bioremediation

Bioremediation is a cheap and simple method, and unlike traditional methods, it does not require special equipment. In fact, one of the most important advantages of this method is its cheapness and simplicity, which has made it a popular method (Alghamdi et al. 2023).

4.4.1 Cost-Effectiveness and Scalability

1. Lower Costs: Bioremediation is generally less expensive than conventional remediation methods, as it requires minimal mechanical intervention and reduces the need for costly disposal processes.

2. Scalable Solutions: This technology can be applied to various scales, from treating small patches of contaminated soil to large-scale water treatment projects, making it a versatile option for different types of environmental clean-ups.

4.4.2 Eco-Friendly Impact

Bioremediation promotes a healthier environment by using natural organisms to clean up contaminants. This method not only improves soil quality and water clarity but also enhances air quality by reducing the release of harmful pollutants. The process typically results in a 95-percent net reduction in chemical concentrations, which facilitates the reuse of land and contributes to sustainable development (Jaiswal et al. 2024; Tedesco et al. 2024).

4.5 Disadvantages

It has disadvantages such as being time-consuming compared to other remediation methods, seasonal changes in microbial activity due to changes in environmental factors, and the problematic application of treatment additives.

Microorganisms only do well when the available waste materials allow them to produce nutrients and energy to grow more cells. When these conditions are not favorable, their destruction capacity decreases. In such cases, genetically engineered microorganisms should be used, although native microorganisms are preferred (Kumar et al. 2011).

4.6 Microbial and Environmental Constraints

The success of bioremediation heavily relies on the presence of suitable microorganisms that can survive in the contaminated environment and have the ability to degrade specific pollutants. However, not all contaminants are biodegradable, and the effectiveness of bioremediation can vary significantly with the type of pollutant. Moreover, microorganisms require specific nutrients and oxygen levels to thrive and effectively break down contaminants, which might not always be available in sufficient quantities at the polluted site (Samanta and Roychoudhury 2024). The physicochemical characteristics of the environment, such as soil type, temperature, and pH also play a crucial role in the effectiveness of bioremediation (Kumar and Chakraborty 2024). Some researchers reported microbial consortium (Rhodococcus KD0901 and Paenarthrobacter sp. KN0901) succeeded in removing Dibutyl phthalate (DBP) and atrazine (ATZ), which led to an increase in plant length and biomass by 33.3% and 273.7%, respectively (Kang et al. 2024). Also, a bacterial consortium was obtained from a contaminated river in São Paulo, Brazil was a good choice for removing copper pollution (Carpio et al. 2014).

4.7 Optimization of Bioremediation

To maximize bioremediation efficiency, microbial consortia need to be optimized. This involves:

4.7.1 Selection of Appropriate Species

The species included in the consortia should possess the necessary metabolic pathways for degrading the targeted contaminants (Li et al. 2023).

4.7.2 Optimization of Consortia Composition (Tizazu et al. 2022)

The optimal ratio of different species in the consortia can significantly impact biodegradation efficiency. It has been reported halotolerant-thermo-alkliphilic microorganisms are suitable microbial consortia. Also, Ma et al. (2024) found that synthetic fungal-bacterial consortium design leads to high removal efficiency and total petrochemical hydrocarbon removal efficiency of 92% was obtained.

4.7.3 Environmental Manipulation (Pallavi et al. 2024)

Adjusting soil pH, temperature, and nutrient availability can enhance the metabolic activity of the consortia. It has been reported neutral pH and temperature equal 40 °C are optimal environmental conditions.

On the other hand, in order for all the mentioned factors to be in the best conditions for the highest effect, each of the factors must be in an optimal state. To achieve this goal, statistical methods such as the response surface method (RSM) and Taguchi approach can be used. These two methods are among the successful statistical techniques in optimizing the test conditions. These two methods are economical, easy and with high accuracy (Mohrazi et al. 2023a, b; Razmi et al. 2021). It was stated in research that RSM design is useful for the optimization of mixed dyes degradation by microbial consortium. It was reported the optimal conditions to removal dye were salinity (30 g/l), pH (9), peptone (5 g/l), inoculum size (1.5 10^8 CFU/ml), agitation (150 rpm) and contact time (72 h) (Alghamdi et al. 2023).

4.8 Conclusion

In this text, an attempt was made to bring new information about the methods of pollutant purification, especially the bioremediation method. In this text, the advantages, disadvantages, challenges, and limitations of this technique were discussed. The development strategy of future remediation technologies is to research green, environmental-friendly biological remediation, combined remediation, in-situ remediation, based on fully equipped rapid approaches, and provide technical support for agricultural soil contamination, industrial enterprises brownfield, mining sites, etc. In fact, this technique is a cost-effective method, cheaply available, which is divided into two categories based on the place of implementation of the method, in-place and out-of-place. One of the inherent abilities of bacteria mentioned in this text was the secretion of destructive enzymes. It was also stated that if the environmental conditions are optimized, the efficiency of this technique will increase. But more serious studies are still needed in this field.

References

Abatenh E, Gizaw B, Tsegaye Z, Wassie M (2017) The role of microorganisms in bioremediation—a review. Open J Environ Biol 2(1):038–046

Adnan M, Xiao B, Ali MU, Xiao P, Zhao P, Wang H, Bibi S (2024) Heavy metals pollution from smelting activities: a threat to soil and groundwater. Ecotoxicol Environ Saf 274:116189

Alghamdi MA, Ayed L, Aljarad MR, Altayeb HN, Chaieb K (2023) Whole genome sequencing analysis and Box-Behnken design for the optimization of the decolourization of mixture textile dyes by halotolerant microbial consortium. Microbiol Res 276:127481

Angon PB, Islam MS, Kc S, Das A, Anjum N, Poudel A, Suchi SA (2024) Sources, effects and present perspectives of heavy metals contamination: soil, plants and human food chain. Heliyon 10:e28357

Azubuike CC, Chikere CB, Okpokwasili GC (2016) Bioremediation techniques–classification based on site of application: principles, advantages, limitations and prospects. World J Microbiol Biotechnol 32:1–18

Bala S, Garg D, Thirumalesh BV, Sharma M, Sridhar K, Inbaraj BS, Tripathi M (2022) Recent strategies for bioremediation of emerging pollutants: a review for a green and sustainable environment. Toxics 10(8):484

Carpio IEM, Machado-Santelli G, Sakata SK, Ferreira Filho SS, Rodrigues DF (2014) Copper removal using a heavy-metal resistant microbial consortium in a fixed-bed reactor. Water Res 62:156–166

Chen L, Zhou M, Wang J, Zhang Z, Duan C, Wang X et al (2022) A global meta-analysis of heavy metal (loid) s pollution in soils near copper mines: evaluation of pollution level and probabilistic health risks. Sci Total Environ 835:155441

Chen W, Qin S, Yang C, Long K, Liang S, Liu H et al (2024) Bioaugmentation using salt-tolerant bacteria in a dual-stage process for high-salinity wastewater treatment: performance, microbial community, and salt-tolerance mechanism. J Water Process Eng 57:104620

Davoodi SM, Miri S, Brar SK, Knystautas E, Martel R (2023) Simulation of novel jellyfish type of process for bioremediation application. Chemosphere 313:137376

Ferreira AD, Zem TMS, Barcellos D, Nóbrega GN, Queiroz HM, Otero XL et al (2024) Assessment of the potential of microbial consortium for the reclamation of mine tailings containing potentially toxic elements. J Environ Chem Eng 12(2):112399

Ghaneei-Bafghi MJ, Feiznia S, Mokhtari AR, Jaafari M, Tavili A, Khodaeian Z (2024) Agricultural soil contamination and degradation near a mining area in an arid region. J Geochem Explor 256:107349

Hu F, Wang P, Li Y, Ling J, Ruan Y, Yu J, Zhang L (2023) Bioremediation of environmental organic pollutants by Pseudomonas aeruginosa: mechanisms, methods and challenges. Environ Res 239(Pt 1):117211

Huybrechts M, Hendrix S, Kyndt T, Demeestere K, Vandamme D, Cuypers A (2021) Short-term effects of cadmium on leaf growth and nutrient transport in rice plants. Plant Sci 313:111054

Jaiswal S, Singh B, Dhingra I, Joshi A, Kodgire P (2024) Bioremediation and bioscavenging for elimination of Organophosphorus threats: an approach using enzymatic advancements. Environ Res 252(Pt 2):118888

Kang Z, Han X, Wang C, Yu H (2024) Artificial microbial consortium for simultaneous removal of dibutyl phthalate and atrazine combined pollutants at low temperatures. Chem Eng J 482:149075

Kaur D, Singh A, Kumar A, Gupta S (2020) Genetic engineering approaches and applicability for the bioremediation of metalloids. In: Plant life under changing environment. Academic, pp 207–235

Kaya C, Uğurlar F, Ashraf M, Hou D, Kirkham MB, Bolan N (2024) Microbial consortia-mediated arsenic bioremediation in agricultural soils: current status, challenges, and solutions. Sci Total Environ 917:170297

Koza NA, Adedayo AA, Babalola OO, Kappo AP (2022) Microorganisms in plant growth and development: roles in abiotic stress tolerance and secondary metabolites secretion. Microorganisms 10(8):1528

Kumar V, Chakraborty P (2024) Microbial bioremediation of metal and radionuclides: approaches and advancement. In: Microbiome-based decontamination of environmental pollutants. Academic, pp 269–292

Kumar A, Bisht BS, Joshi VD, Dhewa T (2011) Review on bioremediation of polluted environment: a management tool. Int J Environ Sci 1(6):1079–1093

Li W, Zhu Y, Li K, Wang L, Li D, Liu N, Huang S (2023) Synergistic remediation of phenanthrene–cadmium co-contaminants by an immobilized acclimated bacterial–fungal consortium and its community response. Chemosphere 336:139234

Liao X, Li Y, Miranda-Avilés R, Zha X, Anguiano JHH, Sánchez CDM et al (2022) In situ remediation and ex situ treatment practices of arsenic-contaminated soil: an overview on recent advances. J Hazard Mater Adv 8:100157

Liu J, Sun S, Zhang H, Kong Q, Li Q, Yao X (2023) Remediation materials for the immobilization of hexavalent chromium in contaminated soil: preparation, applications, and mechanisms. Environ Res 237:116918

Ma XK, Mao D, Chen J, Wang Q, Dong Z, Zhang H et al (2024) Enhancing bioremediation of petroleum sludge-polluted soil: the impact of fungal-bacterial consortium's application strategy. Appl Soil Ecol 195:105267

Mohammadi M, Bayat Z, Hassanshahian M, Mousavi M, Shekarchizadeh F (2024) Microbial community response to biostimulation and bioaugmentation in crude oil-polluted sediments of the Persian Gulf. Environ Res 249:118197

Mohrazi A, Ghasemi-Fasaei R, Mojiri A, Safarzadeh Shirazi S (2023a) Optimization of LDO-pectin synthesis conditions for the removal of metals from wastewater: a comparison of response surface methods and Taguchi approaches. Polymers 15(18):3778

Mohrazi A, Ghasemi-Fasaei R, Mojiri A, Shirazi SS (2023b) Investigating an electro-bio-chemical phytoremediation of multi-metal polluted soil by maize and sunflower using RSM-based optimization methodology. Environ Exp Bot 211:105352

Narayanan M, Ali SS, El-Sheekh M (2023) A comprehensive review on the potential of microbial enzymes in multipollutant bioremediation: mechanisms, challenges, and future prospects. J Environ Manag 334:117532
Ojuederie OB, Babalola OO (2017) Microbial and plant-assisted bioremediation of heavy metal polluted environments: a review. Int J Environ Res Public Health 14(12):1504
Pallavi P, Manikandan SK, Nair V (2024) Optimization and mechanistic study on bioremediation of Cr (VI) using microbial cell immobilized sugarcane bagasse biochar. J Water Process Eng 58:104859
Qian SQ, Liu Z (2000) An overview of development in the soil-remediation technologies. Chem Ind Eng Process 4(10–2):20
Razmi B, Ghasemi-Fasaei R, Ronaghi A, Mostowfizadeh-Ghalamfarsa R (2021) Investigation of factors affecting phytoremediation of multi-elements polluted calcareous soil using Taguchi optimization. Ecotoxicol Environ Saf 207:111315
Rendón-Castrillón L, Ramírez-Carmona M, Ocampo-López C, González-López F, Cuartas-Uribe B, Mendoza-Roca JA (2024) Efficient bioremediation of indigo-dye contaminated textile wastewater using native microorganisms and combined bioaumentation-biostimulation techniques. Chemosphere 353:141538
Rishi S, Kaur I, Naseem M, Gaur VK, Mishra S, Srivastava S et al (2023) Development of immobilized novel fungal consortium for the efficient remediation of cyanide-contaminated wastewaters. Bioresour Technol 373:128750
Saleem S, Mushtaq NU, Shah WH, Rasool A, Rehman RU (2021) Microbial and plant-assisted bioremediation of heavy metal polluted environments. In: Heavy metal toxicity in plants. CRC Press, pp 139–156
Samanta S, Roychoudhury A (2024) Microbial bioremediation of metal and radionuclides: approaches and advancement. In: Microbiome-based decontamination of environmental pollutants. Academic, pp 293–316
Schommer VA, Vanin AP, Nazari MT, Ferrari V, Dettmer A, Colla LM, Piccin JS (2023) Biochar-immobilized Bacillus spp. for heavy metals bioremediation: a review on immobilization techniques, bioremediation mechanisms and effects on soil. Sci Total Environ 881:163385
Shi J, Zhao D, Ren F, Huang L (2023) Spatiotemporal variation of soil heavy metals in China: the pollution status and risk assessment. Sci Total Environ 871:161768
Sravya K, Sangeetha S (2022) Feasibility study on bioremediation techniques to contaminated soils. Mater Today Proc 51:2556–2256
Tedesco P, Balzano S, Coppola D, Esposito FP, de Pascale D, Denato R (2024) Bioremediation for the recovery of oil polluted marine environment, opportunities and challenges approaching the Blue Growth. Mar Pollut Bull 200:116157
Tizazu S, Tesfaye G, Andualem B, Wang A, Guadie A (2022) Evaluating the potential of thermo-alkaliphilic microbial consortia for azo dye biodegradation under anaerobic-aerobic conditions: optimization and microbial diversity analysis. J Environ Manag 323:116235
Wei Y, Shen D, Nicholaus R, Wang Y, Lukwambe B, Zhu J et al (2024) Exogenous compound bacteria enhance the nutrient removal efficiency of integrated bioremediation systems: functional genes and microorganisms play key roles. Environ Res 252(Pt 1):118864
Wu Y, Li X, Yu L, Wang T, Wang J, Liu T (2022) Review of soil heavy metal pollution in China: spatial distribution, primary sources, and remediation alternatives. Resour Conserv Recycl 181:106261
Yahaghi Z, Shirvani M, Nourbakhsh F, Pueyo JJ (2019) Uptake and effects of lead and zinc on alfalfa (Medicago sativa L.) seed germination and seedling growth: role of plant growth promoting bacteria. S Afr J Bot 124:573–582
Yang S, Sun L, Sun Y, Song K, Qin Q, Zhu Z, Xue Y (2023) Towards an integrated health risk assessment framework of soil heavy metals pollution: theoretical basis, conceptual model, and perspectives. Environ Pollut 316:120596

Yu X, Jiang N, Yang Y, Liu H, Gao X, Cheng L (2023) Heavy metals remediation through bio-solidification: potential application in environmental geotechnics. Ecotoxicol Environ Saf 263:115305

Zeng S, Dai Z, Ma B, Dahlgren RA, Xu J (2024) Environmental interactions and remediation strategies for co-occurring pollutants in soil—a review. Earth Crit Zone 1:100002

Zhang YF, Sheng JC, Lu QY (2004) Review on the soil remediation technologies. Gansu Agric Sci Technol 10:36–38

Zhou L, Li S, Li F (2022) Damage and elimination of soil and water antibiotic and heavy metal pollution caused by livestock husbandry. Environ Res 215:114188

Zhou Q, Wu K, Yao L, Chen R, Liu S, Xing H et al (2023) Bio-ecological remediation of freshwater aquaculture environments: a systematic review and bibliometric analysis. Water Biol Secur 3(1):100229

Zhou X, Jiang Y, Leng G, Ling W, Wang J (2024) The addition of Glomalin-related soil protein and functional microbial consortium increased bound PAH residue degradation in soil. Appl Soil Ecol 193:105158

Chapter 5
Synergistic Plant-Microbe Interactions in Biomining and Bioleaching: Implications for Sustainable Resource Recovery

Oyindamola John Samson, Bukola Margaret Popoola, and Afolake Atinuke Olanbiwoninu

Abstract Mining and leaching is a means of recovery of valuable/beneficial metals from low-quality ore. Mining is generally faced with some environmental linked challenges such as greenhouse gas emissions. Also residual liquid waste generated from the leaching process is highly acidic in nature. The synergistic interactions between plants and microbes are an integral part of our terrestrial ecosystem and highly promises a more sustainable alternative to conventional mining methods. Various plant species produce exudates, root exudates attracts beneficial microbes, which form mutualistic relationships with plants. Thereby fostering and enhancing their survival and growth especially when targeted for use in biomining and bioleaching. These bio-based approaches gives an edge over traditional mining methods and conventional mining. Exploring this synergism between plant and microbes minimizes negative environmental impact and the consequent effect of global decline in high-grade ore. By replacing risky mining methods and related mining waste, this innovative biotechnological approach has the potential to revolutionize mining worldwide. The end products can also sustainably lower the usage of agrochemicals and facilitate the production of sustainable and healthful food for agricultural sustainability. This review focuses on the role of synergistic plant-microbe interactions in biomining and bioleaching an implication for sustainable agriculture and resource recovery. The study highlights the mechanism, benefits as well as the limitation of biomining and bioleaching in today's world. While also focusing on the future perspectives and research directions in order to have a sustainable resource recovery system.

O. J. Samson · B. M. Popoola (✉) · A. A. Olanbiwoninu
Department of Microbiology and Biotechnology, Faculty of Natural Sciences, Ajayi Crowther University, Oyo, Oyo State, Nigeria
e-mail: bm.popoola@acu.edu.ng

O. O. Babalola, A. S. Ayangbenro (eds.), *Microbial Allies*,
https://doi.org/10.1007/978-3-031-90530-8_5

5.1 Introduction

As global demand for metals and minerals rises due to technological advancements and population growth, there is an increasing urgency to adopt sustainable resource recovery methods. Traditional mining practices often cause significant environmental harm, including deforestation, soil erosion, water pollution, and greenhouse gas emissions (Mudd 2010; Rawat et al. 2023). In response, researchers and industry professionals are exploring innovative approaches to resource extraction that minimize environmental impact while maximizing efficiency.

Microbial activity plays a pivotal role in sustainable agriculture, serving as a key driver of nutrient cycling and soil fertility enhancement (Suman et al. 2022). Through their metabolic processes, microbes facilitate the decomposition of organic matter, release essential nutrients like nitrogen and phosphorus, and improve soil structure. These activities not only boost plant growth but also contribute to ecosystem resilience by maintaining soil health. As foundational elements of biomining and bioleaching, microbes further demonstrate their versatility in advancing sustainable agricultural practices (Samarasekere 2024).

One promising area of research focuses on the synergistic interactions between plants and microorganisms in biomining and bioleaching processes. These bio-based approaches harness the natural capabilities of certain plants and microbes to accumulate, mobilize, or transform metal compounds, offering a sustainable alternative to conventional mining methods (Barapatre et al. 2021). Biomining and bioleaching generally require less energy and emit fewer greenhouse gases than traditional methods. Furthermore, these processes can help rehabilitate contaminated sites and reduce the necessity for new mine development (Johnson 2014).Plant-microbe interactions in these processes are particularly intriguing. Hyperaccumulator plants, such as Indian mustard (*Brassica juncea*), alpine pennycress (*Thlaspi caerulescens*), and sunflower (*Helianthus annuus*), can store large quantities of metals in their tissues. When combined with metal-mobilizing microbes in the rhizosphere, these plants significantly enhance metal extraction efficiency. Moreover, these partnerships can improve soil health, promote biodiversity, and expedite the rehabilitation of mining sites (Jerez 2017).

From an economic standpoint, sustainable resource recovery via plant-microbe interactions opens new avenues by enabling the extraction of valuable metals from previously untapped sources. This approach aligns with several United Nations Sustainable Development Goals, including responsible consumption and production, climate action, and life on land (United Nations 2015).

Exploring the mechanisms behind plant-microbe interactions in metal extraction, analyzing successful case studies, and addressing potential challenges and future directions in this field are critical. By understanding and optimizing these biological processes, we can move towards a more sustainable and environmentally friendly approach to resource recovery. This will ensure the responsible utilization of the Earth's finite mineral resources for future generations.

5.2 Biomining and Bioleaching

Biomining and bioleaching are innovative biotechnological processes that leverage the capabilities of microorganisms to extract valuable metals from ores and waste materials. These techniques have gained significant attention in recent years due to their potential for sustainable resource recovery and reduced environmental impact compared to traditional mining methods (Brierley and Brierley 2013).

Biomining refers to the use of microorganisms to extract metals from low-grade ores, mine tailings, and electronic waste. This process involves the oxidation of sulfide minerals by acidophilic bacteria and archaea, which solubilize metals and make them available for recovery (Johnson 2014). Bioleaching, a subset of biomining, specifically refers to the extraction of metals from sulfide ores through microbial activity (Martínez-Bellange et al. 2022).

Biomining is a biotechnology process that utilizes microorganisms to extract metals, especially copper and gold, from ores and concentrates. Originally, a straightforward process both biologically and technically, biomining has evolved into a complex technology. Today, it is extensively used in some of the largest industrial stirred tanks and heaps worldwide for mineral bioprocessing (Rawlings and Johnson 2007).

The key difference between the two is that biomining deals with the technique that uses prokaryotes or fungi to extract metals from minerals however, bioleaching make use of bacteria to extract metals from minerals. The infographic below gives a summary of the differences.

PECULIARITIES OF BIMINING AND BIOLEACHING

	BIOMINING	BIOLEACHING
MICRO-ORGANISMS	• Fungi and Prokaryotes	Iron-sulfur bacteria and Archaea
TECHNIQUE INVOLVED	• The technique uses microbes to extract metals from their ore	Solubilization of metals from insoluble metals by the use of microbes
PROCESS OF ACTION	• Microbes can uptake metals in the ore into their cells, the metal can then be extracted from the microbial cells	Bacteria are used for further oxidation of the metal ore, which isolates the metal and allows for the extraction of the metal from the ore

The most commonly used microorganisms in biomining and bioleaching are chemolithoautotrophic bacteria and archaea, such as *Acidithiobacillus ferrooxidans*, *Leptospirillum ferrooxidans*, and *Sulfolobus* species. These microorganisms derive energy from the oxidation of reduced inorganic compounds, such as ferrous iron and reduced sulfur species, while using carbon dioxide as their carbon source (Vera et al. 2013).

5.2.1 Significance of Plant-Microbe Interactions in Biomining and Bioleaching for Sustainable Agriculture

While the role of microorganisms in biomining and bioleaching has been well-established, recent research has highlighted the potential for synergistic plant-microbe interactions to enhance these processes. This is key to increasing plant yields and improvement on plant resistance to stress thereby improving agricultural sustainability and subsequently the circular economy. The integration of plants into biomining and bioleaching systems, known as phytomining or phytoremediation. Plant roots release exudates that can modify the rhizosphere environment, promoting the growth and activity of metal-solubilizing microorganisms. This synergistic interaction can increase the efficiency of metal extraction from ores and contaminated soils (Wan et al. 2012). Plants provide a supportive habitat for microorganisms, increasing their population density and diversity in the rhizosphere. This can lead to more robust and resilient microbial communities capable of sustained metal extraction (Sessitsch et al. 2013) the microorganisms dissolve minerals from metals by creating leachates which can be treated to form nutrients. The nutrients are products of dissolved minerals like magnesium, manganese, iron, cobalt etc. alongside proteins and organic acids leading to biostimulating metabolites (Štyriaková 2023). This can rejuvenate unhealthy plants, raise plants immunity against pathogens, improve fruit sugar content and root systems as well as increase plant yield (Štyriaková 2023). Also the produced nutrients causes increased chlorophyll and thereby stimulating photosynthesis, frost damage is prevented via antifreeze proteins. These leads to the production of more leaves and buds, thereby improving plant quality (Štyriaková 2023).

Some plants, known as hyperaccumulators, can absorb and concentrate metals in their tissues. When combined with microbial bioleaching, this can result in a more comprehensive metal recovery process (van der Ent et al. 2013). The presence of plants can help stabilize soil and reduce erosion at mining sites, mitigating some of the environmental impacts associated with traditional mining practices (Mendez and Maier 2008). Plants contribute to carbon sequestration through photosynthesis, potentially offsetting some of the carbon emissions associated with metal recovery processes (Wan et al. 2012).

The synergistic interactions between plants and microorganisms in biomining and bioleaching represent a promising frontier in sustainable resource recovery. By harnessing these natural partnerships, it may be possible to develop more efficient, environmentally friendly, and economically viable methods for metal extraction in sustainable agriculture. In the following sections, we will explore the mechanisms underlying plant-microbe interactions in biomining and bioleaching, examine the potential benefits and challenges of this approach, and discuss the implications for sustainable resource recovery.

This simplified representation in Fig. 5.1 helps visualize how plants and microbes work together in the soil environment to facilitate biomining and bioleaching processes. The plant provides nutrients and creates a favourable environment for the microbes, which in turn help break down ore particles and make valuable metals more accessible for extraction.

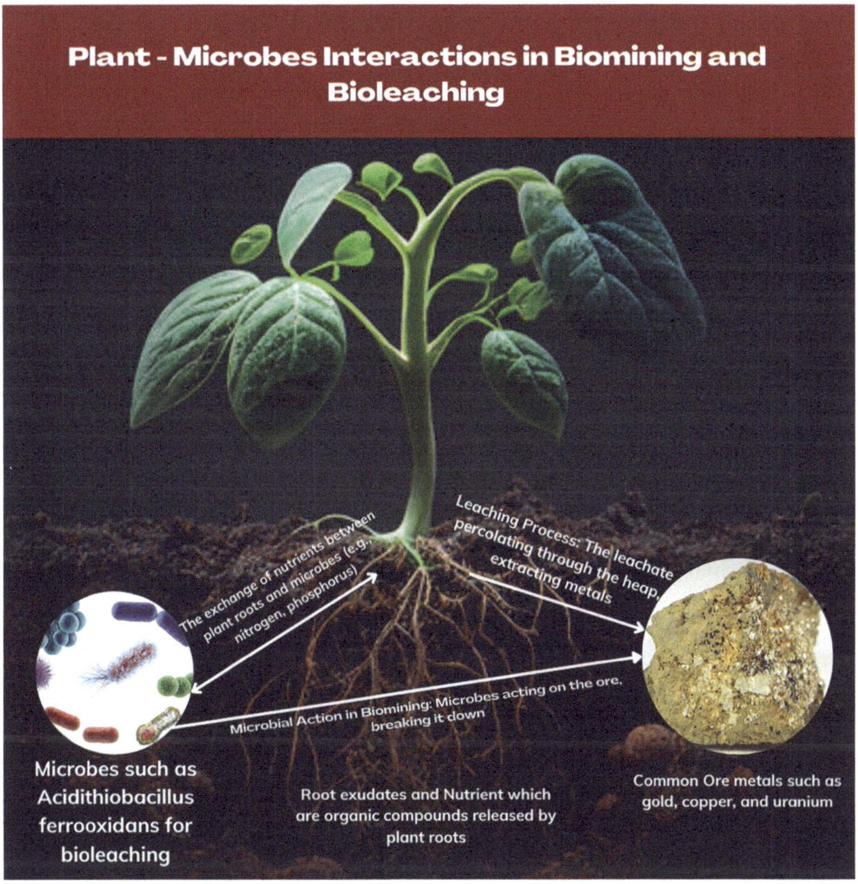

Fig. 5.1 Synergistic plant-microbe interactions in soil: a visualization of biomining and bioleaching processes

5.2.2 *Mechanisms of Biomining and Bioleaching*

5.2.2.1 Microbial Processes Involved

The success of biomining and bioleaching relies on the unique metabolic capabilities of microorganisms, particularly acidophilic bacteria and archaea. These microorganisms employ several mechanisms to solubilize metals from ores and minerals. Chemolithoautotrophic microorganisms, such as *Acidithiobacillus ferrooxidans* and *Leptospirillum ferrooxidans*, oxidize reduced sulfur compounds in metal sulfides. This process generates sulfuric acid, which lowers the pH of the environment and promotes metal solubilization (Rawlings and Johnson 2007). The reaction can be represented as:

$$MS + 2O_2 \rightarrow MSO_4 \left(\text{where M represents a metal}\right)$$

Many biomining microorganisms can oxidize ferrous iron (Fe^{2+}) to ferric iron (Fe^{3+}). Ferric iron is a strong oxidizing agent that can attack metal sulfides, further enhancing metal solubilization (Vera et al. 2013). The reaction can be represented as:

$$4Fe^{2+} + O_2 + 4H^+ \rightarrow 4Fe^{3+} + 2H_2O$$

The generation of sulfuric acid through sulfur oxidation creates an acidic environment that facilitates the dissolution of metal-containing minerals (Johnson 2014). Microorganisms can form biofilms on mineral surfaces, creating microenvironments that enhance metal solubilization through direct contact between cells and the mineral substrate (Bellenberg et al. 2015).

5.2.2.2 Role of Microbial Consortia

While individual microbial species can perform biomining and bioleaching, the use of microbial consortia often results in more efficient and robust metal extraction processes. Microbial consortia offer several advantages. Different microorganisms within a consortium can perform distinct but complementary roles in metal solubilization. For instance, some species may specialize in iron oxidation, while others focus on sulfur oxidation (Johnson 2014). Diverse microbial communities are generally more resilient to environmental fluctuations and stress, ensuring more stable and consistent metal extraction over time (Rawlings and Johnson 2007).

Consortia can more effectively utilize a broader range of substrates present in complex ores, leading to more comprehensive metal recovery (Kaksonen et al. 2018). Microbial species within a consortium can engage in mutually beneficial relationships, such as cross-feeding of metabolites or removal of inhibitory compounds, which can enhance overall metal extraction efficiency (Vera et al. 2013) as well as rejuvenate soil and plant health towards sustainable agricultural practices.

The primary types of microorganisms involved in biomining and bioleaching, along with their mechanism is shown in Fig. 5.2 below. This study explores the use of microbial consortia in ore metal environments, particularly focusing on gold and diamond surfaces (Santomartino et al. 2022). The "bioengineering" panel reveals the role of bioengineering and synthetic biology in improving microbial performance, addressing the limitations of wild-type strains. Sphere colors indicate the suitability for different planetary bodies based on crust composition.

Successful metal recovery examples using microbial consortia include gold recovery in South Africa with *Acidithiobacillus ferrooxidans* and *Leptospirillum ferrooxidans*, copper recovery in Chile with *Acidithiobacillus thiooxidans* and *Acidithiobacillus caldus*, uranium recovery in the United States with *Pseudomonas fluorescens* and *Bacillus subtilis*, and nickel recovery in Australia with *Aspergillus niger* and *Penicillium simplicissimum* (Vardanyan et al. 2018). These cases demonstrate the effectiveness of microbial consortia in metal recovery and the potential for enhancement through bioengineering and synthetic biology.

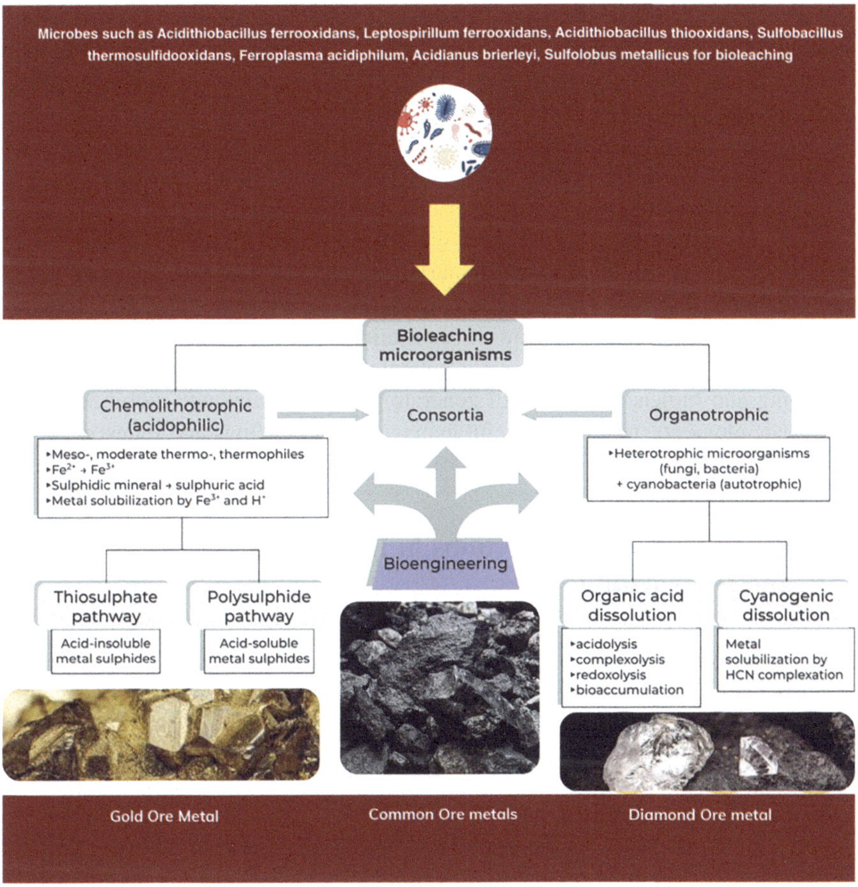

Fig. 5.2 Key biomining and bioleaching microorganisms and mechanisms, with ore metals applicability insights

5.2.2.3 Limitations of Current Techniques

Despite the potential benefits of biomining and bioleaching, these processes face several challenges that limit their widespread adoption. Microbial metal extraction processes are generally slower than traditional hydrometallurgical methods, which can impact economic feasibility (Brierley and Brierley 2013). Certain ores and minerals are resistant to microbial attack, limiting the effectiveness of bioleaching and bioremediation techniques. For instance, sulfide ores containing pyrite (FeS_2) and chalcopyrite ($CuFeS_2$) are less amenable to microbial processing due to their slow oxidative dissolution rates. These ores require high temperatures and specific conditions for microbial activity to be effective, often making traditional mining methods more practical (Johnson 2014).

Another mechanism is the processing of rare earth element (REE) ores, which typically contain minerals such as monazite ($CePO_4$) and bastnäsite ($CeCO_3F$). These minerals are highly resistant to microbial leaching due to their stable chemical structures and the low solubility of the rare earth elements they contain. This resistance necessitates the use of chemical methods, such as acid leaching or solvent extraction, to effectively recover REEs, highlighting the limitations of microbial techniques in certain mining contexts (McNulty et al. 2022).

Maintaining optimal conditions for microbial growth and activity in large-scale operations can be challenging, particularly in heap leaching operations (Watling 2006). Efficient recovery of solubilized metals from leach solutions remains a technical challenge, particularly for low-concentration solutions (Kaksonen et al. 2018). While generally considered more environmentally friendly than traditional mining, biomining and bioleaching can still generate acidic effluents and mobilize potentially toxic elements, requiring careful management (Rawlings and Johnson 2007).

5.3 Plant-Microbe Interactions in Biomining and Bioleaching

The integration of plants into biomining and bioleaching systems represents a promising approach to address some of the limitations of current techniques while enhancing metal extraction efficiency. Plant-microbe interactions in the rhizosphere can significantly influence metal mobilization and recovery through several mechanisms.

5.3.1 Plant Growth-Promoting Rhizobacteria (PGPR)

Plant growth-promoting rhizobacteria (PGPR) are beneficial microorganisms that colonize the rhizosphere and promote plant growth through various mechanisms. In the context of biomining and bioleaching, PGPR can play a crucial role in enhancing metal mobilization and plant tolerance to metal stress.

5.3.1.1 Mechanisms of Action

(a) Production of organic acids: PGPR can produce organic acids, such as citric, oxalic, and gluconic acids, which lower the pH of the rhizosphere and promote metal solubilization (Gadd 2010).

(b) Siderophore production: Some PGPR secrete siderophores, high-affinity iron-chelating compounds that can also bind and mobilize other metals (Rajkumar et al. 2010).

(c) Phytohormone production: PGPR can synthesize plant growth hormones, such as auxins and cytokinins, which enhance plant growth and root development, increasing the plant's ability to explore the soil for metals (Glick 2014).
(d) ACC deaminase production: Some PGPR produce the enzyme 1-aminocyclopr opane-1-carboxylate (ACC) deaminase, which reduces ethylene levels in plants, mitigating the negative effects of metal stress (Glick 2014).

5.3.1.2 Effects on Biomining and Bioleaching

(a) Enhanced metal solubilization: The production of organic acids and siderophores by PGPR can increase the solubilization of metals from ores and minerals, making them more accessible for plant uptake or recovery (Wan et al. 2012).
(b) Improved plant growth: By promoting plant growth and root development, PGPR can increase the surface area available for microbial colonization and metal-microbe interactions (Sessitsch et al. 2013).
(c) Increased metal tolerance: PGPR-mediated production of phytohormones and ACC deaminase can enhance plant tolerance to metal stress, allowing for the use of plants in metal-rich environments (Ma et al. 2016).

5.3.2 Mycorrhizal Fungi

Mycorrhizal fungi form symbiotic associations with plant roots, significantly enhancing the plant's ability to absorb nutrients and water from the soil. In biomining and bioleaching systems, mycorrhizal fungi can aid in metal mobilization and increase plant metal tolerance (Cecchi et al. 2019).

Arbuscular mycorrhizal fungi (AMF) create intracellular structures called arbuscules within plant root cells, facilitating nutrient exchange between the plant and fungus (Smith and Read 2008). Ectomycorrhizal fungi (EMF), on the other hand, form an external sheath around plant roots and a network of hyphae that penetrate between root cells, increasing the root surface area in contact with the soil (Smith and Read 2008).

The extensive hyphal network of mycorrhizal fungi allows plants to access a larger soil volume, potentially increasing metal uptake from metal-rich substrates (Gadd 2010). For instance, AMF have been shown to enhance the uptake of metals like zinc (Zn) and copper (Cu) in plants grown on contaminated soils. Mycorrhizal fungi can bind metals in their cell walls or precipitate them as insoluble compounds, reducing metal toxicity to plants (Meharg 2003). This is evident in plants growing in lead (Pb) and cadmium (Cd) contaminated soils, where AMF help in sequestering these metals, thereby reducing their bioavailability and toxicity.

Additionally, mycorrhizal fungi improve plant nutrition, which can increase plant vigor and metal tolerance (Gadd 2010). This process is complemented by altered gene expression; mycorrhizal colonization can modulate the expression of plant genes involved in metal transport and tolerance, enhancing the plant's ability to cope with metal stress (Ouziad et al. 2005). For example, the presence of AMF has been linked to the upregulation of genes involved in the detoxification and transport of heavy metals in plants growing in contaminated environments, thereby facilitating better growth and survival.

5.3.3 Endophytic Microorganisms

Endophytic microorganisms are bacteria and fungi that live within plant tissues without causing disease. These microorganisms can contribute to biomining and bioleaching processes through several mechanisms:

> Endophytes can colonize various plant tissues, including roots, stems, and leaves, potentially influencing metal distribution within the plant (Sessitsch et al. 2013). Endophytes occupy unique niches within plants, potentially protecting the host from pathogenic microorganisms and abiotic stresses (Hardoim et al. 2015).

5.3.3.1 Potential Benefits for Biomining and Bioleaching

Some endophytes can produce metal-mobilizing compounds, such as organic acids and siderophores, within plant tissues, potentially enhancing metal solubilization and translocation (Ma et al. 2016). Endophytes, microorganisms living within plant tissues, can influence metal distribution, potentially concentrating metals in harvestable parts (Rajkumar et al. 2012a, b). *Pseudomonas fluorescens* has been shown to enhance nickel accumulation in *Alyssum murale*, a nickel-hyperaccumulating plant. Similarly, *Bacillus megaterium* has been used to increase lead uptake in *Brassica juncea*. Another example is *Serratia nematodiphila*, which facilitates cadmium accumulation in *Zea mays*.

These endophytes promote metal uptake and concentration within plants, making them valuable for biomining applications by increasing metal recovery from contaminated soils through plant harvests. Many endophytes exhibit plant growth-promoting traits similar to PGPR, improving plant performance in metal-rich environments (Hardoim et al. 2015). Inoculation of plants with metal-tolerant or metal-mobilizing endophytes could enhance the efficiency of phytomining and phytoremediation processes (Ma et al. 2016).

5.3.4 Synergistic Effects of Plant-Microbe Interactions

The integration of plants and their associated microorganisms in biomining and bioleaching systems can result in synergistic effects that enhance metal recovery and environmental sustainability:

5.3.4.1 Enhanced Metal Solubilization and Leaching

Plant roots create a unique environment in the rhizosphere that can enhance microbial activity and metal solubilization. Root exudates, which include organic acids, amino acids, and carbohydrates, serve as carbon and energy sources for microorganisms, stimulating their growth and metabolic activities (Sessitsch et al. 2013). The rhizosphere supports a diverse community of microorganisms, including bacteria, fungi, and archaea. This diversity can lead to more comprehensive metal solubilization through the combined action of different microbial metabolic pathways (Wenzel 2009). Plant roots can alter soil physical and chemical properties, such as pH, redox potential, and organic matter content, which can influence metal speciation and bioavailability (Kidd et al. 2009).

5.3.4.2 Improved Plant Growth and Stress Tolerance in Agriculture

Plant-associated microorganisms, particularly mycorrhizal fungi and PGPR, can enhance nutrient uptake by plants, improving their overall growth and vigor in metal-contaminated soils (Ma et al. 2016). Microbial production of phytohormones and enzymes like ACC deaminase can help plants maintain normal growth and development under metal stress conditions (Glick 2014). Some plant-associated microorganisms can enhance the plant's antioxidant defense systems, helping to mitigate oxidative stress induced by high metal concentrations (Ma et al. 2016).

5.3.4.3 Potential for Phytoremediation of Contaminated Sites

The combined action of metal-accumulating plants and their associated microorganisms can enhance the extraction of metals from contaminated soils, providing a sustainable approach to site remediation. This is called phytoextraction. (Sheoran et al. 2011). Hyperaccumulating plants and their microbiomes have successfully been used in phytoextraction, stabilizing contaminated soils and preventing the spread of contamination (Mendez and Maier 2008). One notable example is the use of *Brassica juncea* (Indian mustard) combined with *Pseudomonas putida* to extract lead (Pb) from contaminated soils. The microbial partner enhances the plant's ability to absorb and tolerate higher levels of lead, making the extraction process more efficient.

Another success story involves *Thlaspi caerulescens* (Alpine pennycress) and *Rhizobium* species, which have been effectively used to phytoextract zinc (Zn) and cadmium (Cd) from polluted soils. This plant-microbe system not only extracts metals but also reduces their mobility and bioavailability, thus stabilizing the soil.

In addition to metal extraction, plant-microbe systems also facilitate the degradation of organic pollutants. For instance, *Populus* species (poplar trees) in association with *Dehalococcoides* bacteria have been employed to degrade trichloroethylene (TCE) in the rhizosphere, providing a comprehensive approach to site cleanup

(Kuiper et al. 2004). These examples describes the potential of hyperaccumulating plants and their microbial partners in effective phytoextraction and environmental remediation.

5.4 Agricultural Relevance of biomining and Bioleaching

5.4.1 Microbial Technologies for Enhancing Soil Health

Bioleaching-derived minerals hold tremendous promise for improving soil health and fertility through innovative soil amendment practices. Bioleaching, a process that employs microbes to extract valuable elements from ores, releases essential nutrients such as potassium, phosphorus, and trace elements like zinc as by-products. These nutrients are vital for plant growth and soil productivity, making them ideal candidates for agricultural applications (Omotayo and Omotayo 2024). Leveraging the natural capabilities of bioleaching microbes, these processes can lead in sustainable solutions to replenish nutrient-deficient and degraded soils.

The potential of bioleaching-derived minerals extends beyond nutrient supplementation. These by-products can be refined and tailored into soil conditioners that enhance soil structure, water retention, and microbial activity. Their ability to restore soil fertility makes them a crucial tool in addressing land degradation, a challenge faced in many agricultural systems worldwide. This dual-purpose application of bioleaching—industrial resource recovery and agricultural soil enhancement—offers an eco-friendly, cost-effective, and circular approach to managing natural resources while supporting sustainable farming practices (Arya and Kumar 2020).

Incorporation of bioleaching-derived mineral applications within agricultural systems demonstrates the broader impact of microbial technologies. What was once developed for industrial resource extraction can be repurposed to meet the demands of modern agriculture, providing a synergy between environmental sustainability and food security. This innovation emphasized the versatility of microbial processes, illustrating how cutting-edge biotechnologies can transcend their original purpose to address critical challenges in soil remediation and crop productivity (Akintuyi 2024).

5.4.2 Resource Recovery and Agricultural Input

The integration of bio-mined nutrients into sustainable agriculture represents a transformative approach to resource recovery and soil management. Bioleaching processes, traditionally used to extract valuable metals from ores, produce nutrient-rich by-products that can serve as sustainable alternatives to synthetic fertilizers.

These by-products contain essential elements such as potassium, phosphorus, and trace metals in bioavailable forms, making them ideal for agricultural applications. Their environmentally friendly nature further enhances their appeal as a green solution to nutrient supplementation in farming systems (Cozma et al. 2024).

Unlike synthetic fertilizers, which are often energy-intensive to produce and can contribute to environmental degradation, bio-mined nutrients offer a more sustainable option. They are derived through microbial activity, which minimizes chemical inputs and reduces greenhouse gas emissions. Moreover, the bioavailability of these nutrients ensures efficient uptake by plants, potentially decreasing nutrient runoff into water bodies and mitigating eutrophication risks. As a result, integrating bio-mined nutrients into agricultural practices can enhance soil fertility while supporting ecological balance (Chataut et al. 2023).

The dual benefits of resource recovery and agricultural input highlight the potential of biomining as a cornerstone of sustainable farming. By repurposing by-products from industrial processes, this innovation aligns with circular economy principles, promoting the reuse of natural resources. Furthermore, the adoption of bio-mined nutrients can contribute to global food security by improving crop productivity in a manner that is both environmentally responsible and economically viable. This integration describes the critical role of microbial technologies in shaping a sustainable agricultural future (Raabe 2023).

5.4.3 Biomining Innovations in Agroecosystems

The integration of biomining into agroecosystems introduces innovative pathways for sustainable resource management. Biomining, traditionally used for extracting metals, can be adapted to recover valuable nutrients from agricultural or industrial waste. These nutrients, often discarded in conventional systems, can be reintegrated into farming practices as soil amendments or biofertilizers. The process of creating a synergy between biomining and agriculture, such approaches promote resource efficiency and enhance soil health, it addresses critical challenges in modern farming (Corami 2024).

Closed-loop systems exemplify how biomining can complement agricultural practices. In these systems, bioleaching technologies extract nutrients from waste streams, which are then cycled back into agricultural fields to support crop production. This circular approach not only minimizes waste but also optimizes the use of natural resources, reducing dependency on synthetic fertilizers and mitigating the environmental impacts of both mining and agriculture. Additionally, these systems can lower greenhouse gas emissions associated with waste decomposition and industrial fertilizer production, contributing to climate-smart agriculture (Gkika et al. 2024).

The adoption of biomining innovations in agroecosystems underscores the potential of microbial technologies in advancing sustainable development goals. When the combination of linking industrial and agricultural processes in a circular

Table 5.1 Synergy of bioleaching and biomining to agriculture

Action	Biomining/bioleaching	Agricultural application	References
Microbial role	Extracting metals and nutrients from ores.	Mobilizing soil nutrients for plant uptake.	Omotayo and Omotayo (2024)
Output	Metal recovery and nutrient-rich by-products.	Bioavailable nutrients and soil conditioners.	Akintuyi (2024)
Environmental impact	Reducing dependence on conventional mining.	Enhancing sustainable practices and resource reuse.	Raabe (2023)
Potential integration	Use of bioleached nutrients in soil amendments.	Developing nutrient-enriched fertilizers from waste.	Gkika et al. (2024)

economy model, these integrated approaches reduce the environmental footprint while maximizing productivity. Furthermore, the application of bioleaching in agro-ecosystems demonstrates how innovative biotechnologies can provide scalable solutions to pressing global issues such as waste management, soil degradation, and food security, Biomining and agriculture share complementary roles through microbial processes as shown in Table 5.1. Biomining extracts metals and nutrients, producing by-products repurposed as bioavailable soil amendments. This reduces environmental harm while enhancing sustainable farming. Integrating bioleached nutrients into fertilizers fosters a circular economy, aligning industrial recovery with agricultural productivity.

5.5 Sustainable Resource Recovery Implications

5.5.1 Eco-Friendly and Cost-Effective Approach

The synergistic interactions between plants and microbes in biomining and bioleaching processes offer a promising eco-friendly and cost-effective approach to resource recovery. This symbiotic relationship harnesses the natural abilities of both organisms to extract valuable metals from low-grade ores, mine tailings, and electronic waste, presenting a sustainable alternative to traditional mining practices.

One of the primary advantages of this approach is its reduced reliance on harsh chemicals and energy-intensive processes. Conventional mining often involves the use of toxic chemicals such as cyanide and mercury, which can have severe environmental consequences (Gadd 2010). In contrast, plant-microbe interactions utilize natural biological processes to solubilize and accumulate metals, significantly reducing the need for harmful reagents.

The cost-effectiveness of this approach is particularly noteworthy. Traditional mining operations require substantial capital investment in equipment, infrastructure, and energy. Biomining and bioleaching, on the other hand, can be implemented with relatively low initial costs and operational expenses. For instance, a study by Mulligan et al. (2001) demonstrated that bioleaching of copper from low-grade ores could reduce production costs by up to 50% compared to conventional methods.

Moreover, the ability of certain plants to hyperaccumulate metals, combined with the metal-solubilizing capabilities of associated microorganisms, allows for the extraction of metals from sources that would be economically unfeasible using traditional methods. This opens up new opportunities for resource recovery from previously overlooked or abandoned sites (Sheoran et al. 2009).

5.5.2 *Reduced Environmental Impact*

The environmental benefits of plant-microbe interactions in biomining and bioleaching are substantial. Traditional mining practices are often associated with significant ecological disturbances, including deforestation, soil erosion, and water pollution. In contrast, the use of plants and microbes for metal extraction offers a much gentler approach to resource recovery.

Phytoremediation, a key component of this process, involves using plants to remove contaminants from soil and water. This not only extracts valuable metals but also helps to clean up polluted sites. A comprehensive review by Pilon-Smits (2005) highlighted the potential of phytoremediation in restoring ecosystem health while simultaneously recovering valuable resources.

The reduction in greenhouse gas emissions is another crucial environmental benefit. Conventional mining operations are energy-intensive and often rely on fossil fuels, contributing significantly to carbon dioxide emissions. Biomining and bioleaching processes, being largely passive and requiring minimal energy input, have a much lower carbon footprint. A life cycle assessment conducted by Norgate and Jahanshahi (2011) found that bioleaching of copper ores could reduce greenhouse gas emissions by up to 30% compared to conventional methods.

Furthermore, the use of plant-microbe interactions can help mitigate acid mine drainage, a serious environmental problem associated with traditional mining. The ability of certain plants and microbes to stabilize and neutralize acidic mine wastes can prevent the release of toxic metals into surrounding ecosystems (Johnson and Hallberg 2005).

5.5.3 *Potential for* **In-Situ** *biomining and Bioleaching*

One of the most promising aspects of plant-microbe interactions in resource recovery is the potential for *in-situ* biomining and bioleaching. This approach involves treating the ore body or contaminated site directly, without the need for extensive excavation or transportation of materials.

In-situ bioleaching has gained particular attention for its application in recovering metals from low-grade ores that would be uneconomical to mine using conventional methods. The success of in-situ bioleaching in extracting copper from chalcopyrite ores, demonstrating its potential to unlock previously inaccessible mineral resources (Nkuna et al. 2022).

The in-situ approach also offers significant advantages in terms of land disturbance and waste generation. By treating the ore body in place, there is minimal need for large open-pit mines or extensive tailings facilities. This not only reduces the visual impact of mining operations but also minimizes the risk of environmental contamination associated with waste storage (Agboola et al. 2020).

Moreover, in-situ techniques can be applied to remediate and recover metals from abandoned mine sites and contaminated lands. This dual benefit of environmental cleanup and resource recovery makes it an attractive option for addressing legacy mining issues. Hernandez et al. (2023) found that combining in-situ bioleaching with electrokinetics effectively removes heavy metals from mine tailings. This integrated approach enhances the bioleaching process, facilitating the extraction of metals and reducing environmental contamination. The study underscores the potential of this method for sustainable mine waste management.

The potential for in-situ biomining using plants (phytomining) is also being explored. This involves cultivating metal-accumulating plants directly on mineral-rich soils or low-grade ore bodies. The plants, in association with their microbial partners, extract and concentrate metals in their tissues, which can then be harvested and processed. While still in its early stages, phytomining shows promise for the recovery of valuable metals such as nickel and gold from sub-economic deposits (van der Ent et al. 2013).

5.5.4 Future Perspectives and Research Directions

5.5.4.1 Optimization of Plant-Microbe Consortia

The optimization of plant-microbe consortia represents a critical area for future research and development in sustainable resource recovery. While the potential of these synergistic relationships has been demonstrated, there is still much to be learned about how to maximize their efficiency and effectiveness in metal extraction and environmental remediation.

One key focus area is the identification and characterization of hyperaccumulating plants and their associated microbiomes. While over 700 plant species have been identified as metal hyperaccumulators, only a fraction of these have been studied in detail (Reeves et al. 2018). Advanced genomic and metagenomic techniques are now being employed to uncover the genetic basis of metal hyperaccumulation and to identify the microbial communities that support this process.

The engineering of plant-microbe consortia for enhanced metal extraction is another promising avenue of research. This may involve the genetic modification of plants to improve their metal uptake and tolerance, as well as the selection and engineering of microbial strains with enhanced metal-solubilizing capabilities. A review by Lebeau et al. (2008) highlighted the potential of using plant growth-promoting rhizobacteria (PGPR) to enhance phytoremediation efficiency, suggesting that similar approaches could be applied to biomining.

Furthermore, research into the complex interactions between plants, microbes, and the surrounding soil environment is crucial for optimizing metal extraction processes. Understanding how factors such as soil pH, organic matter content, and nutrient availability influence these interactions can help in designing more effective biomining and bioleaching strategies (Rajkumar et al. 2012a, b).

The development of multi-metal recovery systems is another important area for future research. Many ore deposits and contaminated sites contain a mixture of different metals, and developing plant-microbe consortia capable of efficiently extracting multiple metals simultaneously could significantly enhance the economic viability of these approaches (Sheoran et al. 2009).

5.5.4.2 Field-Scale Applications and Challenges

While laboratory studies and small-scale trials have demonstrated the potential of plant-microbe interactions in biomining and bioleaching, scaling up these processes for field-scale applications presents several challenges that need to be addressed.

Recent advancements have significantly addressed the challenge of longer timeframes in biological metal extraction compared to conventional mining methods. Research has focused on optimizing fast-growing, high-biomass hyperaccumulator plants and enhancing microbial metal solubilization techniques. Genetic engineering and selective breeding have produced plants with improved metal uptake and faster growth rates (Gavrilescu 2022).

Additionally, microbial approaches have seen substantial progress; researchers are now employing genetically modified microorganisms and bioaugmentation strategies to accelerate metal solubilization and accumulation (Pande et al. 2022). Innovative bioreactor designs and optimized growth conditions have further reduced processing times. These developments have made biometal extraction more economically feasible and competitive with traditional mining methods (Ali et al. 2013; Wang et al. 2022).

The heterogeneity of field conditions presents another significant challenge. Factors such as soil composition, climate, and native microbial communities can vary widely across a site, affecting the performance of introduced plant-microbe consortia. Developing robust systems that can adapt to these variable conditions is crucial for successful field-scale applications (Weyens et al. 2009; Onajobi et al. 2020).

Managing large-scale plantations for phytomining or phytoremediation also presents logistical challenges. This includes issues related to irrigation, pest management, and harvesting of metal-rich biomass. Innovative approaches, such as the use of remote sensing and precision agriculture techniques, are being explored to address these challenges (van der Ent et al. 2013; Onajobi et al. 2023).

The processing of metal-rich plant biomass is another area that requires further development for field-scale applications. While methods such as ashing and acid leaching have been demonstrated at small scales, more efficient and environmentally friendly methods for extracting metals from harvested biomass need to be developed (Chaney et al. 2005).

Despite these challenges, several field-scale trials have shown promising results. Santos et al. (2014) studied the enhancement of chemical and biological properties of gossan mine wastes by applying amendments and growing *Cistus ladanifer* L. The study found that these interventions significantly improved soil quality, promoting vegetation growth and ecological restoration in contaminated mining areas.

5.5.4.3 Integration with Other Sustainable Mining Practices

The future of sustainable resource recovery lies in the integration of plant-microbe interactions with other innovative mining practices. This holistic approach can maximize resource recovery while minimizing environmental impact.

One promising area is the combination of biomining and bioleaching with urban mining and e-waste recycling. The increasing amount of electronic waste presents both a challenge and an opportunity. Plant-microbe systems could be used to recover valuable metals from e-waste in an environmentally friendly manner. A study by Cui and Zhang (2008) highlighted the potential of bioleaching in recovering metals from printed circuit boards, suggesting that similar approaches could be applied to other types of e-waste.

The integration of biomining with renewable energy sources is another area with significant potential. Solar-powered bioleaching systems, for instance, could provide a sustainable solution for metal recovery in remote areas. Bioremediation offers a sustainable, circular solution for managing mining waste and generating bioenergy. By utilizing microorganisms to detoxify and recover valuable resources, it reduces environmental impact and promotes renewable energy production, aligning with sustainable development goals (Dey et al. 2024).

Combining plant-microbe-based metal recovery with landscape restoration and agroforestry practices also shows promise. This approach, sometimes referred to as "phytomining landscapes," involves designing multifunctional landscapes that provide both metal yield and ecosystem services. For instance, metal-accumulating trees could be used in agroforestry systems, providing both metal yield and other products such as biomass for energy production (van der Ent et al. 2013).

The use of biomining and bioleaching in conjunction with in-situ resource recovery techniques, such as in-situ leaching (ISL) or in-situ recovery (ISR), is another area of interest. These combined approaches could allow for the recovery of metals from deep ore bodies with minimal surface disturbance (Vargas et al. 2020).

5.5.5 Investigating the Synergy Between Plants and Microbes in Biomining and Bioleaching

The benefits and challenges associated with the use of synergistic plant-microbe interactions in biomining and bioleaching for sustainable resource recovery and agriculture (Table 5.2). This approach leverages natural biological processes to extract metals in an eco-friendly and cost-effective manner, minimizing the need for harsh chemicals and reducing energy consumption (Abdellah et al. 2022). The dual capability of these systems to remediate contaminated sites while recovering valuable metals addresses both environmental and economic concerns (Brown and Green 2017). Additionally, the potential for in-situ applications offers a less invasive alternative to traditional mining, reducing land disturbance and waste generation (Akimbekov et al. 2022).

Plant-microbe systems also enable the recovery of metals from low-grade ores and waste, thereby extending global metal reserves and addressing the issue of dwindling high-grade ore deposits (Kalisz et al. 2022; Zhou 2024). However, challenges such as optimizing plant-microbe consortia, scaling up to field-scale applications, and integrating these approaches with existing mining practices remain (Saldana et al. 2023).

The potential impact on the mining industry is significant, with shifts towards environmentally friendly practices that improve public perception and ease regulatory pressures (Akimbekov et al. 2022). The support for the circular economy is also notable, as efficient recovery of metals from waste streams contributes to a more sustainable and resource-efficient economy. Future research should focus on advancing the understanding of genetic and molecular mechanisms, developing more efficient plant-microbe consortia, and improving methods for processing metal-rich biomass (Thavamani et al. 2017). This ongoing research is crucial for addressing the challenges and maximizing the potential of plant-microbe interactions in sustainable resource recovery.

5.6 Conclusion

The exploration of synergistic plant-microbe interactions in biomining and bioleaching has revealed a promising pathway towards more sustainable resource recovery. This approach offers several key advantages over conventional mining methods.

Table 5.2 The exploration of synergistic plant-microbe interactions in biomining and bioleaching

S/N	Strategies	Mechanism of action and potential impact on sustainable resource recovery	References
1.	Eco-friendly and cost-effective metal extraction	Natural biological processes reduce the need for harsh chemicals and energy-intensive operations, leading to lower environmental impact and operational costs.	Abdellah et al. (2022)
2.	Remediation of contaminated sites	Dual capability of cleaning up polluted areas while recovering valuable metals addresses environmental and economic concerns simultaneously.	Brown and Green (2017)
3.	Potential for in-situ applications	Treating ore bodies or contaminated sites directly minimizes land disturbance and waste generation, offering a less invasive alternative to traditional mining.	Akimbekov et al. (2022)
4.	Recovery of metals from low-grade ores and waste	Plant-microbe systems can extract metals from sources that are economically unfeasible using conventional methods, addressing dwindling high-grade ore deposits.	Zhou (2024)
5.	Optimization of plant-microbe consortia	Further research is needed to identify and engineer the most effective combinations of plants and microbes for different metals and environmental conditions.	Thavamani et al. (2017)
6.	Scaling up to field-scale applications	Overcoming issues related to process duration, site heterogeneity, and biomass processing is crucial for commercial viability.	Kalisz et al. (2022)
7.	Integration with existing mining practices	Combining plant-microbe-based approaches with other sustainable mining techniques is key to maximizing their impact in both mining and agricultural industries.	Abdellah et al. (2022)
8.	Transformation of the mining industry	Shifts mining towards environmentally friendly and socially acceptable practices, potentially improving public perception and easing regulatory pressures.	Jang and Topal (2020)
9.	Extension of global metal reserves	Ability to recover metals from low-grade ores, mine tailings, and electronic waste could significantly extend global metal reserves.	Brown and Green (2017)
10.	Environmental remediation and resource recovery	Offers the dual benefit of restoring degraded landscapes and generating economic value.	Akimbekov et al. (2022)
11.	Support for the circular economy	Enables efficient recovery of metals from waste streams and contaminated sites, supporting the transition to a more sustainable and resource-efficient economy.	Saldaña et al. (2023)
12.	Future research focus	Advancing understanding of genetic and molecular mechanisms. Developing efficient and adaptable plant-microbe consortia. Improving methods for processing metal-rich biomass as well as sustainable agriculture processes.	Zhou (2024)

By harnessing natural biological processes, this approach significantly reduces the need for harsh chemicals and energy-intensive operations, leading to lower environmental impact and operational costs. The dual capability of cleaning up polluted areas while recovering valuable metals addresses both environmental and economic concerns simultaneously. The ability to treat ore bodies or contaminated sites directly minimizes land disturbance and waste generation, offering a less invasive alternative to traditional mining practices. Plant-microbe systems can extract metals from sources that would be economically unfeasible using conventional methods, opening up new resources and addressing the issue of dwindling high-grade ore deposits. As global demand for metals continues to grow and easily accessible ore deposits become scarcer, the importance of sustainable resource recovery methods will only increase. Plant-microbe interactions offer a promising solution that aligns with the principles of sustainable agriculture development and circular economy. Since low-quality primary and secondary minerals and trash can be upgraded using ecological bioleaching technology and/or reintroduced into the industrial cycle. Thereby giving rise to new-generation biostimulants with remarkable effects on soil regeneration, plant development, and health.

While challenges remain in scaling up these processes and integrating them with existing mining practices, the potential benefits in terms of environmental protection, resource conservation, and economic opportunity are substantial. With continued research and development, plant-microbe-based approaches could play a pivotal role in shaping the future of both mining and agricultural industries and contributing to a more sustainable global economy.

References

Abdellah WR, Kim JG, Hassan MMA, Ali MAM (2022) The key challenges towards the effective implementation of digital transformation in the mining industry. Geosyst Eng 25(1–2):44–52. https://doi.org/10.1080/12269328.2022.2120093

Agboola O, Babatunde DE, Fayomi OSI, Sadiku ER, Popoola P, Moropeng L, Yahaya A, Mamudu OA (2020) A review on the impact of mining operation: monitoring, assessment and management. Results Eng 8:100181. https://doi.org/10.1016/j.rineng.2020.100181

Akimbekov NS, Digel I, Tastambek KT, Marat AK, Turaliyeva MA, Kaiyrmanova GK (2022) Biotechnology of microorganisms from coal environments: from environmental remediation to energy production. Biology 11(9):1306. https://doi.org/10.3390/biology11091306

Akintuyi OB (2024) Vertical farming in urban environments: a review of architectural integration and food security. Open Access Res J Biol Pharm 10(2):114–126. https://doi.org/10.53022/oarjbp.2024.10.2.0017

Ali H, Khan E, Sajad MA (2013) Phytoremediation of heavy metals—concepts and applications. Chemosphere 91(7):869–881. https://doi.org/10.1016/j.chemosphere.2013.01.075

Arya S, Kumar S (2020) Bioleaching: urban mining option to curb the menace of E-waste challenge. Bioengineered 11(1):640–660

Barapatre A, Burnase N, Das R, Jaiswal S (2021) Harnessing the potential of microbes for rejuvenating soils from mining sites: an initiative for environmental balance and value addition. In: Innovations in sustainable mining: balancing environment, ecology and economy. Springer, Cham, pp 149–181

Bellenberg S, Díaz M, Noël N, Sand W, Poetsch A, Guiliani N, Vera M (2015) Biofilm formation, communication and interactions of leaching bacteria during colonization of pyrite and sulfur surfaces. Res Microbiol 166(9):773–781
Brierley CL, Brierley JA (2013) Progress in bioleaching: part B: applications of microbial processes by the minerals industries. Appl Microbiol Biotechnol 97(17):7543–7552
Brown KM, Green MT (2017) Enhancing bioleaching efficiency through plant-microbe partnerships. Biotechnol Adv 35(1):125–134
Cecchi G, Ceci A, Marescotti P et al (2019) Interactions among microfungi and pyrite-chalcopyrite mineralizations: tolerance, mineral bioleaching, and metal bioaccumulation. Mycol Prog 18:415–423. https://doi.org/10.1007/s11557-018-01466-y
Chaney RL, Angle JS, McIntosh MS, Reeves RD, Li YM, Brewer EP, Chen KY, Roseberg RJ, Perner H, Synkowski EC, Broadhurst CL (2005) Using hyperaccumulator plants to phytoextract soil Ni and Cd. Z Naturforsch C 60(3–4):190–198
Chataut G, Bhatta B, Joshi D, Subedi K, Kafle K (2023) Greenhouse gases emission from agricultural soil: a review. J Agric Food Res 11:100533. https://doi.org/10.1016/j.jafr.2023.100533
Corami A (2024) Sustainable remediation methods for mining wastes. In: Sustainable management of mining waste and tailings. CRC Press, pp 254–288
Cozma P, Bețianu C, Hlihor RM, Simion IM, Gavrilescu M (2024) Bio-recovery of metals through biomining within circularity-based solutions. Processes 12(9):1793. https://doi.org/10.3390/pr12091793
Cui J, Zhang L (2008) Metallurgical recovery of metals from electronic waste: a review. J Hazard Mater 158(2–3):228–256. https://doi.org/10.1016/j.jhazmat.2008.02.001
Dey S, Dash A, Das AP (2024) Bioremediation as a sustainable and circular solution for mining waste management and bioenergy generation. In: Sustainable management of mining waste and tailings. CRC Press, pp 107–119
Gadd GM (2010) Metals, minerals and microbes: geomicrobiology and bioremediation. Microbiology 156(3):609–643. https://doi.org/10.1099/mic.0.037143-0
Gavrilescu M (2022) Enhancing phytoremediation of soils polluted with heavy metals. Curr Opin Biotechnol 74:21–31. https://doi.org/10.1016/j.copbio.2021.10.024
Gkika DA, Chalaris M, Kyzas GZ (2024) Review of methods for obtaining rare earth elements from recycling and their impact on the environment and human health. Processes 12(6):1235. https://doi.org/10.3390/pr12061235
Glick BR (2014) Bacteria with ACC deaminase can promote plant growth and help to feed the world. Microbiol Res 169(1):30–39
Hardoim PR, van Overbeek LS, Berg G, Pirttilä AM, Compant S, Campisano A et al (2015) The hidden world within plants: ecological and evolutionary considerations for defining functioning of microbial endophytes. Microbiol Mol Biol Rev 79(3):293–320
Hernández IA, Morales MM, Morales FJF, Romero LR, Camacho JV (2023) Removal of heavy metals from mine tailings by in-situ bioleaching coupled to electrokinetics. Environ Res 238:117183. https://doi.org/10.1016/j.envres.2023.117183
Jang H, Topal E (2020) Transformation of the Australian mining industry and future prospects. Min Technol 129(3):120–134. https://doi.org/10.1080/25726668.2020.1786298
Jerez CA (2017) Biomining of metals: how to access and exploit natural resource sustainably. Microb Biotechnol 10(5):1191–1193. https://doi.org/10.1111/1751-7915.12792
Johnson DB (2014) Biomining—biotechnologies for extracting and recovering metals from ores and waste materials. Curr Opin Biotechnol 30:24–31
Johnson DB, Hallberg KB (2005) Acid mine drainage remediation options: a review. Sci Total Environ 338(1–2):3–14. https://doi.org/10.1016/j.scitotenv.2004.09.002
Kaksonen AH, Boxall NJ, Gumulya Y, Khaleque HN, Morris C, Bohu T, Cheng KY (2018) Recent progress in biohydrometallurgy and microbial characterisation. Hydrometallurgy 180:7–25
Kalisz S, Kibort K, Mioduska J, Lieder M, Małachowska A (2022) Waste management in the mining industry of metals ores, coal, oil and natural gas—a review. J Environ Manag 304:114239. https://doi.org/10.1016/j.jenvman.2021.114239

Kidd P, Barceló J, Bernal MP, Navari-Izzo F, Poschenrieder C, Shilev S, Monterroso C (2009) Trace element behaviour at the root–soil interface: implications in phytoremediation. Environ Exp Bot 67(1):243–259

Kuiper I, Lagendijk EL, Bloemberg GV, Lugtenberg BJ (2004) Rhizoremediation: a beneficial plant-microbe interaction. Mol Plant-Microbe Interact 17(1):6–15

Lebeau T, Braud A, Jézéquel K (2008) Performance of bioaugmentation-assisted phytoextraction applied to metal contaminated soils: a review. Environ Pollut 153(3):497–522. https://doi.org/10.1016/j.envpol.2007.09.015

Ma Y, Rajkumar M, Zhang C, Freitas H (2016) Beneficial role of bacterial endophytes in heavy metal phytoremediation. J Environ Manag 174:14–25

Martínez-Bellange P, Von Bernath D, Navarro CA, Jerez CA (2022) Biomining of metals: new challenges for the next 15 years. Microb Biotechnol 15(1):186–188

McNulty T, Hazen N, Park S (2022) Processing the ores of rare-earth elements. MRS Bull 47:258–266. https://doi.org/10.1557/s43577-022-00288-4

Meharg AA (2003) The mechanistic basis of interactions between mycorrhizal associations and toxic metal cations. Mycol Res 107(11):1253–1265

Mendez MO, Maier RM (2008) Phytostabilization of mine tailings in arid and semiarid environments—an emerging remediation technology. Environ Health Perspect 116(3):278–283

Mudd GM (2010) The environmental sustainability of mining in Australia: key mega-trends and looming constraints. Resour Policy 35(2):98–115

Mulligan CN, Yong RN, Gibbs BF (2001) Remediation technologies for metal-contaminated soils and groundwater: an evaluation. Eng Geol 60(1–4):193–207. https://doi.org/10.1016/S0013-7952(00)00101-0

Nkuna R, Ijoma GN, Matambo TS, Chimwani N (2022) Accessing metals from low-grade ores and the environmental impact considerations: a review of the perspectives of conventional versus bioleaching strategies. Fortschr Mineral 12(5):506. https://doi.org/10.3390/min12050506

Norgate TE, Jahanshahi S (2011) Low-grade ores—smelt, leach or concentrate? Miner Eng 24(14):1471–1477. https://doi.org/10.1016/j.mineng.2011.05.017

Omotayo AO, Omotayo OP (2024) Potentials of microbe-plant assisted bioremediation in reclaiming heavy metal polluted soil environments for sustainable agriculture. Environ Sustain Indic 22(7):100396. https://doi.org/10.1016/j.indic.2024.100396

Onajobi IB, Idowu EO, Adeyemi JO, Samson OJ, Ogunyinka PI, Fagade OE (2020) In vitro antibacterial activities and molecular characterization of bacterial species isolated from farmlands against selected pathogens. Biotechnol Rep 27:e00513. https://doi.org/10.1016/j.btre.2020.e00513

Onajobi IB, Okon EE, Samson OJ, Aina SA, Adeyemi JO, Adekola HA, Abiona TO (2023) In vitro antibacterial activities and molecular analysis of *Bacillus* species from garden soil against selected pathogens. Biotechnol Rep (Amst) 27:e00513. Copyright© African Scientific Research and Innovation Council

Ouziad F, Hildebrandt U, Schmelzer E, Bothe H (2005) Differential gene expressions in arbuscular mycorrhizal-colonized tomato grown under heavy metal stress. J Plant Physiol 162(6):634–649

Pande V, Pandey SC, Sati D, Bhatt P, Samant M (2022) Microbial interventions in bioremediation of heavy metal contaminants in agroecosystem. Front Microbiol 13:824084. https://doi.org/10.3389/fmicb.2022.824084

Pilon-Smits E (2005) Phytoremediation. Annu Rev Plant Biol 56:15–39. https://doi.org/10.1146/annurev.arplant.56.032604.144214

Raabe D (2023) The materials science behind sustainable metals and alloys. Chem Rev 123(5):2436–2608. https://doi.org/10.1021/acs.chemrev.2c00799

Rajkumar M, Ae N, Freitas H (2010) Endophytic bacteria and their potential to enhance heavy metal phytoextraction. Chemosphere 77(2):153–160

Rajkumar M, Ae N, Prasad MNV, Freitas H (2012a) Potential of siderophore-producing bacteria for improving heavy metal phytoextraction. Trends Biotechnol 28(3):142–149. https://doi.org/10.1016/j.tibtech.2009.12.002

Rajkumar M, Sandhya S, Prasad MNV, Freitas H (2012b) Perspectives of plant-associated microbes in heavy metal phytoremediation. Biotechnol Adv 30(6):1562–1574

Rawat CD, Phian S, Gupta R, Verma H, Kumar M, Kaur J, Rawat VS (2023) Microbial bioprocesses in remediation of contaminated environments and resource recovery. In: Microbial bioprocesses. Academic, pp 225–274. https://doi.org/10.1016/B978-0-323-95332-0.00005-3

Rawlings DE, Johnson DB (2007) The microbiology of biomining: development and optimization of mineral-oxidizing microbial consortia. Microbiology 153(2):315–324

Reeves RD, Baker AJM, Borhidi A, Berazain R (2018) Nickel hyperaccumulation in the serpentine flora of Cuba. Ann Bot 122(3):375–391. https://doi.org/10.1093/aob/mcy092

Saldaña M, Jeldres M, Galleguillos Madrid FM, Gallegos S, Salazar I, Robles P, Toro N (2023) Bioleaching modeling—a review. Materials (Basel, Switzerland) 16(10):3812. https://doi.org/10.3390/ma16103812

Samarasekere PW (2024) Microbial remediation technologies for mining waste management. In: Harnessing microbial potential for multifarious applications. Springer Nature Singapore, Singapore, pp 31–67. https://doi.org/10.1007/978-981-97-1152-9_3

Santomartino R, Zea L, Cockell CS (2022) The smallest space miners: principles of space biomining. Extremophiles 26:7. https://doi.org/10.1007/s00792-021-01253-w

Santos ES, Abreu MM, Macías F, de Varennes A (2014) Improvement of chemical and biological properties of gossan mine wastes following application of amendments and growth of Cistus ladanifer L. J Geochem Explor 147:173–181. https://doi.org/10.1016/j.gexplo.2014.07.007

Sessitsch A, Kuffner M, Kidd P, Vangronsveld J, Wenzel WW, Fallmann K, Puschenreiter M (2013) The role of plant-associated bacteria in the mobilization and phytoextraction of trace elements in contaminated soils. Soil Biol Biochem 60:182–194

Sheoran V, Sheoran AS, Poonia P (2009) Phytomining: a review. Miner Eng 22(12):1007–1019. https://doi.org/10.1016/j.mineng.2009.04.001

Sheoran V, Sheoran AS, Poonia P (2011) Role of hyperaccumulators in phytoextraction of metals from contaminated mining sites: a review. Crit Rev Environ Sci Technol 41(2):168–214

Smith SE, Read DJ (2008) Mycorrhizal symbiosis. Academic

Štyriaková D (2023) Bioleaching eco-mining benefits agricultural sustainability. https://doi.org/10.3030/101071680

Suman J, Rakshit A, Ogireddy SD, Singh S, Gupta C, Chandrakala J (2022) Microbiome as a key player in sustainable agriculture and human health. Front Soil Sci 2:821589. https://doi.org/10.3389/fsoil.2022.821589

Thavamani P, Samkumar RA, Satheesh V, Subashchandrabose SR, Ramadass K, Naidu R, Venkateswarlu K, Megharaj M (2017) Microbes from mined sites: harnessing their potential for reclamation of derelict mine sites. Environ Pollut (Barking, Essex: 1987) 230:495–505. https://doi.org/10.1016/j.envpol.2017.06.056

United Nations (2015) Transforming our world: the 2030 agenda for sustainable development. Resolution adopted by the General Assembly

van der Ent A, Baker AJM, Reeves RD et al (2013) Hyperaccumulators of metal and metalloid trace elements: facts and fiction. Plant Soil 362:319–334. https://doi.org/10.1007/s11104-012-1287-3

Vardanyan N, Sevoyan G, Navasardyan T, Vardanyan A (2018) Recovery of valuable metals from polymetallic mine tailings by natural microbial consortium. Environ Technol 40(26):3467–3472. https://doi.org/10.1080/09593330.2018.1478454

Vargas T, Estay H, Arancibia E, Díaz-Quezada S (2020) In situ recovery of copper sulfide ores: alternative process schemes for bioleaching application. Hydrometallurgy 196:105442. https://doi.org/10.1016/j.hydromet.2020.105442

Vera M, Schippers A, Sand W (2013) Progress in bioleaching: fundamentals and mechanisms of bacterial metal sulfide oxidation—part A. Appl Microbiol Biotechnol 97(17):7529–7541

Wan Y, Luo S, Chen J, Xiao X, Chen L, Zeng G, Liu C, He Y (2012) Effect of endophyte-infection on growth parameters and Cd-induced phytotoxicity of Cd-hyperaccumulator Solanum nigrum L. Chemosphere 89(6):743–750. https://doi.org/10.1016/j.chemosphere.2012.07.005

Wang Y, Narayanan M, Shi X, Chen X, Li Z, Natarajan D, Ma Y (2022) Plant growth-promoting bacteria in metal-contaminated soil: current perspectives on remediation mechanisms. Front Microbiol 13:966226. https://doi.org/10.3389/fmicb.2022.824084

Watling HR (2006) The bioleaching of sulphide minerals with emphasis on copper sulphides—a review. Hydrometallurgy 84(1–2):81–108

Wenzel WW (2009) Rhizosphere processes and management in plant-assisted bioremediation (phytoremediation) of soils. Plant Soil 321(1):385–408

Weyens N, Vangronsveld J, van der Lelie D (2009) Exploiting plant-microbe partnerships to improve biomass production and remediation. Trends Biotechnol 27(8):413–420. https://doi.org/10.1016/j.tibtech.2009.04.006

Zhou Y (2024) Natural resources and green economic growth: a pathway to innovation and digital transformation in the mining industry. Resour Policy 90:104667. https://doi.org/10.1016/j.resourpol.2024.104667

Part II
Advancing Plant Growth and Stress Management

Chapter 6
Enhancing Sustainable Farming with Beneficial Microbes: Soil Health, Crop Growth, and Environmental Safety

Olabode O. Efunwoye and Omowunmi R. Oluwole

Abstract Sustainable farming aim is to propagate crops for production of foods and other products in a manner that preserves human health while the environment is also protected from deleterious effects of conventional farming practices. Soil and plants health is promoted by many beneficial microorganisms that are natural inhabitants of the soil. Microorganisms have been documented to fix nitrogen, solubilize nutrients (such as potassium and phosphorus), and produce plant growth regulators (auxin, gibberellin, indole-3 acetic acid and cytokines). Some soil microbes serve as biocontrol agents in sustainable farming either by antagonistic activities against plants pathogens in competing for nutrients, by parasitism, or by production of substances like antibiotics, siderophores, enzymes and volatile compounds. Others aid plants to resist pathogens by improving the plants immune system. Bioinoculants have been successfully used to improve the yield of crops like maize, millet, cowpea and tomatoes, while there are evidences of these plants growth promoting bacteria (PGPB) aiding crops in pots and on field studies to tolerate abiotic stress conditions like drought and high salinity using various mechanisms, to produce higher growth performances than plants in non-stress conditions. New technological innovations like production of biostimulants, incorporation of nano-particles into bioinoculants and utilization of meta-omics protocols are emerging which have generated greater interests in the use of bioinoculants in sustainable agriculture. However, further studies on the environmental impact and safety of use of these genetically enhanced soil microbes will be required. This chapter aims to provide current information on the use of knowledge on plant-soil microbiome interactions in sustainable agriculture as a potential wealth creation means in commercial food production. It explored the various mechanisms employed by PGPMs in improving agricultural productivity for sustainable food production, and examined recent trends in the studies of soil microbial communities which are helping in developing highly effective bio-inoculants that are economically and environmentally sustainable. Commercial production of bio-inoculants is advocated on wide scale to stimulate accessibility by

O. O. Efunwoye (✉) · O. R. Oluwole
Department of General Studies, Federal College of Agriculture, Akure, Ondo State, Nigeria

O. O. Babalola, A. S. Ayangbenro (eds.), *Microbial Allies*,
https://doi.org/10.1007/978-3-031-90530-8_6

farmers for promoting sustainable agriculture in attaining global food security and environmental safety.

Keywords Bio-inoculants · Microbiome · Sustainable · Agriculture · Phytohormones

6.1 Introduction

The soil is the topmost layer of the earth crust and constitutes the basis of agriculture and of the natural plant communities. The role of plants as producers in the ecosystem makes their existence the basis for the survival of all other land-based organisms in the biosphere (Romero et al. 2023). The soil is a natural habitat to numerous forms of macro and microorganisms. The soil biodiversity is of immense importance to plants and agriculture because of the vast beneficial impacts, on one hand, and deleterious effects on the other hand, of the various biological components of the soil (Nannipieri et al. 2003). This biodiversity plays a crucial role in the myriads of ecosystem functionalities that maintain the ecosystems end products such as agricultural productivity and climate regulation (Doran and Zeiss 2000: Rascovan et al. 2016). Most importantly among the soil microorganisms are bacteria, fungi and Actinomycetes, which play vital roles in many geochemical cycles in the ecosystems (Aguiar-Pulido et al. 2016).

The soil microbes are mostly found around the roots of plants in the region called rhizosphere, while arbuscular mychorrhizal fungi of the division Glomeromycota, live inside the roots of some plants in a symbiotic association (Suman et al. 2022). The presence of microbes in the plant root areas is promoted and influenced by the plant products of photosynthesis-photosynthates, which are released through the plant roots into the soil. These photosynthates contain rhizosphere exudates, shed cells components and mucilage. Composition of the roots exudates are mainly organic acids and sugars with some amount of amino acids, fatty acids, vitamins, hormones, growth factors, and even, antimicrobial compounds (Aguair-Pulido et al. 2016; Suman et al. 2022). The composition of these root exudates are influenced by factors such as the plant species or cultivar, plant age and the developmental stage (Rascovan et al. 2016). The interaction of these soil microbes and the plants grown on the soil may be beneficial, neutral or outright harmful to plants (White et al. 2014). The activities of soil microbes help in maintaining soil health, improving water holding capacity of soil, recycling and availability of essential soil nutrients, filtering of pollutants, root growth, storage of carbon and conservation of biodiversity (Salas-Marina et al. 2011).

The rapid growth of world human population increased the demand for food and food production. As a result, farmers all over the world apply intensive farming practices in enlarged scales which most times leave huge deleterious effects on the soil, environment, soil microbes, animals and man. The large quantities of synthetic fertilizers and agrochemicals have caused detrimental effects on the soil fauna, soil microbiome (set of microorganisms inhabiting a given soil environment), insects and plants, and also generate runoffs which contaminate sub-surface and surface water, leading to climate change occasioned by fluctuations in temperature and rainfall patterns (Rascovan et al. 2016; Glick 2018). This global challenge has given rise to the promotion and advocacy for sustainable agriculture (White et al. 2014). Sustainable farming can be defined as a holistic system of crops farming that maintains stable levels of production and quality in the long term without compromising economic profitability and health of the environment. Sustainable farming ensures that agricultural productivity is not adversely affected while the environment is protected from detrimental effects of agricultural activities (White et al. 2014). This farming system eliminates the use of synthetic fertilizer and agrochemicals that cause harm to crops, soil, animals and humans, by substituting with practices that support soil and plant health, promotes plant productivity and improve general environmental health. The roles of soil microbes in sustainable farming is naturally reserved. The production of Plant Growth-Promoting Microorganisms (PGPM) has improved the practice of sustainable agriculture (Glick 2018). An in-depth study of the wide diversity of beneficial soil microbes will open up a huge and deep goldmine of agricultural wealth for humanity.

6.2 Plant Growth-Promoting Activities of Soil Microorganisms

Plant growth-promoting activities of microorganisms are either direct or indirect. The direct activities of these soil microbes include facilitation of acquisition of nutrients from the environment and modulation of the levels of plant hormones. The indirect activities involve reducing the effects of dangerous plant pathogens (Glick 2018). The benefits of soil microbiome are increasing as more studies are conducted on the roles specific genera of microbes in the soil (Giri et al. 2015). Studies on optimization processes are being extensively conducted by agricultural researchers in order to promote the argument in favour of sustainable farming practices, detailing the effects of common agricultural practices such tilling, intercropping, crop rotation etc. on soil microbiome diversity (Suman et al. 2022). Specific roles of soil microbiome are as follows;

6.2.1 Soil Nutrients Fixation, Mineralization, Solubilization and Mobilization

Soil microbes play important roles in soil enrichment by different mechanisms which include decomposition of organic matter, fixing of nitrogen from the atmosphere, solubilizing essentials elements for easy uptake by plants and aiding the plants to uptake the minerals (Giri et al. 2015; Mostafiz et al. 2012; Suman et al. 2022). Common soil fungi like *Aspergillus* sp., *Penicillum* sp., *Rhizopus* sp., *Trichoderma* sp. and others decompose plants and animal debris to release carbon compounds, sugars, amino acids and many mineral elements into the soil, which provide food and energy source for soil/rhizospheric microbes (Giri et al. 2015). Some bacteria help decompose both allochthonous and autochthonous carbon sources for carbon cycling in the ecosystem (Barea 2015; Maron et al. 2018). Cyanobacteria like *Anabaena* and *Nostoc*, and rhizobacteria (*Rhizobium* sp., *Azotobacter* sp., *Agrobacterium* sp. etc.) help in fixing atmospheric nitrogen in soils as nitrites and nitrates for plants to take up (Mostafiz et al. 2012; Meena and Meena 2017; Matse et al. 2020; Suman et al. 2022). Plant Growth-Promoting rhizobacteria are able to fix atmospheric nitrogen through symbiotic or asymbiotic and associate nitrogen fixing processes. Atmospheric air is rich in nitrogen gas that these microorganisms help to convert into the soil in utilizable forms and reduces dependence on application of nitrogen-rich synthetic fertilizers. Solubilization of minerals such as zinc, phosphorus, and potassium in the soil is associated with genera such as *Flavobacterium, Pseudomonas, Azotobacter, Erwinia* and *Bacillus* (Barea 2015; Suman et al. 2022). These bacteria release the mineral elements from inaccessible forms in compounds into soluble ions in the soil to facilitate plants uptake. This benefit establishes the properties of Plant Growth-Promoting Microorganisms as biofertilizers. The relationships between nitrogen fixing bacteria and legume crops, and arbuscular fungi (mycorrhiza) and phosphorus solubilizing bacteria increases the availability of nitrogen and phosphorus to the plant. The bacteria make these mineral elements available to the plant while the mycorrhiza translocate them to the plant. This natural process prevents pollution in the soil often caused by the prolonged use of phosphorus-containing synthetic fertilizers on the soil (Rascovan et al. 2016). Qarni et al. (2021) in potted plants experiment reported improved uptake of phosphorus by maize plants grown on substrates separately inoculated with bacteria *Staphylococcus* sp., *Bacillus firmus, B. safensis* and *B. licheniforms*, and fungi *Penicillium* sp. and *Pencillium oxalicum*. All the maize plants exhibited higher P uptake with evident better growth performances than the plants grown on un-inoculated substrates. These bacteria produced organic acids which lowered the soil pH for nutrient availability. Organic acids help to solubilize and transform tricalcium phosphate into absorbable soluble monocalcium phosphates. PGP microorganisms solubilize P utilizing mechanisms such as organic acid production, protons release from assimilation of ammonium, direct oxidation pathway, siderophore production and secretion of phosphatase enzymes (Walpola et al. 2022) The potentials of soil microbes as biofertilizers, however, depend on factors such as ability of the PGPMs to compete successfully with other soil microorganisms, soil type, farm management practices and climatic factors (Mostafiz et al. 2012; Gupta et al. 2022).

6.2.2 Production of Plant Growth-Promoting Substances

Plant growth-promoting substances include hormones like ethylene, gibberellin, auxin, and other compounds like indole-3 acetic acid, cytokinins and salicylic acid. These materials regulate plant growth and productivity. *Bacillus* and *Pseudomonas* are PGPB associated with production of these growth regulators. *Azospirillum* is a phytostimulator of phytohormones which directly promotes plant growth. Some bacteria produce the enzyme 1-carboxylic acid-1-aminocyclopropane deaminase, which suppresses production of ethylene in plants (Glick 2018; Suman et al. 2022). Sharma et al. (2014), inoculated seedlings of rice (*Oryza sativa*) with *Pseudomonas putida, P. flourescens* and *Azospirillum lipoferum* (known producers of indole-3-acetic acid) and noted 20% increase in plant height, 10% increase in number of leaves, and an average of 44% increase in grain number against the rice plants that were not inoculated. This is similar to the report of Sharma et al. (2014) on rice. Kalimuthu et al. (2019), reported improved seed germination, increased root and shoot length, and enhanced fresh and dry weight in tomato seedlings from seeds treated with IAA producing *Pseudomonas* sp. VSMKU4050 strain. Indole-3-acetic acid stimulated development of longer root with increased number of root hairs that improved nutrients uptake by the treated tomato seedlings. The phytohormones production stimulating attributes of these PGPMs is a major benefit of these particular soil microbes.

6.2.3 Soil Microbes as Biocontrol Agents

Some soil microbes aid plants growth by inhibiting and controlling the growth of pathogens. Biocontrol activities by these soil microbes against pathogens are achieved by production of volatile compounds, siderophores, antibiotics and enzymes, or by competition for nutrients with pathogens, parasitism, antibiosis, and resistance conferment on plant immune system. In the study of Salas-Marina et al. (2011), *Trichoderma atroviride* inoculated into the roots of pot-grown *Arabidopsis thaliana,* stimulated the biosynthesis of antimicrobials phytoalexin and camalexin, which helps host plants to resist *Fusarium*, *Pseudomonas* and *Pythium* pathogens. *Trichoderma atroviride* induced systemic resistance in *Arabidopsis* by modulating overlapped expression of defense-related genes of salicylic acid and jasmonic acid pathways, and of the gene involved in the synthesis of the antimicrobials phytoalexin and camalexin, This further produced indole-3-acetic acid (IAA) that promoted increase in foliar area, plant biomass and plant growth. In another study, *Bacillus subtilis* produced iturin A, a peptide, which destroys the cell walls of pathogenic fungus *Macrophomina phaseolina* (causal agent of charcoal rot in Mung beans), causing severe mortality in the population of pathogens in the rhizosphere (Hashem et al. 2017). The combine properties of producing PGP substances and biocontrol of plant pathogens are notable features of the genera *Trichoderma*,

Bacillus and *Achromobacter* found in the soil. The use of these bacteria and fungi as inoculants has the potential to improve plant health and resistance to pathogens on the field (Mostafiz et al. 2012; Gupta et al. 2022; Suman et al. 2022).

6.2.4 Tolerance Factor to Abiotic Stress Conditions

Soil microbiome aid plants to develop tolerance to stress or extreme conditions. Soil and environmental factors such as high salinity, drought and exposure to heavy metals hinder plants growth and development leading to decreased yields. *Pseudomonas* spp. in soil are drought-tolerant and confer the ability on host plants. These bacteria produce 1-aminocyclopropane-1-carboxylic acid deaminase and increase antioxidant activity in plants to improve plants fitness by providing protection from oxidative damage induced by drought (Kumar et al. 2014; Chandra et al. 2018). During drought situation, water stress triggers the production of ethylene gas in plants as a stress signal. Continued water stress lead to rapid oxidation of 1-aminocyclopropane-1-carboxylic acid (ACC) from S-adenosine methionine. This reaction results in the production of "stress ethylene", which brings about the death of the plant shoot and root (Danish and Zafar-Ul-Hye 2020). Bacteria, such as soil *Pseudomonas putida* strain Am2, *Variovora paradoxus* strain Bm3, *Pseudomonas putida* strain Bm3, and *P. flourescens* strain FPG 3 (Ali et al. 2014; Ghosh et al. 2018; Gupta and Pandey 2019), produce ACC deaminase enzyme which breaks down the ACC produced in such stress condition into ammonia and α-ketobutyrate, compounds that are beneficial to plant growth (Singh et al. 2015; Mataranyika et al. 2022). Ghosh et al. (2018), reported the reduction in drought stress effects on pea crops inoculated with *Pseudomonas putida* strain Am2 via the activities of ACC deaminase produced by bacteria. Similarly, Gupta and Pandey (2019) noted that *Aneurinibacillus aneurinilyticus* and *Paenibacillus* sp., using similar mechanism, helped *Phaseolus vulgaris* (French beans) to withstand salinity stress. ACC deaminase also helped the plant in root nodulation. Aini et al. (2021), reported that tomato plants inoculated with saline-tolerant PGPB (*Bacillus pumilus* NCT4, *Bacillus firmus* NCT1, *B.* l*icheniformis* LCT4, *B. cereus* LAT3 and *B. safensis* LBM3) grown in saline stress soil showed four times significant increase in plant height, leaf area, dry weight of the plant, and weight of fruit than tomato grown in the same environment without inoculation. These were achieved by the combination effects of the PGPB production of ACC deaminase which lowered ethylene level, and production of metal chelating siderophores which made available to plants the metals needed in the soil. Production of phytohormones by the soil microbes' help plants mitigate against high salinity, conferring on plants resistance against extreme conditions. Some PGPB increase plants defense against environmental stress by producing essential amino acids and hormones. Tomato inoculated with the saline-tolerant PGBR were discovered by Mataranyika et al. (2022) to elicit raised levels of proline and soluble sugars in salty conditions. Proline and soluble sugars promote osmotic adjustment in the cellular water content which shields intracellular macromolecules from dehydration, and also serves as a hydroxyl scavenger thereby eradicating the free radicals from the environment. Plants absorb proline as an adaptive mechanism to salt stress in the environment, as a probable activity of the PGPB in enhancing the metabolic defense mechanisms and aiding plant development in salt stress conditions. Plants may be able to survive heavy metal stress in polluted habitats by the abilities of some microbes to modulate the levels of phytohormones by regulating the production of plant growth hormones like auxins, cytokinins and gibberellins (Devi et al. 2022). In the report of Jing et al. (2014), *Enterobacter* sp. and *Klebsiella* sp. isolated from *Polygonum pubescens* grown on heavy metal contaminated soil helped to improve the growth of rapeseed by producing indole-3-acetic acid, which aided the plant to alleviate zinc, cadmium and lead toxicity. Another study (Bilal et al. 2018), reported a microbial consortium of a fungus and bacterium namely, *Paecilomyces formosus* and *Sphingomonas* sp. produced gibberellins, and the co-inoculation in soybean was reported for alleviating the stress caused by metals aluminum and zinc. The microbial consortium also reduced the concentration of abscisic acid and jasmonic acid (JA) which helps the plant to survive and grow well in stressful conditions. The roles of soil microbiome in aiding plants to tolerate environmental stress is a veritable tool in sustainable farming on a wide array of agricultural lands (Joimel et al. 2018; Yang et al. 2009).

Modification and manipulation of the soil microbiomes have exhibited activities for biofertilization of the soil, reduction in population and the effects of plant pathogens, promoting plant growth and health, maintaining soil health, reducing use of agrochemical inputs, decreasing emission of greenhouse gases, resulting in more sustainable agricultural practices (Anderson et al. 2016; Maron et al. 2018). Glick (2018), noted that in recent years, based on the increased knowledge on the interactions of PGPMs and plants in nature, scientists have deployed the use of PGPMs in horticulture, agriculture and environmental cleanup activities. Some number of strains of PGPMs have been commercialized and are being used successfully in different countries of the world. This has opened the vista to the hidden vast wealth in sustainable farming, if deployed extensively across the globe in order to address the growing demand for food in a sustainable system (Mostafiz et al. 2012).

The ability to tap into the vast hidden wealth of soil microbial diversity for sustainable farming has limitations in practice at the present level of global development. Firstly, need arises to develop a protocol for the culturing, handling, packaging, transporting and storage of the PGPMs. Some of the beneficial endophyte microorganisms (microorganisms living inside the host plant) are obligate symbionts of the host and are difficult to culture outside the host plant. Specific scientific procedures are required to culture and multiply the microbes. PGPMs would require appropriate and adequate handling methods to prevent loss of viability as a result of time travels across the globe for distribution. Secondly, the compatibility and adaptability of bred PGPMs with the natural microbiome of the soil need to be ensured for understanding of the interactions between the two communities of microbes. Alternatively, the manipulation of naturally existing microbes in the soil for sustainable farming is a possibility motivated by increasing knowledge on the plant-microbiome relationships (Glick 2018). Thirdly, most of the existing information on the successful utilization of commercialized PGPMs for sustainable farming are from small-scale laboratory and greenhouse experiments, intensive and extensive implementation for large-scale sustainable farming will require genetic modification of the strains of these soil microbes for optimization of the desirable attributes. This poses the challenge of acceptability by the consuming public and government regulatory agencies of the agricultural products obtained from the crops cultivated sustainably using the genetically modified soil microbe strains. Adequate scientific evidences will be required to be provided on the safety of food produced by this method as well as the impact on the environment (Mostafiz et al. 2012).

Soil microbes have successfully contributed to the sustainable production of wheat, coffee, soybean, oil palm, tomato, beans, Chinese cabbage, lettuce, chilli and potato in various agricultural belts of the world (Meena and Meena 2017; Chandra et al. 2018).

6.3 Bioinoculants and Sustainable Agriculture

Bioinoculants are microorganisms developed for the purpose of application on seeds, plants surfaces or mixing with the soil to culminate in colonization of the rhizosphere or plant endosphere, which confers some benefits on the plants and the soil. Bioinoculants improve plants nutrients uptake, promote plants growth and increase plants defense against abiotic and biotic stresses (Mataranyika et al. 2022). The use of bioinoculants in agriculture as biofertilizers is a recent development that is aimed at improving crop growth in all conditions as a strategy to meet the Strategic Development Goal 2 of the United Nations to attain food security by 2050, in the face of the challenge of rapidly growing world human population. In addition, the use of bioinoculants in agriculture targets a responsible and sustainable food production method that assures of a safe environment. The effectiveness of bioinoculants is majorly dependent on the consortium of microorganisms and the carrier material. The carrier material serves as the habitat for the microbes, providing a safe and suitable environment (Arora and Mishra 2024), and must possess sufficient shelf-life, be biodegradable, have good water-holding capacity and should not be easily contaminated (Bhattacharyya et al. 2020). Liquid bioinoculants are grown in broth cultures containing mineral oils, organic oils, water and polymer-based suspensions. It is user friendly. However, the limitations include short shelf-life and sensitivity to temperature stress in the environment (Arora and Mishra 2024). Another carrier material is kaolin-alginate beads which is used for immobilization of the PGPM. The immobilization of the microorganisms is achieved by encapsulation process. The alginate carriers are natural polymers which serve as matrix for microbial cell beads formed by spray drying or thermal gelation (Deshmukh et al. 2016). Other natural polymers used as bioinoculant carriers are cellulose, starch, carrageenan, gum Arabic, chitosan, proteins such as collagen, keratin and gelatin. Natural polymers are biodegradable, easily available, increase viability of cells and are generally more resilient to environmental stress (Arora and Mishra 2024).

Synthetic materials are also used as carriers for bioinoculants but are less resilient to environmental stress and can create toxicity to microbial cells, if not properly handled. Polyacrylamide and polyurethane are types of synthetic carriers (Juric et al. 2021). The natural and synthetic polymers are both used for cell encapsulation process in preparing bioinoculant carriers (Arora and Mishra 2024). Soil aggregates like peat and biochar, have emerged as effective carriers for agricultural applications due to their abilities to store organic carbon and nutrient retention properties, especially in the use for nodule-forming bacteria (Arora and Mishra 2024). They are cost effective and naturally compatible to these microbes. Variability in its nature and diversity in constituents of peat is a major limitation in its use as carrier. Biochar low water content enables its storage for long periods without requiring sterilization (Singh et al. 2021).

New exciting innovations have emerged in the quest to improve bioinoculants quality and efficiency. The innovations include development of bio-stimulants; mixtures of PGPM, trace elements, plant hormones, enzymes and seaweed extracts (Shubha et al. 2017). Bio-stimulants are ecofriendly and offer immense benefits to plants by influencing their biochemical processes and manipulating physiological mechanisms that favour improved crop yield (Kumar et al. 2018). Another is bio-nanotechnologies innovations which include the beneficial microbes and nanoparticles (Chaudary et al. 2023). Nano-technologies include development of nano-biopesticides, nano-biofertilizers, and nano-herbicides which have been reported as good substitutes for chemical fertilizers, pesticides and herbicides because they are considered safe and environmentally benign (Ayilara et al. 2023). The synergistic application of bioinoculants and nano-particles has stimulated a revolution in agriculture for their beneficial effects of increased crop production, improved soil health and nutrients efficiency, mitigation of adverse soil conditions, and safe ecosystem (Agri et al. 2021; Alharbi et al. 2022).

6.4 Multi-omics: Understainding Plant-Microbe Interactions

Adequate and competent understanding of the plant-microbiome interactions is important to sustainable agriculture because of the huge roles microorganisms play in the biogeochemical processes in the environment, majorly in the soil-plant habitat, considering the wide variation in the biotic and abiotic components of this habitat. Appreciable but relatively limited knowledge has been gained using the conventional methods of studying culturable microbes of the soil origin, while the vast majority of soil microorganisms are unculturable and play far-reaching functions in the soil ecosystems. The multiomic approaches have helped to bridge this knowledge gap while opening a vista into the limitless potentials for sustainable agricultural development to address the global challenge of malnutrition and threat of food insecurity to human population. These technological innovations further broadened the opportunities of the hidden wealth in sustainable agriculture. Selection of microbial consortia for increased effectiveness of bioinoculants have improved tremendously as a result of new technological studies using metagenomics, metatranscriptomics. Metaproteomoics and metabolomics of microbial communities. Integrated multiomics approaches have provided deeper understanding of how soil microorganisms and microbial communities drive plant-microbiome interactions towards plant health, crop productivity and resilience to environmental stresses. The summation of multiomic approaches to plant-microbiome interactions using mathematical and computational modeling is generally referred to as systems biology (Chaudhary and Shukla 2019a, b). Systems biology approaches can be used to change the metabolic networks of microbes in natural environments and to predict the cause and effect of genetic manipulations within the metabolic network. This is an advantage towards enhancing bioinoculant capability (Suyal et al. 2016; Baart and Martens 2012). The different omic approaches and their contributions to knowledge to the structure and functions of plant-microbiome interactions are as follows;

6.4.1 Metagenomics

These are the studies of genetic materials of entire microbial communities focusing on gene contents to profile the microbial community (Aguair-Pulido et al. 2016). This procedure provides full profile of the genomes of an entire microbial community of plant rhizosphere, including microorganisms in the plants-soil interface that cannot be cultured in the laboratory. It is enhanced by metabarcoding that enables studying targeted genomic regions common to all studied microorganisms (Djemiel et al. 2022). Metagenomics have facilitated an understanding of the vast microbial diversities available in the soil environment for the production of novel molecules for sustainable agriculture (Garrido-Oter et al. 2018). Metagenomics utilizes the analyses and sequencing of microbial genomes through DNA extraction, sequencing and bioinformatics to analyze the structure and function of the whole genome. After DNA is extracted from the environmental sample, Next generation sequencing (NGS) methods are used to produce large volumes of data in the form of short reads from which the genomic profile of a microbial community is developed (Aguair-Pulido et al. 2016). Quantitative methods such as quantitative Polymerase Chain Reaction (qPCR), high through-put quantitative PCR (HT-qPCR) and digital droplet PCR (ddPCR) enable achieving the numbers of copies of one or more genes (Djemiel et al. 2022). The Next Generation Sequencing tools includes amplicon sequencing, shot gun sequencing and transposon sequencing or insertion sequencing. Amplicon sequencing involves high-throughput sequencing of marker gene tags (iTAG) which allows the profiling and comparing of spatial distribution, composition and organization of microbial communities. The marker genes target both taxonomic and functional genes in the microbial communities (Trivedi et al. 2021). Tcakz et al. (2020), Marasco et al. (2018), Xu et al. (2018), Walters et al. (2018) and Donn et al. (2015), among several workers, have reported the use of iTAG sequencing to determine the microbiota living on or in pea, grapevine, citrus, corn and wheat respectively. Shot gun sequencing supplies information on the total DNA content of a microbial community in an environmental sample, and has been used to identify microbial genomes linked to plant colonization and plant-microbial interactions (Trivedi et al. 2021; Xu et al. 2018). It has further been used extensively to study microbial diversity and functional genes in various environmental samples. Transposon sequencing has been employed by several researchers to improve the capacities of PGPB. Zamani et al. (2018), utilized mariner transposon to study the oxygen and carbon levels in the fitness of genes disruptions in *Rhizobacter leguminosarum* in a whole genome study. In the same work, Tn5-derived transposon and mariner transposon were employed in studying the whole genome and sensitivity of *Sinorhizobium meliloti* to a nodule-specific cysteine-rich peptide. Functional metagenomics can be used to reshape and readdress microbial activities in the rhizosphere by rhizosphere engineering (Nwachukwu and Babalola 2022). Metagenomics provide genetic details of unculturable soil microbes, give information on their potential functions based on the genomic content, and offer deeper understanding of the response of microbial communities to environmental changes. Although, metagenomics approach provides insight into potential functions of rhizosphere communities, clearer understanding of plant-microbial interactions will only be achieved when metagenomics is complemented with other high-throughput techniques.

6.4.2 Metatranscriptomics

These approaches focus on the particular genes expressed in microbial communities. Metatranscriptomics as molecular methods aid in investigating the functional profiles of the microbial community with varying soil conditions (Aguair-Pulido et al. 2016). The whole genomic profile of a soil microbial community is not indicative of functionality of the community, therefore, metatranscriptomics provides details of the gene expression in a given sample at a particular time, under specific conditions by profiling the total mRNA components. Metatranscriptomics study active microbiome to understand changes in gene expression in response to biotic, abiotic stresses and other environmental disturbances. One major method employed in metatranscriptomic studies is DNA Microarray. This method entails the hybridization of specific oligonucleotide probes that are fixed on a solid surface, a target corresponding to the DNA/RNA soil sample, which is followed by labeling the DNA to help determine the presence or relative abundance of genetic markers. In addition, the method is used to monitor gene expression profiles at individual or community levels (Djemiel et al. 2022; Srivastava et al. 2019). Metatranscriptomic reads are either generally mapped to specialized databases using alignment tools, or assembled into longer fragments called contigs, which is achieved using different software packages such as Trinity (Grabherr et al. 2011), Oases (Schulz et al. 2012), Abyss and RNA-Seq by Expectation Maximization (Li and Dewey 2011). The metatranscriptomic studies of multispecies of microbes unravel the genes that characterize the interactions and strategy for disease-resistance genes. The gene activation process results in disease resistance responses in the plants (Chaudary and Shukla 2019a, b). Detail mechanism and enormous roles of the R-gene are reported in tomato defense. Further, the expression of the Oxo genes in *Ralstonia solanacearum* reduced the symptoms of late blight caused by *Botrytis cinerea* and *Sclerotinia sclerotorium* (Chaudhary and Shukla 2019a, b). The use of metatranscriptomics is less reported than the other omic techniques (Aguair-Pulido et al. 2016).

6.4.3 Metaproteomics

Metaproteomics approaches examine the proteins detected in a microbial community interaction phase as a result of the functionally expressed genes in the microbiome. The studies of metaproteomics give vital knowledge at protein level that is associated with the gene expression of functionally translated regions of the microbial community, showing great potential in linking functional and phylogenetic information of soil microbial communities. Various techniques such as electrophoresis, high performance liquid chromatography, gas chromatography combined with metabolic labeling and mass spectrometry, are used in the study of metaproteomcs (Sander and Joung 2014). These metaproteomics approaches have provided direct insights into phenotypes of microbial communities in the underground (rhizosphere) and above ground (phyllosphere) of plants (Trivedi et al. 2021; Bona et al. 2019). Metaproteomic studies of physiology of microbial communities of rhizospheres established metabolic processes that allow for the selective colonization and adaptation in such rhizosphere regions when compared to the bulk soil (Bona et al. 2019). This is indicative of plant-microbial interactions influencing protein synthesis. Metaproteomics have also been used to distinguish between the functions of specific proteins and noncoding RNAs (Zamani et al. 2018). However, challenges confronting the applications of metaproteomics to studying plant-microbiome interactions include the heterogeneous nature of soil habitats or the dominance of specific few microbial species in certain ecosystems, and the sensitivity of proteins to organic compounds such as complex carbohydrates and lipids, which are endemic in soils. These challenges mitigate against the development of generic protocols for the studies and cause reduced extraction efficiency for proteins from the soil (Djemiel et al. 2022). Metaproteomics provide a powerful tool to understanding the behavior of the active members of the plant-associated microbiome.

6.4.4 Metabolomics

Metabolomics is a comprehensive analysis of all the metabolites released into the immediate environment by the microbial community in a given soil sample (Djemiel et al. 2022). Metabolomics identifies and quantifies all compounds produced and released to the environment by a microbial community. The metabolomics is considered the most direct and efficient indicator of an environment or in the changes in homeostasis (Li and Dewey 2011). Various pathways used by soil microorganism community can be studied using the signature metabolites of such community in a plant-microbiome interaction, and variation in the quantities of such key metabolites may be an indicator to alteration in the metabolic pathways of the soil microbial community (Aguair-Pulido et al. 2016; Djemiel et al. 2022). It is known that different metabolites possess different biological and physicochemical characteristics. *Pseudomonas putida* and *Bacillus subtilis* produce as secondary metabolites putidacin and bacteriocin respectively, which both have antibacterial and antifungal properties. Also, pyrazines are volatile secondary metabolites produced by bacteria of genera *Bacillus, Pseudomonas, Streptomyces* and *Chondromyces*. These metabolites provide potentials for sustainable crop yield by exerting activities against pathogenic species in the rhizosphere of phyllosphere (Chaudhary and Shukla 2019a, b; Lakshmanan et al. 2018). Data generation for metabolomics is significantly different from data generation for other omics techniques because genetic sequencing is not required. Detection, identification and quantification of metabolites of plant-microbiome interactions involve the combination of chromatography methods; liquid for secondary metabolites and gas for primary volatile metabolites, and detection methods such as mass spectrometry and nuclear magnetic resonance (Aguair-Pulido et al. 2016). The technologies produce spectral peaks patterns which are used for identifying and quantifying the metabolites, which are stored in spectral databases for automation analyses and production of metabolomics profiles. Many software programs have been developed to enhance the use and standardization of metabolomics as a high-throughput technique for studying plant-microbiome interactions (Aldridge and Rhee 2014). The qualitative and quantitative measurements microbial metabolites provide a wide and extensive information on the biochemical products of microbial communities that can be used to evaluate their gene functions (Wang et al. 2013). The metabolites can be linked with the transcriptomics and proteomics of the microbes. Metabolomics help in understanding metabolic engineering and its effects in microbial communities, and reveal the cellular processes that help in the control of the phenotypes of microbes (Chaudhary and Shukla 2019a, b). The major drawback in metabolomics is the difficulty in determining whether particular metabolite was produced by the microbes or plant. The best approach at optimizing metabolomics is to combine the data with the other omics data, leading to the integrated multiomic approach. The use of the meta-omics protocols has hugely expanded the potentials of bioinoculants for effectively increasing agricultural productivity in a sustainable manner.

6.5 Optimized Plant-Microbiome Interactions

Multiomics studies have revealed the potentials in altering the genetic composition of the soil microbes or the hosts for enhanced soil and plant health, improved crop productivity and positive responses to environmental stresses. High-yielding crops and high-performing microbes with increased resilience properties are being generated with varying degrees of laboratory, field and commercial scale successes (Djemiel et al. 2022; Trivedi et al. 2021; Bona et al. 2019; Chaudhary and Shukla 2019a, b). Since plants and microbes produce several specialized metabolites to communicate and interact with each other, influencing these processes by manipulating their genetic composition can be used to improve the efficiency of their interaction for positive results. The genes used in producing these metabolites are often located in genomic regions referred to as Biosynthetic Clusters of Genes (BCGs), and newly developed computational algorithms enable the discovery of novel BCGs from environmental samples. A common procedure developed to produce the desired influences in microbiome genomes is called synthetic biology (Trivedi et al. 2021). Synthetic biology involves the use of genetic editing, which is the most recent technology used in this optimization procedure. This technique primarily involves the manipulation of the genetic materials of plant or microbial cells to increase their efficiency (Chaudhary and Shukla 2019a, b). Gene editing employs genetic engineering tools such as Crispr/Cas9, TALENS and zinc finger proteins, to manipulate genes of microbes, rapidly assemble new DNA fragments, and express new BCGs in heterologous hosts to produce novel metabolites of agricultural interests in plant-microbiome interactions (Sugimoto et al. 2019). Synthetic biology tools can be used to engineer multifunctional regulatory genes for efficient assembly and regulation of collective functions in a microbial consortium, while it can also be deployed in combination with multiomics and computational approaches to identify important microbes and metabolites drive metabolic and signaling interactions in plant-microbe interactions (Trivedi et al. 2021; Sugimoto et al. 2019). Genetically modified microorganisms have shown great potentials in improving crop protection during environmental stresses (Rosier et al. 2018). Novel strains have been developed with desirable qualities by modifying the metabolic pathways and enzymatic specificity (Trivedi et al. 2021).

6.6 Conclusion

The rapidly growing human population in the world has increased pressure on safe food production while demanding minimal detrimental effects to the environment in order to ensure continuous production of food. Sustainable farming, using the benefits of the widely distributed and diverse soil microorganisms, has extensive potential to improve plant productivity via nutrients fixing, cycling and mobilization, coupled with plant growth-promoting substances production activities of the soil microbiome. Further, the use of soil microbiome can promote plant health through resistance to pathogens and abiotic stress conditions such as drought, high salinity, soil contamination etc. The emergence of bioinoculants in laboratory, field and commercial forms has offered benefits of increased agricultural productivity and minimize the residues of agricultural chemicals in the environment, thereby promoting responsible and sustainable production of safe food for the increasing population. Recent exciting technological techniques and innovations in soil microbiome studies have deepened understanding of plant-microbe interactions, with the deployment of molecular technology to improve microbial and plants genomes. Multiomics and computational programs approaches have greatly increased crops productivity in environmental stress conditions, which opened the vista of immeasurable wealth in the practice of sustainable farming. Future research is required in understanding the effects of common agronomic practices on the various available and newly developed commercial bioinoculant preparations for life-span, compositional interactions and bio-functionality in delivering the identified agro-ecosystem functions under global changing climatic conditions. Also, the underlying mechanisms of the plant-microbiome interaction in the rhizosphere still require deeper understanding in order to support the establishment of developed strains of PGPMs in the presence of native microbes of the soil.

References

Agri U, Chaudhary P, Sharma A (2021) In vitro compatibility evaluation of agriusable nanochitosan on beneficial plant growth-promoting rhizobacteria and maize plant. Nat Acad Sci Lett 44:555–559. https://doi.org/10.1007/s40009-021-0104

Aguiar-Pulido V, Huang W, Suarez-Ulloa V, Cickovski T, Mathee K, Narasimhan G (2016) Metagenomics, metatranscriptomics, and metabolomics approaches to microbiome analysis. Evol Bioinforma 12(S1):5–16

Aini N, Dwi Yamika WS, Aini LQ, Kurniawan AP (2021) Application of saline tolerant bacteria and soil ameliorants improved growth, yield and nutrient uptake of tomato in saline land. Aust J Crop Sci 15(6):827–834. https://doi.org/10.21475/ajcs.21.15.06.p2903

Aldridge BB, Rhee KY (2014) Microbial metabolomics: innovation, application, insight. Curr Opin Microbiol 19:90–96

Alharbi K, Rashwan E, Mohamed HH, Awadalla A, Omara AE, Hafez EM et al (2022) Application of silica nanoparticles in combination with two bacterial strains improves the growth, antioxidant capacity and production of barley irrigated with saline water in salt-affected soil. Plants (Basel) 11(15):2026. https://doi.org/10.3390/plants11152026

Ali SZ, Sandhya V, Rao LV (2014) Isolation and characterization of drought-tolerant ACC deaminase and exopolysaccharide-producing fluorescent Pseudomonas sp. Ann Microbiol 64:493–502. https://doi.org/10.1007/s13213-013-0680-3

Anderson JAH, Layton R, Pickett J, Shah D (2016) Emerging agricultural biotechnologies for sustainable agriculture and food security. J Agric Food Chem 64:383–393

Arora NK, Mishra J (2024) Next generation microbe-based bioinoculants for sustainable agriculture and food security. Environ Sustain 7:1–4. https://doi.org/10.1007/s42398-024-00308-w

Ayilara MS, Adeleke BS, Akinola SA, Fayose CA, Adeyemi UT, Gbadegesin LA, Omole RK, Johnson RM, Uthman QO, Babalola OO (2023) Biopesticides as a promising alternative to synthetic pesticides: a case for microbial pesticides, phytopesticides, and nanobiopesticides. Front Microbiol 14:1040901, 1–16. https://doi.org/10.3389/fmicb.2023

Baart GJ, Martens DE (2012) Genome-scale metabolic models: reconstruction and analysis. In: Neisseria meningitides. Humana Press/Springer, pp 107–126

Barea JM (2015) Future challenges and perspectives for applying microbial biotechnology in sustainable agriculture based on a better understanding on plant-microbiome interactions. J Soil Sci Plant Nutr 15(2):261–282

Bhattacharyya C, Roy R, Tribedi P, Ghosh A, Ghosh A (2020) Biofertilizers as substitute to commercial agrochemicals. In: Majeti N, Vara P (eds) Agrochemicals detection, treatment and remediation Butterworth-Heinemann, United Kingdom, pp 263–290

Bilal S, Shahzad R, Khan AL, Kang SM, Imran QM, Al-Harrasi A et al (2018) Endophytic microbial consortia of phytohormones-producing fungus *Paecilomyces formosus* LHL10 and bacteria *Sphingomonas* sp. LK11 to *Glycine max* L. regulates physio-hormonal changes to attenuate aluminum and zinc stresses. Front Plant Sci 9:1273

Bona E, Massa N, Novello G, Boatti L, Cesaro P, Todeschini V, Magnelli V, Manfredi M, Marengo E, Mignone F et al (2019) Metaproteomic characterization of the *Vitis vinifera* rhizosphere. FEMS Microbiol Ecol 95:fiy 204

Chandra D, Strivastava R, Glick BR, Sharma AK (2018) Drought-tolerant *Pseudomonas* spp. improve the growth of finger millet (*Eleusine colacarna* (L) Gaertn.) under non-stressed and drought-stressed conditions. Pedosphere 28(2):227–240

Chaudhary T, Shukla P (2019a) Bioinoculant capability enhancement through metabolomics and systems biology approaches. Brief Funct Genomics 18(3):159–168

Chaudhary T, Shukla P (2019b) Bioinoculants for bioremediation applications and disease resistance: innovative perspectives. Indian J Microbiol 59(2):1–8

Chaudhary P, Chen S, Rajput VD, Jaiswal DK (2023) Bioinoculants with nano-compounds to improve soil health: a step toward sustainable agriculture. Front Environ Sci:1–3. https://doi.org/10.3389/fenvs.2023.1270002

Danish S, Zafar-Ul-Hye M (2020) Combined role of ACC deaminaseproducing bacteria and biochar on cereals productivity under drought. Phyton (BAires) 89:217–227. https://doi.org/10.32604/phyton.2020.08523

Deshmukh R, Wagh P, Naik J (2016) Solvent evaporation and spray drying technique for micro- and nanospheres/particles preparation: a review. Dry Technol 34(15):1758–1772

Devi R, Kaur T, Kour D, Hricovec M, Mohan R, Yadav N et al (2022) Microbes-mediated alleviation of heavy metal stress in crops: current research and future challenges. J Appl Biol Biotechnol 10(Suppl 2):25–37

Djemiel C, Dequiedt S, Karimi B, Cottin A, Horrigue A, Bailly A, Boutaleb A, Sadet-Bourgeteau S, Maron P, Prevost-Boure NC, Ranjard L, Terrat S (2022) Potential of meta-omics to provide modern microbial indicators for monitoring soil quality and securing food production. Front Microbiol 13:889788, 1–20. https://doi.org/10.3389/fmicb.2022.889788

Donn S, Kirkegaard JA, Perera G, Richardson AE, Watt M (2015) Evolution of bacterial communities in the wheat crop rhizosphere. Environ Microbiol 17:610–621

Doran JW, Zeiss MR (2000) Soil health and sustainability: managing the biotic components of the soil. Appl Soil Ecol 15(2000):3–11

Garrido-Oter R, Nakano RT, Dombrowski N, Ma KW, Team TA, Mchardy AC et al (2018) Modular traits of the rhizobiales root microbiota and their evolutionary relationship with symbiotic rhizobia. Cell Host Microbe 24:155–167.e5. https://doi.org/10.1016/j.chom.2018.06.006

Ghosh PK, De TK, Maiti TK (2018) Role of ACC deaminase as a stress ameliorating enzyme of plant growth-promoting rhizobacteria useful in stress agriculture: a review. In: Role of rhizospheric microbes in soil: stress management and agricultural sustainability, pp 57–106. https://doi.org/10.1007/978-981-10-8402-7_3

Giri B, Giang PH, Kumari R, Prasad R, Varma A (2015) Microbial diversity in soils. Soil Biol 3:19–54

Glick BR (2018) Soil microbes and sustainable agriculture. Pedosphere 28(1):201–210

Grabherr MG, Haas BJ, Yassour M et al (2011) Full-length transcriptome assembly from RNA-seq data without a reference genome. Nat Biotechnol 29(7):644–652

Gupta S, Pandey S (2019) ACC deaminase producing bacteria with multifarious plant growth promoting traits alleviates salinity stress in French bean (*Phaseolus vulgaris*) plants. Front Microbiol:1–17. https://doi.org/10.3389/fmicb.2019.01506

Gupta A, Singh UB, Sahu PK, Paul S, Kumar A, Malviya D, Singh S et al (2022) Linking soil microbial diversity to modern agricultural practices: a review. Int J Environ Res Public Health 19:3141

Hashem A, Abd-Allah EF, Alqarawi AA, Radhakrishnan R, Kumar A (2017) Plant Defense Approach of *Bacillus subtilis* (BERA 71) against *Macrophomina phaseolina* (Taasi) Goid in mung bean. J Plant Interact 12(1):390–401

Jing YX, Yan JL, He HD, Yang DJ, Xiao L, Zhong T et al (2014) Characterization of bacteria in the rhizosphere soils of *Polygonum pubescens* and their potential in promoting growth and Cd, Pb, Zn Uptake by *Brassica napus*. Int J Phytoremed 16:321–333

Joimel S, Capiaux H, Schwartz C, Hedde M, Lebeau T et al (2018) Effect of geogenic lead on fungal and collembolan communities in garden topsoil. Pedosphere., Elsevier 28(2):215–226

Jurić S, Jurić M, Režek Jambrak A, Vinceković M (2021) Tailoring alginate/chitosan microparticles loaded with chemical and biological agents for agricultural application and production of value-added foods. Appl Sci 11(9):4061

Kalimuthu R, Suresh P, Varatharaju G, Balasubramanian N, Rajasekaran KM, Shanmugaiah V (2019) Isolation and characterization of indole acetic acid (IAA) producing tomato rhizobacterium *Pseudomonas* sp VSMKU4050 and its potential for plant growth promotion. Int J Curr Microbiol App Sci 8(6):443–455

Kumar GP, Ahmed SKM, Desai S, Amalraj ELD, Rasul A (2014) In vitro screening for abiotic stress tolerance in potent biocontrol and plant growth promoting strains of Pseudomonas spp. and Bacillus spp. Int J Bacteriol. Hindawi Publishing Corporation 2014:195946

Kumar S, Kaushik G, Dar MA, Nimesh S, Lopez-Chukenu IR, Villarreal-Chiu JF (2018) Microbial degradation of organophosphate pesticides. A review. Pedosphere 28(2):1–19

Lakshmanan M, Koduru L, Lee DY (2018) Software applications for phenotype analysis and strain design of cellular systems. Emerg Area Bioeng 2:771–791

Li B, Dewey CN (2011) Rsem: accurate transcript quantification from RNA-seq data with or without a reference genome. BMC Bioinform 12(1):323

Marantayika P, Chimwamurombe PM, Venturi V, Uzabakiriho JD (2022) Bacterial bioinoculants adapted for sustainable plant health and soil fertility enhancement in Namibia. Front Sustain Food Syst:1–14. https://doi.org/10.3389/fsufs.2022.1002797

Marasco R, Rolli E, Fusi M, Michoud G, Daffonchio D (2018) Grapevine rootstocks shape underground bacterial microbiome and networking but not potential functionality. Microbiome 6:3

Maron P, Sarr A, Kaisermann A, Leveque J, Mathieu O, Guigue J, Karimi B, Bernard L, Dequiedt S, Terrat S, Chabbi A, Ranjard L (2018) High microbial diversity promotes soil ecosystem functioning. Appl Environ Microbiol 84(9):e02737–17, 13 pages

Matse DT, Huang CH, Huang YM, Yen MY (2020) Effects of coinoculation of Rhizobium with plant growth promoting rhizobacteria on the nitrogen fixation and nutrient uptake of Trifolium repens in low phosphorus soil. J Plant Nutr 43:739–752. https://doi.org/10.1080/01904167.2019.1702205

Meena SK, Meena VS (2017) Importance of soil microbes in nutrients use efficiency and sustainable food production. In: Agriculturally important microbes for sustainable agriculture. Springer Nature Singapore Pte Ltd, pp 3–22

Mostafiz SB, Rahman M, Rahman M (2012) Biotechnology: roles of microbes in sustainable agriculture and environmental health. Internet J Microbiol 10(1):1937–8289

Nannipieri P, Ascher J, Checcherini MT, Landi L, Pietramellara G, Renella G (2003) Microbial diversity and soil functions. Eur J Soil Sci 54:655–670

Nwachukwu BC, Babalola OO (2022) Metagenomics: a tool for exploring key microbiome with the potentials for improving sustainable agriculture. Front Sustain Food Syst 6:886987. https://doi.org/10.3389/fsufs.2022.886987

Qarni A, Billah M, Hussain K, Shah SH, Ahmed W, Alam S, Sheikh AA, Jafri L, Munir A, Malik KM, Khan N (2021) Isolation and characterization of phosphate solubilizing microbes from rock phosphate mines and their potential effect for sustainable agriculture. Sustain For 13:2151, 2–14. https://doi.org/10.3390/su13042151

Rascovan N, Carbonetto B, Perrig DS, Anta GG (2016) Integrated analysis of root microbiomes of soybean and wheat from agricultural fields. Sci Rep 6:28084

Romero F, Hilfiker S, Edlinger A, Held A, Hartman K, Labouyrie M, van der Heijden MGA (2023) Soil microbial diversity promotes crop productivity and agro-ecosystems functioning in experimental microcosms. Sci Total Environ. Elsevier 885(2023):163683

Rosier A, Medeiros FHV, Bais HP (2018) Defining plant growth promoting rhizobacteria molecular and biochemical networks in beneficial plant–microbe interactions. Plant Soil 428(1–2):1–21

Salas-Marina MA, Silva-Flores MA, Uresti-Rivera EE, Castro-Longoria E, Herrera-Estrella A, Casas-Flores S (2011) Colonization of *Arabidopsis* root by *Trichoderma atroviride* promotes growth and enhances systemic disease resistance through jasmonic acid/ethylene and salicylic acid pathways. Eur J Plant Pathol 131:15–26

Sander JD, Joung JK (2014) CRISPR–Cas systems for editing, regulating and targeting genomes. Nat Biotechnol 32:347

Schulz MH, Zerbino DR, Vingron M, Birney E (2012) Oases: robust de novo RNA-seq assembly across the dynamic range of expression levels. Bioinformatics 28(8):1086–1092

Sharma A, Shankdhar D, Sharma A, Shankhdhar SC (2014) Growth promotion of the rice genotypes by pgprs isolated from rice rhizosphere. J Soil Sci Plant Nutr 14(2):505–517

Shubha K, Mukheree A, Kumari M, Tiwari K, Meena VS (2017) Bio-stimulants: an approach towards the sustainable vegetable production. In: Agriculturally important microbes for sustainable agriculture. Springer Nature Singapore Pte Ltd, pp 259–277. https://doi.org/10.1007/978-981-10-5589-8_12

Singh R, Pathak B, Fulekar M (2015) Characterization of PGP traits by heavy metals tolerant Pseudomonas putida and Bacillus safensis strain isolated from rhizospheric zone of weed (Phyllanthus urinaria) and its efficiency in Cd and Pb removal. Int J Curr Microbiol Appl Sci 4:954–975

Singh, Bahsin S, Madan N, Suyal DC, Soni R, Singh D (2021) Bioinoculants for agricultural sustainability. In: Microbiological activity for soil and plant health management. Springer Nature Singapore Ltd, pp 629–641. https://doi.org/10.1007/978-981-16-2922-8_25

Srivastava N, Gupta B, Gupta S, Danquah MK, Sarethy IP (2019) Analyzing functional microbial diversity. In: Das S, Dash HR (eds) Microbial diversity in the genomic era. Academic

Sugimoto Y, Camacho FR, Wang S, Chankhamjon P, Odabas A, Biswas A, Jeffrey PD, Donia MS (2019) A metagenomic strategy for harnessing the chemical repertoire of the human microbiome. Science 366:eaax9176

Suman J, Rakshit A, Ogireddy SD, Singh S, Gupta C, Chandrakala J (2022) Microbiome as a key player in sustainable agriculture and human health. Front Soil Sci 2:Article 821589

Suyal DC, Soni R, Sai S et al (2016) Microbial inoculants as biofertilizer. In: Suyal Chandra D (ed) Microbial inoculants in sustainable agricultural productivity. New Delhi, Springer, pp 311–318

Tkacz A, Pini F, Turner TR, Bestion E, Simmonds J, Howell P, Greenland A, Cheema J, Emms DM, Uauy C et al (2020) Agricultural selection of wheat has been shaped by plant–microbe interactions. Front Microbiol 11:132

Trivedi P, Mattupalli C, Eversole K, Leach JE (2021) Enabling sustainable agriculture through understanding and enhancement of microbiomes—review. New Phyol 230:2129–2147

Walpola BC, Arunkumara KKIU, Yoon MH (2022) Phosphate solubilization by phosphate solubilizing organisms: insight into the mechanisms - review. Korean J Agric Sci 49:1–19. https://doi.org/10.7744/kjoas.20220041

Walters WA, Jin Z, Youngblut N, Wallace JG, Sutter J, Zhang W, Gonzalez-Pena A, Peiffer J, Koren O, Shi Q et al (2018) Large-scale replicated field study of maize rhizosphere identifies heritable microbes. Proc Natl Acad Sci USA 115:7368–7373

Wang H, Yang H, Shivalila CS et al (2013) One-step generation of mice carrying mutations in multiple genes by CRISPR/Cas-mediated genome engineering. Cell 153:910–918

White PJ, Crawford JW, Alvarez MC, Moreno RG (2014) Soil management for sustainable agriculture. Appl Environ Soil Sci 2014:536825, 2 pages

Xu J, Zhang Y, Zhang P, Trivedi P, Riera N, Wang Y, Liu X, Fan G, Tang J, Coletta-Filho HD et al (2018) The structure and function of the global citrus rhizosphere microbiome. Nat Commun 9:4894

Yang J, Kloepper JW, Ryu C (2009) Phizosphere bacteria help plants tolerate abiotic stress. Trends Plant Sci 14(1):1–4

Zamani M, Checcucci A, Fondi M et al (2018) Multi-disciplinary approaches for studying rhizobium–legume symbioses. Can J Microbiol 65(1):1–33

Chapter 7
Plant Growth Promoting Rhizobacteria: An Integration of the Mechanisms Behind Their Agricultural Application

José López-Hernández, Jesús Salvador López-Bucio, Homero Reyes de la Cruz, and José López-Bucio

Abstract Worldwide agriculture is facing an enormous challenge that requires rapid action and sustainable technology. Application of bacteria from the endogenous microbiome encompasses a very promising option to reinforce plant resilience and productivity, due to a number of traits that may lead to beneficial symbioses. Contemporary research aims to understand the molecular mechanisms underlying rhizobacteria-mediated plant growth promotion, and therefore, opens the way for crop improvement. However, no single bacterial trait really determines plant biomass accumulation, but it might be the combination of a number of factors such as the recognition of small-secreted bioactive molecules (i.e. auxins, cyclodipeptides, and *N*-acyl-*L*-homoserine lactones) that acting in concert could modulate root architecture, nutrient acquisition and plant defense. In addition, nutrient solubilization and activation of the molecular mechanisms for nutrient acquisition may not only enhance productivity in fertile soils, but also in challenging soils, such as alkaline soils from desert and salty areas. Altogether, the chemical and nutritional mutualisms lead to highly dynamic plant and bacterial behaviors that may change with time and upon availability of resources, and thus deciphering such complex interplay is required to identify promising isolates for the formulation of inoculants. The aim of this chapter is to summarize and integrate recent data into how plant growth promoting rhizobacteria enhance plant resilience via three main mechanisms, namely chemical plant-bacteria communication, nutritional reinforcement and adaptation to challenging soils.

Keywords Plant growth promoting rhizobacteria · Quorum-sensing · Nutrient acquisition · Root development · Alkaline soils

J. López-Hernández · H. R. de la Cruz · J. López-Bucio (✉)
Instituto de Investigaciones Químico-Biológicas, Universidad Michoacana de San Nicolás de Hidalgo, Morelia, Michoacán, Mexico
e-mail: jbucio@umich.mx

J. S. López-Bucio
Catedrático CONACYT-Instituto de Investigaciones Químico-Biológicas, Universidad Michoacana de San Nicolás de Hidalgo, Morelia, Michoacán, Mexico

O. O. Babalola, A. S. Ayangbenro (eds.), *Microbial Allies*,
https://doi.org/10.1007/978-3-031-90530-8_7

7.1 Introduction

Plants support the food chains worldwide, being approximately 20 crops including cereals such as rice, wheat and maize those that critically attend the growing demand of the human population that just reached 8 billion. Environmental stresses such as cold, drought, or rising temperature, have repercussions on low or null productivity, putting crop performance at risk. In the search for the highest yields per unit of planted area, the use of machinery for the softening of the soil, as well as the application of large quantities of fertilizers and pesticides were implemented for increasing productivity and controlling crop pests. However, this strategy, although effective in its early stages, has caused enormous damage to ecosystems such as soil deterioration, contamination of water bodies and loss of biodiversity (Khush 2001; Cappa et al. 2022; Edlinger et al. 2022). New challenges of optimizing fertilizer usage, and avoiding the degradation of arable soils are vital needs to achieve agricultural sustainability. In recent years the importance of the plant microbiome has attracted considerable attention owing their important roles in growth, development, immunity and abiotic stress adaptation and fundamentally, in achieving sustainable production of agricultural ecosystems (Trivedi et al. 2020; Mesny et al. 2023; Liu et al. 2024).

Bacteria are single-celled organisms that inhabit our planet since at least 3.5 billion years and have been able to adapt to every known environment, from the depths of the oceans to the great deserts (Makhalanyane et al. 2015; Vos et al. 2023). In the late nineteenth and early twentieth centuries, eminent microbiologists Louis Pasteur and Robert Koch highlighted the role of bacteria in causing disease (Opal 2009). Thus, for a long time, these microorganisms were considered dangerous agents. More than 100 years after the discovery of bacteria, the importance of the prokaryote microbiome to drive plant functional processes and to provide adaptive benefits to the species they coexist has started to be recognized (Petipas et al. 2021; Petersen et al. 2023).

Advances in electron microscopy, which make it possible to observe at magnifications thousands of times the original size, have unveiled the shapes and forms of plant growth promoting rhizobacteria, which may include coccoid, round cells and bacilli cells with flagellar structures for movement (Fig. 7.1).

Early studies with legumes such as beans and peanuts, indicated the importance of *Rhizobium* for plant nutrition. These bacteria form specialized structures in roots called nodules where they live, proliferate and provide several benefits including a better nitrogen, phosphorus and iron nutrition through changes in root architecture (Concha and Doerner 2020). In this sense, the nodule represents a niche where plant-bacteria relationships acquire such sophistication enabling plants to survive without the need of chemical fertilizers. The advantages of nodulation were recognized by former civilizations in Mesoamerica, as different groups of farmers combined the planting of beans with corn and squash, obtaining higher yields over the individual planting of each crop (Fonteyne et al. 2023).

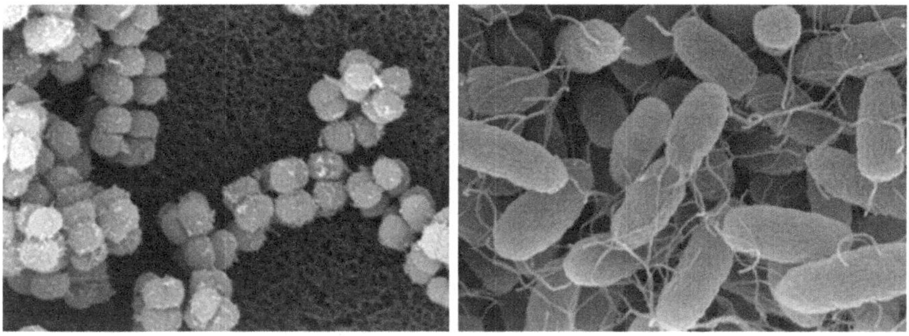

Fig. 7.1 Morphology of plant growth promoting rhizobacteria.
Scanning electron microscopy images of colonies of *Micrococcus luteus*, a Gram-positive coccoid-shaped bacterium and *Achromobacter* sp. 5B1, a Gram-negative bacillus. Magnification = 3000×

Bacteria that inhabit the roots or live inside stems, leaves, flowers and seeds provide countless benefits to their plant partners (Sánchez-Cañizares et al. 2017). Through the advancements of genomic sequencing, metabolomics and proteomics, the molecular mechanisms by which bacterial species promote plant adaptation and productivity have started to be unveiled, and an increasing number of genera including *Pseudomonas*, *Bacillus*, *Arthrobacter*, *Achromobacter*, *Azospirillum* and *Micrococcus* can benefit plants in several ways (Méndez-Gómez et al. 2020, 2021; Barrera-Ortiz et al. 2023; García-Cárdenas et al. 2023; Esparza-Reynoso et al. 2024; Jiménez-Vázquez et al. 2024).

The interactions of bacteria extend the plant functions, thus improving wellness to rhizosphere-engineered crops through properties such as nutrition, hormonal balance, and growth/defense modulation. In this chapter, recent advances into the mechanisms of plant growth regulation in interactions with beneficial bacteria are integrated, including chemical signaling, the configuration of root architecture, nutrient acquisition and adaptation to alkaline stress.

7.2 Chemical Plant-Bacteria Recognition

Plants possess highly sophisticated recognition systems for reacting to beneficial or pathogenic bacteria through detection of structural components of cells and diverse secreted molecules including toxins and bacterial hormones. Lipopolysaccharides (LPS), are major components of the outer-membrane of Gram-negative bacteria that affects growth and immunity in *Arabidopsis* seedlings in interactions with *Azospirillum brasilense* Sp245 (Méndez-Gómez et al. 2020) and *Herbaspirillum* sp. Root189 (Vanacore et al. 2022). LPS induced the TARGET OF RAPAMYCIN (TOR) signaling pathway and promote the activity of meristems, resulting in an increased root growth, biomass production, rosette size, stem cell activity, and number of fruits per plant (Méndez-Gómez et al. 2020).

N-acyl-*L*-homoserine lactones (AHLs), a class of small lipidic compounds control cell proliferation, biofilm formation and virulence in bacteria. AHLs modulate root growth depending on the chemical species and the concentration applied, for which treatment with short (*N*-hexanoyl-*L*-homoserine lactone) and medium-chain (3-oxo-decanoyl-*L*-homoserine lactone) promotes root growth (Pazarlar et al. 2020), whereas molecules with longer acyl moieties such as *N*-dodecanoyl-*L*-homoserine lactone, and 3-oxo-dodecanoyl-*L*-homoserine lactone repress root growth in a dose-dependent manner (Ortiz-Castro et al. 2008, 2011). AHLs also reinforce plant defense acting as elicitors of salicylic acid or jasmonic acid and contribute to resistance against *Pseudomonas syringae* in *Arabidopsis* (Liu et al. 2020) and *Pseudoperonospora cubensis* in cucumber (Pazarlar et al. 2020). Recently, Duan et al. (2023) through combining transcriptomic and hormonal analyses in *Arabidopsis* plants treated with various AHLs showed that the reaction of plants exposed to an AHL mix differs from that of plants exposed to single AHLs, and that AHL combination modifies the metabolism of jasmonates. Indeed, the presence of AHLs lead to an increase in roots and leaves of cyclic adenosine monophosphate (cAMP), an important second messenger in signal transduction pathways orchestrating growth and defense (Zhao et al. 2024).

The biosynthesis of cyclodipeptides (CDPs), such as cyclo(L-Pro-L-Val), cyclo(L-Pro-L-Phe), and cyclo(L-Pro-L-Tyr) is down-regulated by AHLs (Ortiz-Castro et al. 2011), suggesting that AHL and CDP signaling may act antagonistically in bacterial physiology. In an opposite manner to AHLs, which may act as alarm signals for plant cells, CDPs induce an auxin response in the root by acting directly on receptors of the TRANSPORT INHIBITOR RESPONSE 1 (TIR1) family, which lead to an improved biomass production through lateral root and root hair formation in *Arabidopsis*. Ortiz-Castro et al. (2020) showed that the production of cyclodipeptides by *Pseudomonas putida* and *Pseudomonas fluorescens* exert positive effects on the growth of *Arabidopsis*, and these molecules can promote the phosphorylation of the S6 ribosomal protein kinase (S6K), a downstream substrate of the TOR kinase, indicating that CDPs may act as mitogenic factors that contribute to plant metabolic adaptation (Calderan-Rodrigues et al. 2021; González-López et al. 2021).

The chemical recognition systems already present in plant cells can detect toxins and bacterial hormones, an example is coronatine, a toxin produced by the bacterial pathogen *Pseudomonas syringae* that led to the discovery of the jasmonic acid receptor CORONATINE INSENSITIVE 1 (COI). Coronatine is very potent and acts through mimicking defense signaling molecules (Gao et al. 2020), which helps plants to react properly under biotic stress. Regarding those chemicals from plants recognized by bacteria, the photosynthates exported from the shoot to the root and ultimately, secreted to the soil, exert positive chemotaxis towards bacteria, whose proliferation influences the rhizosphere. It is in this region where the root changes its architecture by responding locally to the accumulation of molecules of microbial origin through differential growth, branching patterns and formation of root hairs

(Garnica-Vergara et al. 2016; Esparza-Reynoso et al. 2021; García-Cárdenas et al. 2022).

7.3 Modulation of Root Architecture by Rhizobacteria

The root is the plant organ responsible for taking up water and nutrients. In dicotyledonous plants, the root system is typically a taproot, caused by dominant growth of the main axis that increases in extension throughout the life cycle. From the taproot, branches emerge from an inner cell layer termed pericycle and extend longitudinally, allowing for an increase in the absorption capacity (Du and Scheres 2018; Torres-Martínez et al. 2022). In both the taproot and lateral roots, root hairs are formed from epidermal cells called trichoblasts (Herburger et al. 2022).

Under stress conditions, adventitious roots develop from the stem and leaves, and these are particularly abundant in monocotyledon plants, such as maize, and their proliferation confers strong anchorage to the soil. Both lateral and adventitious roots are formed after germination and are considered post-embryonic structures. The three-dimensional configuration of the root system, its main axes of growth, and its degrees of branching is defined as root architecture (Abbas et al. 2022; Maqbool et al. 2022). It is a concept of great value due to its important applications for agriculture.

Bacteria promote root branching leading to better development and productivity, frequently through the differential production of bioactive substances and secondary metabolites including those mentioned above. Among the best-known genera with probiotic effects, we find species of *Pseudomonas*, *Bacillus*, *Arthrobacter*, *Achromobacter*, *Micrococcus*, *Azospirillum* and *Rhizobium* (Ortiz-Castro et al. 2020; García-Cárdenas et al. 2022, 2023).

Pseudomonas species establish intimate communication with plants and can be pathogens such as *P. aeruginosa* and *P. syringae*, or growth promoters that contribute not only to immunity through systemic acquired resistance (SAR) and induced systemic resistance (ISR) but also increasing phenotypic plasticity (Schikora et al. 2011; Liu et al. 2020; Solis-Ortiz et al. 2022). In relation to root morphogenesis, the inoculation of *A. thaliana* seedlings in Petri dishes with two plant growth promoting species, namely *P. putida* and *P. fluorescens* decreased root apical dominance and promoted the formation of lateral roots (Ortiz-Castro et al. 2020). *Pseudomonas* strains CM11 and WCS417 influence root growth in a largely contrasting manner, for instance, CM11 inoculation leads to the inhibition of primary root growth, whereas WCS417 manifested the opposite trend and promoted root elongation. Despite the contrary effects on extension of the main roots, most plant growth promoting *Pseudomonas* have the common property of inducing root branching (Ortiz-Castro et al. 2020; Li et al. 2022; López-Hernández et al. 2023; Tinoco-Tafolla et al. 2024).

P. aeruginosa and *P. chlororaphis* 3T4-12 led to disease symptoms in plants, termination of growth and overall inhibition of root growth and branching

(Ortiz-Castro et al. 2014; López-Hernández et al. 2023). The deleterious effects of both species diminish by decreasing virulence. In *P. aeruginosa* mutation of the *LasI* gene encoding 3-oxo-*N*-dodecanoyl-*L*-homoserine lactone synthase decreased virulence and increased the production of cyclodipeptides that promote the formation of root hairs and lateral roots, which overall extend the root surface (Ortiz-Castro et al. 2011, 2020). Virulence of *P. chlororaphis* 3T4-12 could be partially reversed by increasing the sucrose concentration in the medium, and then the bacterium could trigger root branching and plant biomass (Tinoco-Tafolla et al. 2024).

The phenotypical changes induced in *A. thaliana* by *Pseudomonas* is through the modulation of an auxin hormonal pathway, as observed upon application of AHLs or CDPs (Ortiz-Castro et al. 2008; Yin et al. 2022). Besides bacterial production of quorum-sensing signals and the release of phytohormones such as auxins account for regulation of root growth, such as *P. putida* PCI2, which produces indole-3-acetic acid upon supplementation of *L*-tryptophan (Pastor et al. 2014). *Bacillus* species increase root branching capacity by modulating either auxin or cytokinin signaling (García-Cárdenas et al. 2023; Barrera-Ortiz et al. 2023).

Bacterial repression of root growth also may be attributed to bacterial auxin production, as demonstrated by the regulation of root architecture by *Micrococcus luteus* LS570 (García-Cárdenas et al. 2022). Auxin accumulation within the *Arabidopsis* root tips inhibited both cell division and elongation, disrupting apical dominance, and thus the formation of lateral roots provides leading branches that replace the missing primary root. These data indicate that the modulation of plant development by rhizobacteria relies on the capacity to produce a wide number of bioactive substances including plant hormones, and compounds that mimic the action of phytohormones (Fig. 7.2). However, most of the molecular players mediating bacterial compound recognition by plant cells remains to be identified.

7.4 Role of Rhizobacteria in Plant Nutrition

The high demand of nutrients for crops such as corn, wheat, rice, barley, berries, and for production of fruits is achieved through the application of chemical fertilizers or organic amendments. Every year, millions of tons of these products are applied by farmers to increase fitness and health, which implies very high monetary investments and strong environmental costs, since excess nutrients may lead to soil salinization and contamination of groundwater, lakes and rivers that ultimately reach the ocean (Chand 2023). For these reasons, it is essential to develop new, effective, economical and environmentally strategies to optimize fertilizer usage.

Roots manifest a highly dynamic behavior to find and take up mineral nutrients. Root secretion of protons, mucilage, organic acids, and amino acids, enables on the one hand to solubilize nutrients, and on the other, to promote the proliferation of bacteria that help the plant in various adaptive functions including nutrient solubilization (Giehl and von Wirén 2014; Wendrich et al. 2020).

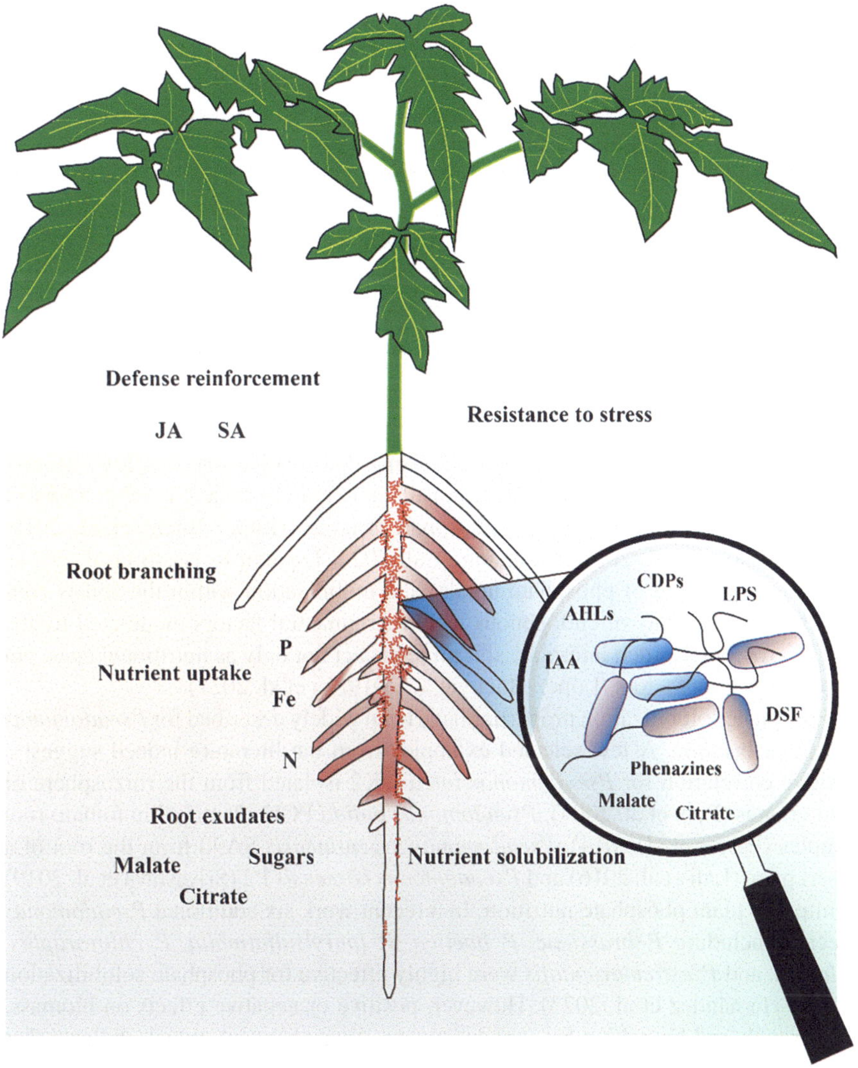

Fig. 7.2 Rhizobacterial traits mediating root growth and plant biomass production. Bacterial spread along the root system leads to a specific cell-to-cell recognition system based on structural cell components and diffusible compounds such as indole-3-acetic acid (IAA), *N*-acyl-*L*-homoserine lactones (AHLs), cyclodipeptides (CDPs), lipopolysaccharides (LPS), and the diffusible signaling factor (DSF). On the other hand, bacterial secretion of organic acids such as citrate and malate, or phenazines enhances nutrient solubilization and root acquisition of minerals such as phosphate (P), nitrate (N) and iron (Fe). By their side, roots secrete sugars and organic acids that act as chemotactic compounds to attract bacteria, and the underlying plant-bacterial interaction promotes root branching, enhances plant defense through salicylic acid (SA) and jasmonic acid (JA), and confers abiotic stress tolerance contributing to plant fitness

There are at least 16 elements of the periodic table with essential functions for plants, which are classified as macro and micronutrients. Macronutrients are present in the cytoplasm of cells in high (millimolar) concentrations, as they participate in a wide diversity of structural and functional processes, such as metabolism, photosynthesis, and production of biomolecules (López-Arredondo et al. 2014; Chen et al. 2023). Micronutrients such as iron, copper and zinc also called trace elements are required in very small concentrations for the fulfillment of their life cycle, acting as cofactors of enzymes and regulatory proteins such as transcription factors (Riaz and Guerinot 2021; Liang 2022; Stanton et al. 2022).

Phosphorus is one of the most limiting elements for agriculture due to its high reactivity with cations such as calcium, magnesium and aluminum (López-Arredondo et al. 2014). It is chemically assimilable as phosphate from organic or inorganic forms that easily precipitate and falls out from water solution (López-Hernández et al. 2023). Roots access phosphate through a combination of biochemical, physiological and molecular mechanisms that involve low and high affinity phosphate transporters and the secretion of malate and citrate by either roots or bacterial cells particularly under phosphate scarcity (López-Bucio et al. 2019; Raya-González et al. 2021; Naumann et al. 2022; Esparza-Reynoso et al. 2024). Thus, the efficiency of phosphate uptake and mobilization within the soil is conveyed by an interplay of endogenous and environmental factors modulated by the root microbiota. In such interplay, soil minerals act not only as nutritional cues, but also as signaling factors (López-Bucio et al. 2019; Li et al. 2021).

Phosphate solubilization properties have been widely described for *Pseudomonas* and other bacteria. A few selected examples from the literature indeed suggest a positive correlation for *Pseudomonas lutea* OK2 isolated from the rhizosphere of wild grasses (Peix et al. 2004), *Pseudomonas putida* PCI2 identified in tomato root samples (Pastor et al. 2014), *Pseudomonas argentinensis* SA90 from the root of a desert palm (Lafi et al. 2016) and *Pseudomonas koreensis* P2 (Srivastava et al. 2019) to improve plant phosphate nutrition. In a recent work six additional *Pseudomonas* species, including *P. brassicae*, *P. baetica*, *P. laurylsulfatiphila*, *P. chlororaphis*, *P. lurida*, and *P. extremorientalis* were highly effective for phosphate solubilization (López-Hernández et al. 2023). However, positive or negative effects on biomass, root growth and branching by each bacterium were observed, which indicate that not all phosphate solubilizing bacteria may be considered beneficial for plants. Indeed, these bacteria may be pathogenic, which makes necessary protocols to clearly separate the soil isolates specially manifesting beneficial traits over deleterious ones.

Phosphate solubilization appears to be under hormonal control, since indole-3-acetic acid (IAA), can upregulate genes encoding enzymes responsible for the synthesis of organic acids that chelate phosphates and make them available for nutrition. An analysis of 841 rhizobacteria identified 15 genera including phosphate solubilizing species, being able of producing IAA. The auxin precursor *L*-tryptophan further enhanced phosphate solubilization due to an increased production of IAA, the drop of pH and release of a high concentration of organic acids such as α-ketoglutarate, *cis*-aconitate, citrate, malate and succinate (Alemneh et al. 2021). On the other

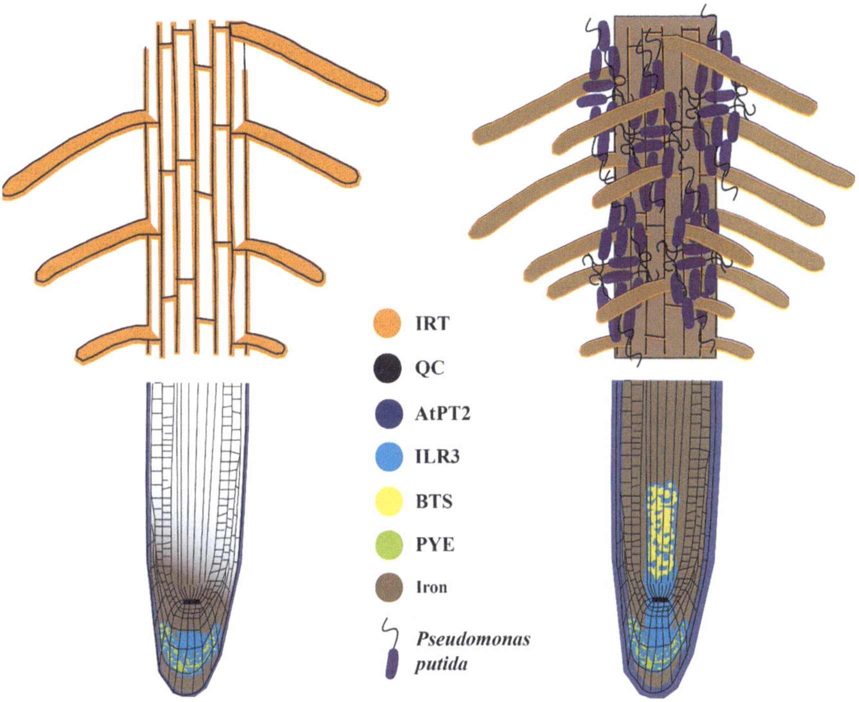

Fig. 7.3 Simultaneous activation in roots of the low phosphate and low iron rescue systems by *Pseudomonas putida*.
Inoculation of roots with *P. putida* leads to induction of master genes involved in phosphate and iron acquisition, including genes encoding the iron transporter IRT1, the phosphate transporter AtPT2, the ubiquitin ligase BRUTUS (BTS), and the transcription factors ILR3 and POPEYE (PYE) that promote iron uptake and its accumulation within roots. The quiescent center is indicated in black, and determines the zonation within the root stem cell niche. Expression domains of genes are indicated by specific colors and iron accumulation within the root tip is indicate in brown color

hand, *Bacillus amyloliquefaciens*, a saline-alkali resistant bacterium was able to solubilize phosphate through producing tartaric acid, citric acid and succinic acid, which may protect plants from heavy metal stress (Zhang et al. 2023).

In a study of plant interaction with *Pseudomonas putida*, Esparza-Reynoso et al. (2024) showed that root colonization by the bacterium in medium with high phosphate induced the expression of genes that are normally activated under phosphate scarcity and are part of the so-called low phosphate rescue system (Fig. 7.3). *Pseudomonas putida* also promoted iron accumulation and expression of low iron-inducible genes, indicating the integration of mechanisms for phosphate and iron acquisition simultaneously. Mutation of master genes for iron nutrition, including the protein ligase BRUTUS (BTS), and the bHLH transcription factors POPEYE (PYE) and IAA-LEUCINE RESISTANT3 (ILR3) compromised the root branching induced by *P. putida*, implying a role for a beneficial bacterium in improving plant nutrition through modulating cell signaling.

An induction of plant genes for nitrate nutrition such as the dual-affinity nitrate transceptor CHL1/AtNRT1/NPF6.3 and the nitrate reductases NIA1 and NIA2 was evidenced in *Arabidopsis* seedlings grown in interaction with *P. aeruginosa ΔlasI* mutant (López-Bucio et al. 2023). This improved nitrate nutrition led to better root branching and biomass production, and strengthening of plant immunity. Indeed, *Arabidopsis chl1-5* and *chl1-12* mutants defective on CHL1 and *nia1nia2* double mutants failed to grow properly under low nitrate availability and to develop robust root branches regardless of bacterial inoculation, indicating that nitrate nutrition determines plant biomass production driven by disruption of the *LasI* quorum-sensing system.

7.5 Rhizobacterial Traits Help Plants to Resist Alkaline Stress

Bacteria may extend plant fitness through several mechanisms. However, it remains to be clarified if application of pure bacterial cultures may be decisive in a greater production of biomass by plants being cultivated in soils with an adverse condition. An example of these soils are salty, alkaline calcareous soils that are widely distributed around the world. Recent reports advanced our knowledge into the mechanisms by which rhizobacteria increase plant resilience in these challenging soils.

Jiménez-Vázquez et al. (2020) isolated *Achromobacter* sp. 5B1 from the soil attached to roots of mesquite trees inhabiting a salty pool within the Chihuahua desert. Inoculation of pure *Achromobacter* sp. 5B1 cultures to *Arabidopsis* seedlings *in vitro* and in soil enhanced plant growth and biomass production, and helped plants to resist both saline and alkaline stress. A very interesting characteristic of this isolate was the property to direct root movements through interfering with the normal gravity response, thus the inoculated roots formed waves, turns and coils by which root branches proliferate and this response was thought to help plants to better explore the soil layers and to take up more water and nutrients. Noteworthy, *Achromobacter* sp. 5B1 helped *Arabidopsis* seedlings to grow in high pH (9.0) medium relieving plants from leaf chlorosis and activating the shoot and root meristems. It was noted that an alkaline condition impairs mitosis in meristems due to decreased auxin production in axenically grown seedlings, and the bacterium could normalize growth and development through enhancing auxin accumulation (Jiménez-Vázquez et al. 2024). Sequencing of the *Achromobacter* sp. 5B1 genome unveiled genes encoding for proteins potentially involved in alkaline stress adaptation influencing processes such as quorum-sensing signaling, auxin biosynthesis and nutrient uptake that may account for an efficient biomass production under stress. In another report, Zhuang et al. (2023) compared the effects of *S. rhizophila* wild-type (WT) and a *rpfF*-knockout mutant strain lacking DSF-QS system in *Brassica napus* L. regarding seed germination, biomass production, antioxidant capacity, and chlorophyll content. DSF enhanced adaptation and survival rate of *S. rhizophila*, and promoted both germination growth of plants under saline-alkaline stress, indicating the potential of quorum-sensing modulation to manage environmental stress in crops.

7.6 Conclusions

Global agriculture is currently under enormous pressure. The demand for food implies the challenge of obtaining the highest yields with the use of the smallest possible area of arable land, in order to prevent further deforestation and loss of biodiversity. In addition to this, the obsolete practices of applying excessive inputs such as fertilizers and pesticides with the subsequent pollution of the environment and damage to pollinators makes it necessary to change the way agriculture is managed. In this sense, the identification of microorganisms with beneficial properties, such as bacteria that promote plant growth, opens up new possibilities for more resilient and sustainable practices.

Our planet Earth is very rich in terms of diversity of environments, temperature, types of soils and water availability. Over the course of the year, seasonal changes in spring, summer, fall, and winter cause substantial dynamism in climate and precipitation, but also in the adaptive responses of all organisms. The abundance and distribution of species is subject to patterns in resource availability that tend to occur seasonally. Moreover, high pressure environments such as deserts, or alkaline/acid soils promote adaptive traits that favor relationships with soil microorganisms that help to better acquire nutrients, solubilize phosphate, or fix nitrogen from the atmosphere, enabling plants to grow despite the highly challenging variations in temperature, sunlight exposure and very low soil nutrient content (Petipas et al. 2021).

For the nutrition of humans, including probiotic bacteria in the diet strengthens the microbiome and helps in different physiological functions, such is the case of yogurt bacteria, comprising mostly lactobacilli. This concept has recently been extrapolated to plants, and thus, the possibility to formulate highly efficient bioinoculants through direct inoculation or applying derived biomolecules to improve plant functions is promising, but very challenging (Mandal et al. 2023; Poppeliers et al. 2023; Ravelo-Ortega et al. 2023). In the rhizosphere, a number of species of protozoa, bacteria and fungi coexist and few of them may be pathogenic, but the vast majority are commensal species that depend on root exudates for their nutrition. Thus, microbes adjust their metabolism to access easily metabolizable sugars (Villalobos-Escobedo et al. 2020; Tinoco-Tafolla et al. 2024). In the case of bacteria, these small, single-celled prokaryotic organisms may exist by thousands around roots and their importance to plants is due to the fact that they produce different types of bioactive substances and secondary metabolites that promote root growth and branching, as well as water and nutrient uptake, leading to better development and productivity (Jiménez-Vázquez et al. 2020; García-Cárdenas et al. 2022).

It should be emphasized that single bacterial traits are unlikely to influence overall plant biomass. However, acting in concert the properties of nutrient solubilization, production of hormones that promote the formation of lateral roots such as auxins, or quorum-sensing metabolites such as AHLs and cyclodipeptides, which strengthen plant immunity, may greatly contribute to crop fitness and resilience. Thus, no single trait but a combination of them, may ensure that a given bacterial isolate may be promising for making plant bio-stimulants and to support their probiotic effects in the long term.

References

Abbas M, Abid MA, Meng Z, Abbas M, Wang P, Lu C, Askari M, Akram U, Ye Y, Wei Y, Wang Y, Guo S, Liang C, Zhang R (2022) Integrating advancements in root phenotyping and genome-wide association studies to open the root genetics gateway. Physiol Plant 174(6):e13787. https://doi.org/10.1111/ppl.13787

Alemneh AA, Cawthray GR, Zhou Y, Ryder MH, Denton MD (2021) Ability to produce indole acetic acid is associated with improved phosphate solubilising activity of rhizobacteria. Arch Microbiol 203(7):3825–3837. https://doi.org/10.1007/s00203-021-02364-w

Barrera-Ortiz S, Balderas-Ruíz KA, López-Bucio JS, López-Bucio J, Flores C, Galindo E, Serrano-Carreón L, Guevara-García AA (2023) A *Bacillus velezensis* strain improves growth and root system development in *Arabidopsis thaliana* through cytokinin signaling. Rhizosphere 28:100815. https://doi.org/10.1016/j.rhisph.2023.100815

Calderan-Rodrigues MJ, Luzarowski M, Monte-Bello CC, Minen RI, Zühlke BM, Nikoloski Z, Skirycz A, Caldana C (2021) Proteogenic dipeptides are characterized by diel fluctuations and target of rapamycin complex-signaling dependency in the model plant *Arabidopsis thaliana*. Front Plant Sci 12:758933. https://doi.org/10.3389/fpls.2021.758933

Cappa F, Baracchi D, Cervo R (2022) Biopesticides and insect pollinators: detrimental effects, outdated guidelines, and future directions. Sci Total Environ 837:155714. https://doi.org/10.1016/j.scitotenv.2022.155714

Chand A (2023) Fertilizer use and ocean eutrophication. Nat Food 4(5):351. https://doi.org/10.1038/s43016-023-00763-4

Chen Z, Wang L, Cardoso JA, Zhu S, Liu G, Rao IM, Lin Y (2023) Improving phosphorus acquisition efficiency through modification of root growth responses to phosphate starvation in legumes. Front Plant Sci 14:1094157. https://doi.org/10.3389/fpls.2023.1094157

Concha C, Doerner P (2020) The impact of the rhizobia-legume symbiosis on host root system architecture. J Exp Bot 71:3902–3921. https://doi.org/10.1093/jxb/eraa198

Du Y, Scheres B (2018) Lateral root formation and the multiple roles of auxin. J Exp Bot 69(2):155–167. https://doi.org/10.1093/jxb/erx223

Duan Y, Han M, Grimm M, Ponath J, Reichelt M, Mithöfer A, Schikora A (2023) Combination of bacterial *N*-acyl homoserine lactones primes *Arabidopsis* defenses via jasmonate metabolism. Plant Physiol 191:2027–2044. https://doi.org/10.1093/plphys/kiad017

Edlinger A, Garland G, Hartman K, Banerjee S, Degrune F, García-Palacios P, Hallin S, Valzano-Held A, Herzog C, Jansa J, Kost E, Maestre FT, Pescador DS, Philippot L, Rillig MC, Romdhane S, Saghaï A, Spor A, Frossard E, van der Heijden MGA (2022) Agricultural management and pesticide use reduce the functioning of beneficial plant symbionts. Nat Ecol Evol 6:1145–1154. https://doi.org/10.1038/s41559-022-01799-8

Esparza-Reynoso S, Ruíz-Herrera LF, Pelagio-Flores R, Macías-Rodríguez LI, Martínez-Trujillo M, López-Coria M, Sánchez-Nieto S, Herrera-Estrella A, López-Bucio J (2021) *Trichoderma atroviride*-emitted volatiles improve growth of *Arabidopsis* seedlings through modulation of sucrose transport and metabolism. Plant Cell Environ 44(6):1961–1976. https://doi.org/10.1111/pce.14014

Esparza-Reynoso S, Ayala-Rodríguez JÁ, López-Bucio J (2024) *Pseudomonas putida* configures *Arabidopsis* root architecture through modulating the sensing systems for phosphate and iron acquisition. Plant Sci 342:112028. https://doi.org/10.1016/j.plantsci.2024.112028

Fonteyne S, Castillo Caamal JB, Lopez-Ridaura S, Van Loon J, Espidio Balbuena J, Osorio Alcalá L, Martínez Hernández F, Odjo S, Verhulst N (2023) Review of agronomic research on the milpa, the traditional polyculture system of Mesoamerica. Front Agron 5:1115490. https://doi.org/10.3389/fagro.2023.1115490

Gao M, Zhang C, Lu H (2020) Coronatine is more potent than jasmonates in regulating *Arabidopsis* circadian clock. Sci Rep 10:12862. https://doi.org/10.1038/s41598-020-69627-2

García-Cárdenas E, Ortiz-Castro R, Ruiz-Herrera LF, Valencia-Cantero E, López-Bucio J (2022) *Micrococcus luteus* LS570 promotes root branching in *Arabidopsis* via decreasing apical dominance of the primary root and an enhanced auxin response. Protoplasma 259:1139–1155. https://doi.org/10.1007/s00709-021-01724-z

García-Cárdenas E, Ortiz-Castro R, Ruiz-Herrera LF et al (2023) *Bacillus sp.* LC390B from the Maize rhizosphere improves plant biomass, root elongation, and branching and requires the phytochromes PHYA and PHYB for phytostimulation. J Plant Growth Regul 42:3056–3070. https://doi.org/10.1007/s00344-022-10770-7

Garnica-Vergara A, Barrera-Ortiz S, Muñoz-Parra E, Raya-González J, Méndez-Bravo A, Macías-Rodríguez L, Ruiz-Herrera LF, López-Bucio J (2016) The volatile 6-pentyl-2H-pyran-2-one from *Trichoderma atroviride* regulates *Arabidopsis thaliana* root morphogenesis via auxin signaling and ETHYLENE INSENSITIVE 2 functioning. New Phytol 209:1496–1512. https://doi.org/10.1111/nph.13725

Giehl RF, von Wirén N (2014) Root nutrient foraging. Plant Physiol 166:509–517. https://doi.org/10.1104/pp.114.245225

González-López O, Palacios-Nava B, Peña-Uribe CA, Campos-García J, López-Bucio J, García-Pineda E, Reyes de la Cruz H (2021) Growth promotion in *Arabidopsis thaliana* by bacterial cyclodipeptides involves the TOR/S6K pathway activation. J Plant Physiol 257:153343. https://doi.org/10.1016/j.jplph.2020.153343

Herburger K, Schoenaers S, Vissenberg K, Mravec J (2022) Shank-localized cell wall growth contributes to *Arabidopsis* root hair elongation. Nat Plants 8:1222–1232. https://doi.org/10.1038/s41477-022-01259-y

Jiménez-Vázquez KR, García-Cárdenas E, Barrera-Ortiz S, Ortiz-Castro R, Ruiz-Herrera LF, Ramos-Acosta BP, Coria-Arellano JL, Sáenz-Mata J, López-Bucio J (2020) The plant beneficial rhizobacterium *Achromobacter* sp. 5B1 influences root development through auxin signaling and redistribution. Plant J 103:1639–1654. https://doi.org/10.1111/tpj.14853

Jiménez-Vázquez KR, López-Hernández J, García-Cárdenas E, Pelagio-Flores R, López-Bucio JS, Téxon AC, Ibarra-Laclette E, López-Bucio J (2024) The plant growth promoting rhizobacterium *Achromobacter* sp. 5B1, rescues *Arabidopsis* seedlings from alkaline stress by enhancing root organogenesis and hormonal responses. Microbiol Res 281:127594. https://doi.org/10.1016/j.micres.2023.127594

Khush G (2001) Green revolution: the way forward. Nat Rev Genet 2:815–822. https://doi.org/10.1038/35093585

Lafi FF, Alam I, Geurts R, Bisseling T, Bajic VB, Hirt H, Saad MM (2016) Draft genome sequence of the phosphate-solubilizing bacterium *Pseudomonas argentinensis* strain SA190 isolated from the desert plant *Indigofera argentea*. Genome Announc 4(6):e01431–e01416. https://doi.org/10.1128/genomeA.01431-16

Li L, Liu KH, Sheen J (2021) Dynamic nutrient signaling networks in plants. Annu Rev Cell Dev Biol 37:341–367. https://doi.org/10.1146/annurev-cellbio-010521-015047

Li Q, Li H, Yang Z, Cheng X, Zhao Y, Qin L, Bisseling T, Cao Q, Willemsem V (2022) Plant growth-promoting rhizobacterium *Pseudomonas* sp.CM11 specifically induces lateral roots. New Phytol 235:1574–1588. https://doi.org/10.1111/nph.18199

Liang G (2022) Iron uptake, signaling, and sensing in plants. Plant Commun 3:100349. https://doi.org/10.1016/j.xplc.2022.100349

Liu F, Zhao Q, Jia Z, Song C, Huang Y, Ma H, Song S (2020) *N*-3-oxo-octanoyl-homoserine lactone-mediated priming of resistance to *Pseudomonas syringae* requires the salicylic acid signaling pathway in *Arabidopsis thaliana*. BMC Plant Biol 20:38. https://doi.org/10.1186/s12870-019-2228-6

Liu Y, Xu Z, Chen L, Xun W, Shu X, Chen Y, Sun X, Wang Z, Ren Y, Shen Q, Zhang R (2024) Root colonization by beneficial rhizobacteria. FEMS Microbiol Rev 48(1):fuad066. https://doi.org/10.1093/femsre/fuad066

López-Arredondo DL, Leyva-González MA, González-Morales SI, López-Bucio J, Herrera-Estrella L (2014) Phosphate nutrition: improving low-phosphate tolerance in crops. Annu Rev Plant Biol 65:95–123. https://doi.org/10.1146/annurev-arplant-050213-035949

López-Bucio JS, Salmerón-Barrera GJ, Ravelo-Ortega G, Raya-González J, León P, Reyes de la Cruz H, Campos-García J, López-Bucio J, Guevara-García AA (2019) Mitogen-activated protein kinase 6 integrates phosphate and iron responses for indeterminate root growth in *Arabidopsis thaliana*. Planta 250:1177–1189. https://doi.org/10.1007/s00425-019-03212-4

López-Bucio J, Ortiz-Castro R, Magaña-Dueñas V, García-Cárdenas E, Jiménez-Vázquez KR, Raya-González J, Pelagio-Flores R, Ibarra-Laclette E, Herrera-Estrella L (2023) *Pseudomonas aeruginosa LasI*-dependent plant growth promotion requires the host nitrate transceptor AtNRT1.1/CHL1 and the nitrate reductases NIA1 and NIA2. Planta 258:80. https://doi.org/10.1007/s00425-023-04236-7

López-Hernández J, García-Cárdenas E, López-Bucio JS, Jiménez-Vázquez KR, de la Cruz HR, Ferrera-Rodríguez O, Santos-Rodríguez DL, Ortiz-Castro R, López-Bucio J (2023) Screening of phosphate solubilization identifies six *Pseudomonas* species with contrasting phytostimulation properties in *Arabidopsis* seedlings. Microb Ecol 86:431–445. https://doi.org/10.1007/s00248-022-02080-y

Makhalanyane TP, Valverde A, Gunnigle E, Frossard A, Ramond JB, Cowan DA (2015) Microbial ecology of hot desert edaphic systems. FEMS Microbiol Rev 39:203–221. https://doi.org/10.1093/femsre/fuu011

Mandal S, Anand U, López-Bucio J, Radha, Kumar M, Lal MK, Tiwari RK, Dey A (2023) Biostimulants and environmental stress mitigation in crops: a novel and emerging approach for agricultural sustainability under climate change. Environ Res 15(233):116357. https://doi.org/10.1016/j.envres.2023.116357

Maqbool S, Hassan MA, Xia X, York LM, Rasheed A, He Z (2022) Root system architecture in cereals: progress, challenges and perspective. Plant J 110(1):23–42. https://doi.org/10.1111/tpj.15669

Méndez-Gómez M, Castro-Mercado E, Peña-Uribe CA, Reyes de la Cruz H, López-Bucio J, García-Pineda E (2020) *Azospirillum brasilense* Sp245 lipopolysaccharides induce target of rapamycin signaling and growth in *Arabidopsis thaliana*. J Plant Physiol 253:153270. https://doi.org/10.1016/j.jplph.2020.153270

Méndez-Gómez M, Barrera-Ortiz S, Castro-Mercado E, López-Bucio J, García-Pineda E (2021) The nature of the interaction *Azospirillum-Arabidopsis* determines the molecular and morphological changes in root and plant growth promotion. Protoplasma 258:179–189. https://doi.org/10.1007/s00709-020-01552-7

Mesny F, Hacquard S, Thomma BP (2023) Co-evolution within the plant holobiont drives host performance. EMBO Rep 24:e57455. https://doi.org/10.15252/embr.202357455

Naumann C, Heisters M, Brandt W, Janitza P, Alfs C, Tang N, Toto Nienguesso A, Ziegler J, Imre R, Mechtler K, Dagdas Y, Hoehenwarter W, Sawers G, Quint M, Abel S (2022) Bacterial-type ferroxidase tunes iron-dependent phosphate sensing during *Arabidopsis* root development. Curr Biol 32:2189–2205.e6. https://doi.org/10.1016/j.cub.2022.04.005

Opal SM (2009) A brief history of microbiology and immunology. Vaccines Biography 2009:31–56. https://doi.org/10.1007/978-1-4419-1108-7_3

Ortíz-Castro R, Martínez-Trujillo M, López-Bucio J (2008) *N*-acyl-*L*-homoserine lactones: a class of bacterial quorum-sensing signals alter post-embryonic root development in *Arabidopsis thaliana*. Plant Cell Environ 31:1497–1509

Ortiz-Castro R, Díaz-Pérez C, Martínez-Trujillo M, del Río RE, Campos-García J, López-Bucio J (2011) Transkingdom signaling based on bacterial cyclodipeptides with auxin activity in plants. Proc Natl Acad Sci USA 108:7253–7258

Ortiz-Castro R, Pelagio-Flores R, Méndez-Bravo A, Ruiz-Herrera LF, Campos-García J, López-Bucio J (2014) Pyocyanin, a virulence factor produced by *Pseudomonas aeruginosa*, alters root development through reactive oxygen species and ethylene signaling in Arabidopsis. Mol Plant-Microbe Interact 27:364–378. https://doi.org/10.1094/MPMI-08-13-0219-R

Ortiz-Castro R, Campos-García J, López-Bucio J (2020) *Pseudomonas putida* and *Pseudomonas fluorescens* influence *Arabidopsis* root system architecture through an auxin response mediated by bioactive cyclodipeptides. J Plant Growth Regul 39:254–265. https://doi.org/10.1007/s00344-019-09979-w

Pastor N, Rosas S, Luna M, Rovera M (2014) Inoculation with *Pseudomonas putida* PCI2, a phosphate solubilizing rhizobacterium, stimulates the growth of tomato plants. Symbiosis 62(3):157–167. https://doi.org/10.1007/s13199-014-0281-3

Pazarlar S, Cetinkaya N, Bor M, Kara RS (2020) *N*-acyl homoserine lactone-mediated modulation of plant growth and defense against *Pseudoperonospora cubensis* in cucumber. J Exp Bot 71:6638–6654. https://doi.org/10.1093/jxb/eraa384

Peix A, Rivas R, Santa-Regina I, Mateos PF, Martínez-Molina E, Rodríguez-Barrueco C, Velázquez E (2004) *Pseudomonas lutea* sp. nov., a novel phosphate-solubilizing bacterium isolated from the rhizosphere of grasses. Int J Syst Evol Microbiol 54(Pt 3):847–850. https://doi.org/10.1099/ijs.0.02966-0

Petersen C, Hamerich IK, Adair KL, Griem-Krey H, Torres Oliva M, Hoeppner MP, Bohannan BJM, Schulenburg H (2023) Host and microbiome jointly contribute to environmental adaptation. ISME J 17:1953–1965. https://doi.org/10.1038/s41396-023-01507-9

Petipas RH, Geber MA, Lau JA (2021) Microbe-mediated adaptation in plants. Ecol Lett 24:1302–1317. https://doi.org/10.1111/ele.13755

Poppeliers SW, Sánchez-Gil JJ, de Jonge R (2023) Microbes to support plant health: understanding bioinoculant success in complex conditions. Curr Opin Microbiol 73:102286. https://doi.org/10.1016/j.mib.2023.102286

Ravelo-Ortega G, Raya-González J, López-Bucio J (2023) Compounds from rhizosphere microbes that promote plant growth. Curr Opin Plant Biol 73:102336. https://doi.org/10.1016/j.pbi.2023.102336

Raya-González J, Ojeda-Rivera JO, Mora-Macias J, Oropeza-Aburto A, Ruiz-Herrera LF, López-Bucio J, Herrera-Estrella L (2021) MEDIATOR16 orchestrates local and systemic responses to phosphate scarcity in *Arabidopsis* roots. New Phytol 229:1278–1288. https://doi.org/10.1111/nph.16989

Riaz N, Guerinot ML (2021) All together now: regulation of the iron deficiency response. J Exp Bot 72:2045–2055. https://doi.org/10.1093/jxb/erab003

Sánchez-Cañizares C, Jorrín B, Poole PS, Tkacz A (2017) Understanding the holobiont: the interdependence of plants and their microbiome. Curr Opin Microbiol 38:188–196. https://doi.org/10.1016/j.mib.2017.07.001

Schikora A, Schenk ST, Stein E, Molitor A, Zuccaro A, Kogel KH (2011) *N*-acyl-homoserine lactone confers resistance toward biotrophic and hemibiotrophic pathogens via altered activation of AtMPK6. Plant Physiol 157(3):1407–1418. https://doi.org/10.1104/pp.111.180604

Solis-Ortiz CS, Gonzalez-Bernal J, Kido-Díaz HA, Peña-Uribe CA, López-Bucio JS, López-Bucio J, Guevara-García AA, García-Pineda E, Villegas J, Campos-García J, Reyes de La Cruz H (2022) Bacterial cyclodipeptides elicit *Arabidopsis thaliana* immune responses reducing the pathogenic effects of *Pseudomonas aeruginosa* PAO1 strains on plant development. J Plant Physiol 275:153738. https://doi.org/10.1016/j.jplph.2022.153738

Srivastava AK, Saxena P, Sharma A, Srivastava R, Jamali H, Bharati AP, Yadav J, Srivastava AK, Kumar M, Chakdar H, Kashyap PL, Saxena AK (2019) Draft genome sequence of a cold-adapted phosphorous-solubilizing *Pseudomonas koreensis* P2 isolated from Sela Lake, India. 3 Biotech 9(7):256. https://doi.org/10.1007/s13205-019-1784-7

Stanton C, Sanders D, Krämer U, Podar D (2022) Zinc in plants: integrating homeostasis and biofortification. Mol Plant 15:65–85. https://doi.org/10.1016/j.molp.2021.12.008

Tinoco-Tafolla HA, López-Hernández J, Ortiz-Castro R, López-Bucio J, Reyes de la Cruz H, Campos-García J, López-Bucio JS (2024) Sucrose supplements modulate the *Pseudomonas chlororaphis-Arabidopsis thaliana* interaction via decreasing the production of phenazines and enhancing the root auxin response. J Plant Physiol 297:154259. https://doi.org/10.1016/j.jplph.2024.154259

Torres-Martínez HH, Napsucialy-Mendivil S, Dubrovsky JG (2022) Cellular and molecular bases of lateral root initiation and morphogenesis. Curr Opin Plant Biol 65:102115. https://doi.org/10.1016/j.pbi.2021.102115

Trivedi P, Leach JE, Tringe SG, Sa T, Singh BK (2020) Plant-microbiome interactions: from community assembly to plant health. Nat Rev Microbiol 18:607–621. https://doi.org/10.1038/s41579-020-0412-1

Vanacore A, Vitiello G, Wanke A, Cavasso D, Clifton LA, Mahdi L, Campanero-Rhodes MA, Solís D, Wuhrer M, Nicolardi S, Molinaro A, Marchetti R, Zuccaro A, Paduano L, Silipo A (2022) Lipopolysaccharide O-antigen molecular and supramolecular modifications of plant root microbiota are pivotal for host recognition. Carbohydr Polym 277:118839. https://doi.org/10.1016/j.carbpol.2021.118839

Villalobos-Escobedo JM, Esparza-Reynoso S, Pelagio-Flores R, López-Ramírez F, Ruiz-Herrera LF, López-Bucio J, Herrera-Estrella A (2020) The fungal NADPH oxidase is an essential element for the molecular dialog between *Trichoderma* and *Arabidopsis*. Plant J 103:2178–2192. https://doi.org/10.1111/tpj.14891

Vos M, Padfield D, Quince C, Vos R (2023) Adaptive radiations in natural populations of prokaryotes: innovation is key. FEMS Microbiol Ecol 99:fiad154. https://doi.org/10.1093/femsec/fiad154

Wendrich JR, Yang B, Vandamme N, Verstaen K, Smet W, Van de Velde C, Minne M, Wybouw B, Mor E, Arents HE, Nolf J, Van Duyse J, Van Isterdael G, Maere S, Saeys Y, De Rybel B (2020) Vascular transcription factors guide plant epidermal responses to limiting phosphate conditions. Science 370(6518):eaay4970

Yin L, Chen X, Chen Q, Wei D, Hu XY, Jia AQ (2022) Diketopiperazine modulates *Arabidopsis thaliana* root system architecture by promoting interactions of auxin receptor TIR1 and IAA7/17 proteins. Plant Cell Physiol 63:57–69

Zhang C, Chen H, Dai Y, Chen Y, Tian Y, Huo Z (2023) Isolation and screening of phosphorus solubilizing bacteria from saline alkali soil and their potential for Pb pollution remediation. Front Bioeng Biotechnol 11:1134310. https://doi.org/10.3389/fbioe.2023.1134310

Zhao X, Li W, Li X, Jia Z, Song S, Zhao Q (2024) The effect of bacterial AHL on the cyclic adenosine monophosphate content in plants according to high-performance liquid chromatography. Molecules 29:1074. https://doi.org/10.3390/molecules29051074

Zhuang X, Liu Y, Fang N, Bai Z, Gao J (2023) Quorum sensing improves the plant growth-promoting ability of *Stenotrophomonas rhizophila* under saline-alkaline stress by enhancing its environmental adaptability. Front Microbiol 14:1155081. https://doi.org/10.3389/fmicb.2023.1155081

Chapter 8
Microbial Biostimulants: Boosting Plant Performance and Stress Tolerance

Adedayo Ayodeji Lanrewaju and Ayomide Emmanuel Fadiji

Abstract Globally, food production and agricultural productivity are threatened by declining arable land for cultivation, rising human activities, and the effects of climate change, which presents regular floods, long-lasting droughts, and abrupt shifts in temperature. In addition, these climatic fluctuations exacerbate disease and insect occurrences, thus lowering agricultural productivity. Owing to these escalating environmental concerns, there is a great need for global collaboration and implementation of sustainable, ecologically secure and environmentally friendly agricultural techniques that enhance plant growth, stress tolerance and overall productivity. Microbial biostimulants (MBs) have emerged as a promising approach to enhance plant growth and development, even in stressful situations. These biostimulants comprise a diverse array of plant growth-promoting microorganisms (PGPMs) used separately or in combination, such as bacteria, fungi, and algae. This profoundly influences the growth, development, and stress responses of plants exposed to different environmental conditions. These microbes employ different mechanisms in mitigating different environmental stresses. However, each mechanism may vary depending on the specific microorganism-plant interaction and environmental conditions utilized. Hence, in this chapter, notable roles and potentials of MBs in revolutionizing plant performance and stress tolerance were highlighted. Also, the mechanisms by which these MBs operate were elucidated while their multifaceted interactions with plants were analyzed and their measures of ameliorating abiotic and biotic stresses. Additionally, the chapter analyzes the major challenges and prospects associated with the emerging adoption of these biostimulants, emphasiz-

A. A. Lanrewaju
Department of Biotechnology and Food Science, Faculty of Applied Science, Durban University of Technology, Durban, South Africa

A. E. Fadiji (✉)
Hawkesbury Institute for the Environment, Western Sydney University, Penrith, NSW, Australia

Food Security and Safety Focus Area, North-West University, Mmabatho, South Africa
e-mail: a.fadiji@westernsydney.edu.au

O. O. Babalola, A. S. Ayangbenro (eds.), *Microbial Allies*,
https://doi.org/10.1007/978-3-031-90530-8_8

ing the need for further research to optimize their application strategies and commercial viability.

Keywords Microbial biostimulants · Climate change · Climate-resistant crops · Modern agriculture · Plant growth improvement · Sustainable agriculture

8.1 Introduction

Global climate data indicate a rise in global temperature since 1970, along with alterations in patterns of rainfall with harmful consequences on agriculture (Lotze-Campen et al. 2012). Unfortunately, these conditions subject crop plants to both abiotic and biotic stresses with a tendency to widen the diversity of pathogens and pests, leading to more frequent and severe disease epidemics (De Wolf and Isard 2007; Garrett et al. 2021). Consequently, present estimates indicated that global agricultural productivity is projected to experience a 50% reduction owing to abiotic stress and a 30% reduction due to biotic stress (Kumar and Verma 2018). Regrettably, the current concerns about agricultural sustainability, such as population growth, climatic variability, and declining soil health for crop production, are projected to increase in the future as a result of climate change and widespread farming practices (Wassmann et al. 2009). Given the ongoing increase in the world's population, the decrease in the available land suitable for farming, and the decline in the genetic capacity of crops, it is necessary to adopt novel agricultural technologies. Therefore, solutions with little environmental effect are essential for ensuring the rising call for food with high nutritional content by enhancing plant tolerance to adverse soil conditions (Szparaga et al. 2019; Mannino et al. 2021). Interestingly, enhancing the capability of plants to withstand climate change-induced pressures is a sustainable approach to ensuring food security without significantly expanding agricultural land and adopting MBs has been considered one of the most effective strategies to accomplish this target (Calvo et al. 2014; Yakhin et al. 2017).

Plants harbour a wide range of microbes in different parts, such as the internal compartments (endosphere), soil attached to their roots (rhizosphere), and the aboveground parts (phyllosphere) (Compant et al. 2019; Babalola et al. 2020). MBs comprise <25% of the total commercial biostimulant products in the global market projected to have significant development, with an anticipated annual growth rate of 10.2% from 2017 to 2025 (Bulgari et al. 2019; Hamid et al. 2021). MBs are advanced technologies that may enhance agricultural productivity by alleviating the adverse impacts of environmental changes while also increasing the nutritional content of crops (Castiglione et al. 2021). They are used on crops to improve the uptake and utilization of various nutrients by plants, thus enhancing plant growth and production, and consist of plant growth-promoting rhizobacteria (PGPR), mycorrhizal and non-mycorrhizal fungi, and bacterial endosymbionts (Enagbonma et al. 2023).

These microorganisms use both direct and indirect processes to enhance plant development in both typical and stressed environments. Specifically, mycorrhizal fungi have a key responsibility in promoting and regulating plant responses to diverse kinds of stress by establishing and sustaining enough activity in the area around plants (Du Jardin 2015; Ali et al. 2017).

Furthermore, MBs improve nutrient absorption, plant immunity and crop tolerance to abiotic stressors and are financially viable and environmentally friendly (Nephali et al. 2020). Microorganisms like *Arthrobacter, Azotobacter, Azospirillum, Bacillus, Pseudomonas,* and arbuscular mycorrhizal fungi (AMF) function as plant growth-promoting bacteria and have been found to improve plant growth and alleviate abiotic stresses in various crops including tomato, potato, soybean, cabbage, broccoli, maize, and rice (Asaf et al. 2018; Kim et al. 2020; Li et al. 2020; Kubi et al. 2021). Mycorrhizal plant-beneficial fungi enhance plant nutrition, trigger substantial alterations in secondary metabolism and increase tolerance to soil and environmental challenges (Rouphael and Colla 2020). In addition, they employ various mechanisms and approaches to ensure plants' survival in unfavourable conditions, such as modifying the cell wall constituents and accumulating high levels of solutes, thus enabling plants to retain water more effectively and enhance their ability to withstand ionic and osmotic stresses (Van Oosten et al. 2017). The capacity of MBs to safeguard plants from environmental pressures is crucial for the continued existence of crops (Nephali et al. 2020), and their usage is a favourable, practical, and eco-friendly approach to lessen the detrimental influences of different stressors.

Therefore, a sustainable approach to help plants cope with biotic and abiotic stresses in the era of climate change is the use of MBs. Although research and technological advancements have shed light on the useful roles of plant-associated microorganisms, these roles have often been oversimplified or have not been effectively replicated in the field. Consequently, instead of fully attaining the goal of microbiome management, MBs have often been used as alternative treatments. Hence, this chapter examined the influence of abiotic and biotic stresses on plant growth and development, the mechanistic impact of MBs in mitigating stress in plants and comprehending the synergy between fungi and bacteria as MBs. In addition, the major constraints and prospects associated with utilizing MBs were highlighted, while the need for further research to maximize their application strategies and commercial viability was emphasized.

8.2 Overview of the Impact of Stresses on Plant Growth and Development

8.2.1 Abiotic Factors

Climate change has mostly resulted from the burning of fossil fuels and the rise in the concentration of harmful greenhouse gases (GHGs) in the atmosphere during the post-industrialization period due to the industrial revolution, often considered as the primary catalyst for the alteration in climatic conditions worldwide (Dutta et al. 2020). Consequently, climate change has led to fluctuations in temperature, precipitation, and air conditions, which have subjected plants to severe and extreme weather conditions with accompanying negative effects on the physical, developmental, cellular, and molecular processes in plants (Chaudhry and Sidhu 2022). Unfortunately, the development and yield of the plants are affected by adverse environmental conditions such as elevated temperatures and CO_2 concentrations (Raza et al. 2019). Plants have a high degree of sensitivity to climate change because their extended lifespan makes it challenging to adjust to fluctuating environmental circumstances (Lindner et al. 2010); hence, temperature change could affect the growth and reproduction of plants as it unfavourably impacts the molecular pathways involved in growth, hence affecting plants' development (Bita and Gerats 2013).

8.2.1.1 Impact of Temperature on Plant Growth and Development

The morphological and physiological activities of plants are influenced by the temperature circumstances in which they are grown. Each plant species has an ideal temperature range for maximum growth, often between 25 and 30 °C (Hatfield and Prueger 2015). Exposure to temperatures higher than the optimal range leads to thermal damage in plants as fluctuations in temperature over the summer might negatively impact the intermolecular interactions essential for regular plant growth, hence affecting the process of fruit formation and development (Bita and Gerats 2013). Prolonged exposure of plants to high temperatures could result in significant changes in metabolic processes and cellular disorganization, leading to decreased growth and development (Rai et al. 2021). Specifically, increased temperature reduces the percentage of germination, viability, and development of the plumule and radicle in plants, which may be ascribed to the reduction in cellular water component leading to a decrease in cell size and overall plant growth (Rodríguez et al. 2005). Furthermore, higher temperatures may lead to a decline in the germination of *Triticum aestivum* and *Pisum sativum* accompanied by a loss in seed dry weight and changes in the nutritional composition of the crop (Dürr et al. 2018). The decrease in seed germination at elevated temperatures is associated with a direct influence on plant metabolism, specifically the activation of specific enzymes within the optimal temperature range (Belmehdi et al. 2018).

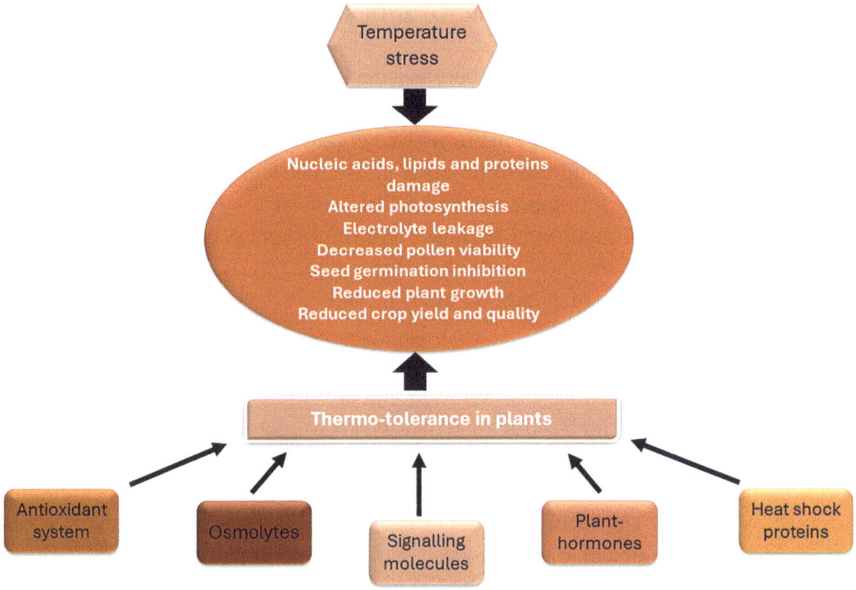

Fig. 8.1 Impact of temperature on the physiological, biochemical and molecular state of the plant. (Adapted from Chaudhry and Sidhu 2022)

Similarly, plant metabolism is negatively affected by low temperatures, resulting in photosynthetic activity reduction, foliar growth and premature senescence above freezing point temperatures (Huner et al. 1993). Temperatures below freezing might impede bud development since the rehydrated and developing buds may be damaged by cold stress (Mishra et al. 2012). Low temperatures may cause dehydration of plant tissues, increasing the concentration of osmolytes inside the cell cytoplasm and, eventually, the collapse of the plasma membrane (Mishra et al. 2012).

Generally, heat stress has a separate and distinct effect on the physiology and metabolism of plants; though, its impact is amplified when paired with other abiotic factors like drought and salt stress. Zinn et al. (2010), observed that severe temperature stress has a higher influence on the reproductive stage of crop species relative to the vegetative stage, leading to a fast decrease in yield owing to pollen infertility. Heat stress influences both the vegetative and reproductive stages of plants; however, the extent of this impact differs across different species (Bita and Gerats 2013). Heat stress induces the destruction of cellular organelles, destabilizes proteins, causes membrane breakdown, and disrupts metabolism. These effects result in the excessive formation of reactive oxygen species (ROS) and contribute to severe oxidative stress in plants (Fig. 8.1) (Bita and Gerats 2013; He et al. 2017). Hence, it is essential to comprehensively fathom the connection between temperature and water to devise more efficient adaptation techniques to mitigate the consequences of heightened temperature extremes linked to a changing climate.

8.2.1.2 Impact of Drought on Plant Growth and Development

Plants undergo drought stress when there is a decline in the quantity of water available in the soil due to water loss via evaporation and transpiration owing to the prevailing climatic conditions (Mishra and Cherkauer 2010; Fadiji et al. 2022b). Drought is a very damaging environmental stress that significantly impacts plant production, growth rate, biomass, and yield (Anjum et al. 2011). Drought stress disrupts normal plant functioning by impacting turgor and water potential (Nawaz et al. 2015), which subsequently leads to changes in the physical, physiological, and molecular characteristics of plants by altering their fresh weight and water status (Vurukonda et al. 2016).

Drought is a significant hindrance to plant development and includes the structural and physical interaction between ROS and abscisic acid (ABA) (Jones 2016). Drought stress induces the buildup of ABA, which triggers the production of H_2O_2 in plants when applied externally. To ensure that water is available in the aboveground sections of a plant, a synchronized series of signals is necessary, extending from the roots to the shoots of the plant (Jones 2016). In stressful conditions, roots are the first organ to detect water presence and transmit signals to the aerial organs via xylem channels in the roots, causing the guard cells to become less rigid and the stomatal opening to shut, so reducing loss of water (Takahashi and Shinozaki 2019).

Drought stress causes stomata to close, resulting in a reduction in the rate of photosynthesis in plants. In addition, decreased water availability causes a decrease in cell volume, which subsequently increases the thickness of cellular contents. This leads to the impairment of proteins and enzymes essential for photosynthesis in plants, resulting in damage (Impa et al. 2012). Drought stress causes stomata to partially close to decrease water loss and prevent the entrance of CO_2, leading to H_2O_2 formation in peroxisomes and mitochondria via the process of photorespiration, ultimately limiting photosynthesis in plants (Miller et al. 2010; Noctor et al. 2014). Hence, future studies should prioritize the adoption of biotechnological and molecular techniques to create genetically modified plants that possess heightened resistance to this specific stressor.

8.2.1.3 Impact of Salinity on Plant Growth and Development

Salt or salinity stress refers to the detrimental impact of excessive mineral salts, such as Na^+ and Cl^-, on plants (Parihar et al. 2015). More than 6% of all land on earth is salinized, which results in 22–33% of all cultivated and irrigated agricultural land having lower crop yield (Ali et al. 2022). According to Sahile et al. (2021), soil salinity will pose a hazard to almost half of arable land by 2050. It is a critical factor which control the growth of plants and is projected to deepen owing to global climate change and increasing irrigation activities (Rengasamy 2010). Regions characterized by aridity with little rainfall, high rates of evaporation and transpiration, elevated temperatures, and limited water supply are more prone to being impacted by salt stress (Nawaz et al. 2010). The presence of water-soluble salts like sodium chloride (NaCl) and sodium carbonate (Na_2CO_3), as well as partly water-soluble ones like calcium chloride ($CaCl_2$), leads to soil salinity (Nawaz et al. 2010). Soil is considered salty when its electrical conductivity (EC) is 4 Ds m^{-1} and its osmotic pressure is 0.2 MPa, which has a significant impact on plant metabolism (Munns and Tester 2008), resulting in chlorosis and necrosis in plants (Acosta-Motos et al. 2017).

Plants are categorized based on their adaptation mechanism to flourish in saline soil as either glycophytes, which are unable to resist salt and eventually die or halophytes which can handle salinity (Acosta-Motos et al. 2017). Most plants are glycophytes, and their development is significantly reduced or totally halted when exposed to salt concentrations of roughly 100–200 mM NaCl while halophytes can tolerate high concentrations of NaCl ranging from 300 to 500 mM. This tolerance may be related to the adoption of improved mechanisms by plants (Flowers and Colmer 2015; Acosta-Motos et al. 2017).

The occurrence of an excessive quantity of salts in water hinders plant development either by reducing the intake of water, resulting in water deficiency in plants, or by flowing into the transpiring stream and damaging cells, leading to abundance of ions in the plant (Parihar et al. 2015). High salt content was observed to cause decreased growth in *Ulva pertusa* and may be linked to lower turgor pressure, which limits cell division (Choi et al. 2010) like the findings on the development of *Glycine max* and *Vicia faba* when exposed to increased salt concentrations (Amirjani 2010; Qados 2011). Furthermore, the reduced plant growth caused by salt stress may be attributed to the adverse impact of salt on the photosynthetic rate and enzyme activity, which in turn influences the production of protein required for growth (Qados 2011). Plants detect an increase in Na^+ ions, which causes them to quickly close their stomata and prevent the growth of leaves within minutes of exposure (Negrão et al. 2017). Subsequently, plants experience early leaf ageing and a decrease in yield owing to an excessive buildup of ions (Negrão et al. 2017). Hence, it is essential to comprehensively grasp the fundamental metabolic and biochemical structures that establish the physiological and phenotypic response to salinity stress.

8.2.1.4 Impact of CO_2 on Plant Growth and Development

According to the IPCC (2014), the atmospheric CO_2 content has significantly risen from the pre-industrial period and is estimated to continue to grow to around 730–1000 ppm by the end of the twenty-first century. The rise in atmospheric CO_2 levels is linked to a sharp increase in average global temperatures caused by the greenhouse effect, which will result in substantial alterations of the global climate (Balasooriya et al. 2018). Research has shown that increased levels of CO_2 have a direct impact on photosynthesis, gaseous exchange, and other developmental activities in plants (Albert et al. 2011; Gray and Brady 2016; Katul et al. 2010). Plants are categorized into C3 and C4 plants based on their photosynthetic ability. The C3 plants such as wheat, rye, oats use the C3 pathway by forming a 3-carbon compound; 3-phosphoglycerate while the C4 plants which include maize, sugarcane produce the 4-carbon compound; oxaloacetate which draws more CO_2 from the atmosphere by suppressing photorespiration having reduced oxygen requirements (Biswal et al. 2018; Talukder et al. 2024). Increased CO_2 levels have a beneficial impact on photosynthesis in C3 plants, leading to enhanced plant development. The increase in photosynthesis due to higher levels of CO_2 is linked to the enhanced concentration of CO_2 at the RuBisCO site that restricts the accessibility of NADPH and ATP for photorespiration, making plants more efficient in assimilating CO_2 (Hay et al. 2017).

Previously, Kaiser et al. (2017) observed that when atmospheric CO_2 levels were increased to 800 μbar, there was a 12% increase in relative carbon gain, a 17% increase in photosynthesis, and a 14% decrease in the loss of photosynthetic induction relative to levels of 400 μbar in *Solanum lycopersicum* owing to an elevated rate of carboxylation and a decreasing rate of oxygenation reaction at RuBisCO. The increased net photosynthetic rate resulting from higher atmospheric CO_2 levels may stimulate lettuce development by providing more carbon skeletons, leading to a greater buildup of carbohydrates (Pan et al. 2020; Pérez-López et al. 2013). Additionally, Pan et al. (2020) found that elevated levels of CO_2 enhance the maximal photosynthetic rate, noticeable quantum efficiency, and carboxylation efficiency in *S. lycopersicum*.

Furthermore, sugarcane, a C4 plant, benefited from higher CO_2 levels in terms of biomass, photosynthesis, leaf area, stalk juice volume, leaf dry weight, and stem dry weight (Misra et al. 2019). Their superior performance may be attributable to their high optimal growth temperature, compensating capacity, very low CO_2 compensation point, and distinctive carbon sequestration method. Increased CO_2 boosted photosynthetic rate, intrinsic water usage efficiency, seed and tiller quantity, spike and stem weight, above-ground biomass, and grain yield in plants, as reported in *Setaria italica* (C4 plant) (Li et al. 2019). Genes involved in response to elevated CO_2 levels in foxtail processes, including cell wall reinforcement, shoot initiation, stomatal conductance, carbon fixation, and glycolysis/gluconeogenesis, were identified through transcriptome analysis (Li et al. 2019). *Cenchrus pedicellatus* also showed decreased stomatal conductance, root-to-shoot ratio, and protein content when exposed to high concentrations of CO_2 (Tom-Dery et al. 2018). Given the interdependence of respiration and photosynthesis, it is crucial to comprehend the metabolic fluxes between chloroplasts and mitochondria when CO_2 levels rise. This understanding is essential for developing strategies to safeguard respiration-related activities, such as nutrient uptake and transport, without compromising them due to excessive carbon fertilization.

8.2.1.5 Impact of Heavy Metals on Plant Growth and Development

Heavy metals in nature exist either in very low concentrations or are sequestered in rocks until they are released, used, and accumulated because of industrial activities. An excessive accumulation of heavy metals in soil may render it hazardous to plants. The primary contributors to heavy metal pollution in soil have been mining activities, exhaust emissions, sewage irrigation, and the utilization of metal products (He et al. 2018). The prevalence and severity of heavy metal pollution in agrosystems have significantly increased over the last decades globally.

Heavy metals are typically conveyed into roots via the plasma membrane using specific transporter proteins and then further transported to the xylem or phloem through extracellular plastid or coplanar pathways, except for cadmium (Cd), which does not appear to have specific transporter proteins (Zhao et al. 2023). Stressed plants exhibit various abnormal symptoms when exposed to excessive heavy metals, including reduced seed germination, reduced plant growth, leaf yellowing, chlorosis, necrosis, and stunted roots. These symptoms are caused by the disruption and intrusion of heavy metals on plant growth, development, and reproduction (Liu et al. 2015). Heavy metals have negative effects on plant physiological metabolisms. They cause the deactivation and denaturation of intracellular enzymes, alter protein structures, and disrupt membrane integrity. As a result, normal plant metabolic activities like photosynthesis, respiration, and transpiration are affected (Liu et al. 2015). The buildup of certain heavy metals negatively influences the texture and pH of the soil. Thereafter, crop development is inhibited by the modified pH and texture of the soil, which negatively impacts various biological activities in plants. According to recent findings, heavy metal stress boosts the secretion of enzymes that scavenge ROS, denatures structures of the cells structurally, and changes the inhibition of cytoplasmic enzymes (Ali et al. 2022). Therefore, an elevated concentration of superoxide anion (O^{2-}) and hydrogen peroxide (H_2O_2) results in more oxidation of lipid membranes, breakdown of biological macromolecules, increased production of unsaturated fatty acids, and cell destruction by oxidative processes (Wani et al. 2017; Feng et al. 2023). Furthermore, prioritizing research on the variety of microorganisms in the rhizosphere and soil is necessary because of their direct impact on the solubility of heavy metals.

8.2.2 *Biotic Factors*

Viruses, bacteria, fungi, nematodes, insects, arachnids and weeds are the living organisms that may induce biotic stress in plants. They directly deplete the nutrients of their host, which may ultimately lead to the death of plants. Biotic stress may have a significant impact due to losses occurring before and after harvesting (Gull et al. 2019). For instance, viruses are obligatory intracellular pathogens that have a wide-ranging impact on almost all biological organisms (Guo et al. 2019a). Plant viral infections portend a critical danger to global agriculture, resulting in substantial reductions in crop quality and production. Common plant viral infections include Potato virus X, Tobacco mosaic virus, Potato virus Y, Cucumber mosaic virus, Tomato yellow leaf curl virus, Cauliflower mosaic virus, Plum pox virus, Brome mosaic virus, and Tomato spotted wilt virus (Scholthof et al. 2011). Fungal pathogens are a significant obstacle to the development of high-quality and abundant crops. The most efficient and uncomplicated means of management are the adoption of fungicides and the utilization of resistant types to reduce the harshness of the disease (Shephard 1987; Dubin and Rajaram 1996).

Furthermore, insects exhibit a remarkable level of biodiversity, surpassing that of other animal species on earth with over a million described species, and they account for more than half of all known living creatures (Chapman 2009). They have successfully evolved to thrive in many terrestrial environments, which is one of the factors contributing to their increased abundance compared to other animals. Only a small percentage (<0.5%) of total insect species are recognized as pests, and some of them could be a direct hazard to humans and agriculture (Sallam 2008). Insects decimate around 20% of the global agricultural food output by leaf-chewing, plant juice absorption, root, fruit, stem, leaf boring and the transmission of numerous plant diseases (Aetiba and Peter 2015; Sallam 2008). In addition, people have chosen agricultural plants that possess traits such as increased productivity, larger size, and greater nutritional value. These plants not only meet human dietary requirements but attract herbivorous insects and offer a favourable condition for their proliferation and maturation. In addition, insect pests possess the capacity to acquire tolerance to poisonous substances or evade the natural or manufactured resistance of plants, therefore exacerbating the issue (Dos Santos et al. 2015; Damalas 2009; Korth et al. 2006). Undoubtedly, pesticides have enhanced the productivity of crops; nonetheless, their widespread and uninterrupted use is accountable for the rapid emergence of resistance in several insects.

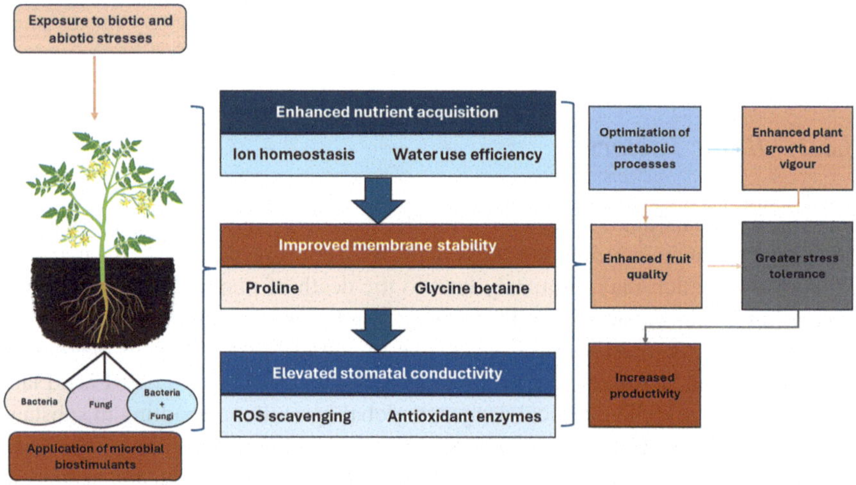

Fig. 8.2 Overview of the major mechanisms adopted by MBs in mitigating biotic and abiotic stresses. (Adapted from Aamir et al. 2020)

8.3 Overview of the Mechanistic Impact of MBs in Mitigating Stress in Plant

Plant growth-promoting microorganisms (PGPM), including rhizobia, mycorrhizae, and plant growth-promoting bacteria, have been shown to improve plant development in both stressed and non-stressed environments. Researchers have studied the capacity of microorganisms to reduce the impact of both biotic and abiotic stresses on plants to a large extent with interesting results (Naamala and Smith 2020). In this section, an overview of the mechanisms employed by PGPM (Fig. 8.2) to reduce the effect of both biotic and abiotic stress on plants is discussed.

8.3.1 Role of MBs in Mitigating Abiotic Stress

Abiotic stresses like drought, salinity, extreme temperatures, and heavy metals can negatively affect plant growth and productivity, as highlighted earlier; hence, this section provides an overview of the impact of MBs in mitigating abiotic stress.

8.3.1.1 Flooding, Water Stagnation, and Heavy Rainfall

Two of the biggest issues resulting from the present climate change are interannual rainfall and seasonal variability in rainfall patterns (Stevenson et al. 2012). About 13% of the earth's surface is prone to flooding, and extreme rainfall frequency and severity are expected to increase globally in the future (Sangiorgio et al. 2020). Intense rains and floods can result in root anoxia, water stagnation and hypoxia. When roots are flooded, they create large amounts of the enzyme ACC synthase, which is known to be implicated in the synthesis of ethylene (Sangiorgio et al. 2020). In order for the enzyme ACC oxidase to catalyse the last phase of ethylene biosynthesis, it needs oxygen. ACC is conveyed to the aerial region of the plant through the xylem, where it is converted to ethylene, causing wilting, chlorosis, leaf necrosis, loss of fruit and flowers and limited growth (Glick 2014).

MBs can assist in stress reduction linked with water stagnation because they inhibit endogenous ethylene through the action of ACC deaminase (Ali and Kim 2018). For instance, tomato seedlings inoculated with *Enterobacter* and *Pseudomonas* strains that produce ACC deaminase exhibit improved germination and decreased anoxia stress (Glick 2014; Sangiorgio et al. 2020). Although *Enterobacter* and *Pseudomonas* spp. are included in the commercial product SumaGrow R (RRR Supply Inc., Munger, MI, United States), it does not specifically guarantee protection against waterlogging stress. Instead, it boosts yield and enhances stress tolerance. While ethylene levels were decreased, plant elongation, chlorophyll content, leaf area, adventitious root development, and biomass were all enhanced. *Pseudomonas* sp. was used on cucumber seeds (Li et al. 2013; Fadiji et al. 2022a), while *Streptomyces* sp. GMKU 336 strain was used on *Vigna radiata* (Jaemsaeng et al. 2018). *S. lydicus* WYEC 108 and *Streptomyces* K61, respectively, are the active components of Mycostop R (Verdera Oy, Espoo, Finland) and Actinovate R (Mycorrhizal Applications, Inc., OR, United States) (Fadiji et al. 2022a). Beneficial bacteria like *Streptomyces* spp. are often employed to counteract biotic stressors; however, their use against abiotic stressors is now lacking in research. Future research is required to fully examine the functions of all commercially available items now on the market and to produce new MBs for the reduction of water stress.

8.3.1.2 Salinity Stress

It has been observed that several MBs that promote plant development can help different crops become more resilient to salt stress. Salinity stress is lessened by plant growth-promoting biostimulants via a variety of synergistic processes, such as improved photosynthesis, enhanced nutrient absorption, phytohormone signalling, and osmotic control. In addition, during salt stress, MBs raise chlorophyll content, boost N, P, and K uptake, reduce Na uptake, stimulate several antioxidants, and modulate hormone secretion (Gul et al. 2022; Ali et al. 2022). According to Miceli et al. (2021), applying MBs improved the salt-tolerance of lettuce and tomato plants. Two distinct commercial MBs, TNC BactorrS13 (containing *Bacillus* spp.) and Flortis Micorrize (containing *A. radiobacter, B. subtilis*, and *Streptomyces* spp.), were applied to the seeds before sowing. The susceptibility of seedlings to salt stress was controlled when MBs were employed to improve plant growth-promoting characteristics, including biomass (dry and fresh), leaf number, and area of unstressed seedlings (Miceli et al. 2021). In a similar vein, MBs treated with *Flortis Micorizze* significantly increased their resistance to salt stress. The inoculation of MBs stands as a workable way to enhance the development of *Solanum lycopersicum* and lettuce while mitigating the adverse effects of high-salinity water in field settings and vegetable nurseries. The application of two *Bacillus* isolates to *S. tuberosum* plants by Gururani et al. (2013) resulted in a notable resistance to abiotic challenges such as salt, drought, and heavy metals, as well as improvements in tuber size, productivity, plant height, and leaf number. According to earlier research, under abiotic stress circumstances, every MB examined had positive impacts on the productive and qualitative characteristics of crops (Ali et al. 2022).

8.3.1.3 Drought Stress

Plants with various MBs may be cultivated to increase drought stress resistance and water usage efficiency. According to Begum et al. (2022), MBs usage is a sustainable strategy for encouraging crop development and tolerance towards water-deficient circumstances. MBs often improve drought stress tolerance by secreting various volatile organic compounds (VOCs), ACC deaminase, a range of exopolysaccharides, and phytohormones (Fadiji et al. 2022a). This aggravates modulation of the osmolytes, modifications in root structure, and expresses genes that respond to stress (Moon and Ali 2022). Mannino et al. (2020) also examined the impact of microbial inocula on *S. lycopersicum* (cv. San Marzano-nano) resilience to drought stress conditions. To confirm the effects of the microbial inocula on drought stress tolerance, biochemical stress markers and hormonal profiles were assessed. The tomato plants were injected with a blend of diverse microbial inocula, including a single AMF species, a blend of two PGPB, a blend of three AMF species, and a commercial inoculum. In order to optimize tolerance and plant performance under such stressful conditions, it is important to identify the optimal genotype combination(s) of plants and microbes. They observed that the reactions of the tomato plant to drought differ subject to the microbial inocula (Mannino et al. 2020). Furthermore, Efthimiadou et al. (2020), investigated the effects of MBs on maize grown under Mediterranean conditions (hot, dry summers and cold, rainy winters) utilizing soil and foliar treatment techniques. These MBs included *B. megatherium*, and *B. subtilis*, and *A. chroococcum*, as well as their combinations. The photosynthetic rate, chlorophyll content, and transpiration all increased after *A. chroococcum* treatment. The greatest maize yields were obtained in the soil region of application by plants treated with *B. megatherium* and a combination of *B. subtilis* and *A. chroococcum* (1:1). MBs are state-of-the-art technologies that can raise agricultural productivity while lessening the detrimental consequences of environmental variations. This is as a result of the increasing request for a sustainable agricultural production approach that will safeguard soil fertility (Castiglione et al. 2021).

8.3.1.4 Temperature Stress

Thermotolerant MBs lessen the impacts of heat stress via the production of several polysaccharides, the development of biofilm on improved water retention and root nodules. Additionally, they generate lipopolysaccharides, EPS, and heat shock proteins that can actively shield plants from the damaging effects of heat stress (Ali and Kim 2019; Kang et al. 2019; Fadiji et al. 2023a). According to Abd El-Daim et al. (2019), the capacity of *B. velezensis* to induce abiotic stress tolerance in wheat is linked to notable molecular and metabolic changes. The seedlings inoculated with *Bacillus* were subjected to cold, drought and heat stressors. *Bacillus* increased the amount of chlorophyll in wheat and the percentage of wheat that survived abiotic stress in all stress scenarios. They concluded that metabolic reprogramming in *Bacillus*-treated seedlings and the increase of numerous common stress metabolites in stressed wheat seedlings were both demonstrated by metabolite analysis using ESI-MS and NMR. Using MBs in an integrated crop management (ICM) approach is a crucial component of contemporary agriculture, enhancing its resilience and sustainability. According to Hamid et al. (2021), MBs support plant development via a number of indirect and direct processes. The acquisition of nutrients, organic acids, siderophores, growth regulators and phytohormones that respond to stress have all been well documented. Through a variety of mechanisms of action, these substances reduce abiotic stressors and stimulate plant defences.

Furthermore, MBs can aid in mitigating the effects of chilling or cold stress. Microorganisms with ice nucleating activity (INAC), which can live on fruits, leaves, and roots, can aid in the formation of ice cores in plants. These bacteria create extracellular polymeric molecules (exopolysaccharides) composed of proteins that act as ice nucleation centres on their cell wall, encouraging the formation of ice crystals (Lee et al. 1995; Fadiji et al. 2024). These microbes are predominantly bacteria, even though ice-nucleating fungi have been reported (Pouleur et al. 1992) and may inhabit the plant both endophytically and epiphytically. Low temperature can restrict plant development; however, MBs that release growth-related hormones (e.g., gibberellins or auxins) or lower ethylene concentration can mitigate its impact on the plant. (Sangiorgio et al. 2020). Several microbes generate indole-3-acetic acid (IAA), an example of auxin (Amara et al. 2015; Fadiji et al. 2023b). For example, *S. marcescens* SRM and *P. dispersa* 1A strains isolated from the Himalayan north-west secrete IAA (Amara et al. 2015). Consequently, wheat seedlings infected with these strains and grown under cold conditions outperformed control seedlings in terms of nutrient absorption and improved yield (Kour and Yadav 2023). Similarly, *Pseudomonas* sp. NARs9 and PGERs-treated wheat seedlings show improved cold tolerance (Mishra et al. 2012). Plant productivity is reduced with exposure to cold stress due to the increment in the synthesis of ethylene (Glick 2014) and *B. phytofirmans* PsJN, which demonstrated ACC deaminase activity, enhanced grapevine cold tolerance by reducing cell membrane damage (Theocharis et al. 2012). Reduced lipid peroxidation and ROS production were observed in bean plants grown in extremely cold climate when injected with psychrophilic ACC deaminase-producing bacteria, such as *Pseudomonas proteolytica, P. fluorescens, P. fragi, B. frigoritolerans*, and *P. chlororaphis* resulting in reduced damage from frost (Tiryaki et al. 2019).

8.3.1.5 Heavy Metals Stress

Several MBs have been shown to enhance crop growth and production while alleviating the harmful impacts of heavy metals. MBs prevent heavy metals from moving around and being saturated in different parts of crops. Through adsorption, precipitation and chelation, they modify metabolism. MBs release EPS compounds with many anion-binding sites, which assist in the biosorption process of removing heavy metals from plant rhizospheres (Ain et al. 2016; Abdel Latef et al. 2020).

MBs technologies have the possibility to be effective substitutes for conventional techniques in the extraction of certain heavy metals from agricultural soils. The potential of heavy metal-stressed spinach plant bioremediators was investigated by Desoky et al. (2020) through their study of heavy metal-resistant bacteria isolated from industrial effluents. Three isolates were chosen for further study due to their high resistance to heavy metals. The isolates were determined to be *Paenibacillus jamilae* DSM, *B. subtilis* subsp. spizizenii DSM and *P. aeruginosa* DSM. The results showed that the application of heavy metal-resistant bacteria enhanced the

Table 8.1 Overview of MBs involved in the mitigation of abiotic stress

Crops	Microorganisms	Mode of action	References
Mitigation of heat stress			
Tomato	*Glomus* sp	Higher scavenging activity of ROS in the roots and leaves and reduction in lipid peroxidation.	Duc et al. (2018)
Wheat	*A. brasilense* and *B. amyloliquefaciens*	Pre-activation of heat shock proteins and reduction of ROS	Abd El-Daim et al. (2014)
Potato	*P. phytofirmans*	Production of 1-Aminocyclopropane-1-carboxylic acid (ACC) and reduction of hydrogen peroxide (H_2O_2)	Sangiorgio et al. (2020)
Soybean	*B. aryabhatthai* SRBO2	Production of abscisic acid	Park et al. (2017)
Mitigation of cold stress			
Wheat	*Pseudomonas* spp., *S. marcescens SRM* and *P. dispersa* 1A	Production of ACC and indole acetic acid (IAA)	Mishra et al. (2008)
Grapevine	*P phytofirmans*	Production of ACC	Theocharis et al. (2012)
Bean plant	*P. fragi, P. fluorescens, P. proteolytica, B. frigoritolerans,* and *P. chlororaphis*	Lipid peroxidation and reduction of ROS	Tiryaki et al. (2019)
Mitigation of salt stress			
Soybean	*P. simiae*	Quinoline and Glycine max 4-nitroguaiacol and promotion of seed germination.	Vaishnav et al. (2016)
Tomato	*L. adecarboxylata* MO1	Production of ACC deaminase and IAA	Khan et al. (2019)
Tomato	*P. fluorescens* UM270	Production of IAA and VOCs	Hernández-León et al. (2015)
Potato	*Bacillus spp.*	IAA production	Tahir et al. (2019)
French bean	*A. aneurinilyticus* and *Paenibacillus sp.*	Production of ACC deaminase, IAA and VOCs	Gupta and Pandey (2019)
Mitigation of water stress			
Cucumber	*Pseudomonas, Burkholderia* and *Acinetobacter* spp.	Gibberellins production	Kang et al. (2015)
Chikenpea	*B. thuringiensis* AZP2	Production of VOCs	Khan et al. (2018a)
Foxtail millet	*Acinetobacter calcoaceticus* EU-LRNA-72 and *Penicillium* sp. EU-FTF-6	Improved accumulation of proline, glycine sugars, and betaine, Increment in chlorophyll content, reduction in lipid peroxidation	Kour et al. (2020c)
Pepper	*B. licheformis* strain K11	Stress-related genes and protein	Ali and Xie (2020)

(continued)

Table 8.1 (continued)

Crops	Microorganisms	Mode of action	References
Millet	*Penicillium* sp. strain EU-DSF-10 and *S. laurentii* EU-LWT3-69 and	Improved accumulation of proline, sugars, and glycine, betaine, alongside decrease in peroxidation of lipid	Kour et al. (2020a)
Mung bean	*Staphylococcus* sp. Acb13, *Staphylococcus* sp. Acb14 and *Staphylococcus* sp. Acb12	Production of ACC deaminase, IAA, and plant growth promotion.	Jayakumar et al. (2020)
Wheat	*P. libanensis* EU-LWNA-33	Production of ammonia, hydrogen cyanide (HCN), ACC and adaptive phosphorus solubilization	Kour et al. (2020b)
Orange	*F. mosseae*	IAA production	Liu et al. (2018)

spinach's tolerance to stressful conditions and increased the plant's dry and fresh weights, transpiration rate, net photosynthesis, chlorophyll content, relative water content, membrane stability and stomatal conductance index relative to control plants. Conversely, stress caused by lead (Pb) and cadmium (Cd) greatly reduced the previously listed features and inhibited the action of both non-enzymatic and enzymatic antioxidants. These results demonstrate how heavy metals affect plant development and growth with how MBs bioremediate stressor chemicals. Summary of other studies on the abiotic stress mitigation abilities of MBs is presented in Table 8.1.

8.3.2 Role of MBs in Mitigating Biotic Stress

The two main processes that underpin plant resistance to pathogens are systemic induced systemic resistance (ISR) and acquired resistance (SAR). While SAR is defined as the plant's adaptive or acquired resistance, ISR is principally facilitated by beneficial microorganisms via root immunity modulation, root colonization, and specific elicitors production like polysaccharides, siderophores, VOCs, enzymes, and plant hormones, among others (Hamid et al. 2021). A broad range of pests and diseases may be effectively combatted by both SAR and ISR resistance mechanisms (Meena et al. 2022). While a number of investigations (Castiglione et al. 2021; Yu et al. 2022) have demonstrated that PGPR-based MBs affect a variety of biochemical, molecular and physiological processes in plants and aids in their continued existence when attacked by pathogens, the fundamental mechanism of action remains unclear.

According to Yu et al. (2022), the outcomes of broad investigations suggest that the development of resistance to various pathogens, such as bacteria, fungi, and viruses, is dependent on a combination of mechanisms which could function concurrently. The process involves the activation of specific defence response genes and enzymes, such as ROS scavengers/antioxidants [catalase (CAT), glutathione peroxidase (GPX), ascorbate peroxidase (APX), superoxide dismutase (SOD) and peroxidase (POD)], buildup of phytohormones [ethylene (ET), jasmonic acid (JA), auxin and salicylic acid (SA)], chitinases, glucanases, osmolytes, sugars, secondary metabolites and pathogenesis-related (PR) proteins. All of these processes have a direct bearing on the multiplication and growth of pathogens (Guo et al. 2019b; Ebrahimi et al. 2020; da Silva et al. 2021; Luo et al. 2022). Within the broad category of PGPR, *Bacillus* sp. has been identified as a highly effective agent in the management of pathogen-induced damage to a variety of plants, including tomato, cotton, tobacco, banana, rice, cucumber, wheat, watermelon (Dimopoulou et al. 2021; Kazerooni et al. 2021; Luo et al. 2022). Rice plants' defence response is strengthened by *B. amyloliquefaciens* (SN13), a bioprotective agent against *R. solani*, the causal agent of sheath blight disease. The colonized plants exhibited changes in the signalling pathways of MAPKinases (phospholipase D and serine-threonine protein kinases) and phytohormone constituent (elevated levels of ABA, GA and SA), which assisted in limiting the spread of the illness (Kaushal et al. 2023).

In addition to these, the production of secondary metabolites (quinazoline), ROS regulators (Ferric reductase, myoinositol, peroxidase precursor and glutathione S-transferase), sugars like b-D-glucopyranose, fructopyranose, and mannitol was also modulated by SN13 treatment (Srivastava et al. 2016). Furthermore, *Gossypium hirsutum*, or cotton, plants inoculated with a mixture of *Bacillus* species showed increased generation of an allelochemical, JA and gossypol, a phytohormone connected to defence, which consequently decreased beetroot armyworm larval eating. Furthermore, treated cotton plants displayed upregulated expression of genes implicated in the generation of allelochemicals, such as jasmonates (allene oxide synthase) (GhAOS), 12-oxo-phytodienoicacid reductase 3 (GhOPR3) and 13-lipoxygenase (GhLOX1) (Zebelo et al. 2016). The tomato bacterial wilt disease (caused by *R. solanacearum*) was significantly suppressed by another strain of rhizobacterium, *B. amyloliquefaciens* SQRT3. According to Chunyu et al. (2017), the use of SQRT3 enhanced the expression of polyphenol oxidase (PPO), POD, stress marker genes such as proteinase inhibitor 2 (PIN2), which is linked to the pathogenesis-related protein-1a (PR-1a), and JA pathway, which is linked to the omsotin-like and SA pathway, which is linked to the ET pathway. Similarly, *B. velezensis* amplified pepper plants' resistance to *B. cinerea* BC1301, the agent that causes grey mould disease, by inducing antioxidants CAT, SOD, SA-mediated defence and POD signalling genes, which include peroxidase, pathogenesis-related protein-1 (PR1), and non-expressor of pathogenesis-related genes 1 (NPR1). However, Jiang et al. (2018) found no difference in the expression of the TIN1 genes and proteinase inhibitor 2 (PIN2). Another study found that *Bacillus amyloliquefaciens* Ba13 has an anti-pathogenic effect on the tomato yellow leaf curl virus (TYLCV) (Guo et al. 2023). It was discovered that injection of this advantageous microorganism enhanced tomato development by increasing the expression of genes linked to systemic resistance, such as pathogenesis-related protein-1, 2, and 3 (PR2, PR3 and PR1), PAL, b-1,3-glucanase, chitinase, PPO, and POD. A summary of other studies on the biotic stress mitigation abilities of MBs is presented in Table 8.2.

Table 8.2 Overview of MBs involved in the mitigation of biotic stress (plant protection)

Crops	Microbes involved	Pathogens	Microbial Elicitor	References
Cucumber	*B. gladioli*	*C. orbiculare*	Production of exopolysaccharides	Park et al. (2008)
Radicchio	*P. fluorescens* WCS374	*F. oxysporum* f. sp. Raphanin	Lipopolysaccharides and Pigment (pseudobactin)	Leeman et al. (1995)
Peanut	*Pseudomonas* sp. BREN6 and *Bacillus* sp. CHEP5 e	*S. rolfsii*	–	Tonelli et al. (2011)
Bean	*B. subtilis* S499	*B. cinerea*	Cyclic lipopeptides (fengicine and surfactin)	Durrant and Dong (2004) and Sangiorgio et al. (2020)
Tomato	*S. liquefaciens* MG1	*A. alternata*	N-acyl homoserine lactones Probably depending on SA and ET	Noyes et al. (2009)
Cucumber and pepper	*B. amyloliquefaciens* IN937a	*X. axonopodis* pv. vesicatoria	VOCs (3-pentanol)	Choi et al. (2014)
Potato	*R. etli* G12	*G. pallida*	Production of lipopolysaccharides	Reitz et al. (2000)
Cabbage	*Paenibacillus* P16	*X. campestris*	–	Ghazalibiglar et al. (2016)

8.4 Understanding the Synergy Between Fungi and Bacteria as MBs

An emerging area in crop biology is the application of microbial consortia, which facilitates plant microbiome engineering. Microbial consortia are constructed blends of microorganisms that can affect desired aspects of plant development and growth (Kong et al. 2018). Plant growth-promoting bacteria and fungi are thought to be the most promising MBs since they can be inoculated to produce the required effects, as demonstrated by Ruzzi and Aroca (2015). In their natural habitats, plants are linked to diverse bacterial and fungal species, and a range of epiphytic, rhizospheric and endophytic microorganisms mediates their reaction to stress. MBs have been developed using this principle to support agricultural plant growth and development in unfavourable environmental conditions.

To enable broad-spectrum utilization for crop development, MBs are made up of complementary beneficial microbial strains with various mechanisms of action. Before being used, a genetically diverse set of strains that can adapt to different environmental circumstances is chosen (Bradáčová et al. 2019). Kong et al. (2018) provided fundamental information about the dynamics, ecology and structure of interacting microbial species. MBs, as defined by CMC-7, can be either a single microbe or a group of microorganisms from many genera, including *Azotobacter* spp., *Azospirillum* spp., *Rhizobium* spp. and mycorrhizal fungi by describing the function of their synthetic microbial consortia and their uses, Therefore, even in the presence of abiotic stress, the combined benefits of fungal and bacterial biostimulants are critical for the regular development and growth of plants. The term "biostimulant" refers to a broader category of microorganism than "biofertilizer," because MBs can be prepared with a single strain of bacteria or fungi, or a mixture of bacteria and fungi, to induce most natural processes, including increasing nutrient uptake, decreasing water use efficiency, mitigating environmental stresses, and promoting plant growth (Ali et al. 2022). Similar to this, a number of variables, such as the crop, soil, product, and application method, affect how effectiveness of MBs (Ali et al. 2022; Fadiji et al. 2022a, 2023b). Furthermore, some recent studies have shown the possibility of MBs consortia as an agricultural probiotic that have to do with a number of processes, including carbon exchange, improving plant nutrient uptake capacity, and mitigating the damaging effects of various stressors (Woo and Pepe 2018).

8.5 Challenges and Prospects in the Adoption of MBs

The process of developing novel MB is ladened with numerous challenges. Firstly, the business registration procedure is unnecessarily difficult because there is currently no unified global regulation (Backer et al. 2018). Furthermore, the efficiency of the product is heavily dependent on the type of plant onto which it is applied, the growth stage and environmental conditions. An assessment of the interaction between microorganisms and the host plant is required in the development of MBs. An essential need for the effectiveness of MBs is the establishment of favourable and enduring plant colonization. Ultimately, it is crucial to develop the optimal formulation that ensures product effectiveness and conservation while also minimizing the impact of environmental and cultural factors (Bashan et al. 2014). Owing to their capability to form spores, MBs containing Gram-positive bacteria enable the manufacture of powder formulations that have long-lasting stability and are resistant to drying (Tabassum et al. 2017).

While several studies have displayed the effectiveness of using microorganisms to enhance plant development in challenging environments, there is a scarcity of MBs solutions known to target the stressors caused by climate change explicitly. Generally, the expensive production costs and the inconsistent effectiveness observed in the field are significant obstacles to the advancement of MBs products (Nadeem et al. 2014). As a result, there are only a few commercially available products, and their use in horticulture is limited. Therefore, prior to using MBs, many criteria must be taken into consideration, such as soil and crop characteristics, which play a crucial role in determining the suitability of microorganisms for use as biostimulant products as no microorganism can be suitable in all ecosystems (Adesemoye et al. 2009) or on every plant (Finkel et al. 2017), hence, it is important to consider the specific soil properties and crop requirements when selecting microbial strains, as different strains may have varying levels of adaptation to different conditions (Ipek et al. 2014).

Furthermore, reduced biostimulant efficacy may result from competition for nutrients and ecological niche occupancy between native microflora and some microbial strains (Jefwa et al. 2008; Strigul and Kravchenko 2006; Ruzzi and Aroca 2015; Bashan et al. 2016). The application method of the MB should reduce the dispersion or mortality of microorganisms caused by abiotic variables such as UV radiation and temperature (Vejan et al. 2016). Microbial strains that possess numerous plant growth-promoting (PGP) qualities are more desirable than strains with just one PGP trait because they can alleviate many stressors at the same time (Tabassum et al. 2017; Etesami and Maheshwari 2018). Incorporating microbial and plant genetic resources into future crop breeding projects should prioritize the plant's ability to form lasting mutual partnerships with beneficial bacteria. This characteristic is strongly associated with stress tolerance, productivity, and resilience, making it extremely desirable. Simultaneously, an in-depth understanding of microbial roles and mechanisms of interaction could facilitate the identification of targeted biostimulants for a certain crop or cultivar under specific cultural conditions (Sangiorgio et al. 2020) just as transcription factors are essential for controlling the expression of stress-responsive genes (Khan et al. 2018b).

Also, an essential approach to harness the potential of MBs completely involves conducting a thorough analysis of the plant microbial biocoenosis using next-generation sequencing (NGS), including RNA/metagenomic/amplicon sequencing optimization (Fadiji and Babalola 2020). This is crucial to monitor the dynamics of microbial functions in real-time continuously and to develop and optimize synthetic communities of microorganisms. Multiple studies indicate that microorganisms obtained from the microbiome of the host plant have greater effectiveness compared to non-native microbial inoculants (Mazzola and Freilich 2017). Therefore, the identification and analysis of the natural microbiome using advanced high-throughput NGS technologies is crucial for effectively choosing MBs. It is possible to discover bacteria that can survive on plants in challenging conditions by combining population association meta-analysis with NGS technology, (Purahong et al. 2018). Indeed, the survival of these organisms may be attributed to positive selection owing to their favourable impact on plant development and protection (Berendsen et al. 2012).

Furthermore, the examination of the microbiome using functional markers, in addition to taxonomic markers, is essential for comprehending and using plant-microbe ecological interactions (Lemanceau et al. 2017). The use of real-time monitoring methods to enhance the efficiency of microbial inoculants may be employed to improve agricultural practices and circumstances. Presently, real-time PCR techniques enable the comprehensive measurement of microbial functional genes and have the potential to be modified for the purpose of monitoring agroecological functions in the future (Bouffaud et al. 2018; Almario et al. 2013).

Also, integrating multiple microorganisms with different plant growth-promoting (PGP) functions to create synthetic microbial communities offers a promising way to enhance the effectiveness and dependability of MBs. However, constructing such a microbial community poses a substantial limitation (Ahkami et al. 2017). The intricate ecological interactions between microorganisms, such as commensalism and competition, must be thoroughly studied to ensure the survival of the beneficial community (Sangiorgio et al. 2020).

8.6 Conclusion

Taken together, environmental stresses have a negative influence on crop growth, physiology, and biochemical processes, which influence crop output and quality. As a result, they pose a danger to world food security. MBs in the root microbiome promote plant development by controlling the secretion of phytohormones, osmolytes, organic acid generation, nutrient absorption, boosting the antioxidant system and up-regulating stress tolerance genes. Nevertheless, to completely comprehend the process by which MBs mitigate stress tolerance, additional molecular study is necessary. Therefore, to address the problems with global food security, investigations into the formulations of more stress-tolerant MBs and their implementation on the farm will be essential.

The choice of MB product composition significantly influences the effectiveness of the formulation. It is important to note that the biostimulant activity of each formulation, the inherent qualities of the chosen inoculants, their relationships with host plants, and the interactions between PGPM strains and organic matrices might vary. This is because biostimulants vary greatly in their effects and efficacy, and they can meet a wider range of demands, including yield preservation, soil fertility, and the rehabilitation of damaged soil. Owing to the increased genetic heterogeneity of PGPMs, several strategies may be taken into account. To begin with, choosing PGPMs that can develop the greatest number of interactions with rhizospheres and plants across a wide range of environmental circumstances might be beneficial. Furthermore, microorganisms with specific activities, such as improving uptake in conditions where nutrients are scarce, restoration of native plant life in degraded soils, to enhancement of plant tolerance to salinity and drought stress, among others, should be explored. Using the aforementioned strategies, the application of MBs may aid in preparing for future challenges in agriculture. Moreover, many earlier relevant laws did not support the use of microorganisms, thereby discouraging studies in this direction. Interestingly, microorganisms have now been included in fertilizer laws for the first time according to the new European Regulation (EU) 2019/1009; nevertheless, the approved positive list is still rather small. The legislative processes for MBs adhere to many legal regulations that need to be harmonized internationally. The risk and efficacy assessment criteria are highly variable, and there is no precise and unified categorization system. Also, due to the complexity of MBs regulation and the need for a harmonized framework, alongside the commercial registration procedure for MBs requires urgent harmonized international law.

References

Aamir M, Rai KK, Zehra A, Dubey MK, Kumar S, Shukla V, Upadhyay RS (2020) Microbial bioformulation-based plant biostimulants: a plausible approach toward next generation of sustainable agriculture. In: Microbial endophytes. Elsevier

Abd El-Daim IA, Bejai S, Meijer J (2014) Improved heat stress tolerance of wheat seedlings by bacterial seed treatment. Plant Soil 379:337–350

Abd El-Daim IA, Bejai S, Meijer J (2019) *Bacillus velezensis* 5113 induced metabolic and molecular reprogramming during abiotic stress tolerance in wheat. Sci Rep 9:16282

Abdel Latef AAH, Zaid A, Abo-Baker A-BA-E, Salem W, Abu Alhmad MF (2020) Mitigation of copper stress in maize by inoculation with *Paenibacillus polymyxa* and *Bacillus circulans*. Plan Theory 9:1513

Acosta-Motos JR, Ortuño MF, Bernal-Vicente A, Diaz-Vivancos P, Sanchez-Blanco MJ, Hernandez JA (2017) Plant responses to salt stress: adaptive mechanisms. Agronomy 7:18

Adesemoye A, Torbert H, Kloepper J (2009) Plant growth-promoting rhizobacteria allow reduced application rates of chemical fertilizers. Microb Ecol 58:921–929

Aetiba N, Peter J (2015) Field evaluation of levo 2.4 sl (Botanical) for the management of insect pests of eggplant (Solanum melongena L.) and okra (Abelmoschus esculentus L. Moench). Kwame Nkrumah University of Science and Technology, Kumasi, Ghana

Ahkami AH, White Iii RA, Handakumbura PP, Jansson C (2017) Rhizosphere engineering: enhancing sustainable plant ecosystem productivity. Rhizosphere 3:233–243

Ain Q, Akhtar J, Amjad M, Haq M, Saqib Z (2016) Effect of enhanced nickel levels on wheat plant growth and physiology under salt stress. Commun Soil Sci Plant Anal 47:2538–2546

Albert KR, Mikkelsen TN, Michelsen A, Ro-Poulsen H, Van Der Linden L (2011) Interactive effects of drought, elevated CO2 and warming on photosynthetic capacity and photosystem performance in temperate heath plants. J Plant Physiol 168:1550–1561

Ali S, Kim W-C (2018) Plant growth promotion under water: decrease of waterlogging-induced ACC and ethylene levels by ACC deaminase-producing bacteria. Front Microbiol 9:1096

Ali S, Kim W-C (2019) A fruitful decade using synthetic promoters in the improvement of transgenic plants. Front Plant Sci 10:493712

Ali S, Xie L (2020) Plant growth promoting and stress mitigating abilities of soil born microorganisms. Recent Pat Food Nutr Agric 11:96–104

Ali S, Khan AL, Ali L, Rizvi TS, Khan SA, Hussain J, Hamayun M, Al-Harrasi A (2017) Enzyme inhibitory metabolites from endophytic Penicillium citrinum isolated from Boswellia sacra. Arch Microbiol 199:691–700

Ali S, Moon Y-S, Hamayun M, Khan MA, Bibi K, Lee I-J (2022) Pragmatic role of microbial plant biostimulants in abiotic stress relief in crop plants. J Plant Interact 17:705–718

Almario J, Moënne-Loccoz Y, Muller D (2013) Monitoring of the relation between 2, 4-diacetylphloroglucinol-producing Pseudomonas and Thielaviopsis basicola populations by real-time PCR in tobacco black root-rot suppressive and conducive soils. Soil Biol Biochem 57:144–155

Amara U, Khalid R, Hayat R (2015) Soil bacteria and phytohormones for sustainable crop production. In: Maheshwari DK (ed) Bacterial metabolites in sustainable agroecosystem. Springer, Cham, pp 87–103

Amirjani M (2010) Effect of salinity stress on growth, mineral composition, proline content, antioxidant enzymes of soybean. Am J Plant Physiol 5:350–360

Anjum SA, Xie X, Wang L, Saleem MF, Man C, Lei W (2011) Morphological, physiological and biochemical responses of plants to drought stress. Afr J Agric Res 6:2026–2032

Asaf S, Hamayun M, Khan AL, Waqas M, Khan MA, Jan R, Lee I-J, Hussain A (2018) Salt tolerance of Glycine max. L induced by endophytic fungus Aspergillus flavus CSH1, via regulating its endogenous hormones and antioxidative system. Plant Physiol Biochem 128:13–23

Babalola OO, Fadiji AE, Enagbonma BJ, Alori ET, Ayilara MS, Ayangbenro AS (2020) The nexus between plant and plant microbiome: revelation of the networking strategies. Front Microbiol 11:548037

Backer R, Rokem JS, Ilangumaran G, Lamont J, Praslickova D, Ricci E, Subramanian S, Smith DL (2018) Plant growth-promoting rhizobacteria: context, mechanisms of action, and roadmap to commercialization of biostimulants for sustainable agriculture. Front Plant Sci 9:402666

Balasooriya HN, Dassanayake KB, Seneweera S, Ajlouni S (2018) Interaction of elevated carbon dioxide and temperature on strawberry (Fragaria × ananassa) growth and fruit yield. Int J Agric Biosyst Eng 12:279–287

Bashan Y, De-Bashan LE, Prabhu S, Hernandez J-P (2014) Advances in plant growth-promoting bacterial inoculant technology: formulations and practical perspectives (1998–2013). Plant Soil 378:1–33

Bashan Y, De-Bashan LE, Prabhu S (2016) Superior polymeric formulations and emerging innovative products of bacterial inoculants for sustainable agriculture and the environment. In: Agriculturally important microorganisms: commercialization and regulatory requirements in Asia. Springer

Begum N, Wang L, Ahmad H, Akhtar K, Roy R, Khan MI, Zhao T (2022) Co-inoculation of arbuscular mycorrhizal fungi and the plant growth-promoting rhizobacteria improve growth and photosynthesis in tobacco under drought stress by up-regulating antioxidant and mineral nutrition metabolism. Microb Ecol 83:971–988

Belmehdi O, El Harsal A, Benmoussi M, Laghmouchi Y, Senhaji NS, Abrini J (2018) Effect of light, temperature, salt stress and pH on seed germination of medicinal plant Origanum elongatum (Bonnet) Emb. & Maire. Biocatal Agric Biotechnol 16:126–131

Berendsen RL, Pieterse CM, Bakker PA (2012) The rhizosphere microbiome and plant health. Trends Plant Sci 17:478–486

Biswal AK, Singh AK, Thakur V, Mangrauthia SK, Ponnuswamy R (2018, October) Breeding strategies to convert C3 plants into C4 plants. In: Advanced molecular plant breeding: meeting the challenge of food security (369). CRC Press

Bita CE, Gerats T (2013) Plant tolerance to high temperature in a changing environment: scientific fundamentals and production of heat stress-tolerant crops. Front Plant Sci 4:273

Bouffaud M-L, Renoud S, Dubost A, Moënne-Loccoz Y, Muller D (2018) 1-Aminocyclopropane-1-carboxylate deaminase producers associated to maize and other Poaceae species. Microbiome 6:1–12

Bradáčová K, Florea AS, Bar-Tal A, Minz D, Yermiyahu U, Shawahna R, Kraut-Cohen J, Zolti A, Erel R, Dietel K (2019) Microbial consortia versus single-strain inoculants: an advantage in PGPM-assisted tomato production? Agronomy 9:105

Bulgari R, Franzoni G, Ferrante A (2019) Biostimulants application in horticultural crops under abiotic stress conditions. Agronomy 9:306

Calvo P, Nelson L, Kloepper JW (2014) Agricultural uses of plant biostimulants. Plant Soil 383:3–41

Castiglione AM, Mannino G, Contartese V, Bertea CM, Ertani A (2021) Microbial biostimulants as response to modern agriculture needs: composition, role and application of these innovative products. Plan Theory 10:1533

Chapman AD (2009) Numbers of living species in Australia and the world. Australian Biolical Resources Study

Chaudhry S, Sidhu GPS (2022) Climate change regulated abiotic stress mechanisms in plants: a comprehensive review. Plant Cell Rep 41:1–31

Choi T-S, Kang E-J, Kim J-H, Kim K-Y (2010) Effect of salinity on growth and nutrient uptake of Ulva pertusa (Chlorophyta) from an eelgrass bed. Algae 25:17–26

Choi HK, Song GC, Yi H-S, Ryu C-M (2014) Field evaluation of the bacterial volatile derivative 3-pentanol in priming for induced resistance in pepper. J Chem Ecol 40:882–892

Chunyu L, Weicong H, Bin P, Yan L, Saifei Y, Yuanyuan D, Rong L, Zheng X, Biao S, Qirong S (2017) Rhizobacterium *Bacillus amyloliquefaciens* strain SQRT3-mediated induced systemic resistance controls bacterial wilt of tomato. Pedosphere 27:1135–1146

Compant S, Samad A, Faist H, Sessitsch A (2019) A review on the plant microbiome: ecology, functions, and emerging trends in microbial application. J Adv Res 19:29–37

Da Silva APS, Olivares FL, Sudré CP, Peres LEP, Canellas NA, Da Silva RM, Cozzolino V, Canellas LP (2021) Attenuations of bacterial spot disease Xanthomonas euvesicatoria on tomato plants treated with biostimulants. Chem Biol Technol Agric 8:1–9

Damalas CA (2009) Understanding benefits and risks of pesticide use. Sci Res Essays 4:945–949

De Wolf ED, Isard SA (2007) Disease cycle approach to plant disease prediction. Annu Rev Phytopathol 45:203–220

Desoky E-SM, Merwad A-RM, Semida WM, Ibrahim SA, El-Saadony MT, Rady MM (2020) Heavy metals-resistant bacteria (HM-RB): potential bioremediators of heavy metals-stressed *Spinacia oleracea* plant. Ecotoxicol Environ Saf 198:110685

Dimopoulou A, Theologidis I, Benaki D, Koukounia M, Zervakou A, Tzima A, Diallinas G, Hatzinikolaou DG, Skandalis N (2021) Direct antibiotic activity of bacillibactin broadens the biocontrol range of *Bacillus amyloliquefaciens* MBI600. Msphere 6. https://doi.org/10.1128/msphere.00376-21

Dos Santos M, Junqueira MR, De Sá VM, Zanúncio J, Serrão J (2015) Effect of silicon on the morphology of the midgut and mandible of tomato leafminer Tuta absoluta (Lepidoptera: Gelechiidae) larvae. Invertebr Surviv J 12:158–165

Du Jardin P (2015) Plant biostimulants: definition, concept, main categories and regulation. Sci Hortic 196:3–14

Dubin H, Rajaram S (1996) Breeding disease-resistant wheats for tropical highlands and lowlands. Annu Rev Phytopathol 34:503–526

Duc NH, Csintalan Z, Posta K (2018) Arbuscular mycorrhizal fungi mitigate negative effects of combined drought and heat stress on tomato plants. Plant Physiol Biochem 132:297–307

Dürr C, Brunel-Muguet S, Girousse C, Larmure A, Larré C, Rolland-Sabaté A, Wagner M-H (2018) Changes in seed composition and germination of wheat (Triticum aestivum) and pea (Pisum sativum) when exposed to high temperatures during grain filling and maturation. Crop Pasture Sci 69:374–386

Durrant WE, Dong X (2004) Systemic acquired resistance. Annu Rev Phytopathol 42:185–209

Dutta P, Chakraborti S, Chaudhuri KM, Mondal S (2020) Physiological responses and resilience of plants to climate change. In: New frontiers in stress management for durable agriculture. Springer Singapore, pp 3–20

Ebrahimi S, Eini O, Koolivand D (2020) Arbuscular mycorrhizal symbiosis enhances virus accumulation and attenuates resistance-related gene expression in tomato plants infected with Beet curly top Iran virus. J Plant Dis Prot 127:341–348

Efthimiadou A, Katsenios N, Chanioti S, Giannoglou M, Djordjevic N, Katsaros G (2020) Effect of foliar and soil application of plant growth promoting bacteria on growth, physiology, yield and seed quality of maize under Mediterranean conditions. Sci Rep 10:1–11

Enagbonma BJ, Fadiji AE, Ayangbenro AS, Babalola OO (2023) Communication between plants and rhizosphere microbiome: exploring the root microbiome for sustainable agriculture. Microorganisms 11:2003

Etesami H, Maheshwari DK (2018) Use of plant growth promoting rhizobacteria (PGPRs) with multiple plant growth promoting traits in stress agriculture: action mechanisms and future prospects. Ecotoxicol Environ Saf 156:225–246

Fadiji AE, Babalola OO (2020) Metagenomics methods for the study of plant-associated microbial communities: a review. J Microbiol Methods 170:105860

Fadiji AE, Babalola OO, Santoyo G, Perazzolli M (2022a) The potential role of microbial biostimulants in the amelioration of climate change-associated abiotic stresses on crops. Front Microbiol 12:829099

Fadiji AE, Orozco-Mosqueda MDC, Santos-Villalobos SDL, Santoyo G, Babalola OO (2022b) Recent developments in the application of plant growth-promoting drought adaptive rhizobacteria for drought mitigation. Plan Theory 11:3090

Fadiji AE, Barmukh R, Varshney RK, Singh BK (2023a) Exploring the connectivity between rhizosphere microbiomes and the plant genes: a way forward for sustainable increase in primary productivity. J Sustain Agric Environ 2:424–443

Fadiji AE, Yadav AN, Santoyo G, Babalola OO (2023b) Understanding the plant-microbe interactions in environments exposed to abiotic stresses: an overview. Microbiol Res 271:127368

Fadiji AE, Olowe OM, Akanmu AO, Ayangbenro AS, Glick BR, Babalola OO (2024) Microbe-mediated amelioration of cold stress in plants: mechanisms and challenges. In: Sa T (ed) Beneficial microbes for sustainable agriculture under stress conditions, pp 263–279. https://doi.org/10.1016/B978-0-443-13193-6.00014-2
Feng D, Wang R, Sun X, Liu P, Tang J, Zhang C, Liu H (2023) Heavy metal stress in plants: ways to alleviate with exogenous substances. Sci Total Environ 897:165397
Finkel OM, Castrillo G, Paredes SH, González IS, Dangl JL (2017) Understanding and exploiting plant beneficial microbes. Curr Opin Plant Biol 38:155–163
Flowers TJ, Colmer TD (2015) Plant salt tolerance: adaptations in halophytes. Ann Bot 115:327–331
Garrett KA, Nita M, De Wolf E, Esker PD, Gomez-Montano L, Sparks AH (2021) Plant pathogens as indicators of climate change. In: Climate change. Elsevier
Ghazalibiglar H, Hampton JG, De Jong EVZ, Holyoake A (2016) Is induced systemic resistance the mechanism for control of black rot in Brassica oleracea by a Paenibacillus sp.? Biol Control 92:195–201
Glick BR (2014) Bacteria with ACC deaminase can promote plant growth and help to feed the world. Microbiol Res 169:30–39
Gray SB, Brady SM (2016) Plant developmental responses to climate change. Dev Biol 419:64–77
Gul SL, Moon Y-S, Hamayun M, Khan SA, Iqbal A, Khan MA, Hussain A, Shafique M, Kim Y-H, Ali S (2022) *Porostereum spadiceum*-AGH786 regulates the growth and metabolites production in *Triticum aestivum* L. under salt stress. Curr Microbiol 79:159
Gull A, Lone AA, Wani NUI (2019) Biotic and abiotic stresses in plants. In: Abiotic and biotic stress in plants. IntechOpen, pp 1–19
Guo H, Gu L, Liu F, Chen F, Ge F, Sun Y (2019a) Aphid-borne viral spread is enhanced by virus-induced accumulation of plant reactive oxygen species. Plant Physiol 179:143–155
Guo Q, Li Y, Lou Y, Shi M, Jiang Y, Zhou J, Sun Y, Xue Q, Lai H (2019b) *Bacillus amyloliquefaciens* Ba13 induces plant systemic resistance and improves rhizosphere microecology against tomato yellow leaf curl virus disease. Appl Soil Ecol 137:154–166
Guo Q, Sun Y, Ji C, Kong Z, Liu Z, Li Y, Li Y, Lai H (2023) Plant resistance to tomato yellow leaf curl virus is enhanced by *Bacillus amyloliquefaciens* Ba13 through modulation of RNA interference. Front Microbiol 14:1251698
Gupta S, Pandey S (2019) ACC deaminase producing bacteria with multifarious plant growth promoting traits alleviates salinity stress in French bean (*Phaseolus vulgaris*) plants. Front Microbiol 10:1506
Gururani MA, Upadhyaya CP, Baskar V, Venkatesh J, Nookaraju A, Park SW (2013) Plant growth-promoting rhizobacteria enhance abiotic stress tolerance in *Solanum tuberosum* through inducing changes in the expression of ROS-scavenging enzymes and improved photosynthetic performance. J Plant Growth Regul 32:245–258
Hamid B, Zaman M, Farooq S, Fatima S, Sayyed R, Baba Z, Sheikh T, Reddy M, El Enshasy H, Gafur A (2021) Bacterial plant biostimulants: a sustainable way towards improving growth, productivity, and health of crops. Sustain For 13:1–24
Hatfield JL, Prueger JH (2015) Temperature extremes: effect on plant growth and development. Weather Clim Extr 10:4–10
Hay WT, Bihmidine S, Mutlu N, Le Hoang K, Awada T, Weeks DP, Clemente TE, Long SP (2017) Enhancing soybean photosynthetic CO_2 assimilation using a cyanobacterial membrane protein, ictB. J Plant Physiol 212:58–68
He B, Guo T, Huang H, Xi W, Chen X (2017) Physiological responses of Scaevola aemula seedlings under high temperature stress. S Afr J Bot 112:203–209
He H, Li Y, He L-F (2018) The central role of hydrogen sulfide in plant responses to toxic metal stress. Ecotoxicol Environ Saf 157:403–408
Hernández-León R, Rojas-Solís D, Contreras-Pérez M, Del Carmen Orozco-Mosqueda M, Macías-Rodríguez LI, Reyes-De La Cruz H, Valencia-Cantero E, Santoyo G (2015) Characterization of the antifungal and plant growth-promoting effects of diffusible and volatile organic compounds produced by *Pseudomonas fluorescens* strains. Biol Control 81:83–92

Huner NP, Öquist G, Hurry VM, Krol M, Falk S, Griffith M (1993) Photosynthesis, photoinhibition and low temperature acclimation in cold tolerant plants. Photosynth Res 37:19–39

Impa S, Nadaradjan S, Jagadish S (2012) Drought stress induced reactive oxygen species and anti-oxidants in plants. In: Abiotic stress responses in plants: metabolism, productivity and sustainability. Springer, pp 131–147

IPCC (2014) Climate change 2014: synthesis report. Contribution of Working Groups I, II and III to the Fifth Assessment Report of the Intergovernmental Panel on Climate Change (Core Writing Team, Pachauri RK, Meyer LA, eds). IPCC, Geneva, pp 151

Ipek M, Pirlak L, Esitken A, Figen Dönmez M, Turan M, Sahin F (2014) Plant growth-promoting rhizobacteria (PGPR) increase yield, growth and nutrition of strawberry under high-calcareous soil conditions. J Plant Nutr 37:990–1001

Jaemsaeng R, Jantasuriyarat C, Thamchaipenet A (2018) Positive role of 1-aminocyclopropane-1-carboxylate deaminase-producing endophytic Streptomyces sp. GMKU 336 on flooding resistance of mung bean. Agric Nat Resour 52:330–334

Jayakumar A, Padmakumar P, Nair IC, Radhakrishnan E (2020) Drought tolerant bacterial endophytes with potential plant probiotic effects from *Ananas comosus*. Biologia 75:1769–1778

Jefwa J, Vanlauwe B, Coyne D, Van Asten P, Gaidashova S, Rurangwa E, Mwashasha M, Elsen A (2008) Benefits and potential use of arbuscular mycorrhizal fungi (AMF) in banana and plantain (Musa spp.) systems in Africa. In: IV international symposium on Banana: international conference on Banana and Plantain in Africa: harnessing international 879, pp 479–486

Jiang C-H, Liao M-J, Wang H-K, Zheng M-Z, Xu J-J, Guo J-H (2018) *Bacillus velezensis*, a potential and efficient biocontrol agent in control of pepper gray mold caused by *Botrytis cinerea*. Biol Control 126:147–157

Jones AM (2016) A new look at stress: abscisic acid patterns and dynamics at high-resolution. New Phytol 210:38–44

Kaiser E, Zhou D, Heuvelink E, Harbinson J, Morales A, Marcelis LF (2017) Elevated CO_2 increases photosynthesis in fluctuating irradiance regardless of photosynthetic induction state. J Exp Bot 68:5629–5640

Kang S-M, Radhakrishnan R, You Y-H, Khan AL, Park J-M, Lee S-M, Lee I-J (2015) Cucumber performance is improved by inoculation with plant growth-promoting microorganisms. Acta Agric Scand Sect B Soil Plant Sci 65:36–44

Kang S-M, Khan AL, Waqas M, Asaf S, Lee K-E, Park Y-G, Kim A-Y, Khan MA, You Y-H, Lee I-J (2019) Integrated phytohormone production by the plant growth-promoting rhizobacterium *Bacillus tequilensis* SSB07 induced thermotolerance in soybean. J Plant Interact 14:416–423

Katul G, Manzoni S, Palmroth S, Oren R (2010) A stomatal optimization theory to describe the effects of atmospheric CO2 on leaf photosynthesis and transpiration. Ann Bot 105:431–442

Kaushal P, Ali N, Saini S, Pati PK, Pati AM (2023) Physiological and molecular insight of microbial biostimulants for sustainable agriculture. Front Plant Sci 14:1041413

Kazerooni EA, Maharachchikumbura SS, Adhikari A, Al-Sadi AM, Kang S-M, Kim L-R, Lee I-J (2021) Rhizospheric *Bacillus amyloliquefaciens* protects *Capsicum annuum* cv. Geumsugangsan from multiple abiotic stresses via multifarious plant growth-promoting attributes. Front Plant Sci 12:669693

Khan N, Bano A, Zandi P (2018a) Effects of exogenously applied plant growth regulators in combination with PGPR on the physiology and root growth of chickpea (*Cicer arietinum*) and their role in drought tolerance. J Plant Interact 13:239–247

Khan S-A, Li M-Z, Wang S-M, Yin H-J (2018b) Revisiting the role of plant transcription factors in the battle against abiotic stress. Int J Mol Sci 19:1634

Khan MA, Asaf S, Khan AL, Ullah I, Ali S, Kang S-M, Lee I-J (2019) Alleviation of salt stress response in soybean plants with the endophytic bacterial isolate *Curtobacterium* sp. SAK1. Ann Microbiol 69:797–800

Kim Y-N, Khan MA, Kang S-M, Hamayun M, Lee I-J (2020) Enhancement of drought-stress tolerance of Brassica oleracea var. italica L. by newly isolated Variovorax sp. YNA59. J Microbiol Biotechnol 30:1500

Kong W, Meldgin DR, Collins JJ, Lu T (2018) Designing microbial consortia with defined social interactions. Nat Chem Biol 14:821–829
Korth KL, Doege SJ, Park S-H, Goggin FL, Wang Q, Gomez SK, Liu G, Jia L, Nakata PA (2006) Medicago truncatula mutants demonstrate the role of plant calcium oxalate crystals as an effective defense against chewing insects. Plant Physiol 141:188–195
Kour D, Yadav AN (2023) Alleviation of cold stress in wheat with psychrotrophic phosphorus solubilizing Acinetobacter rhizosphaerae EU-KL44. Braz. J. Microbiol 54(1):371–383
Kour D, Rana KL, Kaur T, Sheikh I, Yadav AN, Kumar V, Dhaliwal HS, Saxena AK (2020a) Microbe-mediated alleviation of drought stress and acquisition of phosphorus in great millet (*Sorghum bicolour* L.) by drought-adaptive and phosphorus-solubilizing microbes. Biocatal Agric Biotechnol 23:101501
Kour D, Rana KL, Sheikh I, Kumar V, Yadav AN, Dhaliwal HS, Saxena AK (2020b) Alleviation of drought stress and plant growth promotion by *Pseudomonas libanensis* EU-LWNA-33, a drought-adaptive phosphorus-solubilizing bacterium. Proc Nat Acad Sci India Sect B Biol Sci 90:785–795
Kour D, Rana KL, Yadav AN, Sheikh I, Kumar V, Dhaliwal HS, Saxena AK (2020c) Amelioration of drought stress in Foxtail millet (*Setaria italica* L.) by P-solubilizing drought-tolerant microbes with multifarious plant growth promoting attributes. Environ Sustain 3:23–34
Kubi HAA, Khan MA, Adhikari A, Imran M, Kang S-M, Hamayun M, Lee I-J (2021) Silicon and plant growth-promoting rhizobacteria Pseudomonas psychrotolerans CS51 mitigates salt stress in Zea mays L. Agriculture 11:272
Kumar A, Verma JP (2018) Does plant—microbe interaction confer stress tolerance in plants: a review? Microbiol Res 207:41–52
Lee MR, Lee RE Jr, Strong-Gunderson JM, Minges SR (1995) Isolation of ice-nucleating active bacteria from the freeze-tolerant frog, *Rana sylvatica*. Cryobiology 32:358–365
Leeman M, Van Pelt J, Den Ouden F, Heinsbroek M, Bakker P, Schippers B (1995) Induction of systemic resistance against Fusarium wilt of radish by lipopolysaccharides of *Pseudomonas fluorescens*. Phytopathology 85:1021–1027
Lemanceau P, Blouin M, Muller D, Moënne-Loccoz Y (2017) Let the core microbiota be functional. Trends Plant Sci 22:583–595
Li J, Mcconkey BJ, Cheng Z, Guo S, Glick BR (2013) Identification of plant growth-promoting bacteria-responsive proteins in cucumber roots under hypoxic stress using a proteomic approach. J Proteome 84:119–131
Li P, Li B, Seneweera S, Zong Y, Li FY, Han Y, Hao X (2019) Photosynthesis and yield response to elevated CO2, C4 plant foxtail millet behaves similarly to C3 species. Plant Sci 285:239–247
Li Y, Zeng J, Wang S, Lin Q, Ruan D, Chi H, Zheng M, Chao Y, Qiu R, Yang Y (2020) Effects of cadmium-resistant plant growth-promoting rhizobacteria and Funneliformis mosseae on the cadmium tolerance of tomato (Lycopersicon esculentum L.). Int J Phytoremediation 22:451–458
Lindner M, Maroschek M, Netherer S, Kremer A, Barbati A, Garcia-Gonzalo J, Seidl R, Delzon S, Corona P, Kolström M (2010) Climate change impacts, adaptive capacity, and vulnerability of European forest ecosystems. For Ecol Manag 259:698–709
Liu S, Yang R, Pan Y, Ma M, Pan J, Zhao Y, Cheng Q, Wu M, Wang M, Zhang L (2015) Nitric oxide contributes to minerals absorption, proton pumps and hormone equilibrium under cadmium excess in Trifolium repens L. plants. Ecotoxicol Environ Saf 119:35–46
Liu C-Y, Zhang F, Zhang D-J, Srivastava A, Wu Q-S, Zou Y-N (2018) Mycorrhiza stimulates root-hair growth and IAA synthesis and transport in trifoliate orange under drought stress. Sci Rep 8:1–9
Lotze-Campen H, Müller C, Popp A, Füssel H-M (2012) Food security in a changing climate. In: Climate change, justice and sustainability: linking climate and development policy. Springer, pp 33–43
Luo L, Zhao C, Wang E, Raza A, Yin C (2022) Bacillus amyloliquefaciens as an excellent agent for biofertilizer and biocontrol in agriculture: an overview for its mechanisms. Microbiol Res 259:127016

Mannino G, Nerva L, Gritli T, Novero M, Fiorilli V, Bacem M, Bertea CM, Lumini E, Chitarra W, Balestrini R (2020) Effects of different microbial inocula on tomato tolerance to water deficit. Agronomy 10:170

Mannino G, Gentile C, Ertani A, Serio G, Bertea CM (2021) Anthocyanins: biosynthesis, distribution, ecological role, and use of biostimulants to increase their content in plant foods—a review. Agriculture 11:212

Mazzola M, Freilich S (2017) Prospects for biological soilborne disease control: application of indigenous versus synthetic microbiomes. Phytopathology 107:256–263

Meena M, Yadav G, Sonigra P, Nagda A, Mehta T, Swapnil P, Marwal A (2022) Role of elicitors to initiate the induction of systemic resistance in plants to biotic stress. Plant Stress 5:100103

Miceli A, Moncada A, Vetrano F (2021) Use of microbial biostimulants to increase the salinity tolerance of vegetable transplants. Agronomy 11:1143

Miller G, Suzuki N, Ciftci-Yilmaz S, Mittler R (2010) Reactive oxygen species homeostasis and signalling during drought and salinity stresses. Plant Cell Environ 33:453–467

Mishra V, Cherkauer KA (2010) Retrospective droughts in the crop growing season: implications to corn and soybean yield in the Midwestern United States. Agric For Meteorol 150:1030–1045

Mishra PK, Mishra S, Selvakumar G, Bisht SC, Bisht JK, Kundu S, Gupta HS (2008) Characterisation of a psychrotolerant plant growth promoting Pseudomonas sp. strain PGERs17 (MTCC 9000) isolated from North Western Indian Himalayas. Ann Microbiol 58:561–568

Mishra PK, Bisht SC, Bisht JK, Bhatt JC (2012) Cold-tolerant PGPRs as bioinoculants for stress management. In: Maheshwari DK (ed) Bacteria in agrobiology: stress management. Springer, Berlin/Heidelberg, pp 95–118

Misra V, Shrivastava A, Mall A, Solomon S, Singh AK, Ansari MI (2019) Can sugarcane cope with increasing atmospheric CO2 concentration? Aust J Crop Sci 13:780–784

Moon Y-S, Ali S (2022) A fruitful decade of bacterial ACC deaminase biotechnology: a pragmatic approach towards abiotic stress relief in plants. Theor Exp Plant Physiol 34:109–129

Munns R, Tester M (2008) Mechanisms of salinity tolerance. Annu Rev Plant Biol 59:651–681

Naamala J, Smith DL (2020) Relevance of plant growth promoting microorganisms and their derived compounds, in the face of climate change. Agronomy 10:1179

Nadeem SM, Ahmad M, Zahir ZA, Javaid A, Ashraf M (2014) The role of mycorrhizae and plant growth promoting rhizobacteria (PGPR) in improving crop productivity under stressful environments. Biotechnol Adv 32:429–448

Nawaz K, Hussain K, Majeed A, Khan F, Afghan S, Ali K (2010) Fatality of salt stress to plants: morphological, physiological and biochemical aspects. Afr J Biotechnol 9(34):5475–5480

Nawaz F, Ahmad R, Ashraf M, Waraich E, Khan S (2015) Effect of selenium foliar spray on physiological and biochemical processes and chemical constituents of wheat under drought stress. Ecotoxicol Environ Saf 113:191–200

Negrão S, Schmöckel S, Tester M (2017) Evaluating physiological responses of plants to salinity stress. Ann Bot 119:1–11

Nephali L, Piater LA, Dubery IA, Patterson V, Huyser J, Burgess K, Tugizimana F (2020) Biostimulants for plant growth and mitigation of abiotic stresses: a metabolomics perspective. Metabolites 10:505

Noctor G, Mhamdi A, Foyer CH (2014) The roles of reactive oxygen metabolism in drought: not so cut and dried. Plant Physiol 164:1636–1648

Noyes PD, Mcelwee MK, Miller HD, Clark BW, Van Tiem LA, Walcott KC, Erwin KN, Levin ED (2009) The toxicology of climate change: environmental contaminants in a warming world. Environ Int 35:971–986

Pan T, Wang Y, Wang L, Ding J, Cao Y, Qin G, Yan L, Xi L, Zhang J, Zou Z (2020) Increased CO2 and light intensity regulate growth and leaf gas exchange in tomato. Physiol Plant 168:694–708

Parihar P, Singh S, Singh R, Singh VP, Prasad SM (2015) Effect of salinity stress on plants and its tolerance strategies: a review. Environ Sci Pollut Res 22:4056–4075

Park K-S, Kloepper JW, Ryu C-M (2008) Rhizobacterial exopolysaccharides elicit induced resistance on cucumber. J Microbiol Biotechnol 18:1095–1100

Park Y-G, Mun B-G, Kang S-M, Hussain A, Shahzad R, Seo C-W, Kim A-Y, Lee S-U, Oh KY, Lee DY (2017) *Bacillus aryabhattai* SRB02 tolerates oxidative and nitrosative stress and promotes the growth of soybean by modulating the production of phytohormones. PLoS One 12:e0173203

Pérez-López U, Miranda-Apodaca J, Muñoz-Rueda A, Mena-Petite A (2013) Lettuce production and antioxidant capacity are differentially modified by salt stress and light intensity under ambient and elevated CO2. J Plant Physiol 170:1517–1525

Pouleur S, Richard C, Martin J-G, Antoun H (1992) Ice nucleation activity in *Fusarium acuminatum* and *Fusarium avenaceum*. Appl Environ Microbiol 58:2960–2964

Purahong W, Orrù L, Donati I, Perpetuini G, Cellini A, Lamontanara A, Michelotti V, Tacconi G, Spinelli F (2018) Plant microbiome and its link to plant health: host species, organs and Pseudomonas syringae pv. actinidiae infection shaping bacterial phyllosphere communities of kiwifruit plants. Front Plant Sci 9:404926

Qados AMA (2011) Effect of salt stress on plant growth and metabolism of bean plant Vicia faba (L.). J Saudi Soc Agric Sci 10:7–15

Rai A, Rai GK, Dubey R (2021) Heat stress and its effects on plant growth and metabolism. In: Abiotic stress tolerance mechanisms in plants. CRC Press

Raza A, Razzaq A, Mehmood SS, Zou X, Zhang X, Lv Y, Xu J (2019) Impact of climate change on crops adaptation and strategies to tackle its outcome: a review. Plan Theory 8:34

Reitz M, Rudolph K, Schroder I, Hoffmann-Hergarten S, Hallmann J, Sikora R (2000) Lipopolysaccharides of *Rhizobium etli* strain G12 act in potato roots as an inducing agent of systemic resistance to infection by the cyst nematode *Globodera pallida*. Appl Environ Microbiol 66:3515–3518

Rengasamy P (2010) Soil processes affecting crop production in salt-affected soils. Funct Plant Biol 37:613–620

Rodríguez M, Canales E, Borrás-Hidalgo O (2005) Molecular aspects of abiotic stress in plants. Biotecnol Apl 22:1–10

Rouphael Y, Colla G (2020) Biostimulants in agriculture. Front Plant Sci 11:511937

Ruzzi M, Aroca R (2015) Plant growth-promoting rhizobacteria act as biostimulants in horticulture. Sci Hortic 196:124–134

Sahile AA, Khan MA, Hamayun M, Imran M, Kang S-M, Lee I-J (2021) Novel *Bacillus cereus* strain, ALT1, enhance growth and strengthens the antioxidant system of soybean under cadmium stress. Agronomy 11:404

Sallam M (2008) Insect damage: damage on post-harvest. AGSI/FAO: INPhO. Available via http://www.fao.org/inpho/content/compend/text/ch02–01.htm. Accessed

Sangiorgio D, Cellini A, Donati I, Pastore C, Onofrietti C, Spinelli F (2020) Facing climate change: application of microbial biostimulants to mitigate stress in horticultural crops. Agronomy 10:794

Scholthof KBG, Adkins S, Czosnek H, Palukaitis P, Jacquot E, Hohn T, Hohn B, Saunders K, Candresse T, Ahlquist P (2011) Top 10 plant viruses in molecular plant pathology. Mol Plant Pathol 12:938–954

Shephard M (1987) Screening for fungicides. Annu Rev Phytopathol 25:189–206

Srivastava S, Bist V, Srivastava S, Singh PC, Trivedi PK, Asif MH, Chauhan PS, Nautiyal CS (2016) Unraveling aspects of *Bacillus amyloliquefaciens* mediated enhanced production of rice under biotic stress of *Rhizoctonia solani*. Front Plant Sci 7:587

Stevenson S, Fox-Kemper B, Jochum M, Neale R, Deser C, Meehl G (2012) Will there be a significant change to El Niño in the twenty-first century? J Clim 25:2129–2145

Strigul NS, Kravchenko LV (2006) Mathematical modeling of PGPR inoculation into the rhizosphere. Environ Model Softw 21:1158–1171

Szparaga A, Kuboń M, Kocira S, Czerwińska E, Pawłowska A, Hara P, Kobus Z, Kwaśniewski D (2019) Towards sustainable agriculture—agronomic and economic effects of biostimulant use in common bean cultivation. Sustain For 11:4575

Tabassum B, Khan A, Tariq M, Ramzan M, Khan MSI, Shahid N, Aaliya K (2017) Bottlenecks in commercialisation and future prospects of PGPR. Appl Soil Ecol 121:102–117
Tahir M, Ahmad I, Shahid M, Shah GM, Farooq ABU, Akram M, Tabassum SA, Naeem MA, Khalid U, Ahmad S (2019) Regulation of antioxidant production, ion uptake and productivity in potato (*Solanum tuberosum* L.) plant inoculated with growth promoting salt tolerant Bacillus strains. Ecotoxicol Environ Saf 178:33–42
Takahashi F, Shinozaki K (2019) Long-distance signaling in plant stress response. Curr Opin Plant Biol 47:106–111
Talukder P, Sinha B, Biswas S, Ghosh A, Banerjee A, Paul S (2024) A study on the prospect of converting C3 Plants into C4 plants. Biocatal Agric Biotechnol 65:103191
Theocharis A, Bordiec S, Fernandez O, Paquis S, Dhondt-Cordelier S, Baillieul F, Clément C, Barka EA (2012) *Burkholderia phytofirmans* PsJN primes *Vitis vinifera* L. and confers a better tolerance to low nonfreezing temperatures. Mol Plant-Microbe Interact 25:241–249
Tiryaki D, Aydın İ, Atıcı Ö (2019) Psychrotolerant bacteria isolated from the leaf apoplast of cold-adapted wild plants improve the cold resistance of bean (*Phaseolus vulgaris* L.) under low temperature. Cryobiology 86:111–119
Tom-Dery D, Eller F, Jensen K, Reisdorff C (2018) Effects of elevated carbon dioxide and climate change on biomass and nutritive value of Kyasuwa (Cenchrus pedicellatus Trin.). J Appl Bot Food Qual 91:88–95
Tonelli ML, Furlan A, Taurian T, Castro S, Fabra A (2011) Peanut priming induced by biocontrol agents. Physiol Mol Plant Pathol 75:100–105
Vaishnav A, Kumari S, Jain S, Varma A, Tuteja N, Choudhary DK (2016) PGPR-mediated expression of salt tolerance gene in soybean through volatiles under sodium nitroprusside. J Basic Microbiol 56:1274–1288
Van Oosten MJ, Pepe O, De Pascale S, Silletti S, Maggio A (2017) The role of biostimulants and bioeffectors as alleviators of abiotic stress in crop plants. Chem Biol Technol Agric 4:1–12
Vejan P, Abdullah R, Khadiran T, Ismail S, Nasrulhaq Boyce A (2016) Role of plant growth promoting rhizobacteria in agricultural sustainability—a review. Molecules 21:573
Vurukonda SSKP, Vardharajula S, Shrivastava M, Skz A (2016) Enhancement of drought stress tolerance in crops by plant growth promoting rhizobacteria. Microbiol Res 184:13–24
Wani AS, Tahir I, Ahmad SS, Dar RA, Nisar S (2017) Efficacy of 24-epibrassinolide in improving the nitrogen metabolism and antioxidant system in chickpea cultivars under cadmium and/or NaCl stress. Sci Hortic 225:48–55
Wassmann R, Jagadish S, Sumfleth K, Pathak H, Howell G, Ismail A, Serraj R, Redona E, Singh R, Heuer S (2009) Regional vulnerability of climate change impacts on Asian rice production and scope for adaptation. Adv Agron 102:91–133
Woo SL, Pepe O (2018) Microbial consortia: promising probiotics as plant biostimulants for sustainable agriculture. Front Plant Sci 9:428205
Yakhin OI, Lubyanov AA, Yakhin IA, Brown PH (2017) Biostimulants in plant science: a global perspective. Front Plant Sci 7:238366
Yu Y, Gui Y, Li Z, Jiang C, Guo J, Niu D (2022) Induced systemic resistance for improving plant immunity by beneficial microbes. Plan Theory 11:386
Zebelo S, Song Y, Kloepper JW, Fadamiro H (2016) Rhizobacteria activates (+)-δ-cadinene synthase genes and induces systemic resistance in cotton against beet armyworm (*Spodoptera exigua*). Plant Cell Environ 39:935–943
Zhao Y, Wang J, Huang W, Zhang D, Wu J, Li B, Li M, Liu L, Yan M (2023) Abscisic-acid-regulated responses to alleviate cadmium toxicity in plants. Plan Theory 12:1023
Zinn KE, Tunc-Ozdemir M, Harper JF (2010) Temperature stress and plant sexual reproduction: uncovering the weakest links. J Exp Bot 61:1959–1968

Chapter 9
Plant Growth Promotion Induced by Phosphate Solubilizing Endophytes from the Tropics

Isadora Camargo Pedrino, Ana Carolina Vasconcelos Fernandes, Alessandra Pereira Fávero, and Paulo Teixeira Lacava

Abstract Recent advances in agricultural biotechnology have highlighted the potential of microbial inoculants as biofertilizers to enhance plant growth and reduce dependency on agrochemicals. Endophytic microorganisms have gained prominence in agriculture for their ability to solubilize and mineralize nutrients, particularly insoluble mineral phosphates. Endophytes, residing within plant tissues without causing evident disease, represent a promising avenue for sustainable agricultural practices. One key mechanism by which plant-associated endophytes promote plant growth is through the solubilization of inorganic phosphate. Endophytic microorganisms play a crucial role in enhancing nutrient uptake by solubilizing phosphate, thereby increasing its availability for plant absorption. These microorganisms include phosphate-solubilizing bacteria and fungi that actively participate in the phosphorus cycle by secreting organic acids and phosphatases, converting insoluble phosphorus into soluble forms. This process facilitates the efficient utilization of phosphorus by plants. Significant research over the past two decades has contributed extensive knowledge on the natural functions of endophytes. Their ability to colonize internal plant tissues underscores their potential as beneficial tools for enhancing crop productivity. Tropical plants harbor a diverse array of endophytic microorganisms, many of which remain unclassified and may represent novel genera and species. This chapter focuses on examples of endophytes, particularly those capable of enhancing plant growth through phosphate solubilization, associated with tropical plants. This approach serves as a sustainable strategy to bolster crop production while minimizing ecological impact in tropical ecosystems.

I. C. Pedrino · A. C. V. Fernandes · P. T. Lacava (✉)
Laboratory of Microbiology and Biomolecules, Department of Morphology and Pathology, Federal University of São Carlos, São Carlos, Brazil

Biotechnology Graduate Program, Federal University of São Carlos, São Carlos, Brazil
e-mail: ptlacava@ufscar.br

A. P. Fávero
Embrapa Pecuária Sudeste, São Carlos, Brazil

O. O. Babalola, A. S. Ayangbenro (eds.), *Microbial Allies*,
https://doi.org/10.1007/978-3-031-90530-8_9

Keywords Biofertilizers · Endophytic microbes · Phosphate solubilization · Plant growth promotion · Sustainable agriculture

9.1 Introduction

Endophytic microorganisms inhabit plant tissues without causing harm to the host plant. These associations can be symbiotic or mutualistic, providing benefits such as enhanced disease resistance, tolerance to stress conditions like herbivore attacks, improved nutrient availability, and increased biodiversity (Golinska et al. 2015). It is theorized that all plants harbor endophytes, and over evolutionary time, microorganisms have adopted certain metabolic pathways from plants to enable the production of phytochemical compounds (Debbab et al. 2013; Deshmukh et al. 2018; Bibi et al. 2020).

The widespread presence of endophytic microorganisms is widely acknowledged, and their potential use in agriculture continues to be the scientific community's interest (Watts et al. 2023). Unlike rhizospheric members residing on root surfaces or phyllosphere ones colonizing aerial plant parts, endophytes establish closer interactions with the host plant. Under certain circumstances, they notably influence the plant's phenotype, resulting in tangible benefits (Vasileva et al. 2019). An example of this influence of endophytes on plant phenotype can be found in the study by Ting et al. (2008), where it was observed that an endophytic isolate from the roots of *Berangan* cv. Intan, belonging to the genus *Serratia*, led to significant increases in four growth criteria: plant height, pseudostem diameter, root mass, and total number of leaves in healthy seedlings. Sriwati et al. (2023) also describe the potential of endophytic microorganisms to beneficially influence the phenotype of plants. Endophytic isolates were used to analyze their potential for promoting plant growth in tomatoes, as well as their influence on *Fusarium oxysporum* control. They determined that the application of isolates from 5 different bacterial genera to tomato seeds reduced disease occurrence and minimized xylem discoloration length, while also enhancing plant stature and root length. The colonization of soil microorganisms within plants significantly impacts plant growth stimulation, leading to distinct improvements: defense against pests through antibiosis, the release of siderophores aiding nutrient absorption, enhanced disease resistance via systemic induction, soil purification through bioremediation, and production of growth hormones fostering plant development (Adeleke and Babalola 2021).

Plant growth relies on a broad spectrum of essential macronutrients and minerals, with phosphorus playing a pivotal role. Besides fostering root development, phosphorus influences tillering and early flowering and plays vital roles in metabolic activities, notably protein synthesis (Lacava et al. 2021). Despite phosphorus's abundant presence in organic and inorganic forms within the soil, its uptake by plants is greatly reduced because it forms complexes with metallic ions in the soil (Rawat et al. 2021). Thus, exploring alternatives to manage this nutrient is crucial, where phosphorus-solubilizing microorganisms emerge as a viable solution. This facilitates enhanced nutrient utilization by crops, enabling them to access less accessible phosphorus fractions in the soil, consequently reducing the reliance on phosphate fertilizers. The intensive use of agrochemicals has disrupted soil microbiota, underscoring the need for sustainable approaches (Silva et al. 2023; Rawat et al. 2021).

Over the past two decades, there has been a notable increase in scientific production investigating endophytic microorganisms and their environmental impact, suggesting that using these microorganisms as inoculants holds promising potential for the advancement of sustainable agriculture (Azevedo and Quecine 2019; Burragoni and Jeon 2021).

Exploring endophytes in tropical plants can be an opportunity to discover new applications and beneficial compounds in agriculture and biotechnology (Tsipinana et al. 2023). Given the large number of unstudied plants, this field remains a substantial and unexplored territory.

9.2 Endophytic Microorganisms and Their Role in Phosphorus Solubilization

Phosphorus (P) is one of the main essential nutrients that often limit plant growth, especially vital to proper plant development in tropical regions (Santana et al. 2016). In soil, phosphorus exists in various chemical forms, including both organic and inorganic phosphorus (Hanyabui et al. 2020). A portion of the phosphorus present in the soil, derived from plant biomass and microbial substances, is considered available to plants. The forms H_2PO_4 and HPO_4^{2-} are the main sources of available phosphorus for plants (Hanyabui et al. 2020). Phosphorus complexation is a natural phenomenon driven by the presence of ions such as aluminum, calcium, and iron in the soil, which form insoluble compounds with phosphorus. This process also occurs during phosphate fertilization, where the phosphorus from phosphate-based fertilizers complexes with these ions present in the soil. This necessitates the continuous application of these fertilizers in crops (Etesami et al. 2021; Zoboli et al. 2016).

The potential of endophytic microorganisms to solubilize phosphate has been documented (Rawat et al. 2021; Wang et al. 2022). Lucero et al. (2021) investigated the effect of inoculating endophytic phosphate-solubilizing bacteria native to peanuts on soybean and maize plants. The results showed substantial increases in phosphorus content in the aerial parts of inoculated plants compared to non-inoculated controls. These findings suggest that endophytic microorganisms were capable of mobilizing phosphate in the soil, thereby increasing phosphorus availability to host plants. Priyadharsini et al. (2017) investigated the ability of the endophytic fungus *Curvularia* sp., isolated from *Parthenium hysterophorus*, to solubilize phosphate in *Cajanus cajan*. The results demonstrated that the isolate is effective in solubilizing

Table 9.1 Commercial bioinoculants for phosphate solubilization and their applications

Commercial bioinoculant	Microorganisms	Application
Biagro Energia	*Bacillus megaterium* *Bacillus subtilis*	Soybean
BiomaPhos®	*Bacillus subtilis* *Bacillus megaterium*	Corn Soybean
Omsugo™ ECO	*Bacillus subtilis* *Bacillus megaterium*	Sugarcane
Omsugo™ P	*Bacillus subtilis* *Bacillus megaterium*	Soybean
SolubPhos®	*Bacillus megaterium* *Burkholderia pyrrocinia*	Corn Soybean

various phosphate sources, resulting in a significant increase in *Cajanus cajan* growth compared to the non-inoculated control.

During the solubilization process, these microorganisms produce organic and inorganic acids, release protons derived from the processing of ammonium into ammonia, and synthesize siderophores. These mechanisms contribute to the release of complex phosphate, increasing its availability to plants (Tomer et al. 2016).

Microbial-based commercial inoculants aimed at phosphate solubilization are already being applied, as cited in Table 9.1. The microorganisms that make up these bioinoculants were already described in the literature as endophytic microorganisms with the potential for phosphate solubilization.

In the study conducted by Lovecká et al. (2023), endophytic isolates from *Miscanthus giganteus*, identified as *Bacillus megaterium*, stood out for their phosphate solubilization potential. Di et al. (2023) elucidated the phosphorus solubilization potential of *Bacillus subtilis* isolated as an endophyte from sugarcane. The strain demonstrated its capability to enhance plant growth, increase essential nutrient content such as nitrogen, phosphorus, and potassium in sugarcane seedlings, and improve agronomic parameters including photosynthetic rate, root development, and chlorophyll content. Adhikari and Pandey (2019) isolated the genus *Burkholderia* sp. as an endophyte from *Taxus wallichiana*, demonstrating a high potential for phosphate solubilization. This bacterial endophyte exhibited efficacy in phosphate solubilization and phytase production across three different substrates: tricalcium phosphate, aluminum phosphate, and iron phosphate. The experiments resulted in solubilization values of 82.76 ± 1.24 μg/mL for tricalcium phosphate, 59.34 ± 0.54 μg/mL for iron phosphate, and 49.53 ± 0.41 μg/mL for aluminum phosphate, indicating a significant phosphate solubilization capacity.

Given the use of these microbial genera in the formulation of bioinputs and the possibility of finding them as endophytes, there is a promising outlook for the development of new, more efficient, and sustainable phosphate fertilizers.

9.3 Endophytic Microorganisms: A Strategy for Reducing Agrochemical Use in Agrobiological Systems

Agricultural development impacts national economic growth by ensuring food supply and primary resources for non-agricultural sectors. It is evident that efficiency in agricultural production, through the judicious application of production factors, plays a crucial role in the structural reconfiguration of the economy. However, the unregulated and improper application of agrochemicals has led to numerous issues for ecosystems (Mitra et al. 2021). Fertilizers and pesticides possess persistent characteristics in soil and the environment, exerting influence over various biotic and abiotic factors. These substances significantly affect soil, microflora, diverse organisms, the ecosystem, and human health. Due to these undesirable properties, there has been a quest for alternatives, culminating in the advent of sustainable agriculture, which has piqued farmers' interest and attracted considerable attention (Baweja et al. 2020).

In this context, biofertilizers are promising alternatives to conventional agrochemical use. Based on biological compounds, these products offer a more sustainable and ecologically viable approach to promoting plant growth and soil health. In addition to nurturing crops, biofertilizers also have the potential to enhance soil biodiversity, minimizing adverse impacts on the environment (Daniel et al. 2022).

Various microorganisms, such as nitrogen-fixing bacteria, cyanobacteria, and bacteria and fungi capable of phosphate solubilization, are widely employed as biofertilizers. Concurrently, microorganisms responsible for phytohormone production also play a fundamental role in formulating biofertilizers. These agents provide plants with essential compounds to stimulate their growth, such as indole acetic acid (IAA), amino acids, and vitamins, while enhancing soil productivity and fertility and maintaining the viability of crops (Lacava et al. 2021; Daniel et al. 2022; Atieno et al. 2020).

Considering the need to obtain biological agents that provide plants with essential compounds to stimulate their growth more sustainably, the use of endophytic microorganisms has already been documented. The endophytic fungus *Peridormospora indica* has been described as a plant growth-promoting agent, as have fungi from the genus *Trichoderma*, such as *T. hamatum, T. harzianum, T. polysporum,* and *T. viride*, which are being explored for the formulation of biopesticides and biofertilizers (Burrogoni and Jeon 2021). Bacterial endophytes are also being documented as having great potential for promoting plant growth, including genera such as *Rhizobium, Bacillus, Klebsiella, Pseudomonas, Pantoea, Methylobacterium, Acinetobacter, Microccocus,* and *Burkholderia* (Rana et al. 2020a, b).

Endophytic microorganisms are microbes that reside inside plant tissues, such as roots, stems, and leaves, either localized or systemically, without causing diseases in plants (Rajamanikyam et al. 2017; Souza and Santos 2017; Toghueo and Boyom 2019). These microorganisms enhance plant resistance against abiotic stresses like drought, salinity, and extreme temperatures. This enhancement is attributed to the presence of these endophytes, which, once established in plant tissues, trigger gene

Table 9.2 Examples of endophytic microorganisms with the ability to solubilize phosphate and their respective host plants

Endophytic microorganisms	Host plant	References
Penicillium sclerotiorum	*Camellia sinensis*	Nath et al. (2015)
Curvularia geniculata	*Parthenium hysterophorus*	Priyadharsini and Muthukumar (2017)
Aneurinibacillus sp. and Lysinibacillus sp.	*Musa sp.*	Matos et al. (2017)
Penicillium sp. and Aspergillus sp.	*Taxus wallichiana*	Adhikari and Pandey (2019)
Acinetobacter guillouiae	*Triticum aestivum*	Rana et al. (2020a, b)
Bacillus sp.	*Zea mays*	Prihatiningsih et al. (2020)
Pestalotiopsis thailandica, Trichoderma atroviride, Trichoderma brevicrassum and Trichoderma scalesiae	*Zingiberaceaus sp.*	Munir et al. (2022)
Bacillus vallismortis and Bacillus tequilensis	*Stevia rebaudiana and Kalanchoe blossfeldiana*	Nagah et al. (2024)

activation and produce metabolites in response to stress (Wang et al. 2003; Khare and Arora 2015; Lata et al. 2018). Just as these microorganisms are involved in increasing resistance against abiotic stresses, they are also associated with enhancing plant resistance to biotic stresses, inducing systemic resistance by activating genes related to plant defense and inhibiting the growth of phytopathogens through antagonistic interactions (Khare and Arora 2015; Kavroulakis et al. 2007).

Endophytic microorganisms have been described as having potential for phosphate solubilization, as described in Table 9.2.

The ability of endophytes to solubilize phosphate is related to metabolic processes. The release of low molecular weight organic acids, such as malic, succinic, fumaric, citric, gluconic, and 2-ketogluconic acids, reduces the rhizosphere's pH. This leads to the binding of cations responsible for phosphorus precipitation, creating soluble complexes with metal ions from insoluble phosphorus compounds such as Ca-P, Al-P, and Fe-P. Moreover, these acids compete for phosphorus adsorption sites in the soil (Shrivastava et al. 2018).

Another way endophytes solubilize phosphate is by releasing mineral-dissolving compounds generated by the activity of extracellular enzymes that facilitate the biochemical mineralization of phosphate. This process also involves phosphate release during substrate degradation, resulting in the biological mineralization of phosphate (Choudhary et al. 2018). Certain endophytic microorganisms, such as those found in the genera *Aspergillus*, *Bacillus*, *Mucor*, *Penicillium, Pseudomonas,* and *Rhizopus*, can produce phosphatases, enzymes that hydrolyze phosphoesters or phosphoranhydrides and remove phosphorus from organic phosphorus compounds (Mengel et al. 2001). Research has shown that microbial phosphatases are more effective in mineralizing organic phosphorus compounds than those produced by plants (Tarafdar et al. 2001).

Given this, endophytic microorganisms, due to their potential to solubilize phosphate, have been recognized as promising agents for creating phosphate-based biofertilizers. Their positive impact has been observed in various tropical crops, ranging from maize and lettuce to canola, chickpea, barley, common bean, sugarcane, wheat, mung bean, and millet (Shrivastava et al. 2018). This recognition reinforces the perspective that these phosphate-solubilizing microorganisms have the potential to be valuable components in formulating phosphate-based biofertilizers, promoting significant benefits in sustainable agriculture.

9.4 Endophytic Microorganisms from Economically Significant Tropical Plants: Potential for Phosphate Solubilization

9.4.1 Phosphate Solubilization by Endophytic Fungi of **Paspalum** *sp.*

The *Paspalum* sp., belonging to the Poaceae family, encompasses a wide range of grasses comprising over 400 species extensively distributed across various tropical countries. In the Brazilian context, more than 200 of these species are native. Species of this genus demonstrate adaptation to environments with specific characteristics, such as seasonally flooded or hot region soils with low fertility, acidity, and dryness. It is notably recognized for its superior quality as forage, presenting outstanding nutritional properties (Novo et al. 2016).

Several countries, such as Paraguay, Bolivia, Southern Brazil, northern Argentina, and north Uruguay, harbor natural grasslands that serve as habitats for native species of the *Paspalum*. The introduction of these native species through cultivation and genetic enhancement may represent a promising opportunity to expand forage options not only in Brazil but also in neighboring countries (Quarin et al. 1997; Catanzaro et al. 2015; Acunã et al. 2011). The utilization of these native *Paspalum* species in pastures and lawns can reduce the risks of ecological imbalances associated with the introduction of pest insects and diseases, as their natural origin renders them more resilient in this context (Acuña et al. 2011). Therefore, these grasses present themselves as a more sustainable agricultural alternative, offering significant benefits in terms of ecological viability in extensive cultivation systems.

Microorganisms that generate benefits when associated with plants have been the target of isolation, multiplication, formulation, and widespread incorporation as an agricultural practice in several nations. This process has significantly contributed to the growth and productivity of plants (Ahemad and Kibret 2014; Castanheira et al. 2017).

The investigation of endophytic microorganisms capable of phosphate solubilization in plants of the genus *Paspalum* represents a promising field in agricultural research. Due to the adaptability of these grasses to challenging conditions such as acidic, dry, and low-fertility soils, the identification and study of these specific microorganisms may have a notable impact on cultivation practices (Novo et al. 2016; Marcon et al. 2018). Endophytic microorganisms, whether bacterial or fungal, demonstrate an ability to solubilize phosphate present in the soil, resulting in greater availability of this essential nutrient for *Paspalum* plants (Tomer et al. 2016). This critical functionality can reduce the need for industrial phosphate fertilizers, contributing to decreased agricultural production costs. Additionally, this practice helps mitigate potential environmental impacts from excessive chemical substance application in the soil.

Yet, there is limited characterization of fungal diversity and their potential for phosphate solubilization in association with *Paspalum* species. There are some examples of studies on the diversity of endophytic fungi associated with *Paspalum*, such as Depetris et al. (2020, 2022), which reported the fungal diversity of isolates from branches and leaves of *Paspalum dilatatum* and *Paspalum notatum* in pastures subjected to intensive grazing in Argentina, identifying genera like *Acremonium* sp., *Alternaria* sp., *Aspergillus* sp., *Bipolaris* sp., *Cercospora* sp., *Cladosporium* sp., *Curvularia* sp., *Fusarium* sp., *Penicillium* sp., and *Trichoderma* sp., and Sikes et al. (2017), who isolated endophytic fungi from *Paspalum notatum* at the Archbold Biological Station in the United States, finding genera such as *Curvularia* sp., *Fusarium* sp., *Helotiadeles* sp., and *Trichoderma* sp. Nevertheless, none of these studies explored the phosphate solubilization potential associated with the endophytic fungi species, and no other reports were found in the literature regarding phosphate solubilization by endophytic fungi of the genus *Paspalum*.

Our research investigated the phosphate solubilization capacity of endophytes from two species of *Paspalum* isolated from leaves and roots. In this context, Pedrino (2023) conducted a study involving the isolation, biochemical characterization, identification, and analysis of the genetic diversity of the endophytic fungal community associated with *Paspalum atratum* and *P. notatum* sourced from the Active Germplasm Bank of *Paspalum* at Embrapa Southeast Livestock (São Carlos, SP, Brazil).

The results showed that out of the 385 endophytic fungal isolates evaluated for *in vitro* phosphate solubilization capacity, approximately 54.80% of the isolates solubilized inorganic phosphate, with 37.14% exhibiting high phosphate solubilization potential (Figs. 9.1 and 9.2). The endophytic strains were identified through sequencing of the ITS region of rDNA. The genera identified as having the highest phosphate solubilization potentials were *Aspergillus* sp., *Colletotrichum* sp., *Fusarium* sp., *Microdochium* sp., *Neosartorya* sp., *Penicillium* sp., and *Phoma* sp. as endophytes of *P. atratum*, and *Aspergillus* sp., *Ceratobasidium* sp., *Colletotrichum* sp., *Curvularia* sp., *Fusarium* sp., *Microdochium* sp., *Penicillium* sp., and *Talaromyces* sp. as endophytes of *P. notatum*.

While information on the phosphate solubilization capabilities of endophytic fungi in Paspalum remains scarce, the preliminary findings suggest encouraging results. Studying these microorganisms within *Paspalum* species is crucial as it unveils potential agricultural alternatives. Understanding their phosphate solubilization potential holds significance in seeking sustainable alternatives to chemical use.

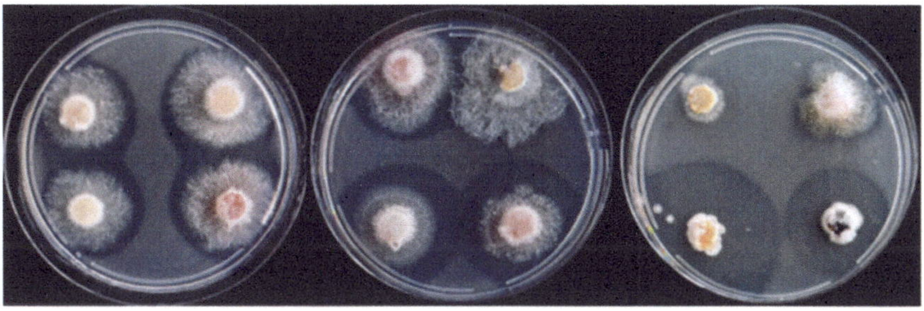

Fig. 9.1 *In vitro* assessment of insoluble phosphate solubilization activity by endophytic fungi. (Source: Author's record)

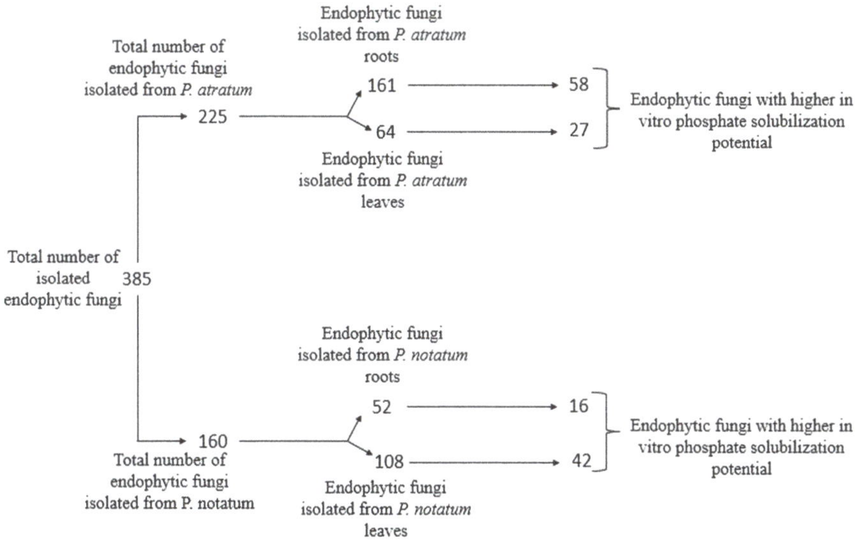

Fig. 9.2 Schematic representation of the selection of endophytic fungi based on their potential for inorganic phosphate solubilization and their plant origin from *Paspalum atratum* and *P. notatum*. (Source: Authored diagram)

This research paves the way for innovative agricultural practices, offering sustainable solutions, reducing reliance on chemical fertilizers, and fostering environmentally conscious cultivation methods.

9.4.2 *Phosphate Solubilization by Endophytic Bacteria of* Paspalum *sp.*

In the context of the diversity of endophytic bacteria in *Paspalum*, they are considered subjects of study, along with fungi, due to their role in plant phosphorus availability.

Our research investigated the phosphate solubilization capacity of endophytes from *Paspalum vaginatum* isolated from leaves and roots. In this context, Fernandes (2023) conducted a study involving the isolation, biochemical characterization, identification, and analysis of the genetic diversity of the endophytic bacteria community associated with *P. vaginatum*, sourced from the Active Germplasm Bank of *Paspalum* at Embrapa Southeast Livestock (São Carlos, SP, Brazil). The results demonstrated that among the 72 endophytic bacterial strains assessed for in vitro phosphate solubilization capacity, approximately 38.90% solubilized inorganic phosphate, with 35.71% exhibiting high phosphate solubilization potential. The endophytic strains were characterized through sequencing of the 16S rDNA region. The genera identified as having the highest phosphate solubilization potentials were *Enterobacte*r sp. and *Pantoea* sp.

Other works were developed by our research group addressing phosphorus-solubilizing bacteria and endophytic bacteria from *Paspalum* sp. De Moraes (2021) isolated 73 endophytic bacteria strains from the leaves and roots of *Paspalum compressifolium*, *P. rojasii*, and *P. lenticulare*, of which 52.94% of strains showed the potential of solubilize phosphate *in vitro*. The genera identified as having the highest phosphate solubilization potentials were *Agrobacterium* sp., *Enterobacter* sp., *Flavobacterium* sp., *Pantoea* sp., and *Pseudomonas* sp.

De Paula (2019) isolated endophytic bacteria from *Paspalum atratum* and *Paspalum notatum*, originating from leaves and roots. Among the 62 bacterial strains isolated, the capacity for phosphate solubilization stood out, with an expressive 67.74% showing significant potential for this activity. Among the genera identified with the highest rates of phosphate solubilization, *Bacillus* sp., *Enterobacter* sp., *Microbacterium* sp., *Micrococcus* sp., *Pantoea* sp., *Pseudomonas* sp., and *Rhizobium* sp. were prominent.

In a subsequent development, de Paula et al. (2022) conducted *in vivo* assays, analyzing endophytic bacteria associated with *Paspalum atratum* and their ability to promote plant growth with different phosphate sources. The results revealed that the most promising isolates belong to the genera *Bacillus* sp., *Enterobacter* sp., and *Pseudomonas* sp., highlighting these lineages' relevance in promoting plant development.

9.4.3 Endophytic Bacteria from Coffea arabica *with Phosphorus Solubilization Potential*

Coffee, an essential agricultural commodity, is cultivated in approximately 80 tropical countries. Originating from tropical regions of South America, Africa, and Asia, the *Coffea* genus encompasses at least 125 species distributed across Africa, Madagascar, Comoros Islands, Mascarene Islands (La Réunion and Mauritius), tropical Asia, and Australia (Krishnan et al. 2017). Two species, *Coffea arabica* and *Coffea canephora*, stand out in beverage production. They are valued for their distinct characteristics: the former for its sweet flavor and the latter for its high caffeine content. This shrub or small tree, a member of the Rubiaceae family, produces fruits known as "coffee cherries," displaying a red or purple coloration. The seeds, notable for their richness in caffeine, have broad applications in beverages, flavorings, and cosmetics (Tsai et al. 2023).

The abundance of endophytic bacteria in coffee plants is well-documented. However, there has been limited exploration of how this bacterial diversity correlates with their ability to enhance plant growth (Pratiwi et al. 2020). The identification and detailed study of microorganisms associated with coffee have highlighted advantageous properties, pointing to their potential as sustainable alternatives to agrochemical inputs (Andrade et al. 2023). These endophytic microorganisms exhibit the capacity to enhance plant growth and show significant potential in the biological control of pests linked to coffee cultivation (Duong et al. 2020).

In Andrade's (2019) study, which aligns with our research focus, the capability for phosphate solubilization by the endophytic bacteria inhabiting Coffea arabica was examined, encompassing both conventional and organic cultivation environments. In evaluating phosphate solubilization capacity by endophytic bacterial isolates, 342 strains were examined *in vitro*, revealing that approximately 64.33% of these isolates demonstrated competence in inorganic phosphate solubilization. The identified genera included *Arthrobacter, Bacillus, Cronobacter, Enterobacter, Erwinia, Klebsiella, Kosakonia, Kurthia, Lysinibacillus, Microbacterium, Pantoea, Pseudomonas,* and *Rhizobium* (Andrade 2019; Andrade et al. 2023).

The most effective endophytic strains, previously selected based on phosphate solubilization tests, were inoculated into coffee seedlings. After 180 days from the initial inoculation, all treatments with inoculated plants, except one, showed a decrease in aerial height (AH) relative to the control group. Regarding root length (RL), all treatments showed a significant average increase of 77.76%. In contrast, stem diameter (SD), aerial dry weight (ADW), and dry weight of the root system (DWRS) did not reveal significant differences between treatments and the control group. Analyses of macro and micronutrients in leaf tissue did not indicate notable differences between treatments and the control. Factor analysis accounted for 99% of the total variance in the observed morphological characteristics. These findings highlight the significant enhancement in coffee seedling growth facilitated by the examined bacterial strains, particularly through efficient phosphate solubilization (Andrade 2019).

Pratiwi et al. (2020) conducted a study to assess the phosphate solubilization potential of bacteria isolated from the roots of *Coffea arabica*. The isolates were tested on different phosphate sources at three distinct time points (24 h, 48 h, and 72 h) (Pratiwi et al. 2020).

Endophytic bacteria were successfully isolated from *Coffea arabica* plants, totaling seven isolates. Each of these bacteria demonstrated the unique ability to solubilize different types of phosphate. The findings indicated that after 48 h of incubation, the SW.P5 isolate from the endophytic roots of *C. arabica* exhibited the highest phosphate solubilization capacity, reaching a concentration of up to 4.55 $\mu g.mL^{-1}$, with statistical significance ($p < 0.05$). In contrast, the SS.P3 and SW.P5 isolates demonstrated lower phosphate solubilization levels compared to the EB14 isolates, which achieved a concentration of 12.54 $\mu g.mL^{-1}$ in the same incubation period. The ability of these bacteria to solubilize phosphate is attributed to the production of organic acids, which decrease the medium's pH. Furthermore, when tricalcium phosphate (TCP) was utilized as a phosphorus source in the culture medium, the bacteria reached a peak phosphate concentration of 764.7 $\mu g.mL^{-1}$. The SS.P3 and SW.P5 isolates were identified as the most proficient in phosphate solubilization, with Rahnella aquatilis B35 (99.9%) and Kluyvera intermedia TPY16 (99.9%) being the respective identities (Pratiwi et al. 2020).

9.4.4 *Bacterial Endophytes of* Jatropha curcas *with Phosphorus Solubilization Potential*

Originating from the tropical and subtropical regions of Central and South America, *Jatropha curcas* was introduced and cultivated in various parts of the world due to its potential as an oil-producing plant (Freitas 2015). The species is recognized for its remarkable adaptability to water stress and degraded soils, standing out for its ability to enhance the structure and quality of the environment in which it is cultivated. This cultivar plays a significant role in the energy, environmental, and socioeconomic domains, and it has been identified as a promising energy crop (Ewunie et al. 2021).

The understanding of *Jatropha curcas* ability to survive in conditions of limited nutrients and adverse environments, as well as the recommendation of its cultivation for bioremediation and soil recovery, is widely disseminated (Mohanty et al. 2017). These issues are intrinsic to its use as a second-generation biofuel source (Abhilash et al. 2011). However, the reason for this remarkable adaptation by this plant has not been fully elucidated, with hypotheses pointing to the role of endophytes present in these plants (Mohanty et al. 2017; Machado et al. 2024).

A study by Mohanty and colleagues (2017) aimed to characterize the cultivable endophytic bacteria of *Jatropha curcas*, especially regarding their potential for promoting growth in corn plants (*Zea mays*). Among the various tests performed, the ability to solubilize phosphate was one of the focus areas. Researchers characterized 15 endophytic isolates for their phosphate solubilization potential, and all isolates yielded positive results. The variation in phosphatase activity, measured in milligrams of PO_4 released per day, ranged from 0.107 to 0.298. The identified genera of the isolates were *Bacillus* sp., *Brevibacillus* sp., *Paenibacillus* sp., *Rhizobium* sp., *Sphingomonas* sp., *Staphylococcus* sp., and *Teribacillus* sp. (Mohanty et al. 2017).

In our study, Machado et al. (2020) conducted a characterization of 72 endophytic bacteria found in association with *Jatropha curcas*. Among the isolates, it was found that 43% displayed proficiency in solubilizing inorganic phosphate. The taxonomy of these endophytic strains revealed identifications in the genera *Bacillus, Citrobacter, Curtobacterium, Enterococcus, Klebsiella, Microbacterium,*

Promicromonosporaceae, Sanguibacter, and *Serratia*. Specifically, the endophytic genera *Bacillus, Citrobacter, Curtobacterium, Klebsiella,* and *Serratia* presented notable phosphate solubilization indices (Machado et al. 2020).

Also, Machado (2019) experimented to assess the impacts of inoculating corn seeds with different endophytic bacterial strains previously isolated from *J. curcas* on the growth parameters of plants. The experimental design included a control group, receiving only a culture medium and various treatments associated with a specific bacterial strain. The plants were cultivated in a greenhouse, receiving irrigation and nutrients as needed. Quantitative assessments were carried out at 30 and 60 days, encompassing measurements such as stem height, stem diameter, shoot dry mass, and root system dry mass. The findings demonstrated that plants treated with the T1 strain (EPM2- *Serratia marcescens*) showed markedly superior performance compared to other strains tested and the control group. This finding suggests that the mentioned strain had a positive and distinctive impact on the development of corn plants.

9.4.5 Endophytic Microorganisms from Tropical Mangrove Forests: Phosphorus Solubilization Potential

The mangroves, coastal ecosystems found in tropical and subtropical regions are characterized by the presence of trees and woody shrubs adapted to saline and waterlogged conditions, such as the genera *Rhizophora, Avicennia,* and *Ceriops*, which have developed aerial roots to facilitate gas exchange (Alongi 2018). These forests thrive in muddy protected areas, where stable conditions promote plant establishment and vigorous growth, protecting against adversities like intense waves and thus fostering robust mangrove development (Friess et al. 2019; Alongi 2018).

It is relevant to note that mangroves reach their maximum prosperity in muddy and protected environments, but their presence is not exclusively limited to these areas. Mangroves can also be identified in coastal ecosystems characterized by rocky and sandy shores, highlighting the remarkable adaptive ability of these forests to survive and thrive in various coastal environments (Biswas and Biswas 2019; Friess et al. 2019).

The notable adaptive ability observed in mangrove ecosystems may be correlated with endophytic microbiota associated with these environments. Due to the high nutrient demand at the primary trophic level in these ecosystems, the crucial role played by such microorganisms stands out, establishing a direct correlation between their presence and the nutritional requirements inherent to these habitats. In this context, endophytes emerge as a subject of interest, given their ability to enhance plant growth and sustain in diverse environmental situations, with phosphorus solubilization being particularly pertinent in this scenario (Quecine et al. 2014; Rukshana-Begum and Tamilselvi 2016).

Studies highlighting the capabilities of endophytic microorganisms collected from tropical mangroves and their beneficial impacts on phosphorus solubilization support the proposed approach (Sebatianes et al. 2017). In this context, Castro et al. (2018) studied endophytic bacteria from the mangrove species *Rhizophora mangle, Laguncularia racemosa*, and *Avicennia nitida* on the Brazilian coast. They obtained 115 bacterial isolates, and the strains were identified for their phosphorus solubilization potential through a quantitative test using $Ca_3(PO4)_2$ substrate in the culture medium. The outcomes were measured by calculating the halo's diameter (cm) relative to the colony's size (cm) to produce a solubilization index (SI). All 115 tested strains exhibited a halo formation in the phosphate solubilization assay, showing the inherent capacity of each strain to dissolve inorganic phosphate. Notably, elevated rates were recorded in the *Pantoea* genera, with *Pantoea agglomerans* standing out, while lower rates were found in the *Bacillus, Ochrobactrum, Sphingosinicella,* and *Xanthomonas* genera, ranging between 1.2 and 2.6 (Castro et al. 2018).

Sawant and Rodrigues (2020) studied the solubilization potential of microorganisms isolated from 12 plant species in the mangrove area of Chorao Island, one of the largest islands in the state of Goa, India. Of the 378 plant segments collected, 217 endophytic fungi were obtained, identified as 37 fungal taxa, representing 15 distinct genera. Of the 37 fungal taxa evaluated, 24 fungal species showed positive activity in solubilizing the phosphorus source in the medium. The analysis focused on the size of clear halo zones on culture plates to evaluate phosphorus solubilization potential in a solid medium. Results revealed a variation in the Phosphorus Solubilization Index (PSI) from 2 to 3.8 cm, with the highest index recorded in *Drechslera* sp. (3.8 cm). In the quantitative phosphorus solubilization test, fungal endophytes demonstrated a reduction in the culture filtrate pH, indicating their ability to solubilize phosphorus. The total amount of phosphorus solubilized by endophytic isolates ranged from 9.70 to 0.1 μg/mL, with *Drechslera* sp. standing out with the highest solubilization (9.70 μg/mL). In liquid medium, a significant reduction in pH (from 3 to 5.7) was observed, indicating the impact of these fungi on phosphorus solubilization (Sawant and Rodrigues 2020).

9.4.6 Solubilization of Phosphate by Endophytes in the Tropical Savanna

Savannas are recognized as extensive biomes in tropical and subtropical regions, characterized by a plant composition that includes grasses, shrubs, and scattered trees. This vegetation demonstrates adaptation to periods of drought, exposure to fire, and nutrient-poor soils with acidic characteristics (Borghetti et al. 2023).

The Brazilian savanna, the cerrado, is Latin America's second-largest biome, covering approximately 200 million hectares. Recognized as a global hotspot for biodiversity conservation due to its remarkable diversity (Sano et al. 2019), this biome is located in the northeastern part of Paraguay, eastern Bolivia, and a significant portion of the Brazilian territory, constituting about 22% of it. The Cerrado harbors a considerable variety of 4400 endemic plant species, including trees, herbaceous plants, shrubs, and vines (Myers et al. 2000)

The limited availability of phosphorus in acidic soils, such as those found in the Brazilian cerrado, and the abundant biodiversity of this biome establish a notable dichotomy, emphasizing the importance of efficient strategies for supplying this vital component in these ecosystems (Silva et al. 2020). In this context, plants play a central role in phosphorus bioavailability, establishing symbiotic associations with microorganisms that can provide these compounds and improve plant resilience to abiotic stresses (Bogas et al. 2020). These interactions are particularly crucial as plants face constant challenges from stressors such as acidic soils, exposure to fire, and water scarcity (Lacava et al. 2021).

In this context, endophytic microorganisms have demonstrated their ability to induce systemic plant resistance against abiotic stresses and are pivotal in the solubilization of phosphorus (Bogas et al. 2020). Consequently, their thematic relevance remains substantial in scientific investigations (de Sousa et al. 2017). In our research group, Torres et al. (2022) studied 66 endophytic fungi isolated from *S. adstringens* and *S. lycocarpum* St. Hill, collected in the Brazilian Tropical Savanna Reserve at the Federal University of São Carlos, Brazil. Healthy leaves and stems of both plants underwent surface sterilization. Following incubation of the plant material, endophytic fungi were isolated and molecularly identified. In vitro evaluation of these strains revealed their capacity to solubilize inorganic phosphate. The strains exhibited promising potential, with average phosphate solubilization indices ranging from 1.46 to 1.93 for those isolated from *S. adstringens* and 1.19–2.61 for those from *S. lycocarpum* (Torres et al. 2022).

Noriler et al. (2018) conducted a study addressing the functional diversity of microorganisms associated with promoting plant growth in the endemic species of the Brazilian cerrado, *Butia purpurascens*. One of the research objectives was to investigate the ability of this plant's root and rhizosphere microbiota to solubilize phosphate. Among the 166 bacterial isolates and 46 fungal isolates evaluated, all bacterial and some fungal isolates demonstrated potential to solubilize calcium phosphate ($CaHPO_4$). While no isolate demonstrated all tested functional traits, 72% exhibited the ability to solubilize $CaHPO_4$. The results highlighted that isolate of the genus Bacillus, particularly *Bacillus subtilis*, showed an adequate capacity for solubilizing calcium phosphate.

9.5 Nanotechnology: Innovations in Developing New Agricultural Bioinputs for Enhanced Phosphorus Solubilization

Nanotechnology involves materials that have unique characteristics due to quantum confinement effects or highly reactive surfaces, both of which are specific to the nanoscale. Compared to the macroscopic scale, the nanoscale brings about distinct and innovative material properties due to the smaller size, increased surface-to-weight ratio, and unique structure of nanomaterials (Roduner 2006; Gutiérrez et al. 2011). Nanomaterials exhibit high reactivity, enhanced bioavailability/bioactivity, adhesive capabilities, and improved surface interactions as advantageous traits (Gutiérrez et al. 2011). Nanoparticles can be made from a single element (such as silver) or a combination of elements, including those found in oxides (TiO2, SiO2, ZnO) (Dinesh et al. 2012; de Moraes et al. 2021).

The challenge of phosphorus scarcity in agriculture has directed attention towards nanotechnology. This field demonstrates the ability to enhance the availability of low-bioavailability elements like phosphorus and zinc, highlighting nanofertilizers as a promising alternative. Hydroxyapatite nanoparticles (HANP) have been studied for their viability as a phosphorus fertilizer. Their characteristic sustained phosphate release and increased mobility of this element compared to triple superphosphate and bulk hydroxyapatite have sparked interest (de Moraes et al. 2021).

These nanoparticles have been proposed as innovative, highly efficient "green" phosphorus fertilizers for modern agriculture, aiming to improve efficacy and reduce adverse environmental impacts (Liu et al. 2014; Wang et al. 2016). Additionally, these nanofertilizers are thought to exert a lesser effect on water quality because of their nanoscale size, large reactive surface area, and minimal tendency to leach (Liu et al. 2014; Wang et al. 2016).

In agriculture, nanotechnology is utilized across various fields, from seed technology to the development of nanobased fertilizers, herbicides, water resource management, nanoscale delivery systems, biosensing devices, agricultural machinery, and livestock science (Koul 2019). Cutting-edge agricultural systems utilizing nanoscale materials facilitate efficient nutrient uptake by crops, accurate and regulated delivery of substances to specific plant parts, early disease identification, and environmental conservation (Suman et al. 2010; Tarafdar et al. 2013; Bhattacharyya et al. 2016). Nanoengineered materials have been applied with agricultural inputs such as nano-fertilizers and nano-pesticides, allowing for the targeted release of substances to plants at lower application rates, thereby reducing soil residue and safeguarding the environment (Varma et al. 2017).

Contemporary agricultural techniques employ nanoencapsulations containing specific bacterial strains known as "smart seeds." These reduce sowing rates, ensure field stand adequacy, and enhance crop yields. Additionally, such smart seeds can be distributed in fields and programmed to germinate according to ideal soil temperature, humidity, and pH conditions (Chinnamuthu and Boopathi 2009; El-Ramady et al. 2018).

Nanostructures represent a modern and promising approach to enhancing the stability and effectiveness of beneficial microorganisms in biofertilizers, acting as vehicles for their precise delivery (Vandergheynst et al. 2007; Ghormade et al. 2011). Embedding microbial organisms within either inorganic or organic nanoparticles is a fundamental part of this method, facilitating the controlled delivery of microbes to specific areas at predetermined intervals (Timmusk et al. 2018).

The interactions between microorganisms and nanoparticles are complex and multifaceted, influenced by charge, chemical composition, and particle size. Various forces, including electrostatic attraction and chemical reactions on the bacterial surface (Palmqvist et al. 2015) mediate adhesion between nanoparticles and microorganisms. These interactions play a crucial role in the stability and effectiveness of biofertilizers, affecting the delivery capacity and response of microorganisms in plant tissues (Shukla et al. 2015).

The indirect effects of nanoparticles on the plant-microorganism association are remarkable, increasing bacterial viability and growth rate, as well as reinforcing plant defenses against adverse conditions (Ghalamboran and Ramsden 2012; Karunakaran et al. 2013; Abdel Latef et al. 2018). For instance, studies have shown that nanoparticles like CuO and ZnO can stimulate the production of beneficial compounds by microorganisms, while AgNPs can positively influence nitrogen fixation in some bacterial species (Dimkpa et al. 2012; Yang et al. 2013). However, it is essential to recognize that these effects can vary and be unpredictable, depending on the particular properties of the microorganisms and nanoparticles involved.

Therefore, investing in nanoparticles for phosphate solubilization is a strategic and innovative choice due to their potential to revolutionize agriculture. These nanometric structures present unique properties that allow controlled and sustained release of phosphorus, a crucial element for plant growth (de Moraes and Lacava 2022). Their ability to enhance the availability of this nutrient, especially through nanoparticles like hydroxyapatite, offers a promising alternative to conventional fertilizers (de Moraes et al. 2021). Additionally, nanoparticles show the potential to reduce nutrient leaching, minimize negative environmental impacts, and promote more sustainable and efficient agricultural practices. This innovative approach highlights the fundamental role of nanoparticles in optimizing phosphorus supply, significantly improving agricultural productivity and environmental preservation (de Moraes et al. 2023; de Moraes and Lacava 2022).

9.6 Conclusion

Research on biological activities of interest has formed the foundation for advancing various applications in biotechnological and agricultural domains. Research on biologically relevant activities has underpinned the advancement of various applications in biotechnological and agricultural domains. Microorganisms, including endophytes, display extensive genetic and metabolic diversity that remains incompletely explored. Available literary analyses indicate that tropical plants harbor many endophytic microorganisms, many lacking in-depth studies (Lacava et al. 2022). Phosphate-solubilizing endophytes are significant for their contribution to plant development and their potential viability for commercialization in biofertilizer production. Exploring the capability and effectiveness of these microorganisms in phosphate solubilization could aid in identifying strains with significant potential for biofertilizer applications (Lacava et al. 2021). Such microbial applications have the potential to replace or reduce reliance on soluble phosphate fertilizers, thereby optimizing the utilization of natural resources or modifying soil conditions (Bogas et al. 2024).

Acknowledgments We thank the financial support of the São Paulo Research Foundation, FAPESP (Proc. No. 2015/10974-8 and Proc. No. 2020/11315-6).

References

Abdel Latef AAH, Srivastava AK, El-sadek MSA et al (2018) Titanium dioxide nanoparticles improve growth and enhance tolerance of broad bean plants under saline soil conditions. Land Degrad Dev 29:1065–1073. https://doi.org/10.1002/ldr.2780

Abhilash PC, Jamil S, Singh N (2011) Transgenic plants for enhanced biodegradation and phytoremediation of organic xenobiotics. Biotechnology Advances 29(6):792–802. https://doi.org/10.1016/j.biotechadv.2011.06.005

Acuña CA, Blount AR, Quesenberry KH, Kenworthy KE, Hanna WW (2011) Tetraploid bahiagrass hybrids: breeding technique, genetic variability and proportion of heterotic hybrids. Euphytica 179:227–235. https://doi.org/10.1007/s10681-010-0276-y

Adeleke BS, Babalola OO (2021) Roles of plant endosphere microbes in agriculture—a review. J Plant Growth Regul 41:1411–1428. https://doi.org/10.1007/s00344-021-10406-2

Adhikari P, Pandey A (2019) Phosphate solubilization potential of endophytic fungi isolated from *Taxus wallichiana* Zucc. roots. Rhizosphere 9:2–9. https://doi.org/10.1016/j.rhisph.2018.11.002

Ahemad M, Kibret M (2014) Mechanisms and applications of plant growth promoting rhizobacteria: current perspective. J King Saud Univ Sci 26:1–20. https://doi.org/10.1016/j.jksus.2013.05.001

Alongi DM (2018) Impact of global change on nutrient dynamics in mangrove forests. Forests 9(10):596. https://doi.org/10.3390/f9100596

Alori ET, Glick BR, Babalola OO (2017) Microbial phosphorus solubilization and its potential for use in sustainable agriculture. Front Microbiol 8:971. https://doi.org/10.3389/fmicb.2017.00971

Andrade PHM (2019) Análise da diversidade genética e potencial biotecnológico da comunidade bacteriana associada a *Coffea arabica* de cultivo convencional e orgânico. PhD thesis, Universidade Federal de São Carlos

Andrade PHM, Machado PC, Paula AF, Paganin ACL, Rezende GS, Matheucci JRE, Carvalho LM, Freire CCM, Cunha AF, Lacava PT (2023) 16S metabarcoding analysis reveals the influence of organic and conventional farming practices on bacterial communities from the rhizospheric of *Coffea arabica* L. Braz J Biol 83:e274070. https://doi.org/10.1590/1519-6984.274070

Atieno M, Herrmann L, Nguyen HT, Phan HT, Nguyen NK, Srean P, Lesueur D (2020) Assessment of biofertilizer use for sustainable agriculture in the Great Mekong Region. J Environ Manag 275:111300. https://doi.org/10.1016/j.jenvman.2020.111300

Azevedo JL, Quecine MC (2019) Biodiversity and biotechnological applications of microorganisms associated with tropical plants. In: Azevedo JL, Quecine MC (eds) *Diversity and Benefits of Microorganisms from the Tropics*. Springer, Cham. https://doi.org/10.1007/978-3-319-55804-2_1

Baweja P, Kumar S, Kumar G (2020) Fertilizers and pesticides: their impact on soil health and environment. In: Giri B, Varma A (eds) Soil health. Soil biology, vol 59. Springer, Cham. https://doi.org/10.1007/978-3-030-44364-1_15

Bhattacharyya A, Duraisamy P, Govindarajan M, Buhroo AA, Prasad R (2016) Nanobiofungicides: emerging trend in insect pest control. In: Advances and applications through fungal nanobiotechnology. Springer, pp 307–319

Bibi SN, Gokhan Z, Rajesh J, Mahomoodally MF (2020) Fungal endophytes associated with mangroves—chemistry and biopharmaceutical potential. S Afr J Bot 134:187–212. https://doi.org/10.1016/j.sajb.2019.12.016

Biswas S, Biswas A (2019) A study correlate perceived stress and coping strategies among nurses. Open J Nurs 9(6):547–557. https://doi.org/10.4236/ojn.2019.96044

Bogas AC, Torres FL, Sousa CP, Lacava PT (2020) Bioactivity of endophytes from the Brazilian tropical savannah. Acta Sci Microbiol 3:15–22. https://doi.org/10.31080/ASMI.2020.03.0670

Bogas AC, Cruz FPN, Lacava PT, Sousa CP (2024) Endophytic fungi: an overview on biotechnological and agronomic potential. Braz J Biol 84:e258557. https://doi.org/10.1590/1519-6984.258557

Borghetti F, Barbosa E, Ribeiro L, Ribeiro JF, Maciel E, Walter BMT (2023) Fitogeografia das savanas sul-americanas. Heringeriana 17:e918014. https://doi.org/10.17648/heringeriana.v17i1.918014

Burragoni SG, Jeon J (2021) Applications of endophytic microbes in agriculture, biotechnology, medicine, and beyond. Microbiol Res 245:126691. https://doi.org/10.1016/j.micres.2020.126691

Castanheira NL, Dourado AC, Pais I, Semedo J, Scotti-Campos P, Borges N, Fareleira P (2017) Colonization and beneficial effects on annual ryegrass by mixed inoculation with plant growth promoting bacteria. Microbiol Res 198:47–55. https://doi.org/10.1016/j.micres.2017.01.009

Catanzaro M, Bonasora M, Speranza P, Medina-Nicolas M, Valls J, Rua GH (2015) *Paspalum chilense* (Poaceae, Paspaleae): a new species from southern South America. Phytotaxa 197:245–256. https://doi.org/10.11646/phytotaxa.197.4.2

Castro FS, De Souza AQL, De Souza ADL et al (2018) Overview on biodiversity, chemistry, and biotechnological potential of microorganisms from the Brazilian Amazon. In: Azevedo JL, Quecine MC (eds) *Diversity and Benefits of Microorganisms from the Tropics*. Springer, Cham. https://doi.org/10.1007/978-3-319-55804-2_5

Chinnamuthu C, Boopathi P (2009) Nanotechnology and agroecosystem. Madras Agric J 96:17–31. https://doi.org/10.29321/MAJ.10.100436

Choudhary M, Ghasal PC, Yadav RP, Meena VS, Mondal T, Bisht JK (2018) Towards plant-beneficiary rhizobacteria and agricultural sustainability. In: Role of rhizospheric microbes in soil: nutrient management and crop improvement. Springer Nature Singapore, pp 1–46

Daniel AI, Fadaka AO, Gokul A, Bakare OO, Aina O, Fisher S, Klein A (2022) Biofertilizer: the future of food security and food safety. Microorganisms 10:e1220

de Moraes ACP (2021) Potencial biotecnológico de bactérias endofíticas e rizosféricas de *Paspalum* spp. e nanopartículas de sílica para promoção de crescimento vegetal. PhD thesis, Universidade Federal de São Carlos

de Moraes ACP, Lacava PT (2022) Use of silicon and nano-silicon in agro-biotechnologies. In: Silicon and nano-silicon in environmental stress management and crop quality improvement: progress and prospects. Elsevier, pp 55–65

de Moraes ACP, Ribeiro LDS, de Camargo ER, Lacava PT (2021) The potential of nanomaterials associated with plant growth-promoting bacteria in agriculture. 3 Biotech 11:318. https://doi.org/10.1007/s13205-021-02870-0

de Moraes ACP, Lisboa PHG, Ribeiro LS, Camargo ER, Lacava PT (2023) Plant growth-promoting microbes and nanoparticles: biotechnological potential in agrobiological systems. In: Microbial inoculants recent progress and applications. Elsevier, pp 183–206

de Paula AF (2019) Identificação taxonômica, potencial agrícola e biotecnológico de bactérias endofíticas e rizosféricas associadas a *Paspalum atratum* e *P. notatum*. PhD thesis, Universidade Federal de São Carlos

de Paula AF, Cruz FPN, Dinato NB et al (2022) Endophytic and rhizospheric bacteria associated with *Paspalum atratum* and its potential for plant growth promotion with different phosphate sources. Front Plant Sci 13:884716. https://doi.org/10.3389/fpls.2022.884716

de Sousa CP, Serrano NFG, Lacava PT (2017) Endophytic microorganisms of the tropical savannah: a promising source of bioactive molecules. In: Diversity and benefits of microorganisms from the tropics. Springer, pp 57–70

Debbab A, Aly AH, Proksch P (2013) Mangrove derived fungal endophytes—a chemical and biological perception. Fungal Divers 61:1–27. https://doi.org/10.1007/s13225-013-0243-8

Depetris MB, Acuña CA, Gutierrez S, Marcón F, Felitti SA (2020) Fungal endophyte diversity in *Paspalum* and effects on plant growth. Grass Forage Sci 75:316–325. https://doi.org/10.1111/gfs.12494

Depetris MB, Acuña CA, Gutierrez S et al (2022) First report of *Cercospora dianellicola* as an endophyte of *Paspalum notatum* (Poaceae). Australas Plant Dis Notes 17:5. https://doi.org/10.1007/s13314-022-00447-1

Deshmukh SK, Gupta MK, Prakash V, Saxena S (2018) Endophytic fungi: a source of potential antifungal compounds. J Fungi 4:77. https://doi.org/10.3390/jof4030077

Di Y-n et al (2023) Identification and characterization of Bacillus subtilis B9: a diazotrophic plant growth-promoting endophytic bacterium isolated from sugarcane root. J Plant Growth Regul 42:1720–1737. https://doi.org/10.1007/s00344-023-10112-3

Dimkpa CO, Zeng J, McLean JE et al (2012) Production of indole3-acetic acid via the indole-3-acetamide pathway in the plant-beneficial bacterium *Pseudomonas chlororaphis* O_6 is inhibited by ZnO nanoparticles but enhanced by CuO nanoparticles. Appl Environ Microbiol 78:1404–1410. https://doi.org/10.1128/AEM.07424-11

Dinesh R, Anandaraj M, Srinivasan V, Hamza S (2012) Engineered nanoparticles in the soil and their potential implications to microbial activity. Geoderma 173–174. https://doi.org/10.1016/j.geoderma.2011.12.018

Duong J, Chu CD, Tuot D et al (2020) Completeness and quality of text paging for subspecialty consult requests. Postgrad Med J 96(1139):456–460. https://doi.org/10.1136/postgradmedj-2019-137186

Elias E, Okoth S, Smaling EMA (2016) Exploring the potential for biochar application to improve soil fertility and productivity in Ethiopia. Environ Sys Res 5:10. https://doi.org/10.1186/s40068-016-0050-9

El-Lattief EA (2016) Effect of bio and organic fertilizers on growth, yield, and quality of potato under sandy soil conditions. Int J Adv Res 4(3):378–388

El-Ramady H, El-Ghamry A, Mosa A, Alshaal T (2018) Nanofertilizers vs. biofertilizers: new insights. EBSS 2:51–72. https://doi.org/10.21608/jenvbs.2018.3880.1029

Etesami H, Jeong BR, Glick BR (2021) Contribution of arbuscular mycorrhizal fungi, phosphate-solubilizing bacteria, and silicon to P uptake by plant. Front Plant Sci. https://doi.org/10.3389/fpls.2021.699618

Ewunie GA, Lekang OI, Morken J, Yigezu ZD (2021) Characterizing the potential and suitability of Ethiopian variety Jatropha curcas for biodiesel production. Heliyon 7(3):e06492. https://doi.org/10.1016/j.heliyon.2021.e06492

Fernandes ACV, Lisboa PHG, de Andrade PHM, Vigna BBZ, Nogueira SR, Fávero AP, Lacava PT (2023) Endophytic bacteria associated with *Paspalum vaginatum* Swartz with potential for plant growth promotion and biocontrol. BJAER 6:1071–1089. https://doi.org/10.34188/bjaerv6n2-010

Friess DA, Rogers K, Lovelock CE et al (2019) The state of the world's mangrove forests: past, present, and future. Annu Rev Environ Resour 44:89–115. https://doi.org/10.1146/annurev-environ-101718-033302

Freitas DR, Campos JMS, Marcondes MI et al (2015) Dry yeast for lactating dairy cows. Arquivos Brasileiros de Medicina Veterinária e Zootecnia 67(1):211–220. https://doi.org/10.1590/1678-4162-7585

Ghalamboran MR, Ramsden JJ (2012) Viability of *Bradyrhizobium japonicum* on soybean seeds enhanced by magnetite nanoparticles during desiccation. World Acad Sci Eng Technol 63:198–203

Ghormade V, Deshpande M, Paknikar KM (2011) Perspectives for nano-biotechnology enabled protection and nutrition of plants. Biotechnol Adv 29:792–803. https://doi.org/10.1016/j.biotechadv.2011.06.007

Golinska P, Wypij M, Agarkar G, Rathod D, Dahm H, Rai M (2015) Endophytic actinobacteria of medicinal plants: diversity and bioactivity. Anton Leeuw Int J G 108:267–289. https://doi.org/10.1007/s10482-015-0502-7

Gutiérrez FJ, Mussons ML, Gatón P, Rojo R (2011) Nanotechnology and food industry. In: Scientific, health and social aspects of the food industry. Tech, Croatia Book Chapter, pp 95–128

Hanyabui E, Apori O, Frimpong K, Atiah K, Abindaw T, Ali M, Asiamah J, Byalebeka J (2020) Phosphorus sorption in tropical soils. AIMS Agric Food 5:599–616. https://doi.org/10.3934/agrfood.2020.4.599

Karunakaran G, Suriyaprabha R, Manivasakan P et al (2013) Effect of nanosilica and silicon sources on plant growth promoting rhizobacteria, soil nutrients and maize seed germination. IET Nanobiotechnol 7:70–77. https://doi.org/10.1049/iet-nbt.2012.0048

Kavroulakis NS, Zervakis GI, Ehaliotis C, Haralampidis K, Papadopoulou KK (2007) Role of ethylene in the protection of tomato plants against soil-borne fungal pathogens conferred by an endophytic *Fusarium solani* strain. J Exp Bot 58:3853–3864. https://doi.org/10.1093/jxb/erm230

Khare EK, Arora NK (2015) Effects of soil environment on field efficacy of microbial inoculants. In: Plant microbes symbiosis: applied facets. Springer, pp 353–381

Koul O (2019) Nano-biopesticides today and future perspectives. Academic. https://doi.org/10.1016/C2017-0-03028-8

Krishnan K, Lau BY, Ewall G et al (2017) MECP2 regulates cortical plasticity underlying a learned behaviour in adult female mice. Nat Commun 8:14077. https://doi.org/10.1038/ncomms14077

Lacava PT, Machado PC, Andrade PH (2021) Phosphate solubilization by endophytes from the tropical plants. In: Sustainable development and biodiversity. Springer, pp 207–226

Lacava PT, Bogas AC, Cruz FPN (2022) Plant growth promotion and biocontrol by endophytic and rhizospheric microorganisms from the tropics: a review and perspectives. Front Sustain Food Syst 6:796113. https://doi.org/10.3389/fsufs.2022.796113

Lata R, Chowdhury S, Gond SK, White JF Jr (2018) Induction of abiotic stress tolerance in plants by endophytic microbes. Lett Appl Microbiol 66:268–276. https://doi.org/10.1111/lam.12855

Liu N, Chen X, Zhang J, Schwank JW (2014) A review on TiO_2-based nanotubes synthesized via hydrothermal method: formation mechanism, structure modification, and photocatalytic applications. Catal Today 225:34–51. https://doi.org/10.1016/j.cattod.2013.10.090

Lovecká P et al (2023) Plant growth-promoting endophytic bacteria isolated from Miscanthus giganteus and their antifungal activity. Microorganisms 11:2710. https://doi.org/10.3390/microorganisms11112710

Lucero CT, Lorda GS, Anzuay MS et al (2021) Peanut endophytic phosphate solubilizing bacteria increase growth and P content of soybean and maize plants. Curr Microbiol 78:1961–1972. https://doi.org/10.1007/s00284-021-02469-x

Machado JA, Silva JS (2019) Quantiles via moments. J Econom 213(1):145–173. https://doi.org/10.1016/j.jeconom.2019.04.009

Machado PC, de Andrade PHM, Sousa CP, Souza CWO, Lacava PT (2020) In vitro characterization of endophytic bacteria associated with physic nut (*Jatropha curcas* L.) and their potential for plant-growth promotion and biocontrol. Braz J Dev 6:88572–88589. https://doi.org/10.34117/bjdv6n11-326

Machado PC, de Andrade PHM, Matheucci Junior E, Carazzolle MF, de Carvalho LM, da Cunha AF, Lacava PT (2024) Characterization of rhizobacteria associated with physic nut (*Jatropha curcas* L.) by metagenomic analysis. BJAER 7:173–190. https://doi.org/10.34188/bjaerv7n1-014

Marcon F, Urbani MH, Quarin CL, Acuña CA (2018) Agronomic characterization of *Paspalum atratum* Swallen and *P. lenticulare* Kunth. Trop Grassl-Forrajes Trop 6:70–81. https://doi.org/10.17138/tgft(6)70-81

Matos ADM, Gomes ICP, Nietsche S, Xavier AA, Gomes WS, dos Santos JA, Pereira MCT (2017) Phosphate solubilization by endophytic bacteria isolated from banana trees. An Acad Bras Ciênc 89(4):2945–2954. https://doi.org/10.1590/0001-3765201720160111

Mengel K, Kirkby EA, Kosegarten H, Appel T (2001) Phosphorus. In: Principles of plant nutrition. Springer, pp 453–479

Mitra B, Chowdhury AR, Dey P, Hazra KK, Sinha AK, Hossain A, Meena RS (2021) Use of agrochemicals in agriculture: alarming issues and solutions. In: Input use efficiency for food and environmental security. Springer, pp 85–122

Mohanty KB, Mohapatra A, Nayak B (2017) Parameter extraction of PV module using NLS algorithm with experimental validation. IJECE 7(5):2392–2400. https://doi.org/10.11591/ijece.v7i5.pp2392-2400

Munir E et al (2022) Isolation and characterization of phosphate solubilizing activity of endophytic fungi from Zingiberaceous species. OnLine J Biol Sci 22:149–156. https://doi.org/10.3844/ojbsci.2022.149.156

Myers N, Mittermeier RA, Mittermeier CG, da Fonseca GAB, Kent J (2000) Biodiversity hotspots for conservation priorities. Nature 403:853–858

Nagah A et al (2024) Endophytic Bacillus vallismortis and Bacillus tequilensis bacteria isolated from medicinal plants enhance phosphorus acquisition and fortify Brassica napus L. vegetative growth and metabolic content. Front Plant Sci 15:1324538. https://doi.org/10.3389/fpls.2024.1324538

Nath R, Sharma GD, Barooah M (2015) Plant growth promoting endophytic fungi isolated from tea (Camellia sinensis) shrubs of Assam, India. Appl Ecol Environ Res 13(3):877–891. https://doi.org/10.15666/aeer/1303_877891

Noriler AS, Savi DC, Aluizio R, Palacio-Cortez AM, Possiede YM, Glienke C (2018) Bioprospecting and structure of fungal endophyte communities found in the Brazilian biomes, Pantanal and Cerrado. Front Microbiol 9:1526–1532. https://doi.org/10.3389/fmicb.2018.01526

Novo PE, Valls JFM, Galdeano F, Honfi AI, Espinoza F, Quarin CL (2016) Interspecific hybrids between *Paspalum plicatulum* and *P. oteroi*: a key tool for forage breeding. Sci Agric 73:356–362. https://doi.org/10.1590/0103-9016-2015-0218

Palmqvist NGM, Bejai S, Meijer J et al (2015) Nano titania aided clustering and adhesion of beneficial bacteria to plant roots to enhance crop growth and stress management. Sci Rep 5:10146. https://doi.org/10.1038/srep10146

Pedrino IC (2023) Isolamento e caracterização de fungos endofíticos associados a *Paspalum atratum* Swallen e *Paspalum notatum* Fluggé com potencial biotecnológico e agrícola para promoção de crescimento vegetal e biocontrole. Capstone Project, Universidade Federal de São Carlos

Pratiwi T, Marsidin S, Syarif H, Yahya Y (2020) The effectiveness of the directed project-based learning model in improving the learning outcomes of class VIII students at junior high school. In: Proceedings of the 2nd International Conference Innovation in Education (ICoIE 2020). Atlantis Press. https://doi.org/10.2991/assehr.k.201209.250

Prihatiningsih N et al (2020) Isolation and characterization of the endophytic bacteria, and their potential as maize diseases control. Biodivers J Biol Divers 21(5). https://doi.org/10.13057/biodiv/d210506

Priyadharsini P, Muthukumar T (2017) The root endophytic fungus *Curvularia geniculata* from *Parthenium hysterophorus* roots improves plant growth through phosphate solubilization and phytohormone production. Fungal Ecol 27:69–77

Priyadharsini C, Thangarajathi S (2017) Effect of selected variables on regular school teachers' attitude towards inclusive education. Int J Spec Educ 32(4):1–12

Quarin CL, Valls JFM, Urbani MH (1997) Cytological and reproductive behaviour of *Paspalum atratum*, a promising forage grass for the tropics. Trop Grassl 31:114–116

Quecine MC, Araújo WL, Marcon J et al (2014) Chitinolytic activity of endophytic Streptomyces and potential for biocontrol. Lett Appl Microbiol 47(6):486–491. https://doi.org/10.1111/j.1472-765X.2008.02463.x

Rajamanikyam M, Vadlapudi V, Amanchy R, Upadhyayula SM (2017) Endophytic fungi as novel resources of natural therapeutics. Braz Arch Biol Technol 60:e17160542. https://doi.org/10.1590/1678-4324-2017160542

Rana KL, Kour D, Kaur T et al (2020a) Endophytic microbes: biodiversity, plant growth-promoting mechanisms and potential applications for agricultural sustainability. Antonie Van Leeuwenhoek 113:1075–1107. https://doi.org/10.1007/s10482-020-01429-y

Rana KL, Kour D, Kaur T et al (2020b) Endophytic microbes from diverse wheat genotypes and their potential biotechnological applications in plant growth promotion and nutrient uptake. Proc Natl Acad Sci India Sect B Biol Sci 90:969–979. https://doi.org/10.1007/s40011-020-01168-0

Rawat P, Das S, Shankhdhar D, Shankhdhar SC (2021) Phosphate-solubilizing microorganisms: mechanism and their role in phosphate solubilization and uptake. J Soil Sci Plant Nutr 21:49–68. https://doi.org/10.1007/s42729-020-00342-7

Roduner E (2006) Size matters: why nanomaterials are different. Chem Soc Rev 35:583–592. https://doi.org/10.1039/b502142c

Rukshana-Begum S, Tamilselvi S (2016) Effect of selected variables on regular school teachers' attitude towards inclusive education. Int J Spec Educ 31(2):1–12

Sankaran M (2019) Role of plant growth-promoting rhizobacteria in sustainable agriculture. In: Kumar A, Singh VK, Tripathi DK, Sharma S, Chauhan DK (eds) *Plant Growth Promoting Rhizobacteria for Sustainable Stress Management*. Springer, Singapore. https://doi.org/10.1007/978-981-13-7556-2_2

Sano EE, Rodrigues AA, Martins ES, Bettiol GM, Bustamante MM, Bezerra AS, Bolfe EL (2019) Cerrado ecoregions: a spatial framework to assess and prioritize Brazilian savanna environmental diversity for conservation. J Environ Manag 232:818–828. https://doi.org/10.1016/j.jenvman.2018.11.108

Santana EB, Marques ELS, Dias JCT (2016) Effects of phosphate-solubilizing bacteria, native microorganisms and rock dust on *Jatropha curcas* L. growth. Genet Mol Res 15. https://doi.org/10.4238/gmr.15048729

Sarnaik D, Kanade G, Patil C (2020) Bioremediation of heavy metals using microbial biosorbents: A review. Environ. Sci Pollut Res Int 27:1–14. https://doi.org/10.1007/s11356-020-08168-3

Sawant SS, Rodrigues BF (2020) Phosphate solubilization by fungal endophytes isolated from mangrove plants in India. Kavaka 55:71–76. https://doi.org/10.36460/Kavaka/55/2020/71-76

Sebastianes FLS, Azevedo JL, Lacava PT (2017) Diversity and biotechnological potential of endophytic microorganisms associated with tropical mangrove forests. In: Diversity and benefits of microorganisms from the tropics. Springer, pp 37–56

Shrivastava M, Srivastava PC, D'Souza SF (2018) Phosphate-solubilizing microbes: diversity and phosphates solubilization mechanism. In: Role of rhizospheric microbes in soil volume 2: nutrient management and crop improvement. Springer, pp 137–165

Shukla SK, Kumar R, Mishra RK et al (2015) Prediction and validation of gold nanoparticles (GNPs) on plant growth promoting rhizobacteria (PGPR): a step toward development of nano-biofertilizers. Nanotechnol Rev 4:439–448. https://doi.org/10.1515/ntrev-2015-0036

Sikes BA, Paszalek J, de Leon NE, Hawkes CV (2017) Regardless of N-substrate, multiple fungal root endophytes isolated from pastures outgrow and outcompete those isolated from undisturbed sites. Pedobiologia 63:52–58. https://doi.org/10.1016/j.pedobi.2017.05.006

Silva IR et al (2020) Phosphorus availability in acid soils of the Cerrado and its implications for plant nutrition. Plant Soil 442:279–301. https://doi.org/10.1007/s11104-019-04150-5

Silva LID, Pereira MC, Carvalho AMXD, Buttrós VH, Pasqual M, Dória J (2023) Phosphorus-solubilizing microorganisms: a key to sustainable agriculture. Agriculture 13:462. https://doi.org/10.3390/agriculture13020462

Souza BS, Santos TT (2017) Endophytic fungi in economically important plants: ecological aspects, diversity and potential biotechnological applications. J Bioeng Food Sci 4:113–126. https://doi.org/10.18067/jbfs.v4i2.121

Sriwati R et al (2023) Endophytic bacteria as biological agents to control Fusarium wilt disease and promote tomato plant growth. Physiol Mol Plant Pathol 125:101994

Suman V, Prasad R, Jain VK, Varma A (2010) Role of nanomaterials in symbiotic fungus growth enhancement. Curr Sci 99:1189–1191

Tarafdar JC, Yadav RS, Meena SC (2001) Comparative efficiency of acid phosphatase originated from plant and fungal sources. J Plant Nutr Soil Sci 164:279–282. https://doi.org/10.1002/1522-2624(200106)164:33.0.CO;2-L

Tarafdar JC, Sharma S, Raliya R (2013) Nanotechnology: interdisciplinary science of applications. Afr J Biotechnol 12:219–226. https://doi.org/10.5897/AJB12

Timmusk S, Seisenbaeva G, Behers L (2018) Titania (TiO_2) nanoparticles enhance the performance of growth-promoting rhizobacteria. Sci Rep 8:617. https://doi.org/10.1038/s41598-017-18939-x

Ting ASY, Meon S, Kadir J et al (2008) Endophytic microorganisms as potential growth promoters of banana. BioControl 53:541–553. https://doi.org/10.1007/s10526-007-9093-1

Toghueo RMK, Boyom FF (2019) Endophytic fungi from Terminalia species: a comprehensive review. J Fungi 5:43. https://doi.org/10.3390/jof5020043. PMID:31137730

Tomer S, Suyal DC, Goel R (2016) Biofertilizers: a timely approach for sustainable agriculture. In: Plant-microbe interaction: an approach to sustainable agriculture. Springer, pp 375–395

Torres FL, Bogas AC, Andrade PHM, de Sousa CP, Lacava PT (2022) Bioprospection and biotechnological potential of endophytic fungi associated to plants of Brazilian savannah. Concilium 22:256–272. https://doi.org/10.53660/CLM-114-133

Tsai C-Y, Yeh Y-H, Tsai L-H, Chou EC-L (2023) Correction: Tsai et al. The efficacy of transvaginal ultrasound-guided BoNT-A external sphincter injection in female patients with underactive bladder. Toxins 15(6):399. https://doi.org/10.3390/toxins15060399

Tsipinana S, Husseiny S, Alayande K, Raslan M, Amoo S, Adeleke R (2023) Contribution of endophytes towards improving plant bioactive metabolites: a rescue option against red-taping of medicinal plants. Front Plant Sci 14:1248319. https://doi.org/10.3389/fpls.2023.1248319

Vandergheynst J, Scher H, Guo HY, Schultz D (2007) Water-in-oil emulsions that improve the storage and delivery of the biolarvacide *Lagenidium giganteum*. BioControl 52:207–229. https://doi.org/10.1007/s10526-006-9021-9

Varma A, Uma S, Khanuja M (2017) Role of nanoparticles on plant growth with special emphasis on *Piriformospora indica*: a review. In: Nanoscience and plant-soil systems. Springer, pp 387–403

Vasileva EN, Akhtemova GA, Zhukov VA, Tikhonovich IA (2019) Endophytic microorganisms in fundamental research and agriculture. Ekol Genet 17:19–32. https://doi.org/10.17816/ecogen17119-32

Wang Y (2004) The role of rhizosphere microorganisms in phytoremediation of heavy metal contaminated soils. Biotechnolo Adv 22(2):135–143. https://doi.org/10.1016/j.biotechadv.2003.08.009

Wang WG, Barak T, Vinocur B, Shoseyov O, Altman A (2003) Abiotic resistance and chaperones: possible physiologi- cal role of SP1, a stable and stabilizing protein from Populus. In: Plant biotechnology 2002 and beyond. Kluwer Academic Publication, pp 439–443

Wang Y, Li H, Wang Y (2016) The role of rhizosphere microorganisms in phytoremediation of heavy metal contaminated soils. Biotechnol Adv 34(5):1131–1140. https://doi.org/10.1016/j.biotechadv.2016.07.003

Wang Z, Zhang H, Liu L et al (2022) Screening of phosphate-solubilizing bacteria and their abilities of phosphorus solubilization and wheat growth promotion. BMC Microbiol 22:296. https://doi.org/10.1186/s12866-022-02715-7

Watts D, Palombo EA, Jaimes Castillo A, Zaferanloo B (2023) Endophytes in agriculture: potential to improve yields and tolerances of agricultural crops. Microorganisms 11:1276. https://doi.org/10.3390/microorganisms11051276

Yang Y, Wang J, Xiu Z, Alvarez PJJ (2013) Impacts of silver nanoparticles on cellular and transcriptional activity of nitrogen-cycling bacteria. Environ Toxicol Chem 32:1488–1494. https://doi.org/10.1002/etc.2230

Zapata F, Axmann H (1995) Transfer of nitrogen fixed by legumes to cereals in intercropping systems. Plant and Soil 174:51–58. https://doi.org/10.1007/BF00032241

Zoboli O, Zessner M, Rechberger H (2016) Supporting phosphorus management in Austria: potential, priorities, and limitations. Sci Total Environ 565:313–323. https://doi.org/10.1016/j.scitotenv.2016.04.171

Part III
Plant Protection and Pathogen Management

Chapter 10
Biocontrol Strategies: Microbial Defenders Against Plant Pathogens

Abdelaaziz Farhaoui, Mohammed Khadiri, Mohammed Taoussi, Rachid Ezzouggari, Oumaima Benaissa, Mohammed Radi, Zineb Belabess, and Rachid Lahlali

Abstract The escalating threat posed by plant pathogens to global agricultural productivity requires the development of sustainable and environmentally friendly control strategies. This chapter provides an in-depth review of microbial biocontrol agents as formidable defenders against plant pathogens. The introduction highlights the critical importance of plant disease control and lays the foundation for understanding biocontrol in agriculture. The chapter explores the diversity of microbial biocontrol agents, encompassing fungi, bacteria, yeast, and viruses, with emphasis on their potential contributions. An important part is devoted to the discovery of the mechanisms of action employed by these microbial defenders. Detailed discussions of competition, parasitism, antibiosis, and induced systemic resistance (ISR) provide valuable insights into how these agents suppress pathogen growth and enhance plant immunity. Choosing the appropriate microbial inoculants is a critical factor determining the success of a biocontrol program, followed by the challenge of deploying them effectively in the field while ensuring consistent performance. This chapter reviews various microbial bioformulations and their application methods, such as seed treatment, soil drenching, and foliar sprays. It examines challenges in microbial biocontrol, including environmental factors and commercialization barriers. Emerging trends and technological advances are also discussed, offering insights into improving the effectiveness of these biocontrol strategies. The conclusion highlights the potential of microbial biocontrol as a sustainable component of modern agriculture. This chapter serves as a guide for researchers, practitioners, and policymakers interested in using microorganisms to reduce pathogen impact and enhance agricultural production.

A. Farhaoui · M. Khadiri · M. Taoussi · R. Ezzouggari · O. Benaissa · M. Radi · R. Lahlali (✉)
Phytopathology Unit, Department of Plant Protection, Ecole National of Agriculture Meknes, Meknes, Morocco
e-mail: rlahlali@enameknes.ac.ma

Z. Belabess
Plant Protection Laboratory, Regional Center of Agricultural Research of Meknes, National Institute of Agricultural Research, Meknes, Morocco

O. O. Babalola, A. S. Ayangbenro (eds.), *Microbial Allies*,
https://doi.org/10.1007/978-3-031-90530-8_10

Keywords Plant pathogen · Microbial biocontrol agents · Biopesticide · Bioformulation

10.1 Introduction

Considering the continuous expansion of the global population, projected to reach 9.7 billion individuals by 2050 (Webb et al. 2020; Lubchenco and Haugan 2023), a substantial augmentation in food production becomes imperative. The Food and Agriculture Organization of the United Nations (FAO) stipulates a minimum 50% rise in agricultural food production by this timeframe (FAO 2017). Confronted with this escalating demand for food and considering the limitations associated with the restricted availability of additional agricultural land, significant emphasis is placed on optimizing yield per unit area and mitigating losses in crop yield (Pradhan et al. 2015; FAO 2017). A noteworthy proportion of crop losses occurs annually both before and after the harvest season, attributed to infestations by various pathogens, encompassing viroids, viruses, bacteria, fungi, and oomycetes. These plant pathogens are recognized for their persistent attacks, resulting in direct and indirect losses estimated at approximately $40 billion globally (Jamiolkowska 2020). Thus, improving the efficiency of strategies to manage plant pathogens is anticipated to play a fundamental role in meeting the growing global food demand (Valin et al. 2014).

Preventive measures are effective options for managing various pathogens. These methods mainly include farming practices, specifically crop rotations and the utilization of plant varieties exhibiting resistance to diseases (Van Bruggen et al. 2016). Chemical pesticides are employed for both preventive and curative purposes to control plant pathogens (Ul Haq et al. 2020). In contemporary agriculture, chemical pesticides play a crucial role in contributing to the improvement of crop quality and yield, prolonging the shelf life of agricultural products, and improving food safety (e.g., by reducing fungal toxins) (Kawasaki and Lichtenberg 2015; Raymaekers et al. 2020). In conjunction with intensified fertilizer use, crop variety selection, and irrigation, the substantial 20-fold increase in pesticide usage between 1960 and 1990 contributed to the doubling of agricultural production during this period (Oerke 2006). This trend has persisted in recent decades (Raymaekers et al. 2020). While their positive contribution to the effective management of plant pathogens and pests is indisputable, worries emerge concerning the adverse impacts of chemical pesticides on the environment and human health. This is especially evident in terms of the contamination of natural environments and their toxicity to non-target beneficial organisms, giving rise to genuine concerns (Rani et al. 2021; Araújo et al. 2023; Khan et al. 2023). Moreover, the intensive application of chemical pesticides has frequently led to the emergence of resistant pathogens and pests (Damalas 2016; Farhaoui et al. 2023b). Due to increasing legislative stringency, access to effective chemical pesticides continues to decline (Raymaekers et al. 2020). In many countries around the world, most of these synthetic molecules have either been permanently banned for sale or will be banned soon (Lecomte et al. 2016; Farhaoui et al. 2023b). Complementary to stricter guidelines concerning the use of pesticides, the European Union (EU) promotes a management approach that minimizes the use of pesticides, such as the integrated pest management program established by Directive 2009/128/EC (Durel et al. 2015). This approach relies on the integration of suitable measures to reduce the expansion of pest populations, with an emphasis on promoting healthy cultivation while minimizing disturbance to agroecosystems. It also encourages natural mechanisms to combat plant diseases and pests (Raymaekers et al. 2020). Many countries strive for a more sustainable approach to protecting plants (Jindal et al. 2013; Farhaoui et al. 2023b), Emphasizing prevention and the biological control of diseases and pests, the population of pathogens, though not entirely eradicated, is anticipated to stay sufficiently low to prevent the development of notable symptoms (Heimpel and Mills 2017). In its strictest sense, biological control entails the utilization of natural organisms, encompassing microorganisms, to manage plant pests or diseases (Heimpel and Mills 2017). Biological control serves as an eco-friendly alternative for effectively managing various plant diseases and minimizing human exposure to harmful chemicals and their residues (Sharma et al. 2017). Biocontrol organisms, abundantly present in nature, comprise fungi, bacteria, yeast, protozoa, and viruses. They can effectively manage plant diseases either through direct or indirect means (Al-Ani et al. 2020). The first involves antagonistic mechanisms such as parasitism and antibiosis. The microbial biological control agent (MBCA) interacts indirectly with the plant pathogen via the host plant, inducing host tissue resistance. This induction encompasses both systemic acquired resistance (SAR) and induced systemic resistance (ISR) (Babalola et al. 2023; Sulaiman and Bello 2023). Therefore, the host is subjected to various molecular and biochemical defense processes, serving as a protective mechanism against diverse plant pathogens. Additionally, various biocontrol agents stimulate plant growth by improving water and nutrient absorption or by producing metabolites like hormones that maintain plant fitness (Zaker 2016; Farhaoui et al. 2023a). The complex interactions among plant pathogens, MBCAs, and plants occur through numerous processes. Thus, recognition biocontrol mechanisms constitute an important challenge (Farzand et al. 2019; Zubair et al. 2019). Comprehending the mechanisms explaining the protective effect of a MBCA can facilitate the optimization of biological control by establishing ideal conditions for the interaction between the biocontrol agent, the host and the plant pathogen. This contributes to the development of appropriate formulations and the adoption of suitable application techniques to improve plant health, thus contributing to increased agricultural production (Jambhulkar et al. 2016). The success in discovering new microbial inoculants with effective control against diverse leaf and root diseases in recent decades has spurred increased interest in utilizing biological control methods for diverse plant pathogens (Desai et al. 2002). Inoculating soil or seeds with microbial inoculants can lead to significant modifications in the composition of the native microbial community, which is crucial for ensuring the safety of introducing MBCAs into the microenvironment of the plants to be treated (Jambhulkar et al. 2016). To ensure the efficacy of MBCAs under field conditions, an appropriate delivery method is crucial when applying a well-formulated microbial inoculant (Kennedy 1999). Several formulations of beneficial microbes, ratified by regulatory authorities globally, are accessible for plant pathogen control. Consequently, microbial antagonists targeting leaf and root diseases have become increasingly significant in recent decades. The delivery system of microbial inoculants primarily relies on factors like the nature of pathogen to be controlled, the crop stage, the severity and characteristics of the disease, and the prevailing climatic conditions (Jambhulkar et al. 2016). The use of microbial inoculants on seeds provides protection against seed- and soil-transmitted plant pathogens, simultaneously boosting plant growth and vigor, thereby enhancing crop yield (Jambhulkar and Sharma 2013). Biological control of soilborne pathogens can also be accomplished by directly applying microbial inoculants through methods such as root dip, flooding, drip, or spraying (Gašić and Tanović 2013). Effective and stable bioformulations, combined with an appropriate delivery system, are essential to promote the adoption of bioformulations by farmers.

The present review paper aims to provide a broader and updated overview and discussion of beneficial microorganisms exploited in the biological control of plant pathogens to improve agricultural production. Understanding the characteristics, including mechanisms of action, of antagonistic microbes is essential to adopt the most appropriate formulation and delivery system. Therefore, this chapter is segmented into 3 main sections. The first part provides a detailed- overview of the types of microbial biocontrol agents. The second part delves into the mechanisms employed by MBCAs in crop protection against phytopathogens. The third part presents several microbial bioformulations available commercially and their modes of implementation in the field.

10.2 Types of Microbial Biocontrol Agents

The rhizosphere harbors microbes capable of suppressing plant pathogens, offering natural protection to plants through mechanisms like competition for nutrients and space, as well as inducing plant resistance against biotic stress. Many of these microbes produce antibiotic compounds, enzymes, and other metabolites with the ability to control various plant pathogens. Microorganisms possessing these characteristics can be harnessed as antagonists in agriculture (Ab Rahman et al. 2018). Many microorganisms can both inhibit plant pathogens and stimulate plant growth. They colonize plant surfaces, establishing plant-microbe interactions (Boro et al. 2022). Fungi, bacteria, protozoa, nematodes, and microarthropods are among the microbes that have the ability to control phytopathogens and improve yields (Köhl et al. 2019). Figure 10.1 illustrates the main kinds of microbial biocontrol agents and their general modes of action. Many commercially available microbial products are employed worldwide to combat pathogens responsible for significant losses to cultivated crops.

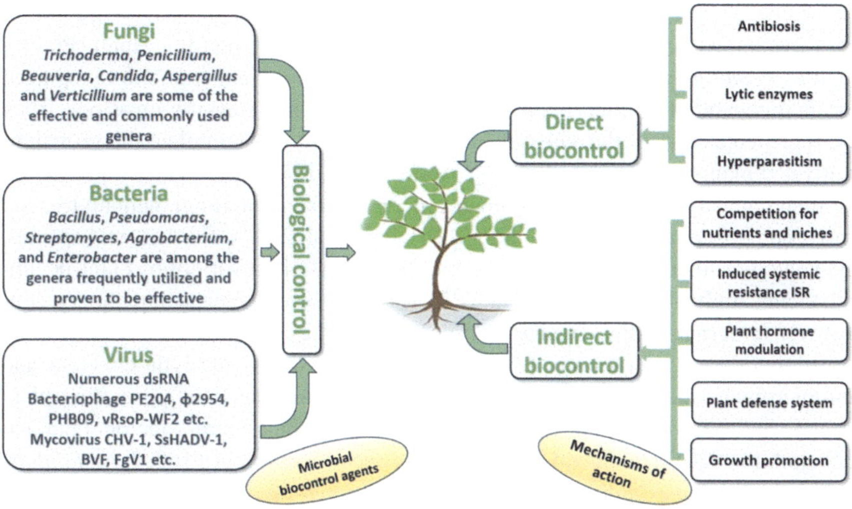

Fig. 10.1 Key microbial biocontrol agents and their diverse mechanisms in combatting plant pathogens

10.2.1 Fungi as Biocontrol Agent

In addition to promoting nutrient absorption and nitrogen utilization by plants, many fungi possess biocontrol abilities. They contribute to the fight against pests and various plant pathogens affecting different plant parts like roots, foliage, and fruits. These fungi provide protection against diseases by employing mechanisms like mycoparasitism, competition for resources with pathogens, antibiosis, inducing systemic resistance in the host plant, and mycovirus-mediated cross-protection (MMCP) (Singh and Giri 2017). This makes fungal species ideal MBCAs for controlling a wide range of crop plant diseases (Savita and Sharma 2019). Widely recognized fungal biocontrol agents include *Trichoderma* species, yeasts, ectomycorrhizas, and arbuscular mycorrhizal fungi (AMF). With advances in genetic engineering and biotechnology, it is now feasible not only to incorporate useful fungal genes into the genomes of host plants but also to amplify the expression of these genes to enhance their effects (Ghorbanpour et al. 2018). A review conducted by Thambugala et al. (2020) presents a comprehensive compilation of antagonistic fungal agents used against fungal plant pathogens, in line with contemporary taxonomic concepts, and elucidates their phylogenetic relationships. This review meticulously describes approximately 300 fungal biocontrol agents across 13 classes and 113 genera, with explicit details on the targeted plant pathogens and resulting diseases. Their findings highlight the *Trichoderma* genus as holding the greatest biocontrol potential, encompassing 25 antagonistic species used to combat various plant pathogens. Additionally, nine other fungal genera (*Aspergillus*, *Alternaria*, *Fusarium*, *Candida*, *Penicillium*, *Pythium*, *Pichia*, *Verticillium*, and

Talaromyces) have been recognized as important in this context, each comprising at least five known antagonistic species. Most plant growth-promoting fungi (PGPF), such as *Penicillium*, *Aspergillus*, *Trichoderma*, and *Fusarium* spp., are recognized for their capability to elicit plant immune responses during enemy invasions. They are regarded as one of the safest means to induce ISR and stimulate the growth of cultivated plants (Jogaiah et al. 2018). Additionally, PGPFs are also acknowledged for their capacity to mitigate the impacts of various bacteria, fungi, nematodes, and viruses on host plants while inducing ISR (Divya et al. 2017). *Trichoderma* species, filamentous soil fungi recognized for their usefulness in a variety of plant health-enhancing applications (Kumar et al. 2021), employ several means of pathogen control. *Trichoderma* strains engage in processes such as soil and root colonization, preventing the proliferation of plant pathogens by synthesizing lytic enzymes and antimicrobial metabolites, generating plant defense mechanisms, promoting plant growth, and improving plant tolerance to biotic and abiotic stress factors (Zin and Badaluddin 2020). The fungal antagonists *T. pseudoharzianum* T160, *T. pseudoharzianum* T113, *T. pseudoharzianum* T129, and *T. asperelloides* T136 release secondary metabolites exhibiting antimicrobial properties against phytopathogenic bacteria, including *Xanthomonas* compestris and *Ralstonia solanacearum* (Khan et al. 2020). Table 10.1 lists some common secondary metabolites and the *Trichoderma* isolates that produce them. Fungal species from other genera, including *Aspergillus*, *Saccharomyces*, *Gliocladium*, and *Penicillium*, exhibited the capability to release secondary metabolites, such as toxins and antibiotics (Prajapati et al. 2020), which exert a biocontrol effect against a wide range of fungal phytopathogens like *Rhizoctonia*, *Fusarium*, *Phytophthora*, *Botrytis*, *Pythium*, and *Alternaria* (Tariq et al. 2020). *Funneliformis mosseae*, the arbuscular mycorrhizal fungus, exhibited a significant antibiotic effect against *F. oxysporum*, causing alfalfa root rot (Wang et al. 2019).

Several yeasts (e.g., *Candida* and *Aureobasidium*) have been reported to display antagonistic effects against phytopathogenic fungi (Grzegorczyk et al. 2017; Ezzouggari et al. 2024). Since yeasts are natural inhabitants of vines and grapes, research has been conducted to explore the biocontrol potential of yeasts in wine production (Calvo-Garrido et al. 2013). *Saccharomyces* and non-*Saccharomyces* yeasts could be potential MBCAs against fungi belonging to the *Botrytis*, *Aspergillus*, and *Penicillium* genera that affect wine grapes and table grapes. Numerous yeast species remain widely unexplored to date, both in fundamental research and for their potential commercial exploitation. Yeasts constitute a vast and untapped reservoir with significant potential for exploring strains with new metabolic traits, adhesiveness, new antimicrobial traits, etc., necessary to understand the antagonistic effects of these microorganisms and their use as MBCAs (Di Canito et al. 2021). Di Canito et al. (2021) suggest that the use of yeasts in preventing infections represents a new approach to preserving the post-harvest quality of wine and table grapes. The antagonistic potential of yeast strains was first observed in *Saccharomyces cerevisiae*. Some yeast strains can impede the proliferation of susceptible strains of the same species by releasing proteins called "killer toxins" (São-José et al. 2017). Since its discovery, this potential has been extensively studied for the development

Table 10.1 Some secondary metabolites secreted by *Trichoderma* strains with an antagonistic effect

Trichoderma strain	Secondary metabolite produced	Target pathogen	Biocontrol mechanisms	References
Trichoderma harzianum *T. viride*	Cyclooctanol Pyrone	*Alternaria alternata*	Results in harm to the mycelial structure of the pathogen	Boro et al. (2022)
T. erinaceum IT-58 *T. gamsii* IT-62, *T. harzianum* P11 *T. afroharzianum* P8	Flavonoids Polyphenols	*Pythium myriotylum*	Inhibition of hyphal growth	Tchameni et al. (2020)
T. atroviride PHYTAT7 *T. asperellum* S11 *T. longibrachiatum* S12	Siderophores Chitinase Amylase Protease enzymes	*Rhizoctonia solani*	Inhibition of hyphal growth	Sallam et al. (2021)
T. asperellum *T. longibrachiatum*	peptaibol compounds	*R. solani* *Fusarium graminearum, F. moniliforme, F. culmorum, F. oxysporum Alternaria spp.*	Induction of systemic resistance in host plants	Rahimi Tamandegani et al. (2020)

of yeast-based biopesticides. The antifungal effects of yeasts are not founded on a single mechanism but also implicate competition for nutrients and space, mycoparasitism, as well as the release of lytic enzymes and volatile secondary metabolites such as CO_2, ethyl acetate, and ethanol (Liu et al. 2013). Examples of bioformulations based on yeast or yeast-like fungi as the active ingredient include BoniProtect® (*Aureobasidium pullulans*), Shemer® (*Metschnikowia fructicola*), Candifruit® (*Candida sake*), and Aspire® (*Candida oleophila*). The significant biocontrol potential of *Pichia anomala* and other antagonistic yeasts has not yet been exploited in the form of products available for the biocontrol of plant pathogens (Oro et al. 2014). Broadly documented successful cases of utilizing yeast strains for the biocontrol of fruit spoilage caused by phytopathogenic fungi (Ferraz et al. 2019). The genomes of a large number of non-conventional yeast species have been fully sequenced, and this number continues to grow (Pandit et al. 2022). Therefore, the anticipated new techniques for genetic analysis and their subsequent modifications in yeasts, along with genomic and post-genomic analysis before and after such changes, will serve as a platform to comprehend the molecular means behind the simple and complex biological attributes. This knowledge is expected to be valuable in developing new eco-compatible applications.

10.2.2 Bacteria as Biocontrol Agent

Among the most studied MBACs are bacteria, which are microorganisms that use many mechanisms to limit the development of plant diseases. These action mechanisms encompass the production of potent antimicrobials like antibiotics or hydrolytic enzymes that target the cell wall, the occupation of infection sites and competitive exclusion of the plant pathogen, and induction of host plant resistance (Berendsen et al. 2012). Bacteria widely studied as antagonists against various plant pathogens include species in the genera *Bacillus*, *Pseudomonas*, *Alcaligenes*, *Arthrobacter*, *Enterobacter*, *Erwinia*, *Serratia*, *Stenotrophomonas*, *Rhizobium*, *Agrobacterium*, *Streptomyces* and *Xanthomonas*. Many antagonistic bacterial strains have already been registered as commercial products and placed on the market. Currently, within the EU, 13 bacterial MBCAs are registered as biopesticides for managing bacterial and fungal diseases (Bonaterra et al. 2022). Some examples of bacteria used as biopesticides are highlighted in Table 10.3. In addition to their antagonistic actions against plant pathogens, the introduction of certain bacteria to the soil to improve plant development and growth is well established. Certain bacterial species produce phytohormones, making several nutrients accessible to plants and breaking down several toxic compounds (Tariq et al. 2020). Several rhizobacteria isolates have been identified as endophytes penetrating the plant system via roots. Various rhizobial strains exhibit antagonistic effects and growth-inhibiting action against soil-borne pathogenic fungi, including *Sclerotium*, *Macrophomina*, *Rhizoctonia,* and *Fusarium* (Das et al. 2017). Rhizobia exert their antagonistic effect by releasing antimicrobial compounds, antifungal enzymes, hydrocyanic acid (HCN), and siderophores in conditions where iron is limited (Farhaoui et al. 2023a). Kanouni et al. (2018) demonstrated the inhibitory effects on mycelial growth exhibited by several Rhizobia strains against the phytopathogens *Humicola* sp., *Penicillium* sp., *Alternaria* sp., and *Cladosporium* sp. Many other Plant growth promoting bacteria (PGPBs) naturally appear as endophytic microorganisms. Generally, PGPB exerts an antagonistic effect through several mechanisms including releasing allelochemicals like siderophores, competition, and induction of plant resistance (Reddy 2014). The antifungal effects of many bacteria have been well elucidated. *Bacillus amyloliquefaciens* exhibited significant inhibitory activity against the hyphae of *F. semitectum, Helminthosporium oryzae*, and *Curvularia lunata*, affecting rice (Saechow et al. 2018). Choub et al. (2022) reported that several species of PGPBs, including *B. velezensis, B. subtilis, B. amyloliquefaciens, B simplex,* and *B. licheniformis*, demonstrated the ability to release antibiotic compounds that impede fungal phytopathogens and compete for space and nutrients with them. *Streptomyces* isolates, obtained from tomato rhizospheric soils, demonstrated their efficiency in the biological control of *F. oxysporum* f. sp. *lycopersici,* the causal agent of *Fusarium* wilt. Ethyl acetate extracted from *Streptomyces carpaticus* strain EGY-S7 exhibited significant antifungal effect against the targeted pathogen. Additionally, seed treatment with ethyl acetate extracts was associated with a substantial diminution in disease severity, as well as a significant increase in

Table 10.3 A broad description of some common microbial biopesticides available worldwide, their targets, and the corresponding delivery systems

Bioagent(s)	Target organism / disease	Delivery system	Product	Developing agency
Bacillus subtilis	*Fusarium* spp., *R. solani*, *Alternaria* spp.	Mix with chemical fungicides when added to the slurry for seed treatment	Epic	Gustafon Inc., Dallas, TX
B. subtilis GB03	*Aspergillus*, *Rhizoctonia*	Drenching during sowing and transplanting	Kodiac, companion	Growth Products, USA
B. subtilis GB34	*Fusarium*, *Rhizoctonia*, *Phytophthora*, *Pythium*.	Apply through drenching during transplanting and sowing	GB34	Gustafon, USA
Agrobacterium radiobacter 84	*Agrobacterium tumefaciens*	Spray	Galtrol	AgroBioChem, USA
Pseudomonas fluorescens A506	Frost damage, fire blight, bunch rot	Spray at blooming flowers and fruiting	Frostban, Blightban A506	Plant Health Technologies
P. syringae ESC-10	*Geotricum candidum,Mucor pyriformis, botrytis cinera F.*	Apply pellets to fruit postharvest through drenching or spraying	Bio-save 100 Bio-save 1000	EcoScience Corp, Orlando, Florida
P. cepacia	*Fusarium* spp., *R. solani*, *Pythium* spp.	Foliar spray, soil application, seed treatment	Intercept	Soil Technologies, Fairfield, IA
P. fluorescens A506	Frost damage, fire blight, bunch rot,	Spray at blooming flowers and fruiting	Frostban, Blightban A506	Plant Health Technologies
Trichoderma harzianum T-22	Soil-borne pathogens	A powder that is mixed with water and applied as a soil drench, granules mixed with soil or potting medium	Root shield or BioTrek T-22G	BioWorks, Inc., USA
T. polysporum (ATCC20475) and *T. harzianum* (ATCC20476)	Root rot tale-all and wilt	Spray, mixing with potting substrate, as paste painting on tree wounds	Binab T	Bio-Innovation AB, Sweden Henry Doubleday Research Association, UK
Trichoderma spp.	Soil-borne pathogens	Seed dressing	Monitor SD	M/s Agriland Biotech Pvt Ltd., Baroda, India
Trichoderma spp.	Soil-borne plant pathogens	Soil application	Monitor WP	M/s Agriland Biotech Pvt Ltd., Baroda, India
T. harzianum strain T-39	*Botrytis cinera*	Spray	Trichodex	BioWorks, Inc., USA

(continued)

Table 10.3 (continued)

Bioagent(s)	Target organism / disease	Delivery system	Product	Developing agency
T. harzianum	*Fusarium, Pythium, Sclerotium, Rhizoctonia*	Seed treatment	Trichostar	GBPUAT, Pantnagar
Candida oleophila	*Penicillium* spp., *Botrytis* spp.	Pasharvest to fruit as drip, spray, or drench	Aspire	Ecogen, Inc., Langhorne, PA
Fusarium oxysporum (nonpathogenic)	*F. moniliforme* , *F. oxysporum*	Soil incorporation or seed treatment	Biofox C	SIAPA, Bologna, Italy
Gliocladium virens GL-21	*Pythium* spp., *R. solani*, Damping-off, root rot,	Granules incorporated in soil	SoilGard	Thermo Trilogy, Columbia, MD
Coniothyrium minitans	*Sclerotinia minor and* and *S. sclerotiorum*	Spray	Contans	Prophyta Biologischer Pflanzenschutz, Germany
Burkholderia cepacia	*Pythium, R. solani, Fusarium* spp.	Use on seeds or in an aqueous suspension for drip irrigation	Deny	Stine Microbial Products
Fusarium oxysporum (nonpathogenic)	*F. oxysporum*	In drip to rock wool, incorporate in potting mix; in rows	Fusaclean	Natural Plant Protection, France
Gliocladium. Spp	*Pythium, Rhizoctonia, Fusarium, Sclerotium*	Drenching and seed treatment,	Gliostar	GBPUAT, Pantnagar

tomato plant biomass (Abdel-Aziz et al. 2021). Fira et al. (2018) documented that species belonging to the genus *Bacillus* are known for their capability to release secondary metabolites, including lipopeptide antibiotics, bacteriocins, siderophores, and polyketides with antagonistic effects. Bacillomycin D, a potent polypeptide antibiotic, is one of the metabolites secreted by the commercially available bacterial strain *B. amyloliquefaciens* FZB42, with significant biocontrol ability against *F. graminearum* infecting barley and wheat (Gu et al. 2017). Fengycin BS155, a cyclic lipopeptide with a low molecular weight, released by *B. subtilis* strain BS155, was identified by Zhang and Sun (2018) as the primary antimicrobial metabolite effective against the rice blast pathogen *Magnaporthe grisea*. It has been demonstrated that this compound alters the structure of the cell wall of *M. grisea*, leading to a disruption of organelle function, oxidative stress, disturbance of mitochondrial membrane activities, and chromatin condensation, leading to the death of mycelium cells (Boro et al. 2022). Many bacteria within the plant microbiome exhibit antifungal effects by releasing specific metabolites (Pang et al. 2021). Fluorescent *pseudomonads* are omnipresent in plant environments and exhibit numerous features conducive to their effectiveness in reducing plant diseases. These characteristics comprise good ecological adaptability, robust antagonistic activity against numerous plant pathogens, and a high capacity to induce resistance in the host plant (Bonaterra et al. 2022). Numerous *Pseudomonas* species are recognized for their capacity to inhabit the phyllosphere, rhizosphere, and endosphere. Utilizing various plant exudates as nutrients and displaying a rapid growth rate, these bacteria effectively compete with numerous plant pathogens for resources and space within the plant environment (Oso et al. 2019). Cabrefiga et al. (2007) reported that *P. fluorescens* EPS62 effectively reduced the infection by *Erwinia amylovora*, the causative agent of fire blight disease, due to its high capacity to colonize plant tissues. This is attributed to its higher growth rate and nutrient assimilation efficiency compared to those of the phytopathogen. *Pseudomonas* spp. are also distinguished by their ability to release bioactive compounds like cyclic lipopeptides, enzymes, and antibiotics with a crucial ecological role. Compounds such as pyrrolnitrin, phloroglucinols, phenazines, pyoluteorin, and dialkylresorcinols have been documented as antimicrobial metabolites secreted by *Pseudomonas* spp. and having a powerful antimicrobial effect (Haas and Keel 2003). For example, phenazines released by *P. chlororaphis* Pcho10 exhibit a potent inhibitory effect against *F. graminearum* causing wheat head blight (Hu et al. 2014). Other pertinent bacterial species can be employed in the development of microbial biopesticides. Examples of such species, which mitigate infections caused by plant pathogenic fungi and bacteria, include *Lactobacillus* spp. and *Pantoea* spp. Several strains of *Pantoea* spp. have MBCA capabilities due to their ubiquity and the release of antimicrobial metabolites. Currently, biopesticides derived from *Pantoea* spp. are registered and commercially accessible in the United States, New Zealand, and Canada. They exert antagonistic effects via diverse mechanisms, like the synthesis of antimicrobial metabolites, competition, and/or induction of host plant defense. Studies have demonstrated that *Pantoea* spp. strains target a broad spectrum of plant pathogens, especially oomycetes, fungi, and bacteria, by producing several antimicrobial metabolites, including phenazines, pantocins, microcins (Walterson and Stavrinides 2015). Lactic acid bacteria are promising candidates as antagonists due to certain strains being widely recognized as biopreservatives for fruits and vegetables (Trias et al. 2008; Lahmamsi et al. 2024). Lactic acid bacteria exhibit an antagonistic effect through the synthesis of one or more antimicrobial metabolites, like diacetyl, carbon dioxide, hydroxide peroxide, organic acids, and protein metabolites such as bacteriocins. These bacteria also possess the ability to prevent plant pathogens by preoccupying plant tissues susceptible to infection, competing for nutrients and space, and inducing resistance in the host plant. For example, *L. plantarum* strains TC92 and PM411 demonstrated significant efficacy in inhibiting fire blight in apples and pears (Roselló et al. 2013).

10.2.3 Viruses as Biocontrol Agent

Many viruses have demonstrated their potential for biocontrol against phytopathogens. Infection of phytopathogenic fungi by mycoviruses is associated with hypovirulence, leading to a reduction in the pathogenicity of these fungi against host plants (Abid et al. 2018). While mycoviral dsRNAs have been reported in *F. poae, F. graminearum, F. oxysporum,* and *F. solani*, their incorporation into a hypovirulent host has only been detected in a few strains, including *F. graminearum* DK21. Viruses transmitted by hypovirulent isolates of plant-parasitic fungi have been extensively examined for their probable use as MBCAs and comprehending the mechanisms behind fungal pathogenesis. Numerous cases of hypovirulence associated with mycoviruses have been elucidated, such as the mitoviruses infecting *Sclerotinia homoeocarpa* and *Ophiostoma novo-ulmi* and the mycoviruses infecting *Botrytis cinerea, F. graminearum,* and *Diaporthe ambigua* (Sharma et al. 2018). Mycoviruses are natural suppressors of pathogenic fungi, like *Ophiostoma* novo-ulmi, *Sclerotinia* sp., *Gaeumannomyces graminis*, and *Cryphonectria parasitica*, causing Dutch elm disease, *sclerotinia* rot, powdery mildew, and chestnut blight, respectively. Hence, a wide range of microbe parasites, such as bacteria, viruses, and fungi, exhibit considerable antagonistic effects against an equally wide range of plant-pathogenic fungi (Elad and Freeman 2002).

Bacteriophages, also known as phages, are viruses that specifically target bacteria and are widespread in the biosphere. They rely on host bacteria to complete their life cycle. Virulent lytic phages are particularly interesting for the suppression of plant pathogenic bacteria because they can lyse host cells and are highly specific to the host with a very restricted target spectrum. To apply bacteriophages in biological control, it is crucial to choose those that do not follow a lysogenic life cycle and do not trigger the transduction of bacterial genes. In such a case, resistance can be developed in host bacteria due to the incorporation of fragments of previous host bacterial genes into the genetic background of the bacteriophage (Barua and Nath 2018). Recent studies have shown that the podovirus PP16, by infecting isolates of *Pectobacterium carotovorum* responsible for blackleg and soft rot in potatoes, significantly reduces bacterial infection in vitro and in vivo (Voronina et al. 2019). Due to their high host specificity, using a cocktail of bacteriophage particles is generally considered to enhance biocontrol potential. Thus, the application of a cocktail consisting of 6 species of bacteriophages was associated with a reduction in the soft rot of potatoes, with rates of 64% and 61% for the severity and occurrence of the disease, respectively. These rates are significantly lower when applying a single virus species (Carstens et al. 2019).

10.3 Modes of Action of MBCA

Plant diseases typically originate from the interaction among the three main components of the disease-causing triangle, including the pathogen, the environment, and the host plant. When incorporating MBCAS into this triangle, it results in disease control. Comprehending the mode of action is crucial to grasp how MBCAs function in inhibiting plant diseases. The predominant mechanisms of antagonism encompass both direct and indirect biocontrol processes, leading to the suppression of plant pathogens (Parveen et al. 2016). The action of MBCAs is commonly classified into two main categories, direct and indirect biocontrol (Fig. 10.1). Direct biocontrol involves antibiosis, the production of lytic enzymes, siderophores, and hyperparasitism. On the other hand, indirect biocontrol encompasses antagonism through compounds that improve antagonistic action or the release of an endogenous neurotransmitter without possessing a receptor. As the name suggests, this mode does not involve direct interaction between the MBCA and the phytopathogen (Pal and Gardener 2006). This method comprises competition for nutrition and space, stimulation of plant growth, and the induction of systemic resistance in plants. The different modes of action are not mutually exclusive. The same antagonistic species may possess several of them. These various modes of action can act independently, successively, or simultaneously (Nega 2014).

10.3.1 Direct Action Mechanisms of MBCA on the Pathogen

10.3.1.1 Antibiosis

Antibiosis is the most studied antagonistic mechanism among MBCAs (M'Henni 2020; Boro et al. 2022). It is characterized by the secretion of metabolites at low concentrations targeting the pathogen. These molecules can be more or less specific, including volatile compounds, antifungals, antibacterials, or other toxic substances with a direct antagonistic effect on target organisms (M'Henni 2020). Generally, these metabolites have a broad spectrum of activity against various phytopathogens (Rettinassababady and Jeyalakshmi 2014). Many fungi and bacteria are producers of these toxic compounds. For bacterial MBCAs, species of *B. subtilis*, *Serratia plymuthica*, and *P. fluorescens* are well-known for their production of various metabolites involved in the suppression of several soilborne phytopathogens (Farhaoui et al. 2023a). Among these metabolites, phenazine, phloroglucinol 2,4-diacetyl, and pyrrolnitrine are produced by the species *P. fluorescens* (Vacheron et al. 2016). Species of the genus *Bacillus* are known for their production of lipopeptides (fengycins, surfactins, iturins, and bacillomycins) that inhibit the growth of various phytopathogens (Ezrari et al. 2021; Farhaoui et al. 2022, 2023a).

Regarding fungal BCAs, species of the genus *Trichoderma* are the most studied due to their significant production of diverse bioactive secondary metabolites. With inhibitory effects, they play a role in the suppression of several soilborne pathogens (Rhouma et al. 2018). Among these metabolites, polyketides, alkaloids, terpenoids, and peptaibols, the most important being 6-pentyl-alpha-pyrone (6PAP), viridins, gliovirins, glisoprenins, and gliotoxins, exhibit toxicity at concentrations dependent on the compound and the target (McMullin et al. 2017).

Secreted metabolites act through antibiosis, operating at multiple levels such as inhibiting the germination, growth, and/or sporulation of the pathogen. These compounds can also distort the pathogen's hyphae, alter colony appearance, and contribute to the production of specific forms such as pseudo-parenchyma (M'Henni 2020).

10.3.1.2 Parasitism

Mycoparasitism is a trophic relationship established by a microorganism at the expense of a fungus or oomycete. The intervention process of MBCAs through mycoparasitism is highly complex, unfolding in several successive steps that are generally very specific and vary depending on the pathogen species (Narayanasamy 2013). The stages of this process include (1) Chemotropism and recognition, (2) penetration, (3) wrapping around the cell wall, and (4) wall lysis (Boro et al. 2022). Mycoparasitism involves several chemical interactions, such as the production of lectins in the early stages of chemotropism and recognition between the mycoparasite (MBCA) and the pathogen (M'Henni 2020). Penetration relies on the production of lytic enzymes such as chitinases and β-glucanase by MBCAs to hydrolyze chitin and β-1,3-glucan, which are the main constituents of the cell wall in most fungi and oomycetes (Guzmán-Guzmán et al. 2023). Wall alteration leads to cell lysis, thus inhibiting the development of the pathogen. This mechanism of action is frequently employed by fungi such as *Gliocladium* spp. and *Trichoderma* spp. (Guzmán-Guzmán et al. 2023). Species of the *Trichoderma* genus are the most extensively studied due to their widely reported mycoparasitic properties (Guzmán-Guzmán et al. 2023). Notably, *T. virens* is considered an aggressive mycoparasite against several fungal pathogens (Guzmán-Guzmán et al. 2023). Other species such as *T. asperellum* and *T. spirale* exhibit mycoparasitic ability, completely inhibiting the mycelial growth of *Phytophthora palmivora* (M'Henni 2020). Additional studies have demonstrated the role of β-1,6-glucanases produced by *Trichoderma* spp. in inhibiting *P. citrophthora* (Druzhinina et al. 2011).

10.3.1.3 Lytic Enzymes

Enzymes function not only as biocatalysts in vital metabolic reactions but also play a crucial role in the survival of microorganisms in a precise niche. Several MBCAs release enzymes to hydrolyze proteins, chitin, cellulose, and hemicellulose, thereby suppressing phytopathogens (Panicker and Sayyed 2022). Several fungal and bacterial species have demonstrated their ability to release specific enzymes that target particular phytopathogens dwelling in the proximity of the rhizosphere (Mishra et al. 2020b). Lytic enzymes can modify or inhibit the synthesis of the cell wall, perforate the cell membrane, or degrade the cell wall of phytopathogens (Mota et al. 2017). *Lysobacter enzymogenes* has been identified as a significant antagonist agent in agriculture. *L. enzymogenes* exhibits a biocontrol effect against numerous phytopathogens such as bacteria, oomycetes, and fungi. Its antagonistic effect is conferred by its ability to produce abundant lytic enzymes such as cellulases, glucanases, and chitinases, which can degrade the cell walls of various fungi, including *R. solani, Sclerotinia sclerotiorum, Penicillium italicum, Colletotrichum gloeosporioides, P. capsica, P. nicotianae*, and *A. alternata* (Chen et al. 2020; Odhiambo 2021). Several reports highlight the role of proteases derived from soil bacteria in the biocontrol of plant-pathogenic bacteria and fungi. An antagonistic isolate of *B. amyloliquefaciens*, isolated from the jute rhizosphere, exhibited protease-mediated biocontrol action against *A. alternata, F. semitectum,* and *Macrophomina phaseolina* (Majumdar and Chakraborty 2017).

10.3.1.4 Siderophores

Siderophores are chelators secreted by microorganisms to capture iron when they are in an iron-deficient environment, an essential element for their survival (Drechsel and Winkelmann 2022). They play a crucial role in biocontrol against phytopathogens through competition. Among the most well-known siderophores are the pyoverdines produced by various species of the *Pseudomonas* genus (Díaz-Pérez et al. 2023). They also contribute to the stimulation of plant growth by reducing oxidative stress. Thus, studies conducted by Segarra et al. (2010) demonstrated the role of siderophores produced by *T. asperellum* T34 in controlling *F. oxysporum*, resulting in a reduction in tomato infestation and the promotion of root plant growth. Other studies have shown the role of siderophores produced by *T. hamatum* GD12 against *S. sclerotiorum* and their involvement in stimulating lettuce plant growth (Shaw et al. 2016). The siderophores produced by *Lysobacter antibioticus* HS124 also exhibited significant antifungal activity against *P. capsici*, which infects peppers (Hyun and Choi 2014).

10.3.2 Indirect Action Mechanisms of MBCAs

10.3.2.1 Nutrient and Spatial Competition

From a microbial standpoint, plant soils are environments where nutrients are limited, leading to competition among microorganisms (Pankhurst and Lynch 2005). MBCAs have the ability to grow rapidly, colonize plant infection sites, and consequently restrict the development of pathogens (Mesguida et al. 2023). This activity helps to reduce the incidence and severity of diseases, especially those caused by soilborne phytopathogens such as *Pythium* spp and *Phytophthora* spp. (Narayanasamy 2013). Competition among microorganisms for carbon, nitrogen, and other nutrients in the rhizosphere can be a mechanism of indirect control of pathogens. Certain bacterial microorganisms, such as *Pseudomonas* spp., have the ability to chelate iron sources in the form of ferric ions (Fe^3+) in the rhizosphere by producing siderophores, thereby reducing the availability of iron crucial for the pathogen's development (Kramer et al. 2020). *Trichoderma* spp. are also known for their ability to sequester ferric ions through siderophores (Woo et al. 2023). Studies conducted by Segarra et al. (2010) have demonstrated the significant role of siderophores produced by *T. asperellum* in antagonism against *F. oxysporum* f. sp. *lycopersici*. Among the *Aspergillus* genus species, non-pathogenic *A. flavus* has proven effective as a MBCA against *A. flavus* affecting cotton crops. This antagonistic effect is attributed to its significant nutritional and spatial competitive capabilities (Ehrlich and Mack 2014).

10.3.2.2 Induction of the Resistance Reaction in the Plant

The induction of resistance in the host plant is among the most exploited mechanisms in biological control against plant pathogens. It can be local or systemic depending on the nature and source of elicitors (abiotic or biotic) (M'Henni 2020). It can be induced by a variety of non-pathogenic microorganisms, saprophytes, and rhizobacteria such as *B. pumilus, B. subtilis, B. amyloliquefaciens*, and *B. thuringiensis* (Elad et al. 2007). Defense reactions in the host plant through the antagonistic effect of certain MBCAs induce the synthesis of different compounds by the plant. These compounds act directly on the development of the pathogen or form barriers that reduce infection (Akbar et al. 2023). Studies on several species belonging to the *Trichoderma* genus correlate their root-colonizing ability with their significant capacity to stimulate defense in host plants. This interaction between the MBCA and the plant induces the accumulation of peroxidases, chitinases, glucanases, phenylalanines, and hydroperoxide lyases, thereby activating defense reaction synthesis pathways and inducing the accumulation of phytoalexins (Narayanasamy 2013).

The pathway for synthesizing host resistance reactions begins with the expression of SAR, a plant defense mechanism activated in response to pathogen infection. The SAR typically results in the production of pathogenesis-related proteins (PR), including various enzymes such as chitinases and β-1,3-glucanases (Durrant and Dong 2004). Concurrently, there is an elevation in the concentration of salicylic acid (SA), both locally at the infection site and systemically in distant plant tissues (Kim and Lim 2023). Subsequently, the induction of ISR is initiated, marked by the synthesis of jasmonic acid and/or ethylene (Rettinassababady and Jeyalakshmi 2014). It is noteworthy that the participation of SA in ISR has been confirmed in specific plant-pathogen-rhizobacteria interactions. (Kamle et al. 2020). The expression of ISR, similar to SAR, occurs in three stages, including (1) Plant perception of active molecules produced by PGPR or elicitation of the phenomenon, (2) Transmission of a systemic signal within the plant, and (3) Expression of the defense mechanism(s) strictly induced to inhibit pathogen penetration into host tissues (Ongena and Thonart 2006). Several bacterial determinants of ISR have been characterized, and they are summarized in Table 10.2.

Table 10.2 Some bacterial determinants inducing the systemic reaction by MBCAs

Bacterial species	Bacterial determinants	Host plant	References
Pseudomonas fluorescens CHA0	Antibiotics (DAPG)	Arabidopsis	Iavicoli et al. (2003)
P. fluorescens WCS374	Siderophore	Radish	Leeman et al. (1995)
	Iron regulated factor		
	Lipopolysaccharide		
P. fluorescens WCS417	Lipopolysaccharide	Carnation	Peer and Schippers (1992)
P. fluorescens WCS374	Lipopolysaccharide	Tomato	Duijff et al. (1997)
	Iron regulated factor	Radish	Leeman et al. (1995)
	Lipopolysaccharide		
	Lipopolysaccharide	Arabidopsis	Van Wees et al. (1997)
P. putida WCS 358	Siderophore	Arabidopsis	Meziane et al. (2005)
	Lipopolysaccharide		
P. putida BTP1	Z,3-hexenal	Bean	Ongena et al. (2004)
Bacillus mycoides	Peroxidase, chitinase et β-1,3- glucanase	Sugar beet	Bargabus et al. (2004)
B. subtilis GB03 and IN937a	2,3-butanediol	Arabidopsis	Ryu et al. (2004)

10.4 Application Techniques

Recent developments in agricultural practices, transitioning from the use of synthetic pesticides to the adoption of organic farming, have highlighted the significance of harnessing microorganisms that perform analogous functions (Chakraborty et al. 2023). Biological control through the use of antagonistic microbial inoculants is a promising alternative to traditional chemical methods for crop protection. This approach is highly attractive due to its potential to significantly decrease dependence on chemical pesticides (Šunjka and Mechora 2022). Biopesticides presently constitute around 10% of the global pesticide market. This increase is attributed to restrictions on chemical pesticides, the emergence of pesticide resistance, and advancements in residue management (Marrone 2024). More than 3000 tons of biopesticide are produced worldwide each year, and this number is increasing rapidly (Chakraborty et al. 2023). Biopesticides exert a substantial impact on sustainable agricultural management, owing to their commendable traits aligning with both the principles of green chemistry and the threefold concept of sustainable development (Chakraborty et al. 2023). Throughout the entire process of agricultural development, biopesticides have played a crucial role in safeguarding plants and enhancing crop quality and productivity (Farhaoui et al. 2023b). This section provides a summary of the recent understanding of various commercially available microbial bioformulations and explores their practical applications in the field.

Various microorganisms have demonstrated their ability—in vitro—to effectively control phytopathogens and have been proposed as efficient MBCAs against crop-infecting pathogens (Farhaoui et al. 2023a, b, c; Khadiri et al. 2023). However, large-scale application of MBCAs encounters numerous challenges, which rely on progress in interdisciplinary research, including advancements in mass production and formulation techniques (Berg 2009). Over the past decades, enhancements have been made in application techniques and formulation technology, leading to enhanced crop protection performance (Singh et al. 2019; Šunjka and Mechora 2022). Numerous microorganisms are presently being investigated and utilized as biopesticides. Microbial biopesticides primarily consist of microorganisms such as fungi, bacteria, viruses, or protozoa. While each active component in microbial biopesticides is specifically tailored to target particular diseases, they exhibit versatility in combating a diverse range of plant diseases (Gupta and Dikshit 2010). Commonly utilized MBCAs include *Bacillus* spp., *Trichoderma* spp., *P. fluorescence*, *Agrobacterium radiobacter*, *Ampelomyces quisqualis*, *Coniothyrium atoxigenic Aspergillus niger,* and nonpathogenic *Fusarium* (Mishra and Arora 2016). These MBCAs can be administered to crops through various methods, such as spores, live organisms, or deceased organisms (Mishra et al. 2020a). A formulated microbial product comprises one or more MBCAs mixed with ingredients providing an initial nutrient source to the antagonists, protecting them from various environmental conditions, and enhancing the effectiveness and shelf life of the product (Jambhulkar and Sharma 2014). The commercial usage of MBCA requires an inoculum with high and enduring cell viability, as well as easy transportability. Additionally, comprehensive research is essential for the large-scale multiplication of the biocontrol agent. This investigation includes the identification of a less expensive and suitable medium, the determination of the type of formulation, the selection of filler material, the establishment of delivery systems, and the identification of optimal storage conditions to ensure a viable shelf life. Application guidelines are then formulated, taking into account all these aspects of bioformulation (Jambhulkar et al. 2016). Microbial inoculants are formulated as either liquid or dry formulations. Powdered formulations are generally utilized, containing concentrated organisms in either dry or wet powders. Application methods vary based on the powder composition, encompassing direct soil application, suspension in water, or dusting onto seeds. The predominant method for formulating granular products involves the amalgamation of MBCA and ingredients with the granules (Mishra et al. 2018). Developing an accurate delivery system is crucial to ensure the effectiveness of bioformulations in real field conditions (Berg 2009). The delivery system for MBCAs primarily relies on factors such as the specific pathogen targeted, the crop's growth stage, the characteristics and intensity of the disease, and the climatic conditions prevalent in the area (Desai et al. 2002). Microbial inoculum is administered through various techniques based on the mode of infection and the survival nature of the phytopathogen. An optimal formulation should enable the delivery of active MBCAs, ensuring they are applied at the right time and in the appropriate place. This is typically achieved through irrigation water, seed treatment, soil application, and root dip (Jambhulkar et al. 2016). Numerous formulations authorized by regulatory authorities worldwide are accessible for application in the management of plant diseases (Desai et al. 2002). Table 10.3 lists some microbial agents registered as biopesticides, their target diseases/organisms, and the corresponding delivery systems.

10.4.1 Seed Treatment

Seed treatment presents a convenient method for administering bacterial and fungal biocontrol agents, delivering a valuable and similarly effective approach to controlling plants against phytopathogens when compared to chemical seed treatment. Additionally, physiological seed treatments, like seed priming, have been employed to accelerate seed germination and enhance seedling survival (Kokalis-Burelle 2002). Seed treatment with MBCAs has proven to be a highly efficient strategy, especially in managing soil-borne pathogens (Mishra et al. 2018). Inspired by the pharmaceutical industry, seed coating was initially introduced to cereal seeds in the 1930s, gaining widespread commercial use in the 1960s. Over the past decade, there has been a substantial rise in research focused on seed treatment, driven by the quest for environmentally friendly technologies to combat diseases transmitted through seeds and soil (Moumni et al. 2023). Currently, seed treatment with MBCAs is widely used by crop and horticultural industries (Rocha et al. 2019). Coating seeds with microbial inoculum has the potential to deliver microbial agents in the right quantity, at the right time, and in the right place (Jambhulkar et al. 2016). The application of biological formulations to seeds is highly effective in facilitating the delivery of microbial inoculum to the spermosphere of plants (Mishra et al. 2018). Seed treatment with MBCAs allows beneficial micro-organisms to proliferate in the soil, protecting seedlings from pathogen attack during emergence and early growth stages. This, in turn, contributes to the establishment of healthy crop plants and ultimately improves overall crop yield (Moumni et al. 2023). Beneficial microorganisms introduced to the soil through seed treatment can also enhance both germination and plant growth (Rocha et al. 2019).

Fundamentally, the seed coating process with MBCAs entails situating the seeds within a container (like a cement mixer or a rotating drum), where a binder (such as an adhesive compound), a filler (if necessary for bulk), and active ingredients (MBCAs) are combined (Accinelli et al. 2018). Fillers can be made from single components or a combination of materials, with lime, talc, or peat being the most commonly used options (Hameeda et al. 2010; Rocha et al. 2019). Certain substances, like alginate, can serve dual purposes as both a binder and a filler (Lally et al. 2017). Lately, chitosan and biochar have found application as fillers/carriers for coating seeds with plant-beneficial microbes (Glodowska et al. 2017). Binders, whether synthetic or natural polymers like gum arabic (Rocha et al. 2019) or the polysaccharide Pelgel (Ugoji et al. 2006), methylcellulose (Farhaoui et al. 2023c), or carboxymethylcellulose (Zhou et al. 2017), are typically employed in the coating process to enhance the adhesion of exogenous materials to the seeds (Pedrini et al.

2017). Certain adhesives, such as xanthan gum and gum arabic, are utilized to extend the viability of beneficial plant microbes when applied to seeds (Jambhulkar et al. 2016). The selection of the suitable type and concentration of filler and binder is critical for the viability of the applied microbial inoculant and its effectiveness in protecting seeds against damping off (Rocha et al. 2019).

An industrial film-coating process has been utilized to apply plant-beneficial microbes to seeds, offering protection for various crops against pathogens. This technique has demonstrated effectiveness in preventing damping off through the application of *Trichoderma* spp. On cucumber and radish seeds (Cliquet and Scheffer 1997). In another seed treatment process, seed biopriming is an approach that involves treating seeds with bacterial or fungal biocontrol agents, followed by incubation in warm and humid conditions until just before radical emergence. This technique has the potential to enhance seedling establishment, as well as overall plant vigor in the field (Jambhulkar et al. 2016). Bioprimed sunflower seeds with UTPf76 and UTPf86 strains of *P. fluorescens* reduced seedling damping-off and improved seedling growth (Moeinzadeh et al. 2010). Chandra Nayaka et al. (2010) documented that the application of a conidial suspension of *T. harzianum* on maize seeds resulted in a significant reduction in the incidence of *F. verticillioides* and fumonisin. Mishra and Pandey (2010) demonstrated that priming field pea seeds with *B. subtilis*, *P. aeruginosa*, and *P. fluorescens* led to a reduction in the incidence of *Uromyces fabae* by approximately 20% under field conditions. Seed encapsulation is a specialized seed-coating process that involves enveloping both the seeds and microbial inoculants with a gelatinous or polymeric gel matrix, thereby prolonging the survival of the microbial inoculants on the seeds (Hubballi et al. 2023). This delivery method has a notable advantage in terms of user-friendliness and environmental safety. It achieves this by securely sealing the MBCAs until they are released during germination (Jambhulkar et al. 2016). Jambhulkar et al. (2016) reported a significant rise in vegetative growth *when P. putida* CC-FR2-4 and *B. subtilis* CC-pg104, encapsulated in alginate supplemented with humic acid, were inoculated into *Lactuca sativa* L. seeds.

Applying MBCAs to seeds is expected to significantly increase in the future, capturing substantial attention from seed marketers aiming to deliver beneficial microbes to plants as close to the target as possible. Moreover, the organisms introduced through this technique possess the selective advantage of being the initial colonizers of plant roots (Jambhulkar et al. 2016).

10.4.2 Soil Application

Multiple attempts have been undertaken to control diverse soil-borne plant pathogens by introducing natural substrates inhabited by MBCAs into the soil (Sesan and Csep 1992). The soil harbors both pathogenic and beneficial microbes. The application of microbial inoculants to the soil has the potential to enhance the population dynamics of MBCAs, thereby suppressing the proliferation and activity of phytopathogens at the infection site (Jambhulkar et al. 2016). Soil application of a peat-based formulation containing *P. fluorescens* Pf1 at a rate of 2.5 kg of formulation mixed with 25 kg of farmyard manure (FYM) demonstrated effective control of chickpea wilting (Mishra et al. 2018). Treatment of the soil with FYM enriched with *T. viride* and *P. fluorescens* significantly reduced sesame root rot caused by *Macrophomina* and decreased seedling mortality by 72.39% compared to the control (Khamari et al. 2022). Likewise, the application of FYM enriched with the Th3 strain of *T. harzianum* to the soil before sowing chickpeas effectively reduced *R. solani* root rot and significantly increased the yield (Jambhulkar et al. 2015). Soil drenching is the preferred method when biocontrol agents are overly susceptible to desiccation. By employing this method, the MBCA can establish a substantial population, rendering essential nutrients inaccessible to soil pathogens and effectively diminishing the progression of the disease (Jambhulkar et al. 2016). Soil drenching with a product based on the *T. harzianum* strain T-22 and the *T. virens* strain G-41 T significantly reduced the severity of yellowing and rhizome rot induced by *F. oxysporum* f.sp. *zingiberi* in ginger, resulting in an increased harvest yield compared to the untreated control (Mersha and Ibarra-Bautista 2022). Application of *P. cepacia* strain N24 at a rate of 500 ml per m^2 in seedbeds within greenhouses was observed to be effective in suppressing wilting of sunflower plants (Hebbar et al. 1991). Soil drenching with a suspension containing *T. viride* effectively reduced the damping off of cowpeas caused by *Colletotrichum truncatum* (brown blotch) (Bankole and Adebanjo 1998). Drenching the soil with aqueous suspensions of microbial inoculants may not result in a uniform distribution of the control agents throughout the treated soil.

10.4.3 Foliar Spray

Numerous works have been made to administer liquid formulations containing MBCAs onto plant leaves for the management of foliar pathogens. The effectiveness of foliar application is markedly influenced by the microclimate within the plant canopy. The plant canopy exhibits diverse concentrations of nutrients, including organic acids, sugars, and amino acids, which are exuded via lenticels, stomata, wounds, and hydathodes. These factors play a significant role in affecting the survival and activity of MBCAs in the phylloplane (Jambhulkar et al. 2016). Through foliar spraying, plants can receive treatment throughout the season, either targeting specific growth phases to stimulate plant growth and enhance resilience or addressing particular climatic conditions to combat phytopathogens. Application frequencies of microbial inoculant sprays may vary depending on the type of plant pathogen and the efficacy of the spray. Successful spraying is generally achieved through the timely application of an effective microbial formulation and the use of appropriate spray equipment (Preininger et al. 2018). In spray applications, it is advisable to utilize microorganisms capable of effectively colonizing leaf tissues and thriving in difficult environmental conditions. This is crucial because biocontrol agents are subject to diverse abiotic stresses, like UV radiation and drought, after foliar application (Marian and Shimizu 2019). Additives contained in foliar sprays, such as emulsifiers, stickers, adjuvants, and spreaders facilitate the adhesion of microbial inoculum to plant tissues (Jambhulkar et al. 2016). Foliar spraying of tomato plants with *B. subtilis* strain G4L1 significantly decreased the incidence of bacterial wilt, decreased the population of the pathogen in the stems, and significantly increased yield compared to the control (Fu et al. 2020). Under field conditions, foliar spraying of canola with a suspension containing *B. amyloquifaciens* (BS6), *B. amyloquifaciens* (E16), *P. chlororaphis* (PA-23), and *Pseudomonas spp.* (DF41) proved highly effective in controlling *S. sclerotiorum*, the causative agent of stem rot (Fernando et al. 2007). Recently, cost-effective, potent microbial bioformulations designed for foliar spray with good stability properties have become available and are successfully used worldwide. However, the development of biological control for foliar diseases lags behind that of soil-borne diseases. This discrepancy can be attributed to the availability and ease of application of chemicals targeting foliar diseases, as well as the relatively less impressive results obtained with MBCAs compared to those achieved by widely used synthetic chemicals (Jambhulkar et al. 2016).

10.4.4 Root Dip

Phytopathogens can be transmitted through soil or seeds, parasitizing the plant by penetrating through the roots. Therefore, protecting the rhizospheric environment through prior colonization by MBCAs may stop the installation of a host-parasite relationship. The roots of the seedlings can be treated with cells or conidial suspensions of MBCAs. This can be done by drenching the MBCAs in the nursery bed or by immersing roots in a suspension of MBCA before transplanting. This approach is particularly effective for rice and several vegetable crops (Singh and Zaidi 2002). Rabindran and Vidhyasekaran (1996) observed a reduction in sheath blight severity and an increase in yield by dipping rice seedlings in a *P. fluorescens* (PfALR1)-based formulation before transplanting. Similarly, Prashant P. Jambhulkar and Sharma (2013) demonstrated a decrease in bacterial leaf blight of rice by dipping paddy seedlings in a suspension containing *P. fluorescens* for 2 h before transplanting. Dipping of tomato seedlings' roots in a suspension (109 CFU mL-1) of *Streptomyces* sp. NEAU-HV9 before transplantation showed effective control of bacterial wilt induced by *Ralstonia solanacearum,* with an efficacy rate of 82% (Ling et al. 2020). Dipping the roots of 4-month-old pepper plants in suspensions of *B. amyloliquefaciens* RS11 (108 CFU mL^{-1}) for 30 min before planting significantly decreased the incidence of *verticillium* wilt (Guenoun et al. 2019).

10.5 Conclusion

The growth of environmental concerns and resulting regulations severely limit the use of chemical pesticides worldwide, creating an urgent need for safer and environmentally friendly solutions. Microbial biological control agents have emerged as a promising strategy in this regard, offering an ecological approach to controlling plant pathogens. However, despite their potential, MBCAs are not without significant challenges. The variable efficacy of MBCAs is one of the major obstacles to their widespread adoption (Lahlali et al. 2022). Their performance can be influenced by a myriad of environmental factors, making it difficult to guarantee consistent results in the field. Moreover, the development of stable and effective formulations for microbial bioagents requires a deep understanding of the complex interactions between the MBCA, formulation components, and the environment. These challenges demand ongoing research and innovation efforts to improve the efficacy and practicality of MBCAs.

The commercialization of biocontrol products is expanding, but is hindered by complex and varying regulations worldwide (Bejarano and Puopolo 2020). This regulatory diversity creates significant barriers to the adoption of MBCAs, particularly in regions like the European Union. Harmonizing specific regulations for biopesticides could stimulate their widespread adoption and facilitate their integration into sustainable agricultural practices.

Furthermore, challenges posed by climate change add an additional layer of complexity. Climate variations can compromise the efficacy of MBCAs and require constant adaptation of biocontrol strategies (Kang and Banga 2013). Future research should therefore focus on improving the production, formulation, and delivery of MBCAs, as well as developing simpler and more universal regulations to foster their adoption.

Finally, it is crucial to recognize that MBCAs are not a one-size-fits-all solution and also present practical and socio-economic challenges. The potentially higher cost of MBCAs compared to traditional chemical pesticides, as well as farmer acceptance and adoption, require special attention. Additionally, the need for careful planning and monitoring of interactions between MBCAs and other pest or disease management methods underscores the complexity of the biological control landscape in agriculture

Acknowledgements This research was funded by the Phytopathology Unit within the Department of Plant Protection and Environment at the Ecole Nationale d'Agriculture de Meknès (Morocco).

References

Ab Rahman SFS, Singh E, Pieterse CM, Schenk PM (2018) Emerging microbial biocontrol strategies for plant pathogens. Plant Sci 267:102–111

Abdel-Aziz MS, Ghareeb MA, Hamed AA, Rashad EM, El-Sawy ER, Saad IM, Ghoneem KM (2021) Ethyl acetate extract of Streptomyces spp. isolated from Egyptian soil for management of *Fusarium oxysporum*: the causing agent of wilt disease of tomato. Biocatal Agric Biotechnol 37:102185

Abid M, Khan MA, Mushtaq S, Afzaal S, Haider MS (2018) A comprehensive review on mycoviruses as biological control agent. World J Biol Biotechnol 3:187–192

Accinelli C, Abbas HK, Shier WT (2018) A bioplastic-based seed coating improves seedling growth and reduces production of coated seed dust. J Crop Improv 32:318–330. https://doi.org/10.1080/15427528.2018.1425792

Akbar MU, Aqeel M, Shah MS, Jeelani G, Iqbal N, Latif A, Elnour RO, Hashem M, Alzoubi OM, Habeeb T (2023) Molecular regulation of antioxidants and secondary metabolites act in conjunction to defend plants against pathogenic infection. S Afr J Bot 161:247–257

Al-Ani LKT, Aguilar-Marcelino L, Fiorotti J, Sharma V, Sarker MS, Furtado EL, Wijayawardene NN, Herrera-Estrella A (2020) Biological control agents and their importance for the plant health. In: Microbial services in restoration ecology. Elsevier, pp 13–36

Araújo MF, Castanheira EM, Sousa SF (2023) The buzz on insecticides: a review of uses, molecular structures, targets, adverse effects, and alternatives. Molecules 28:3641

Babalola OO, Olanrewaju OS, Amoo AE, Ajilogba CF, Chukwuneme FC, Ojuederie OB, Omomowo OI (2023) The functionality of plant-microbe interactions in disease suppression. J King Saud Univ-Sci 35(8):102893

Bankole SA, Adebanjo A (1998) Efficacy of some fungal and bacterial isolates in controlling wet rot disease of cowpea caused by *Pythium aphanidermatum*. Plant Prot Trop 11:37–43

Bargabus RL, Zidack NK, Sherwood JE, Jacobsen BJ (2004) Screening for the identification of potential biological control agents that induce systemic acquired resistance in sugar beet. Biol Control 30:342–350

Barua P, Nath PD (2018) Bacteriophages: a potential next generation biocontrol tool for plant disease management. Int J Curr Microbiol App Sci 7:1103–1112

Bejarano A, Puopolo G (2020) Bioformulation of microbial biocontrol agents for a sustainable agriculture. In: De Cal A, Melgarejo P, Magan N (eds) How research can stimulate the development of commercial biological control against plant diseases. Springer, Cham, pp 275–293

Berendsen RL, Pieterse CM, Bakker PA (2012) The rhizosphere microbiome and plant health. Trends Plant Sci 17:478–486

Berg G (2009) Plant–microbe interactions promoting plant growth and health: perspectives for controlled use of microorganisms in agriculture. Appl Microbiol Biotechnol 84:11–18. https://doi.org/10.1007/s00253-009-2092-7

Bonaterra A, Badosa E, Daranas N, Francés J, Roselló G, Montesinos E (2022) Bacteria as biological control agents of plant diseases. Microorganisms 10:1759

Boro M, Sannyasi S, Chettri D, Verma AK (2022) Microorganisms in biological control strategies to manage microbial plant pathogens: a review. Arch Microbiol 204:666. https://doi.org/10.1007/s00203-022-03279-w

Cabrefiga J, Bonaterra A, Montesinos E (2007) Mechanisms of antagonism of *Pseudomonas fluorescens* EPS62e against *Erwinia amylovora*, the causal agent of fire blight. Int Microbiol 10:123

Calvo-Garrido C, Elmer PAG, Viñas I, Usall J, Bartra E, Teixidó N (2013) Biological control of *botrytis bunch* rot in organic wine grapes with the yeast antagonist C *andida sake* CPA-1. Plant Pathol 62:510–519. https://doi.org/10.1111/j.1365-3059.2012.02684.x

Carstens AB, Djurhuus AM, Kot W, Hansen LH (2019) A novel six-phage cocktail reduces *Pectobacterium atrosepticum* soft rot infection in potato tubers under simulated storage conditions. FEMS Microbiol Lett 366:fnz101

Chakraborty N, Mitra R, Pal S, Ganguly R, Acharya K, Minkina T, Sarkar A, Keswani C (2023) Biopesticide consumption in India: insights into the current trends. Agriculture 13:557

Chandra Nayaka S, Niranjana SR, Uday Shankar AC, Niranjan Raj S, Reddy MS, Prakash HS, Mortensen CN (2010) Seed biopriming with novel strain of *Trichoderma harzianum* for the control of toxigenic *Fusarium verticillioides* and fumonisins in maize. Arch Phytopathol Plant Protect 43:264–282. https://doi.org/10.1080/03235400701803879

Chen DM, Yang HJ, Huang JG, Yuan L (2020) *Lysobacter enzymogenes* LE16 autolysates have potential as biocontrol agents—Lysobacter sp. autolysates as biofungicide. J Appl Microbiol 129:1684–1692

Choub V, Won S-J, Ajuna HB, Moon J-H, Choi S-I, Lim H-I, Ahn YS (2022) Antifungal activity of volatile organic compounds from *Bacillus velezensis* CE 100 against *Colletotrichum gloeosporioides*. Horticulturae 8:557

Cliquet S, Scheffer RJ (1997) Influence of culture conditions on growth and survival of conidia of *Trichoderma* spp. coated on seeds. Biocontrol Sci Tech 7:171–182. https://doi.org/10.1080/09583159730875

Damalas CA (2016) Safe food production with minimum and judicious use of pesticides. In: Selamat J, Iqbal SZ (eds) Food safety. Springer, Cham, pp 43–55

Das K, Prasanna R, Saxena AK (2017) Rhizobia: a potential biocontrol agent for soilborne fungal pathogens. Folia Microbiol (Praha) 62:425–435. https://doi.org/10.1007/s12223-017-0513-z

Desai S, Reddy MS, Kloepper JW (2002) Comprehensive testing of biocontrol agents. In: Biological control of crop diseases. CRC Press, pp 387–420

Di Canito A, Mateo-Vargas MA, Mazzieri M, Cantoral J, Foschino R, Cordero-Bueso G, Vigentini I (2021) The role of yeasts as biocontrol agents for pathogenic fungi on postharvest grapes: a review. Foods 10:1650

Díaz-Pérez SP, Solis CS, López-Bucio JS, Valdez Alarcón JJ, Villegas J, Reyes-De La Cruz H, Campos-Garcia J (2023) Pathogenesis in *Pseudomonas aeruginosa* PAO1 biofilm-associated is dependent on the pyoverdine and pyocyanin siderophores by quorum sensing modulation. Microb Ecol 86:727–741. https://doi.org/10.1007/s00248-022-02095-5

Divya M, Patro T, Rani YS, Triveni U, Anuradha N (2017) Detection of resistant entries of little millet against banded blight disease in cited by rhizoctonia solani kuhn. Front Crop Imp 5(1):74–75

Drechsel H, Winkelmann G (2022) Iron chelation and siderophores. In: Transition metals in microbial metabolism. CRC Press, pp 1–49

Druzhinina IS, Seidl-Seiboth V, Herrera-Estrella A, Horwitz BA, Kenerley CM, Monte E, Mukherjee PK, Zeilinger S, Grigoriev IV, Kubicek CP (2011) Trichoderma: the genomics of opportunistic success. Nat Rev Microbiol 9:749–759

Duijff BJ, Gianinazzi-Pearson V, Lemanceau P (1997) Involvement of the outer membrane lipopolysaccharides in the endophytic colonization of tomato roots by biocontrol *Pseudomonas fluorescens* strain WCS417r. New Phytol 135:325–334

Durel L, Estrada-Peña A, Franc M, Mehlhorn H, Bouyer J (2015) Integrated fly management in European ruminant operations from the perspective of directive 2009/128/EC on sustainable use of pesticides. Parasitol Res 114:379–389. https://doi.org/10.1007/s00436-014-4273-2

Durrant WE, Dong X (2004) systemic acquired resistance. Annu Rev Phytopathol 42:185–209. https://doi.org/10.1146/annurev.phyto.42.040803.140421

Ehrlich KC, Mack BM (2014) Comparison of expression of secondary metabolite biosynthesis cluster genes in *Aspergillus flavus*, A. parasiticus, and A. oryzae. Toxins 6:1916–1928

Elad Y, Freeman S (2002) Biological control of fungal plant pathogens. In: Kempken F (ed) Agricultural applications. Springer Berlin Heidelberg, Berlin, Heidelberg, pp 93–109

Elad Y, Williamson B, Tudzynski P, Delen N (eds) (2007) Botrytis: biology, pathology and control. Springer Netherlands, Dordrecht

Ezrari S, Mhidra O, Radouane N, Tahiri A, Polizzi G, Lazraq A, Lahlali R (2021) Potential role of rhizobacteria isolated from citrus rhizosphere for biological control of citrus dry root rot. Plants 10:872

Ezzouggari R, Bahhou J, Taoussi M, Seddiqi Kallali N, Aberkani K, Barka EA, Lahlali R (2024) Yeast warriors: exploring the potential of yeasts for sustainable citrus post-harvest disease management. Agronomy 14:288

FAO (2017) The future of food and agriculture: trends and challenges. FAO

Farhaoui A, Adadi A, Tahiri A, El Alami N, Khayi S, Mentag R, Ezrari S, Nabil R, Mokrini F, Belabess Z, Lahlali R (2022) Biocontrol potential of plant growth-promoting rhizobacteria (PGPR) against *Sclerotiorum rolfsii* diseases on sugar beet (*Beta vulgaris* L.). Physiol Mol Plant Pathol 119:101829. https://doi.org/10.1016/j.pmpp.2022.101829

Farhaoui A, El Alami N, Khadiri M, Ezrari S, Radouane N, Baala M, Tahiri A, Lahlali R (2023a) Biological control of diseases caused by *Rhizoctonia solani* AG-2-2 in sugar beet (*Beta vulgaris* L.) using plant growth-promoting rhizobacteria (PGPR). Physiol Mol Plant Pathol 124:101966

Farhaoui A, Tahiri A, Khadiri M, El Alami N, Lahlali R (2023b) Fungal root rots of sugar beets: a review of common causal agents and management strategies. In: Gesunde Pflanz. Springer, pp 1–30

Farhaoui A, Tahiri A, Khadiri M, Radouane N, Alami NE, Lahlali R (2023c) First report of *Fusarium tricinctum* causing root rot on sugar beet (*Beta vulgaris* L.) in Morocco. J Plant Pathol 105:1757–1757. https://doi.org/10.1007/s42161-023-01507-w

Farzand A, Moosa A, Zubair M, Khan AR, Massawe VC, Tahir HAS, Sheikh TMM, Ayaz M, Gao X (2019) Suppression of *Sclerotinia sclerotiorum* by the induction of systemic resistance and regulation of antioxidant pathways in tomato using fengycin produced by *Bacillus amyloliquefaciens* FZB42. Biomolecules 9:613

Fernando WGD, Nakkeeran S, Zhang Y, Savchuk S (2007) Biological control of *Sclerotinia sclerotiorum* (Lib.) de Bary by Pseudomonas and *Bacillus* species on canola petals. Crop Prot 26:100–107

Ferraz P, Cássio F, Lucas C (2019) Potential of yeasts as biocontrol agents of the phytopathogen causing cacao witches' broom disease: is microbial warfare a solution? Front Microbiol 10:1766

Fira D, Dimkić I, Berić T, Lozo J, Stanković S (2018) Biological control of plant pathogens by *Bacillus* species. J Biotechnol 285:44–55

Fu H-Z, Marian M, Enomoto T, Hieno A, Ina H, Suga H, Shimizu M (2020) Biocontrol of tomato bacterial wilt by foliar spray application of a novel strain of endophytic *Bacillus* sp. Microbes Environ 35:ME20078

Gašić S, Tanović B (2013) Biopesticide formulations, possibility of application and future trends. Pestic Fitomed 28:97–102

Ghorbanpour M, Omidvari M, Abbaszadeh-Dahaji P, Omidvar R, Kariman K (2018) Mechanisms underlying the protective effects of beneficial fungi against plant diseases. Biol Control 117:147–157

Glodowska M, Schwinghamer T, Husk B, Smith D (2017) Biochar based inoculants improve soybean growth and nodulation. Agric Sci 8:1048–1064

Grzegorczyk M, Żarowska B, Restuccia C, Cirvilleri G (2017) Postharvest biocontrol ability of killer yeasts against *Monilinia fructigena* and *Monilinia fructicola* on stone fruit. Food Microbiol 61:93–101

Gu Q, Yang Y, Yuan Q, Shi G, Wu L, Lou Z, Huo R, Wu H, Borriss R, Gao X (2017) Bacillomycin D produced by *Bacillus amyloliquefaciens* is involved in the antagonistic interaction with the plant-pathogenic fungus Fusarium graminearum. Appl Environ Microbiol 83:e01075–e01017. https://doi.org/10.1128/AEM.01075-17

Guenoun K, Chattaoui M, Bouri M, Rhouma A, Naghmouchi K, Raies A (2019) Biological control of growth promoting rhizobacteria against verticillium wilt of pepper plant. Biologia (Bratisl) 74:237–250. https://doi.org/10.2478/s11756-018-00169-9

Gupta S, Dikshit AK (2010) Biopesticides: an ecofriendly approach for pest control. J Biopestic 3:186

Guzmán-Guzmán P, Kumar A, de Los Santos-Villalobos S, Parra-Cota FI, Orozco-Mosqueda M del C, Fadiji AE, Hyder S, Babalola OO, Santoyo G (2023) *Trichoderma* species: our best fungal allies in the biocontrol of plant diseases—a review. Plants 12:432

Haas D, Keel C (2003) Regulation of antibiotic production in root-colonizing *Pseudomonas* spp. and relevance for biological control of plant disease. Annu Rev Phytopathol 41:117–153

Hameeda B, Harini G, Rupela OP, Kumar Rao JVDK, Reddy G (2010) Biological control of chickpea collar rot by Co-inoculation of antagonistic bacteria and compatible rhizobia. Indian J Microbiol 50:419–424. https://doi.org/10.1007/s12088-011-0083-8

Hebbar P, Berge O, Heulin T, Singh SP (1991) Bacterial antagonists of sunflower (*Helianthus annuus* L.) fungal pathogens. Plant Soil 133:131–140. https://doi.org/10.1007/BF00011907

Heimpel GE, Mills NJ (2017) Biological control. Cambridge University Press

Hu W, Gao Q, Hamada MS, Dawood DH, Zheng J, Chen Y, Ma Z (2014) Potential of *Pseudomonas chlororaphis* subsp. *aurantiaca* strain Pcho10 as a biocontrol agent against *Fusarium graminearum*. Phytopathology 104:1289–1297. https://doi.org/10.1094/PHYTO-02-14-0049-R

Hubballi M, Rajamanickam S, Mawar R, Tuladhar R, Singh A, Sayyed RZ, Nakkeeran S (2023) PGPM: fundamental, bioformulation, commercialization, and success at farmer's field. In: Mawar R, Sayyed RZ, Sharma SK, Sattiraju KS (eds) Plant growth promoting microorganisms of arid region. Springer Nature Singapore, Singapore, pp 257–301

Hyun I-H, Choi W (2014) Phytophthora species, new threats to the plant health in Korea. Plant Pathol J 30:331

Iavicoli A, Boutet E, Buchala A, Métraux J-P (2003) Induced systemic resistance in *Arabidopsis thaliana* in response to root inoculation with *Pseudomonas fluorescens* CHA0. Mol Plant-Microbe Interact 16:851–858. https://doi.org/10.1094/MPMI.2003.16.10.851

Jambhulkar PP, Sharma P (2013) Promotion of rice seedling growth characteristics by development and use of bioformulation of *Pseudomonas fluorescens*. Indian J Agric Sci 83:136–142

Jambhulkar PP, Sharma P (2014) Development of bioformulation and delivery system of *Pseudomonas fluorescens* against bacterial leaf blight of rice (*Xanthomonas oryzae* pv. *oryzae*). J Environ Biol 35:843

Jambhulkar PP, Sharma P, Meghwal ML (2015) Additive effect of soil application with *Trichoderma* enriched FYM along with seed treatment and drenching with *Trichoderma* formulation for management of wet root rot caused by *Rhizoctonia solani* in chickpea. J Pure Appl Microbiol 9:1–8

Jambhulkar PP, Sharma P, Yadav R (2016) Delivery systems for introduction of microbial inoculants in the field. In: Singh DP, Singh HB, Prabha R (eds) Microbial inoculants in sustainable agricultural productivity. Springer India, New Delhi, pp 199–218

Jamiolkowska A (2020) Natural compounds as elicitors of plant resistance against diseases and new biocontrol strategies. Agronomy 10:173

Jindal V, Dhaliwal GS, Koul O (2013) Pest management in 21st century: roadmap for future. Biopestic Int 9:22

Jogaiah S, Abdelrahman M, Tran LP, Ito S (2018) Different mechanisms of *Trichoderma virens* -mediated resistance in tomato against Fusarium wilt involve the jasmonic and salicylic acid pathways. Mol Plant Pathol 19:870–882. https://doi.org/10.1111/mpp.12571

Kamle M, Borah R, Bora H, Jaiswal AK, Singh RK, Kumar P (2020) Systemic acquired resistance (SAR) and induced systemic resistance (ISR): role and mechanism of action against phytopathogens. In: Hesham AE-L, Upadhyay RS, Sharma GD, Manoharachary C, Gupta VK (eds) Fungal biotechnology and bioengineering. Springer, Cham, pp 457–470

Kang MS, Banga SS (2013) Combating climate change: an agricultural perspective. CRC Press

Kanouni L, Larous L, Mezaache-Aichour S (2018) Inhibitory effect of rhizobia isolated from several leguminous against phytopathogenic fungi. Annu Res Rev Biol 22:1–16

Kawasaki K, Lichtenberg E (2015) Quality versus quantity effects of pesticides: joint estimation of quality grade and crop yield. In: Agricultural and Applied Economics Association's 2015 AAEA annual meeting. San Francisco, CA, United States

Kennedy AC (1999) Bacterial diversity in agroecosystems. In: Invertebrate biodiversity as bioindicators of sustainable landscapes. Elsevier, pp 65–76

Khadiri M, Boubaker H, Askarne L, Ezrari S, Radouane N, Farhaoui A, El Hamss H, Tahiri A, Barka EA, Lahlali R (2023) Bacillus cereus B8W8 an effective bacterial antagonist against major postharvest fungal pathogens of fruit. Postharvest Biol Technol 200:112315

Khamari B, Hasmi SK, Samal KC, Sahoo JP, Senapati AK, Ranasingh N, Beura SK, Samal T (2022) Impact of microbial rivals and natural alterations on root decay and plant development in sesame. Indian Phytopathol 75:1075–1083. https://doi.org/10.1007/s42360-022-00552-2

Khan RAA, Najeeb S, Mao Z, Ling J, Yang Y, Li Y, Xie B (2020) Bioactive secondary metabolites from *Trichoderma* spp. against phytopathogenic bacteria and root-knot nematode. Microorganisms 8:401

Khan BA, Nadeem MA, Nawaz H, Amin MM, Abbasi GH, Nadeem M, Ali M, Ameen M, Javaid MM, Maqbool R, Ikram M, Ayub MA (2023) Pesticides: impacts on agriculture productivity, environment, and management strategies. In: Aftab T (ed) Emerging contaminants and plants. Springer, Cham, pp 109–134

Kim T-J, Lim G-H (2023) Salicylic acid and mobile regulators of systemic immunity in plants: transport and metabolism. Plants 12:1013

Köhl J, Kolnaar R, Ravensberg WJ (2019) Mode of action of microbial biological control agents against plant diseases: relevance beyond efficacy. Front Plant Sci 10:845

Kokalis-Burelle N (2002) Biological control of tomato diseases. In: Biological control of crop diseases. CRC Press, pp 239–276

Kramer J, Özkaya Ö, Kümmerli R (2020) Bacterial siderophores in community and host interactions. Nat Rev Microbiol 18:152–163

Kumar J, Ramlal A, Mallick D, Mishra V (2021) An overview of some biopesticides and their importance in plant protection for commercial acceptance. Plants 10:1185

Lahlali R, Ezrari S, Radouane N, Kenfaoui J, Esmaeel Q, El Hamss H, Belabess Z, Barka EA (2022) Biological control of plant pathogens: a global perspective. Microorganisms 10:596

Lahmamsi H, Ananou S, Lahlali R, Tahiri A (2024) Lactic acid bacteria as an eco-friendly approach in plant production: current state and prospects. Folia Microbiol (Praha). https://doi.org/10.1007/s12223-024-01146-3

Lally RD, Galbally P, Moreira AS, Spink J, Ryan D, Germaine KJ, Dowling DN (2017) Application of endophytic *Pseudomonas fluorescens* and a bacterial consortium to *Brassica napus* can increase plant height and biomass under greenhouse and field conditions. Front Plant Sci 8:2193

Lecomte C, Alabouvette C, Edel-Hermann V, Robert F, Steinberg C (2016) Biological control of ornamental plant diseases caused by *Fusarium oxysporum*: a review. Biol Control 101:17–30

Leeman M, Van Pelt JA, Den Ouden FM, Heinsbroek M, Bakker PAHM, Schippers B (1995) Induction of systemic resistance by *Pseudomonas fluorescens* in radish cultivars differing in susceptibility to fusarium wilt, using a novel bioassay. Eur J Plant Pathol 101:655–664. https://doi.org/10.1007/BF01874869

Ling L, Han X, Li X, Zhang X, Wang H, Zhang L, Cao P, Wu Y, Wang X, Zhao J (2020) A *Streptomyces* sp. NEAU-HV9: isolation, identification, and potential as a biocontrol agent against *Ralstonia solanacearum* of tomato plants. Microorganisms 8:351

Liu J, Sui Y, Wisniewski M, Droby S, Liu Y (2013) Utilization of antagonistic yeasts to manage postharvest fungal diseases of fruit. Int J Food Microbiol 167:153–160

Lubchenco J, Haugan PM (2023) A sustainable ocean economy for 2050: approximating its benefits and costs. In: Lubchenco J, Haugan PM (eds) The blue compendium. Springer, Cham, pp 681–714

M'Henni YB (2020) Gestion de la maladie de dépérissement du pommier: criblage in vitro et in planta des activités protectrices d'une collection de microorganismes contre les Pythiacées et caractérisation chimique du principal actif produit par l'isolat A. westerdijkiae A7. PhD thesis, Université Paris-Saclay

Majumdar S, Chakraborty U (2017) Optimization of protease production from plant growth promoting *Bacillus amyloliquefaciens* showing antagonistic activity against phytopathogens. Int J Pharm Bio Sci 8:635–642

Marian M, Shimizu M (2019) Improving performance of microbial biocontrol agents against plant diseases. J Gen Plant Pathol 85:329–336. https://doi.org/10.1007/s10327-019-00866-6

Marrone PG (2024) Status of the biopesticide market and prospects for new bioherbicides. Pest Manag Sci 80:81–86. https://doi.org/10.1002/ps.7403

McMullin DR, Renaud JB, Barasubiye T, Sumarah MW, Miller JD (2017) Metabolites of *Trichoderma* species isolated from damp building materials. Can J Microbiol 63:621–632. https://doi.org/10.1139/cjm-2017-0083

Mersha Z, Ibarra-Bautista MA (2022) Evaluation of selected biological and chemical treatments against soil-borne diseases of ginger in high tunnel production. Horticulturae 8:870

Mesguida O, Haidar R, Yacoub A, Dreux-Zigha A, Berthon J-Y, Guyoneaud R, Attard E, Rey P (2023) Microbial biological control of fungi associated with grapevine trunk diseases: a review of strain diversity, modes of action, and advantages and limits of current strategies. J Fungi 9:638

Meziane H, Van Der Sluis I, Van Loon LC, Höfte M, Bakker PAHM (2005) Determinants of *Pseudomonas putida* WCS358 involved in inducing systemic resistance in plants. Mol Plant Pathol 6:177–185. https://doi.org/10.1111/j.1364-3703.2005.00276.x

Mishra J, Arora NK (2016) Bioformulations for plant growth promotion and combating phytopathogens: a sustainable approach. In: Bioformulations: for sustainable agriculture. Springer, pp 3–33

Mishra RK, Pandey KK (2010) Effect of application of PGPR and neemazal on management of pea rust (Uromyces fabae). J Basic Appl Microbiol 12:115–119

Mishra RK, Bohra A, Kamaal N, Kumar K, Gandhi K, Gk S, Saabale PR, Sj SN, Sarma BK, Kumar D, Mishra M, Srivastava DK, Singh NP (2018) Utilization of biopesticides as sustainable solutions for management of pests in legume crops: achievements and prospects. Egypt J Biol Pest Control 28:3. https://doi.org/10.1186/s41938-017-0004-1

Mishra J, Dutta V, Arora NK (2020a) Biopesticides in India: technology and sustainability linkages. 3 Biotech 10:210. https://doi.org/10.1007/s13205-020-02192-7

Mishra P, Mishra J, Dwivedi SK, Arora NK (2020b) Microbial enzymes in biocontrol of phytopathogens. In: Arora NK, Mishra J, Mishra V (eds) Microbial enzymes: roles and applications in industries. Springer Singapore, Singapore, pp 259–285

Moeinzadeh A, Sharif-Zadeh F, Ahmadzadeh M, Tajabadi FH (2010) Biopriming of sunflower (*"Helianthus annuus"* L.) seed *with 'Pseudomonas fluorescens'* for improvement of seed invigoration and seedling growth. Aust J Crop Sci 4:564–570

Mota MS, Gomes CB, Souza IT, Moura AB (2017) Bacterial selection for biological control of plant disease: criterion determination and validation. Braz J Microbiol 48:62–70

Moumni M, Brodal G, Romanazzi G (2023) Recent innovative seed treatment methods in the management of seedborne pathogens. Food Secur 15:1365–1382. https://doi.org/10.1007/s12571-023-01384-2

Narayanasamy P (2013) Mechanisms of action of bacterial biological control agents. In: Biological management of diseases of crops. Springer Netherlands, Dordrecht, pp 295–429

Nega A (2014) Review on concepts in biological control of plant pathogens. J Biol Agric Healthc 4:33–54

Odhiambo B (2021) Antimicrobial compounds biosynthesis and biocontrol mechanisms of Lysobacter enzymogenes. Pan Afr Sci J 1:85–102

Oerke E-C (2006) Crop losses to pests. J Agric Sci 144:31–43

Ongena M, Thonart P (2006) Resistance induced in plants by non-pathogenic microorganisms: elicitation and defense responses. In: Floriculture, ornamental and plant biotechnology: advances and topical issues. Global Science Books, London, pp 447–463

Ongena M, Duby F, Rossignol F, Fauconnier M-L, Dommes J, Thonart P (2004) Stimulation of the lipoxygenase pathway is associated with systemic resistance induced in bean by a nonpathogenic *Pseudomonas* strain. Mol Plant-Microbe Interact 17:1009–1018. https://doi.org/10.1094/MPMI.2004.17.9.1009

Oro L, Feliziani E, Ciani M, Romanazzi G, Comitini F (2014) Biocontrol of postharvest brown rot of sweet cherries by *Saccharomyces cerevisiae* Disva 599, Metschnikowia pulcherrima Disva 267 and *Wickerhamomyces anomalus* Disva 2 strains. Postharvest Biol Technol 96:64–68

Oso S, Walters M, Schlechter RO, Remus-Emsermann MN (2019) Utilisation of hydrocarbons and production of surfactants by bacteria isolated from plant leaf surfaces. FEMS Microbiol Lett 366:fnz061

Pal KK, Gardener BM (2006) Biological control of plant pathogens. Plant Health Instr 2:1117–1142

Pandit MA, Kumar J, Gulati S, Bhandari N, Mehta P, Katyal R, Rawat CD, Mishra V, Kaur J (2022) Major biological control strategies for plant pathogens. Pathogens 11:273

Pang Z, Chen J, Wang T, Gao C, Li Z, Guo L, Xu J, Cheng Y (2021) Linking plant secondary metabolites and plant microbiomes: a review. Front Plant Sci 12:621276

Panicker S, Sayyed RZ (2022) Hydrolytic enzymes from PGPR against plant fungal pathogens. In: Sayyed RZ, Singh A, Ilyas N (eds) Antifungal metabolites of rhizobacteria for sustainable agriculture. Springer, Cham, pp 211–238

Pankhurst CE, Lynch JM (2005) Biocontrol of soil-borne plant diseases. In: Encyclopedia of soils in the environment. Elsevier, pp 129–136

Parveen S, Wani AH, Bhat MY, Koka JA (2016) Biological control of postharvest fungal rots of rosaceous fruits using microbial antagonists and plant extracts—a review. Czech Mycol 68(1):41–66

Pedrini S, Merritt DJ, Stevens J, Dixon K (2017) Seed coating: science or marketing spin? Trends Plant Sci 22:106–116

Peer R, Schippers B (1992) Lipopolysaccharides of plant-growth promoting Pseudomonas sp. strain WCS417r induce resistance in carnation to Fusarium wilt. Neth J Plant Pathol 98:129–139. https://doi.org/10.1007/BF01996325

Pradhan P, Fischer G, Van Velthuizen H, Reusser DE, Kropp JP (2015) Closing yield gaps: how sustainable can we be? PLoS One 10:e0129487

Prajapati S, Kumar N, Kumar S, Maurya S (2020) Biological control a sustainable approach for plant diseases management: a review. J Pharmacogn Phytochem 9:1514–1523

Preininger C, Sauer U, Bejarano A, Berninger T (2018) Concepts and applications of foliar spray for microbial inoculants. Appl Microbiol Biotechnol 102:7265–7282. https://doi.org/10.1007/s00253-018-9173-4

Rabindran R, Vidhyasekaran P (1996) Development of a formulation of *Pseudomonas fluorescens* PfALR2 for management of rice sheath blight. Crop Prot 15:715–721

Rahimi Tamandegani P, Marik T, Zafari D, Balázs D, Vágvölgyi C, Szekeres A, Kredics L (2020) Changes in peptaibol production of *Trichoderma* species during in vitro antagonistic interactions with fungal plant pathogens. Biomolecules 10:730

Rani L, Thapa K, Kanojia N, Sharma N, Singh S, Grewal AS, Srivastav AL, Kaushal J (2021) An extensive review on the consequences of chemical pesticides on human health and environment. J Clean Prod 283:124657

Raymaekers K, Ponet L, Holtappels D, Berckmans B, Cammue BP (2020) Screening for novel biocontrol agents applicable in plant disease management—a review. Biol Control 144:104240

Reddy PP (2014) Plant growth promoting rhizobacteria for horticultural crop protection. Springer India, New Delhi

Rettinassababady C, Jeyalakshmi C (2014) Bio-fungicides: the best alternative for sustainable food security and ecosystem. In: Kharwar RN, Upadhyay RS, Dubey NK, Raghuwanshi R (eds) Microbial diversity and biotechnology in food security. Springer India, New Delhi, pp 401–411

Rhouma A, Ben Salem I, M'Hamdi M, Boughalleb-M'Hamdi N (2018) Antagonistic potential of certain soilborne fungal bioagents against Monosporascus root rot and vine decline of watermelon and promotion of its growth. Nov Res Microbiol J 2:85–100

Rocha I, Ma Y, Souza-Alonso P, Vosátka M, Freitas H, Oliveira RS (2019) Seed coating: a tool for delivering beneficial microbes to agricultural crops. Front Plant Sci 10:1357

Roselló G, Bonaterra A, Francés J, Montesinos L, Badosa E, Montesinos E (2013) Biological control of fire blight of apple and pear with antagonistic *Lactobacillus plantarum*. Eur J Plant Pathol 137:621–633. https://doi.org/10.1007/s10658-013-0275-7

Ryu C-M, Farag MA, Hu C-H, Reddy MS, Kloepper JW, Paré PW (2004) Bacterial volatiles induce systemic resistance in Arabidopsis. Plant Physiol 134:1017–1026

Saechow S, Thammasittirong A, Kittakoop P, Prachya S, Thammasittirong SN-R (2018) Antagonistic activity against dirty panicle rice fungal pathogens and plant growth-promoting activity of *Bacillus amyloliquefaciens* BAS23. J Microbiol Biotechnol 28(9):1527–1535

Sallam N, Ali EF, Seleim MAA, Khalil Bagy HMM (2021) Endophytic fungi associated with soybean plants and their antagonistic activity against *Rhizoctonia solani*. Egypt J Biol Pest Control 31:54. https://doi.org/10.1186/s41938-021-00402-9

São-José C, Santos MA, Schmitt MJ (2017) Viruses of wine-associated yeasts and bacteria. In: König H, Unden G, Fröhlich J (eds) Biology of microorganisms on grapes, in must and in wine. Springer, Cham, pp 133–154

Savita, Sharma A (2019) Fungi as biological control agents. In: Giri B, Prasad R, Wu Q-S, Varma A (eds) Biofertilizers for sustainable agriculture and environment. Springer, Cham, pp 395–411

Segarra G, Casanova E, Avilés M, Trillas I (2010) *Trichoderma asperellum* strain T34 controls Fusarium wilt disease in tomato plants in soilless culture through competition for iron. Microb Ecol 59:141–149

Sesan T, Csep N (1992) Prevention of white rot (*Sclerotinia sclerotiorum* (Lib.) de Bary) of sunflower and soybean by the biological control agent *Coniothyrium minitans* Campbell. Bull OILB SROP Fr

Sharma M, Tarafdar A, Ghosh R, Gopalakrishanan S (2017) Biological control as a tool for eco-friendly management of plant pathogens. In: Adhya TK, Mishra BB, Annapurna K, Verma DK, Kumar U (eds) Advances in soil microbiology: recent trends and future prospects. Springer Singapore, Singapore, pp 153–188

Sharma M, Guleria S, Singh K, Chauhan A, Kulshrestha S (2018) Mycovirus associated hypovirulence, a potential method for biological control of *Fusarium* species. Virus Dis 29:134–140. https://doi.org/10.1007/s13337-018-0438-4

Shaw S, Le Cocq K, Paszkiewicz K, Moore K, Winsbury R, De Torres ZM, Studholme DJ, Salmon D, Thornton CR, Grant MR (2016) Transcriptional reprogramming underpins enhanced plant growth promotion by the biocontrol fungus *Trichoderma hamatum* GD12 during antagonistic interactions with *Sclerotinia sclerotiorum* in soil. Mol Plant Pathol 17:1425–1441. https://doi.org/10.1111/mpp.12429

Singh I, Giri B (2017) Arbuscular mycorrhiza mediated control of plant pathogens. In: Varma A, Prasad R, Tuteja N (eds) Mycorrhiza—nutrient uptake, biocontrol, ecorestoration. Springer, Cham, pp 131–160

Singh US, Zaidi NW (2002) Current status of formulation and delivery of fungal and bacterial antagonists for disease management in India. In: Microbial biopesticide formulations and application. Biological Control, Bangalore, pp 168–179

Singh A, Bhardwaj R, Singh IK (2019) Biocontrol agents: potential of biopesticides for integrated pest management. In: Giri B, Prasad R, Wu Q-S, Varma A (eds) Biofertilizers for sustainable agriculture and environment. Springer, Cham, pp 413–433

Sulaiman MA, Bello SK (2023) Biological control of soil-borne pathogens in arid lands: a review. J Plant Dis Prot. https://doi.org/10.1007/s41348-023-00824-7

Šunjka D, Mechora Š (2022) An alternative source of biopesticides and improvement in their formulation—recent advances. Plants 11:3172

Tariq M, Khan A, Asif M, Khan F, Ansari T, Shariq M, Siddiqui MA (2020) Biological control: a sustainable and practical approach for plant disease management. Acta Agric Scand Sect B—Soil Plant Sci 70:507–524

Tchameni SN, Cotârleț M, Ghinea IO, Bedine MAB, Sameza ML, Borda D, Bahrim G, Dinică RM (2020) Involvement of lytic enzymes and secondary metabolites produced by *Trichoderma* spp. in the biological control of *Pythium myriotylum*. Int Microbiol 23:179–188. https://doi.org/10.1007/s10123-019-00089-x

Thambugala KM, Daranagama DA, Phillips AJ, Kannangara SD, Promputtha I (2020) Fungi vs. fungi in biocontrol: an overview of fungal antagonists applied against fungal plant pathogens. Front Cell Infect Microbiol 10:604923

Trias R, Badosa E, Montesinos E, Bañeras L (2008) Bioprotective *Leuconostoc* strains against *Listeria monocytogenes* in fresh fruits and vegetables. Int J Food Microbiol 127:91–98

Ugoji EO, Laing MD, Hunter CH (2006) An investigation of the shelf-life (storage) of *Bacillus* isolates on seeds. S Afr J Bot 72:28–33

Ul Haq I, Sarwar MK, Faraz A, Latif MZ (2020) Synthetic chemicals: major component of plant disease management. In: Ul Haq I, Ijaz S (eds) Plant disease management strategies for sustainable agriculture through traditional and modern approaches. Springer, Cham, pp 53–81

Vacheron J, Moënne-Loccoz Y, Dubost A, Gonçalves-Martins M, Muller D, Prigent-Combaret C (2016) *Fluorescent Pseudomonas* strains with only few plant-beneficial properties are favored in the maize rhizosphere. Front Plant Sci 7:1212

Valin H, Sands RD, Van Der Mensbrugghe D, Nelson GC, Ahammad H, Blanc E, Bodirsky B, Fujimori S, Hasegawa T, Havlik P, Heyhoe E, Kyle P, Mason-D'Croz D, Paltsev S, Rolinski S, Tabeau A, Van Meijl H, Von Lampe M, Willenbockel D (2014) The future of food demand: understanding differences in global economic models. Agric Econ 45:51–67. https://doi.org/10.1111/agec.12089

Van Bruggen AH, Gamliel A, Finckh MR (2016) Plant disease management in organic farming systems. Pest Manag Sci 72:30–44. https://doi.org/10.1002/ps.4145

Van Wees SCM, Pieterse CMJ, Trijssenaar A, Van 'T Westende YAM, Hartog F, Van Loon LC (1997) Differential induction of systemic resistance in *Arabidopsis* by biocontrol bacteria. Mol Plant-Microbe Interact 10:716–724. https://doi.org/10.1094/MPMI.1997.10.6.716

Voronina MV, Bugaeva EN, Vasiliev DM, Kabanova AP, Barannik AP, Shneider MM, Kulikov EE, Korzhenkov AA, Toschakov SV, Ignatov AN, Miroshnikov KA (2019) Characterization of *Pectobacterium carotovorum* subsp. *carotovorum* bacteriophage PP16 prospective for biocontrol of potato soft rot. Microbiology 88:451–460. https://doi.org/10.1134/S0026261719040118

Walterson AM, Stavrinides J (2015) Pantoea: insights into a highly versatile and diverse genus within the Enterobacteriaceae. FEMS Microbiol Rev 39:968–984

Wang X, Ding T, Li Y, Duan T (2019) Effects of an arbuscular mycorrhizal fungus and a rhizobium species on Medicago sativa wilt and Fusarium oxysporum root rot. Acta Pratacul Sin 28:139–149

Webb P, Benton TG, Beddington J, Flynn D, Kelly NM, Thomas SM (2020) The urgency of food system transformation is now irrefutable. Nat Food 1:584–585

Woo SL, Hermosa R, Lorito M, Monte E (2023) *Trichoderma*: a multipurpose, plant-beneficial microorganism for eco-sustainable agriculture. Nat Rev Microbiol 21:312–326

Zaker M (2016) Natural plant products as eco-friendly fungicides for plant diseases control-A review. Agriculturists 14:134–141

Zhang L, Sun C (2018) Fengycins, cyclic lipopeptides from marine *Bacillus subtilis* strains, kill the plant-pathogenic fungus Magnaporthe grisea by inducing reactive oxygen species production and chromatin condensation. Appl Environ Microbiol 84:e00445–e00418. https://doi.org/10.1128/AEM.00445-18

Zhou J, Deng B, Zhang Y, Cobb AB, Zhang Z (2017) Molybdate in rhizobial seed-coat formulations improves the production and nodulation of alfalfa. PLoS One 12:e0170179

Zin NA, Badaluddin NA (2020) Biological functions of *Trichoderma* spp. for agriculture applications. Ann Agric Sci 65:168–178

Zubair M, Hanif A, Farzand A, Sheikh TMM, Khan AR, Suleman M, Ayaz M, Gao X (2019) Genetic screening and expression analysis of psychrophilic *Bacillus* spp. reveal their potential to alleviate cold stress and modulate phytohormones in wheat. Microorganisms 7:337

Chapter 11
Battlefronts in Agriculture: Understanding Plant-Pathogen Interactions

Rachid Lahlali, Mohamed Kouighat, Mohammed Baala, Youness Ighissane, Mohammed Khadiri, Rachid Ezzouggari, Moussa El Jarroudi, and Abdelaaziz Farhaoui

Abstract As the demand for global food production continues to surge, combating plant pathogens becomes increasingly imperative. This review delves deeply into the intricate dynamics of plant-pathogen interactions, intending to provide a thorough understanding of the challenges and strategies in agricultural science. We explore the underlying mechanisms governing these interactions, scrutinizing the tactics employed by both plants and pathogens in an ongoing evolutionary battle. By assessing the impact of plant diseases on global food security, we outline essential plant disease management strategies, ranging from traditional methods to emerging biological control approaches, emphasizing the necessity of integrating these methods for effective plant pathogen management within the framework of sustainable agriculture. Furthermore, we shed light on recent advancements in molecular tools and omics technologies, showcasing groundbreaking research that is reshaping our comprehension of these interactions. Our analysis underscores the critical requirement to develop sophisticated plant disease models incorporating nuanced climate projections for effective management. Additionally, we examine

R. Lahlali (✉) · M. Baala · Y. Ighissane · R. Ezzouggari · A. Farhaoui
Phytopathology Unit, Department of Plant Protection, Ecole National of Agriculture Meknes, Meknes, Morocco
e-mail: rlahlali@enameknes.ac.ma

M. Kouighat
Research Unit of Plant Breeding and Plant Genetic Resources Conservation, Regional Agricultural Research Center of Meknes, National Institute of Agricultural Research, Rabat, Morocco

M. Khadiri
Phytopathology Unit, Department of Plant Protection, Ecole National of Agriculture Meknes, Meknes, Morocco

Laboratoire de Biotechnologies Microbiennes et Protection des Végétaux, Faculté des Sciences, Université Ibn Zhor, Agadir, Morocco

M. El Jarroudi
Water, Environment and Development Unit, SPHERES Research Unit, Department of Environmental Sciences and Management, University of Liège, Arlon, Belgium

O. O. Babalola, A. S. Ayangbenro (eds.), *Microbial Allies*,
https://doi.org/10.1007/978-3-031-90530-8_11

the potential of technological breakthroughs such as CRISPR/Cas genome editing and omics methodologies to enhance plant resistance. Nonetheless, there are still hurdles in pinpointing additional metabolites implicated in pathogen-host interactions, potentially leading to a more profound comprehension of these intricate relationships. This exploration of one of the challenges of modern agriculture is a valuable resource for researchers, policymakers, and practitioners seeking innovative solutions to safeguard the global food supply.

Keywords Plant-pathogen interactions · Plant immunity · Disease management strategies · Sustainable agriculture

11.1 Introduction

Contemporary farming methods, characterized by extensive cultivation of single-crop fields chosen primarily for consumer preferences rather than disease resilience, create an ideal environment for pathogenic organisms (Kumar and Murali 2023). These result in significant losses in agricultural systems. Although many biotic constraints are familiar and addressed with varying degrees of success, the emergence of new plant pathogens and the rising prevalence of virulent races, strains, or pathotypes are becoming increasingly apparent. Controlling plant diseases is increasingly challenging due to the rising occurrence of emerging diseases, which ultimately results in a decrease in agricultural production potential (Campos et al. 2021). The significant diseases affecting crop plants, caused by these pathogenic organisms, pose a substantial risk to worldwide food supplies (Jones 2013). In the face of this, new strategies of control are needed, and an understanding of how pathogens adapt and appropriate adaptive mechanisms during plant infection, as well as the exploitation of the diversity of mechanisms that plants possess to control their resistance/susceptibility to plant diseases, is essential to disease prevention and thus improve agricultural yields. Many fungal and bacterial phytopathogens acquire their nourishment from living hosts, necessitating intimate parasitic interaction between these entities. The molecular players of both the host and the pathogen vie for control over the outcome of their interaction. In pursuit of this, phytopathogens release effector proteins aimed at altering the physiology of the host cell to facilitate parasitism (Naveed et al. 2020). Recently, substantial progress has been made in comprehending the molecular mechanisms behind the ongoing battle between plant pathogens and their host plants. The mutual evolution of phytopathogens and their hosts has led to extensively adapted microbial invasion tactics and plant defense responses (Wang et al. 2022). The response of plants to biotic stresses is intricate. To defend against invading plant pathogens, plants deploy both constitutive preformed and induced defense reactions, offering multiple levels of protection (Naveed et al. 2020). Constitutive defense involves chemical or physical obstacles that impede the attachment and entry of the majority of phytopathogens into plants (Chowdhary and Tank 2023). Following the zigzag model of plant-pathogen interactions, induced defense operates on two levels. The initial level is referred to as pathogen-associated molecular pattern (PAMP) triggered immunity (PTI), where plant pattern recognition receptors (PRRs) detect conserved structures or molecules from phytopathogens, initiating defensive reactions. In order to circumvent PTI, multiple phytopathogens generate effector proteins within plant cells, where they interfere with defense mechanisms. The host detects these effectors via resistance (R) genes, prompting a quicker and more vigorous defensive reaction known as effector-triggered immunity (ETI). If one or more phytopathogen effectors manage to bypass PTI, they can successfully infect susceptible plants. In situations where effective R proteins are lacking, ETI becomes overwhelmed, leading to effector-triggered susceptibility (ETS). The continuous alternations between ETI and ETS arise from the ongoing coevolution of plant pathogen effectors and corresponding R genes in the host (Sekhwal et al. 2015). Although its practical applicability has been questioned, the zigzag model is still widely regarded as the most concise plant defense model for pedagogical objectives. Its popularity may be attributed to its elegant representation of the three main outcomes of pathogen-plant interaction: elicitor-triggered non-race-specific defense (PTI), cultivar or race-specific gene-for-gene resistance (ETI), and race-specific phytopathogen virulence and plant vulnerability (ETS) (Naveed et al. 2020). Gaining insights into the molecular mechanisms of plant-pathogen interactions could lead to the development of innovative genetic engineering approaches of disease resistance, including RNA silencing, genetic editing, and other techniques. Recently, advancements in high-throughput sequencing of both plant and phytopathogen genomes have provided new options for gaining deeper insights into the mechanisms governing parasitic processes, signal transduction, and host defense responses. For example, the designation of resistance gene analogs (RGAs) in plants, such as receptor-like proteins (RLPs), receptor-like kinases (RLKs) and nucleotide-binding site leucine-rich repeats (NBS-LRRs), presents a significant opportunity for improving the resistance of plant varieties (Sekhwal et al. 2015). This chapter delves into the latest knowledge of plant-pathogen interactions, focusing on recent advances aimed at improving our understanding of these complex relationships. It further examines plant defense mechanisms, the evolutionary dynamics of the plant immune system, and essential strategies for managing plant diseases. Additionally, we explore how current knowledge of plant-pathogen interactions can be leveraged to develop robust, broad-spectrum resistance in plants.

11.2 Fundamentals of Plant-Pathogen Interactions

11.2.1 Definition and Types of Plant Pathogens

Plant pathogens are microorganisms that cause diseases in plants. They can belong to various groups, such as viruses, bacteria, fungi, etc. Each type of pathogen has specific structures and modes of infection that allow them to colonize plants and cause damage.

11.2.1.1 Fungi

Plant pathogenic fungi can infect different plant organs, such as leaves, stems, or roots. They attack plant tissues and can cause symptoms such as spots, rots, or deformations, representing a highly diverse group of plant pathogens with a significant impact on agriculture (Nazarov et al. 2020). Fungi exhibit a wide range of lifestyles and genetic flexibility, allowing pathogens to quickly invade new hosts and develop resistance to fungicides or to overcome plant resistance mediated by R genes (Sekhwal et al. 2015). Consequently, combating fungal diseases will remain a major challenge in future phytopathology. To develop new strategies for controlling phytopathogenic fungi, there is a need to deepen our understanding of fungal virulence mechanisms, the evolution of pathogenic traits, the molecular basis of host adaptation, and particularly how crop resistance can be engineered more sustainably (Dang et al. 2024). Additionally, fungi are fascinating organisms that have developed an arsenal of strategies to colonize living plants. Studying the molecular mechanisms underlying these complex interactions will not only advance our understanding of microbes but also provide valuable insights into cellular processes in plants.

Infection of plants by phytopathogenic fungi occurs through a mechanism known as the disease cycle, which varies among species but shares common stages (inoculation, adhesion, germination, and invasion) (Dang et al. 2024). The developmental cycle of a fungus comprises a vegetative phase of growth and nutrition, as well as a reproductive phase. Spores originate from the vegetative phase, during which they develop a mycelial network to colonize the plant and obtain nutrients. Typically, the presence of these filaments signifies the onset of contamination. Several fungi produce spores that ensure species dispersion. These spores, produced in large quantities, germinate immediately if environmental conditions are conducive. Moreover, these spores constitute the initial inoculation source (Selvakumar et al. 2018) due to their reduced metabolism and being surrounded by protective walls that confer high resistance. They can remain dormant for several years, awaiting favorable conditions for germination (Wyatt et al. 2013). Under natural ambient conditions, spores are suspended in the air or soil. Through sedimentation, spores settle on all surfaces: plant debris, working material, soil, and spontaneous plants. These surfaces thus become a reservoir of spores (Dang et al. 2024).

11.2.1.2 Bacteria

Plant pathogenic bacteria can infect plant tissues and multiply inside. They can cause symptoms such as lesions, rots, or plant deformations. Bacterial infection cannot induce a disease unless certain conditions exist: host/bacteria compatibility, specific environmental factors, and well-defined plant physiology (Geeta and Mishra 2018). The pathogenic bacterium must initially penetrate the plant through natural openings, with stomata being the most commonly utilized. Additionally, penetration can occur through lenticels in fruits and trees, hydathodes, trichomes, or nectaries in flowers (Dai et al. 2020). Alternatively, penetration can occur through a wound, which may result from mechanical abrasion due to weather conditions or crop maintenance activities such as soil work or pruning. Two types of bacteria are distinguished:

Necrotrophic Bacteria: So-called "necrotrophic" bacteria cause necroses on leaves and wilting on branches. These symptoms are generally caused by toxins produced by the bacteria. These bacterial species have very destructive pathogenesis strategies leading to extensive necroses, tissue macerations, and plant rotting. They kill host plant cells and utilize their contents to support their growth. Additionally, they can induce apoptosis in host cells instead of breaking down plant cell walls (Geeta and Mishra 2018).

Biotrophic Bacteria: These are bacteria that feed on living plants. These bacteria establish a long-term feeding relationship with the living cells of their hosts, rather than killing host cells as part of the infection process (Tian et al. 2020).

Generally, bacteria develop between host cells and only invade a few cells. Through their feeding activities, they create a nutrient sink to the infection site, so that the host is disadvantaged but not killed (Geeta and Mishra 2018). This type of parasitism can lead to serious economic losses in cultivated plants and, in natural environments, weaken the plant. This highly sophisticated type of parasitism maintains the host alive as a long-term food source.

11.2.1.3 Viruses

Viruses are very small pathogens that infect plants by entering their cells and using their machinery to reproduce. They can cause various symptoms such as spots, deformations, or necroses of tissues (Farooq et al. 2021). Plant viruses are obligatory parasites; their transmission mode is via vectors through different mechanisms such as non-persistent, semi-persistent, circulative persistent, non-propagative, and circulative persistent. It is therefore necessary to study these mechanisms to understand the interactions between the plant and viral particles. Most viruses are transmitted by insect vectors (Shi et al. 2021), but some of them are also transferred mechanically (TMV, PVY), through seeds (Soybean mosaic virus) (Bowers Jr and Goodman 1979), pollen (alfalfa cryptic virus-1) (Card et al. 2007), fungi (BNYVV) (Singh et al. 2008), and nematodes (Grapevine fanleaf virus) (Andret-Link et al. 2004). The virus-infected plant shows a decrease in net photosynthesis and alterations in carbohydrate metabolism (Zanini et al. 2021), in the case of CMV (Almási et al. 2001). Plants expressing viral symptoms, which are alterations in primary metabolites such as starch and soluble sugars and decreased photosynthesis and also increased respiration, are observed (Zanini et al. 2021). The Geminiviridae virus family can encode proteins that not only modify the host RNA silencing pathway but also have proteins that modulate cellular function and programming (Shi et al. 2021).

11.2.2 Plant Defense Mechanisms

Events leading to the induction of resistance in plants can occur in three main stages: signal recognition, production of molecular signals, and establishment of defense (Mittler et al. 2022).

Recognition When a plant is exposed to a pathogen or an elicitor, it can recognize the presence of this agent through specific receptors present on the surface of cells. This recognition can occur through the detection of molecular patterns associated with pathogens, such as specific oligosaccharides present in the cell wall (Mittler et al. 2022).

Production of Molecular Signals Once the pathogen or elicitor is recognized, the plant responds by producing molecular signals, also known as defense hormones. These signals include compounds such as jasmonic acid, salicylic acid, ethylene, and phytoalexins. These hormones play a key role in coordinating the plant's defense response (Verma et al. 2023).

Establishment of Defense Mechanisms In response to molecular signals, the plant activates various defense mechanisms to protect itself against the pathogen. These mechanisms include thickening of the cell wall to strengthen the physical barrier, production of PR (Pathogenesis-Related) proteins with antimicrobial properties, and other defense responses specific to each pathogen. The coordination of these different responses enables the plant to effectively defend itself against pathogens and maintain its health (Kumar et al. 2020).

11.2.2.1 Physical Barriers

Passive plant defenses, such as physical barriers, are generally sufficient to protect the plant against many pathogen attacks. It is noted as a method of passive defense whereby plants construct a sort of physical barrier, which prevents the aggressor from penetrating the plant; this is the first line of defense (Wang et al. 2021). In several instances, the initial observable defense mechanism is the papilla, a cell wall structure formed by the host plant at the site where the fungus or oomycete attempts

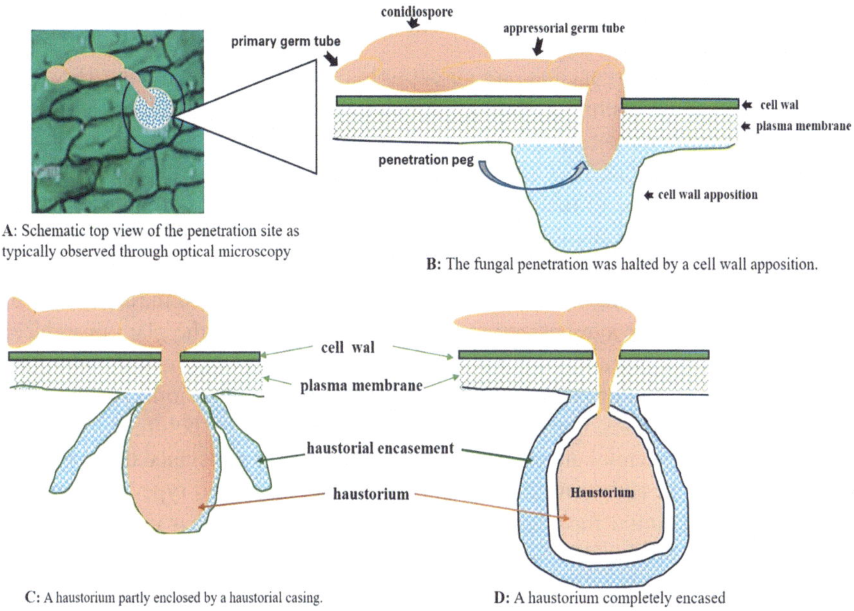

Fig. 11.1 Cell wall-associated structures frequently observed at sites where plants interact with various fungal pathogens (**a**) Schematic top view of the penetration site as typically observed through optical microscopy (**b**) The fungal penetration was halted by a cell wall apposition (**c**) A haustorium partly enclosed by a haustorial casing (**d**) A haustorium completely encased

to penetrate (Fig. 11.1). Papillae are evident in various plant species, with notable studies conducted in interactions with grain-powdery mildew. Histological and chemical examinations reveal the intricate composition of papillae, incorporating components such as callose, proteins, and diverse phenolic and inorganic compounds, including opal silica and, occasionally, reactive oxygen species (Wang et al. 2021).

The deposition of substances like wax and cutin thickens cuticles, effectively preventing pathogens such as viruses, bacteria, and fungi from penetrating and developing. This process also plays a role in the signaling cascade that enables the establishment of secondary defense mechanisms (Verma et al. 2023). Other factors come into play, such as the quantity and shape of wax crystals. Pubescence is the layer of hair on the epidermis of stems, leaves, roots, and even fruit (Kumar et al. 2020). The stomata of leaves and the lenticels of stems, their number, size, and angle of opening, can also close to prevent penetration by fungal spores and bacteria. However, sometimes certain microorganisms manage to bypass these. They can infect the plant by using natural openings like stomata or by exploiting natural or artificial wounds. Some microorganisms also produce hydrolytic enzymes that allow them to degrade the plant's cuticle or cell wall, thereby enabling them to penetrate inside (Verma et al. 2023).

11.2.2.2 Chemical Defenses

Plants, being sessile organisms, are susceptible to attacks by predators. However, some plants produce chemical compounds that aid in preventing viral, bacterial, and fungal infections. Among these compounds are terpenes, derived from isoprene or 2-methyl-buta-1,3-diene. Low molecular weight terpenes are particularly effective in repelling insects due to their volatility (Toffolatti et al. 2021). Phytoalexins and phytoanticipins are chemical substances produced by plants in response to pathogenic attacks. They act as chemical barriers to protect plants against infections (Tiku 2020). Phytoalexins are antimicrobial compounds produced in response to an infection. They can be synthesized from precursors present in the plant or induced by the activation of specific metabolic pathways. Phytoalexins work by inhibiting the growth and spread of pathogens (Jan et al. 2021). Phytoanticipins, on the other hand, are primary or secondary metabolites synthesized even before infection, in preparation for a potential attack. They enhance the plant's resistance by modifying its metabolism and activating defense mechanisms. These two types of chemical substances are essential for plant survival as they strengthen their immune system and prevent infections (Tiku 2020).

11.2.2.3 Molecular Defenses

Plants have developed an effective defense system against pathogens, allowing resistance to many microbial aggressors. Part of this success is due to the existence in each plant cell of a wide range of recognition systems.

11.2.2.3.1 Pathogen Recognition

The investigation into the chronological stages of initiation of different defense reactions in plants following an attack by pathogens has revealed that plant recognition of pathogens begins with the perception of "signal" molecules. These molecules can initiate a cascade of events that ultimately lead to the expression of defense reactions in the plant (Abdul Malik et al. 2020) (Fig. 11.2). Two types of recognition can be proposed:

- Non-specific recognition for general elicitors.
- Specific recognition involves the product of the plant's resistance gene and the product of the corresponding pathogen's virulence gene.

When a plant is exposed to pathogens or stress factors, it can trigger a defense response to protect itself. One of the defense reactions is the hypersensitive reaction, which is characterized by the necrosis of plant cells around the infection site. The transduction of stress signals is the process by which stress signals are detected by the plant and transmitted, leading to the activation and expression of defense genes (Dalio et al. 2021). General elicitors are molecules that can trigger a defense

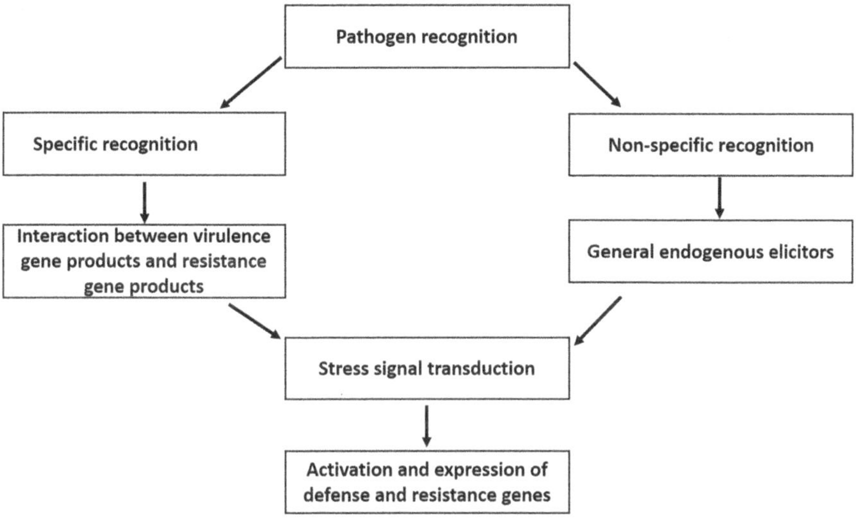

Fig. 11.2 Process of genetic control in plants against pathogens

response in different plants without gene-for-gene specificity (Abdul Malik et al. 2020). They can be of biotic origin (from living organisms) or abiotic (from non-living factors). There are also specific elicitors, such as avirulence proteins, which are recognized by particular cultivars of plants. These specific elicitors are often involved in specific interactions between plants and pathogens. These defense and mutual recognition mechanisms between plants and microorganisms are important for plant survival and ecosystem balance (Abdul Malik et al. 2020). When the pathogen signal is perceived via a potential receptor, extremely complex intracellular signaling steps are set up. This signal transduction cascade leading to the hypersensitive reaction is associated with molecular modifications (Dalio et al. 2021). When numerous elicitors are perceived, one of the first responses observed is the change in membrane permeability and the alkalinization of the external medium due to ion flux. These changes in ion fluxes lead to acidification of the cytoplasm and are manifested by a significant increase in calcium ions (Ca^{2+}) and hydrogen ions (H^{+}), as well as the release of NO_{3-} or Cl^{-} and then K^{+}. These ion movements trigger depolarization of the plasma membrane, the amplitude and duration of which depend on the pathogen (Pathak et al. 2020). Calcium influxes are considered a major signaling event. Indeed, calcium plays a key role as a secondary messenger in the cell signaling processes triggered by pathogen recognition. Calcium transmits the perceived signal from the cell surface to the inside of the cell by causing variations in calcium concentration in the cytosol (Pathak et al. 2020). In plants, there is a wide variety of calcium sensors, including a specific family unique to the plant kingdom called "calcium-dependent protein kinase" (CDPK) (Tong et al. 2021). In all model plant species whose genomes have been sequenced, at least 20 genes encoding CDPKs are found. The channels and transporters involved in ion fluxes are not yet well understood. Membrane calcium channels involved in the signaling process have been characterized in tomato, parsley, and Arabidopsis (Pathak et al. 2020).

11.2.2.3.2 Signal Transduction by Hormones

Several hormones play crucial roles in the gene expression process activated during attacks by biotic stressors, for example:

Salicylic Acid (SA) is naturally produced by plants in response to attack by different pathogens. It plays an essential role in strengthening local resistance and establishing the systemic acquired resistance system (Lukan and Coll 2022). SA signaling is closely related to the production of reactive oxygen species (ROS) (Lukan and Coll 2022). ROS are capable of inducing the accumulation of SA and also contribute to increasing its production. Additionally, SA can itself induce ROS production by acting on the phosphorylation state of upstream signaling elements in ROS production or by inhibiting enzymes responsible for ROS detoxification, such as catalase, which degrades H_20_2 (Mohamed et al. 2020). The ability of SA to inhibit catalase activity leads to an increase in hydrogen peroxide levels in vivo. This induces oxidative stress at sites attacked by the pathogen or during treatment with an elicitor (Mohamed et al. 2020).

Ethylene (ET) is a gaseous plant hormone essential for plant growth and development. It is involved in several stages of plant life cycles, such as seed germination, the formation of absorptive roots, and fruit ripening (Reid 1995). Furthermore, ET plays a key role in regulating plant responses to various stresses, whether abiotic or biotic, resulting from attack by pathogens (Shekhawat et al. 2022). Biosynthesis of ET begins with the conversion of S-adenosyl methionine into 1-aminocyclopropane carboxylic acid (ACC) by ACC synthase. Then, ACC is transformed into ethylene, carbon dioxide, and cyanide by ACC oxidase (Pech et al. 2010). Studies have shown that the expression of ACC synthase and ACC oxidase genes is induced in plants following inoculation with pathogens. The role of ET in plant defense can vary depending on the type of pathogen and plant species. In response to a necrotrophic pathogen, ET can contribute to the production of antimicrobial compounds or modulation of the plant's immune response to combat infection (Shekhawat et al. 2022).

11.3 Dynamics of Plant-Pathogen Interactions

The dynamic interaction between plants and pathogens is crucial for shaping ecosystems and profoundly impacting global food security. Understanding the dynamic nature of plant-pathogen interactions has become increasingly important in light of ongoing climate change and the expanding human population.

11.3.1 Recognition and Signaling Between Plants and Pathogens

The interaction between plants and pathogens has been ongoing for centuries, leading to a continuous evolution in the recognition and signaling mechanisms employed by plants in response to pathogens. The plant-pathogen interaction is a permanent and continuously evolving relationship, where each participant continually adapts to changes in the other to survive and thrive. This mutual adaptation is crucial for the long-term dynamics of coexistence between pathogens and plants (Kumara et al. 2022). The ability of plants to recognize and respond to invading pathogens has significant implications for their survival (Panigrahi and Satapathy 2020). The upcoming section explores the mechanisms through which plants and pathogens recognize and interact with each other, as well as the molecular exchanges that influence the outcomes of these interactions.

11.3.1.1 Molecular-Level Recognition

The molecular recognition of conserved PAMPs by plant PRRs is a crucial component of plant-pathogen interactions. Recent advances have shed light on the structural basis of PAMP recognition by PRRs, highlighting the distinctiveness and versatility of these interactions (Patnaik et al. 2024). Additionally, another study suggests that chaperone proteins play a significant role in regulating the activation of PRRs and the detection of ligands. This adds complexity to the recognition process (Zhang et al. 2021).

11.3.1.2 Recognition Activated by an Effector

To manipulate host cellular processes and evade plant immune responses, pathogens utilize effector proteins. In response, plants have evolved intracellular nucleotide-binding leucine-rich repeat (NLR) receptors to detect the presence of these effectors, thereby initiating ETI (Thordal-Christensen 2020). Research has uncovered innovative mechanisms underlying the recognition of effectors by NLR receptors, including the participation of decoy domains and integrated decoys, which contribute to specificity in recognition (Ngou et al. 2021).

11.3.1.3 Pathways for Signal Transduction

When plants detect invaders, they initiate communication systems that coordinate defense reactions. Research has identified components and regulators of communication pathways, such as receptor-like cytoplasmic kinases (RLCKs) and supporting proteins. These discoveries offer insights into the precise transmission of signals

(Sun and Zhang 2020). Furthermore, advancements in techniques for observing living cells have made it possible to study the temporal patterns of signaling events, shedding light on the complexities of plant responses (De Bruyne et al. 2014).

11.3.1.4 Communication and Defense Priming

Plants can transmit information about their interaction with pathogens to adjacent cells and systemic organs, thereby priming them for heightened defense reactions. Recent investigations have identified mobile signals, such as short RNAs and peptides, as important mediators of systemic defense priming (Baldrich et al. 2019). In addition, studies have demonstrated that epigenetic alterations, such as DNA methylation and histone modifications, play a role in the establishment of durable defense memory, highlighting the importance of epigenetic regulation in plant immunity (Ramos-Cruz et al. 2021).

11.3.2 Infection Strategies Employed by Pathogens

To successfully invade and thrive in their plant hosts, pathogens have developed a range of infection strategies. Understanding how pathogens enter plant tissues and impact host physiology is crucial for developing disease control methods (Khan et al. 2021). Pathogens can penetrate plant surfaces and establish infections using both direct and indirect methods. Certain pathogens have specialized structures and enzymes that allow them to directly breach the chemical defenses of plants. Meanwhile, bacteria and viruses rely on openings or wounds to enter the host's environment (Maheswary et al. 2021).

11.3.2.1 Direct Penetration

Pathogens, including bacteria, fungi, and viruses, pose a significant threat to plant health and agricultural productivity worldwide. One of the primary strategies employed by these pathogens to establish infection is through direct penetration of plant tissues (Maheswary et al. 2021). There are several mechanisms for direct penetration through which pathogens breach the physical barriers of plants and gain access to the host's internal tissues, leading to disease development.

11.3.2.1.1 Mechanical Damage and Wounding

Mechanical damage or wounds on plant surfaces provide entry points for pathogens. Pathogens can exploit these vulnerabilities by secreting enzymes or toxins that degrade the plant's cell wall, allowing them to penetrate the host tissue. For example, the necrotrophic fungus *Botrytis cinerea* secretes cell wall-degrading enzymes, such as polygalacturonases, to break down pectin present in the plant cell walls (Alkan et al. 2015). This enzymatic degradation weakens the cell wall, facilitating the pathogen's entry into the plant tissue (Silva et al. 2023).

11.3.2.1.2 Appressorium Formation

Certain pathogens, such as the rice blast fungus *Magnaporthe oryzae*, utilize specialized infection structures called appressoria to directly penetrate plant surfaces. Appressoria is formed in response to physical cues, such as surface hydrophobicity or hardness (Shi et al. 2023). The fungus generates enormous turgor pressure within the appressorium, enabling it to mechanically rupture the plant cuticle and penetrate the underlying epidermal cells (Alkan et al. 2015).

11.3.2.1.3 Direct Uptake of Nutrients

Some pathogens, particularly parasitic nematodes, directly penetrate plants to obtain nutrients for their survival and reproduction (Kumar and Yadav 2020). Nematodes possess a stylet, a needle-like feeding structure, which they use to puncture plant cells and extract nutrients. For example, the root-knot nematode *Meloidogyne incognita* injects cell wall-degrading enzymes and effector proteins into plant cells via its stylet, leading to cell wall modifications that facilitate nutrient uptake (Alkan et al. 2015).

11.3.2.2 Secretion of Pathogenic Molecules

The secretion of pathogenic molecules during interactions with plants plays a crucial role in the communication between plant pathogens and their hosts, ultimately leading to the development of diseases (Timilsina et al. 2020). One key method employed by pathogens when infecting plants is the type III secretion system (T3SS), a needle-shaped structure that directly injects effector proteins into plant cells. This mechanism is essential for the virulence of pathogens like *Pseudomonas syringae* and *Xanthomonas* spp., allowing them to control host processes, suppress plant defenses, and establish infections. By delivering effectors through the T3SS, bacteria can disrupt plant signaling pathways, alter host metabolism, and trigger disease symptoms, facilitating their survival and proliferation within the host plant (Alkan et al. 2015). The type IV secretion system (T4SS) is essential in how bacterial pathogens interact with plants alongside the T3SS. For instance, certain phytopathogens like *Agrobacterium tumefaciens* employ the T4SS to transport virulence factors, such as DNA, into plant cells, leading to tumor formation (Yu et al. 2021). This mechanism also aids in transferring material between bacteria and plants, thereby influencing adaptation and the emergence of new pathogenic characteristics. Through leveraging the T4SS, phytopathogens can adjust plant hormone signaling, modify gene expression, and control host processes to promote infection and disease progression (Timilsina et al. 2020). Furthermore, fungal pathogens utilize secretion systems, such as the type II secretion system (T2SS), to release virulence factors and establish connections with host plants. The T2SS plays a role in releasing enzymes that break down plant cell walls, toxins, and enzymes that degrade cell walls. This mechanism allows fungi, like *Botrytis cinerea* and *Fusarium spp.*, to break down components of plant cell walls and enter host tissues (Chlebek et al. 2022). By secreting molecules through the T2SS, fungi can lead to tissue necrosis, facilitate the spread of diseases, and improve their ability to acquire nutrients from host plants, thus increasing their pathogenicity and virulence (Timilsina et al. 2020).

11.3.2.3 Manipulation of Host Physiology

Pathogenic microorganisms have evolved a variety of strategies to manipulate the body's responses and establish infections successfully. For example, some pathogens release proteins known as effector proteins that help regulate the host's reactions. This allows the pathogen to avoid detection and counteract the host's defense mechanisms (Kaur et al. 2022). These effectors can target various components of the plant's system, such as receptor proteins, signaling molecules, and transcription factors, to promote the growth and colonization of host tissues by pathogens. (Toruño et al. 2016). Moreover, pathogens can manipulate the functioning of the host's body to create an environment suitable for their growth and reproduction. They not only attack the plant's system directly but also alter the hormonal balance in plants. This is achieved by secreting substances that mimic plant hormones or interfere with the production and communication of these hormones (Timilsina et al. 2020). Pathogens can induce changes in the host's physiology that benefit their survival and spread by manipulating levels of plant hormones, such as auxins, cytokinins, and gibberellins (Akhtar et al. 2020). Furthermore, certain viruses can alter how host plants metabolize to access nutrients and create structures that facilitate their absorption (Zimmerman et al. 2020). Pathogens can alter the metabolic processes of their hosts, thereby controlling nutrient distribution and leading to deficiencies and imbalances within infected plants (Kaur et al. 2022). Infections can significantly influence the growth and vitality of plants, thereby affecting their overall health and productivity. Denancé et al. (2013) demonstrated that weakening the plant's defenses and altering hormone signals could render it more susceptible to infections and less capable of managing stress. Fones and Preston (2013) observed that depleting the plant's resources and experiencing shortages can impede its growth, development, and reproductive capacity, ultimately resulting in reduced crop yield and quality. The process by which pathogens alter a host's physiology is multifaceted and contributes to the onset of plant diseases. Understanding how infections compromise the host's defenses and modify its physiology is essential for devising effective disease management strategies to safeguard plant health and ensure global food security.

11.4 Impact of Plant-Pathogen Interactions on Agriculture

Plants play a vital role in providing more than 80% of food for humans and livestock (Barnum et al. 2021). Food security, which ensures consistent access to adequate, safe, affordable, and nutritious food, is essential to maintaining healthy and productive societies (Savary et al. 2019). However, plant diseases pose a serious threat to global food security and environmental sustainability. They lead to a reduction in primary productivity and biodiversity, thus having a detrimental impact on the environmental and socio-economic conditions of the affected regions (Singh et al. 2023). These diseases threaten farmers' income by increasing production costs and altering the quality of agricultural products. Direct annual losses due to plant pathogens are estimated at 13–22%, equivalent to US$220 billion (Savary et al. 2019). These losses directly affect food security, regional economies, and various other socio-economic aspects (Van Dijk et al. 2021). Beyond direct costs, the implementation of control measures also incurs indirect costs associated with the treatment of phytopathogens (Welsh et al. 2021). Moreover, food obtained from plants infected by pathogens can serve as vectors for human pathogens and microbial toxins, thereby affecting global human health (Singh et al. 2023). Plant diseases can also disrupt food supply chains, leading to local or regional shortages. Particularly in regions where agriculture is an economic pillar, small farms are most affected, leading to increased financial difficulties and economic and social instability (Savary et al. 2019).

Plants are exposed to a variety of pathogens, including bacteria, fungi, viruses, and nematodes, each with its own modes of attack, infection strategies, and tissue targets (Tripathi et al. 2022). Case studies of major crop diseases are crucial to understanding their devastating impact on agricultural production. These analyses allow researchers and farmers to characterize the pathogens responsible, evaluate their spread and their impact on yields, and develop effective management strategies. In Table 11.1, we have listed examples of major diseases threatening field crops.

11.5 Advancements in Understanding Plant-Pathogen Interactions

Interactions between plants and pathogens are a crucial area of research to ensure food security and environmental sustainability. These complex interactions involve a series of molecular, biochemical, and physiological mechanisms that influence plant health and pathogen virulence. Over the years, significant advances have been

Table 11.1 Case studies on major crop diseases and their effects

Crop	Major disease	Study case	Reference
Wheat	Wheat rust	Wheat leaf rust, caused by *Puccinia triticina*, is a fungal disease that affects the stems, leaves, and grains of wheat, barley, and rye. Crop losses estimated up to 55% have been reported for wheat cultivars susceptible to leaf and stem rust	Abebe (2021)
Rice	Rice blast	Rice blast, caused *by Pyricularia oryzae* Cavara, can infect the aerial tissues of rice plants at any stage of growth and result in a complete loss of the harvest. Yield losses due to this pathogen vary between 20% and 70%.	NurulNahar et al. (2020)
Tomato	Tomato mildew	Tomato mildew, caused by *Alternaria solani,* can lead to up to 80% yield loss in severe infections	Panno et al. (2021)
Potato	Potato blight	*Phytophthora infestans* is an oomycete or water mold, a microorganism similar to a fungus that causes severe disease in potatoe and tomato known as late blight or potato blight. Due to the severe development of the disease, yields can be reduced by 5–78%.	Schmey et al. (2024)
Rapeseed	Rapeseed Alternaria leaf spot	*Alternaria* spp. is considered the most virulent pathogen on all brassicaceous plants, with estimated yield losses ranging between 47% and 70%.	Mourou et al. (2023)
Onion	Onion downy mildew	Onion downy mildew, caused by *Pseudoperonospora plantaginis*, resulted in a seed yield reduction of up to 30%.	Chauhan et al. (2023)
Broad bean	Broad bean anthracnose	Anthracnose, caused by *Colletotrichum spp.*, a hemibiotrophic pathogen, is a destructive disease affecting legumes and many other crops worldwide, resulting in yield losses ranging from 30% to 100%.	Pandey et al. (2023)
Chickpea	Chickpea ascochyta blight	Chickpea ascochyta blight, a disease caused by the fungus *Ascochyta rabiei,* can lead to total yield losses or a significant reduction in the quality of susceptible and untreated cultivars. In Australia, for example, it is responsible for average losses of 4.8 million AUD per year	Foresto et al. (2023)
Strawberry	Anthracnose diseases in strawberry	Fruit Anthracnose in strawberries is caused by multiple strains belonging to the Acutatum species complex, which includes 44 species of *Colletotrichum*, resulting in significant crop losses of up to 70% in commercial production fields planted with susceptible strawberry cultivars.	Aljawasim et al. (2023)
Olive	Olive verticillium wilt	*a* Verticillium wilt, caused by the soil-borne fungus *Verticillium dahliae Kleb,* results in a devastating reduction of 75–89% in yield of susceptible olive production.	Rhouma et al. (2023)

(continued)

Table 11.1 (continued)

Crop	Major disease	Study case	Reference
Date palm	Date palm fusarium wilt	*F. oxysporum f.* sp. *albedinis* (Foa) is responsible for vascular fusariosis of date palms, commonly known as Bayoud. In general, Bayoud can lead to significant yield losses, reaching up to 100% in the most severe cases.	Alotaibi et al. (2023)
Apple tree	Apple scab	Apple scab, caused by *Venturia inaequalis*, can result in yield losses of up to 70%.	Khajuria et al. (2023)

made in understanding these interactions, largely due to the emergence of innovative molecular tools and techniques. In this paragraph, we will review recent advances in this field, focusing on more conventional methods, such as culture-based detection strategies, immunological and nucleic acid-based methods, as well as more advanced methods, such as biosensors and high-throughput sequencing techniques.

11.5.1 Methods Based on Optical and Spectral Tools

Non-invasive optical and spectral detection methods have become essential in plant disease monitoring. Unlike visual symptom detection, these methods can identify latent infections and provide objective data (Zubler and Yoon 2020). Stressed or diseased plants emit a distinct spectral signature, allowing early detection of physiological changes before visual symptoms appear (Tanner et al. 2022). Advanced techniques, such as drone and satellite imaging, provide large-scale monitoring, while the use of optical sensors enables real-time, non-invasive detection (Venbrux et al. 2023). However, data interpretation remains complex and often requires machine-learning algorithms (Zubler and Yoon 2020). Although these techniques can detect biotic stress, they cannot always specifically identify the pathogen. Therefore, they are often combined with other methods for precise plant disease management.

11.5.2 Methods Based on Culturing

Culture-based methods are widely used to detect and identify plant pathogens. They consist of cultivating and isolating microorganisms on selective growth media, followed by morphological, biochemical, molecular, or immunological tests to confirm their (Tanner et al. 2022). These methods offer advantages such as simplicity, reliability, and the distinction between viable and nonviable organisms (Venbrux et al. 2023). However, they have disadvantages, including the time required to grow organisms and confirm their identity, as well as the inability to culture certain microorganisms. Despite these limitations, culture-based techniques remain widely used, often in combination with DNA-based methods, for the detection and confirmation of plant pathogens (Venbrux et al. 2023).

11.5.3 Methods Based on Immunological Tests

Immunoassays to detect plant pathogens use specific antibodies linked to epitopes of microbial antigens. These tests can be based on polyclonal antibodies, mixing several antibodies, or monoclonal antibodies, specific to a single epitope (Fang and Ramasamy 2015). Although polyclonal antibodies can cause false positive results due to cross-reactivity, monoclonal antibodies provide higher specificity and reproducibility (Krivitsky et al. 2021). Immunological methods, such as ELISA and LFIA, are rapid, and often used clinically, but may require sample pretreatment to improve sensitivity (Venbrux et al. 2023). ELISA tests are particularly common, offering a variety of formats and possible automation, although their use requires delicate storage and handling of antibodies. LFIAs are portable, inexpensive, and provide rapid results, but are limited in sensitivity for some samples. Aptamers, an alternative to antibodies, show strong potential with their ease of production and low cost, although their use is still limited in the detection of plant pathogens (Krivitsky et al. 2021).

11.5.4 Methods Based on Nucleic Acid and Sequencing

DNA or RNA sequences are used to detect pathogenic microorganisms, but DNA purity is essential for accurate results. Different DNA extraction methods are available, while quantitative PCR (qPCR) and droplet digital PCR (ddPCR) offer increased sensitivity and specificity (Venbrux et al. 2023). Isothermal amplification, such as LAMP and RPA, is a promising alternative to PCR, requiring less DNA purification and providing high sensitivity (Hu et al. 2021). CRISPR-Cas-based detection systems offer high specificity and sensitivity, but face challenges such as multiplexity and sensitivity to genetic mutations (Venbrux et al. 2023). By integrating these techniques, it is possible to improve the speed and sensitivity of plant pathogen detection tests. DNA sequencing is an effective method for identifying microorganisms, providing greater accuracy and reproducibility than conventional techniques. Technological advances have enabled the development of next-generation sequencers, offering a range of options from high-throughput long reads at varying costs. Two main approaches are used for multiplex detection of plant pathogens: metagenome sequencing and amplicon sequencing (Hu et al. 2021). Metagenomics involves sequencing all of the DNA in a sample, while amplicon sequencing targets specific marker genes. Although metagenomics offers identification down to the strain level and detection of nonculturable organisms, it presents challenges such as high cost and limited sensitivity for low-abundance species (Venbrux et al. 2023). In contrast, amplicon sequencing is less expensive and offers increased sensitivity, although its taxonomic resolution is limited (Hu et al. 2021). Biosensors, combining biorecognition elements and physicochemical transducers, are a promising alternative for the detection of plant pathogens, offering speed, ease of use, and sensitivity (Venbrux et al. 2023). These diverse technologies open up new and effective perspectives for the monitoring and management of plant diseases.

Advances in molecular tools, omics technologies, and emerging fields have significantly enriched our understanding of plant-pathogen interactions. By combining these multidisciplinary approaches, researchers can identify new virulence factors, develop disease control strategies, and improve crop resilience to pathogen threats. These advances are essential to address the growing challenges related to food security and environmental sustainability in the context of climate change and the globalization of agricultural trade.

11.6 Strategies for Managing Plant Diseases

11.6.1 Traditional Methods

11.6.1.1 Crop Rotation

Growing crops with varying economic significance on the same land, including fallow periods, can be defined as cultivating plants from the same family or with similar characteristics consecutively (Shah et al. 2021). In contrast to a monoculture system, which often lacks a specific plan, crop rotation involves growing plant species with different characteristics in an organized manner on the same piece of land (Shah et al. 2021). Implementing crop rotation is a crucial practice in plant disease management, aiming to reduce the initial inoculum and control the incidence of diseases (Atay 2023). For annual plants, crop rotation is particularly effective in managing certain plant pathogens in cultivation. Moreover, crop rotation is essential in combating soil and stubble-borne diseases, making it a critical consideration in disease management strategies (Dixon and Tilston 2010).

Furthermore, some plants not only serve as unsuitable hosts for pathogens but also actively suppress their growth (Yang et al. 2023). To establish a successful crop rotation program, understanding various elements is imperative. These elements include the duration of pathogen survival in the soil, the range of host plants, methods for pathogen spread or reintroduction into fields, and strategies for managing other potential sources of pathogens.

11.6.1.2 Sanitation

Sanitation is a critical method in the classical management of plant diseases. Plant pathogens can persist in the soil for extended periods through resting structures like chlamydospores, microsclerotia, oospores, or sclerotia, even in the absence of a living host or plant debris. Thus, it is crucial to remove plant debris from growing areas whenever possible or to accelerate residue breakdown. Sanitation involves activities aimed at preventing the spread of pathogens by removing diseased and infected plant parts, decontaminating tools and equipment, and ensuring proper hand hygiene (Erdel et al. 2023). Weeds and volunteer plants should be eradicated as they can serve as hosts for pathogens and increase the relative humidity around the crop canopy, creating favorable conditions for many pathogens. Plowing under infected crop debris is an effective sanitation measure to control certain soilborne pathogens because tillage exposes infected plant materials to direct sunlight, which can kill some pathogens. For example, removing diseased plants and the immediate soil around their canopy can reduce the spread of diseases like lettuce drop or white mold caused by *Sclerotinia sclerotiorum* (Baysal-Gurel et al. 2012).

Tools used in farming should be disinfected or at least cleaned when moving between different plants or fields. Methods such as heat treatment, ultraviolet (UV) treatment, and chlorine treatment can be employed for this purpose. Preventative measures are essential to avoid pathogen contamination. Field sanitation, when combined with other methods, can yield a desirable outcome. It is the first step in the management of plant diseases in an integrated system (Farhaoui et al. 2023b).

11.6.1.3 Soil Solarization

Soil solarization is an environmentally friendly, pre-planting technique that uses solar energy to manage soilborne bacteria, fungi, oomycetes, and nematodes. This method involves covering the production bed with transparent plastic sheets after thorough irrigation. The plastic traps solar radiation, heating the upper soil layers effectively (Mihajlović et al. 2017). It is most suitable for regions and seasons with abundant sunshine and high temperatures. Soil solarization works by direct thermal destruction, altering the soil's microbial population and eliminating most pests. During hot summer months, it can raise soil temperatures sufficiently to kill many important pathogens, including *Sclerotinia* species, *Streptomyces scabies*, *Verticillium dahliae*, *Agrobacterium tumefaciens*, certain *Fusarium* species, and nematodes (Baysal-Gurel et al. 2012). The effectiveness of soil solarization depends on factors such as soil moisture, soil temperature, mulching material, and the duration of heating. Ethylene-vinyl acetate plastics, for example, have been shown to increase soil temperature more than polyethylene films (Panth et al. 2020). Given the high disposal costs of these materials, biodegradable alternatives like paraffin-wax emulsion mulch are preferable (Zheng et al. 2005).

Researchers have suggested pre-irrigation to balance soil thermal equilibrium, though more studies are needed to fully understand the relationship between soil moisture and temperature. Soil solarization is particularly cost-effective for organic growers, as it requires no special skills or technology. Despite its benefits, the method's dependence on climatic conditions and the time required limit its application in crop production. However, combining soil solarization with allelopathic plants (e.g., *Brassica* cover crops and *Allium* species) can enhance its efficacy (Panth et al. 2020).

11.6.1.4 Biofumigants and Soil Amendments

Biofumigation uses crops from the *Brassicaceae* family, such as mustard, broccoli, and cabbage, to control soilborne pathogens. These crops produce glucosinolates that release isothiocyanates (ITC), toxic to many soil organisms like *Rhizoctonia solani* and *Phytophthora nicotianae* (Baysal-Gurel et al. 2018). This method can also enhance beneficial soil microbes, creating competition that suppresses harmful pathogens. Allium species, like garlic and onions, produce similar pesticidal compounds (Panth et al. 2020). Broccoli, used as green manure, has been shown to reduce pathogen populations (Baysal-Gurel et al. 2012). However, the effectiveness of biofumigation depends on factors like soil type, mulching materials, and climatic conditions. While promising, biofumigation may sometimes increase disease severity or cause phytotoxicity (Panth et al. 2020).

Organic soil amendments improve soil conditions and suppress pathogens. Composts and enriched liquids contain biocidal compounds effective against pathogens like *R.* solani and *V. dahliae* (Shafique et al. 2016). These amendments enhance beneficial microorganisms, promoting competition that suppresses pathogens. Compost extracts with rhizobacteria and *Trichoderma* can enhance plant growth and disease resistance (Welke 2005). Biochar, a biomass decomposition by-product, also improves soil properties and supports beneficial microbes (Jaiswal et al. 2019). The effectiveness of organic amendments depends on factors like soil type, amendment amount, and quality (Panth et al. 2020). Integrating organic amendments with other methods, like soil solarization and biofortification, can enhance disease suppression (Katan 2017). However, research is needed to optimize amendment use and avoid potential phytotoxicity.

In addition to the methods discussed, several other classical techniques are widely used in the management of plant diseases. These include mechanical control methods, such as hand-picking pests (Khan 2021) and using physical barriers to protect plants (Kashyap et al. 2021), as well as proper irrigation and drainage practices (Krishnamoorthy and Subramaniam 2023) to prevent the conditions that favor pathogen proliferation. These methods, combined with others, contribute to a comprehensive and sustainable approach to plant disease management.

11.6.1.5 Resistant Varieties

Plant disease resistance refers to the ability of a plant variety to limit the growth and development of specific pests or pathogens, as well as the damage they cause, offering a cost-effective and efficient approach to plant disease management. Plants exhibit varying degrees of resistance to pathogenic attacks, with resistant plants effectively fending off such assaults, while susceptible ones are easily affected (Akbar et al. 2023). This resistance is facilitated by specific genes, such as R-genes, which encode materials that respond to particular pathogenic assaults (Khan et al. 2020).

There are two main types of resistance: vertical and horizontal, which serve as opposing protective mechanisms against various pathogens. Vertical resistance involves genes conferring resistance to specific strains or types of pathogens, whereas horizontal resistance is polygenic and provides broad-spectrum protection (Akbar et al. 2023). These distinctions between horizontal and vertical resistance are outlined in Table 11.2.

Horizontal resistance in plants, as described by Keane (2012), is a polygenic trait involving multiple genes that confer protection against various strains of pathogens. This form of resistance operates in a nonspecific manner, providing defense against all types of pathogens. When a plant is under attack, its horizontal resistance mechanism activates against all groups of pathogens, as noted by Kaur et al. 2021. In contrast, vertical resistance, primarily described as monogenic with occasional oligogenic traits, acts specifically against certain genes or strains of pathogens, akin to a gene-for-gene interaction (Khan et al. 2020). While vertical resistance proves effective against one group of pathogens, it may not be as successful against another group (Lin et al. 2022). Horizontal resistance demonstrates effectiveness against all races of pathogens, although it may not offer complete immunity. This resistance mechanism slows down the damage caused by pathogens and is particularly efficient against necrotrophs, which kill host tissue before penetrating it. Activating its mechanism when pathogens attempt to harm the host tissue, horizontal resistance plays a crucial role in plant defense (Khan et al. 2020). Vertical resistance, on the other hand, exhibits high efficacy against biotrophs, which cannot survive outside living cells. By leveraging hypersensitivity, vertical resistance triggers the necrosis of its plant cells when attacked by biotrophs, thus preventing the biotrophs from

Table 11.2 The main distinctions between horizontal and vertical resistance in plants (Khan et al. 2020)

Criteria	Horizontal resistance	Vertical resistance
Nature of gene action	Polygenic	Monogenic/oligogenic
Pathotype specificity	Non-specific	Specific
Efficiency	Against all races	Against specific races
Effectivity	Nectrophs	Biotrophs
Stage of expression	Increases with maturity	Same from seedling to maturity
Chance of epidemic	Less/minimal	Present

surviving outside living cells and making vertical resistance effective against them (Kaur et al. 2021). As plants mature, horizontal resistance strengthens and becomes a durable form of defense. In contrast, vertical resistance remains consistent from seedling to maturity but generally lacks the enduring nature of horizontal resistance. Horizontal resistance's ability to act against all races of pathogens renders plants unaffected by changes or mutations in the pathogen's race (Keane 2012). However, any alterations within the pathogen or an attack by a new pathogen can pose a significant threat to the plant, potentially leading to epidemics (Dent 2000).

11.6.2 Chemical Control Methods

Various chemicals are available to control plant diseases by inhibiting the growth of or killing disease-causing pathogens. These chemicals, including bactericides for bacteria, fungicides for fungi, and nematicides for nematodes, can be applied to seeds, foliage, flowers, fruit, or soil (Baweja et al. 2020). They operate based on different principles of disease control.

Eradicants are designed to eliminate pathogens present in the soil, on seeds, or vegetative propagative organs such as bulbs, corms, and tubers (Narayan et al. 2022). Protectants create a chemical barrier between the plant and the pathogen, thereby preventing infections. Therapeutic chemicals are employed to combat infections that are already in progress (Narayan et al. 2022).

11.6.2.1 Chemical Control Techniques

Foliage Sprays and Dusts involve the application of chemical or organic substances onto the leaves of plants to control pests and diseases, or promote their growth (Deliopoulos et al. 2010). This method entails spraying or dusting the foliage with a fine mist or powder to ensure complete coverage of the entire plant surface (Narayan et al. 2022). It serves as an effective means to safeguard plants from insects, mites, fungi, and other harmful organisms that can impede foliage growth (Narayan et al. 2022).

Seed Treatment is the process of applying substances to seeds to protect them from diseases, pests, or environmental stresses (Mastouri et al. 2010). Typically conducted before planting, this treatment aims to ensure healthy seedlings and maximize crop yield. It involves coating seeds with fungicides, insecticides, or other protective agents to prevent disease transmission and minimize the risk of seed-borne infections (Cardarelli et al. 2022). Seed treatment can also enhance seed germination, improve plant vigor, and increase resistance to adverse conditions (Cardarelli et al. 2022).

Soil Treatment encompasses various techniques aimed at improving soil fertility, structure, and health to create an optimal environment for plant growth (Tahat et al. 2020). It includes adding organic matter such as compost or manure to enhance soil structure and nutrient content. Additionally, soil treatments may involve applying fertilizers, soil amendments, or microbial inoculants to boost soil fertility and promote beneficial microbial activity (Cardarelli et al. 2022).

Treatment of Tree Wounds treating tree wounds involves proper care and management to facilitate healing and prevent further damage (Negut et al. 2018). This process may entail cleaning the wound area, removing damaged or diseased tissue, and applying protective substances like wound dressings or sealants. The objective is to support the tree's natural healing process, prevent infections, and encourage healthy tissue growth (Holland et al. 2021).

Control of Post-harvest Diseases post-harvest diseases refer to infections or deterioration of crops or produce after harvesting. Effectively controlling post-harvest diseases is crucial for maintaining the quality and shelf life of agricultural products. This can be achieved through proper storage conditions, temperature and humidity control, and the use of fungicides or other preservatives (Elik et al. 2019). Implementing robust post-harvest disease control measures helps minimize losses and ensures consumers receive fresh and safe products (Rajapaksha et al. 2021).

Disinfestation of Warehouses it is a vital step in ensuring the quality and safety of stored products (Hamel et al. 2020). Warehouses are susceptible to infestations by pests like insects, rodents, or microorganisms, which can damage or contaminate stored goods. Disinfestation involves employing appropriate methods such as fumigation, heat treatment, or insecticide application to eliminate or control these pests. Regular monitoring and maintenance of warehouses are essential to prevent infestations and uphold the integrity of stored products (Elik et al. 2019).

Chemicals utilized for plant disease management can be classified into several types based on their mode of action, spectrum of activity, and the specific diseases they target (Hirooka and Ishii 2013). Here is an overview of the various types of chemicals commonly employed in plant disease control:

- Fungicides: Among the most commonly used chemicals, fungicides function by inhibiting fungal growth or eradicating fungal spores responsible for plant diseases. They can be categorized as contact fungicides, which remain on the plant surface, or systemic fungicides, which are absorbed by the plant (Farhaoui et al. 2023b).
- Bactericides: These chemicals are specifically designed to combat bacterial diseases in plants. Bactericides target the bacteria causing the disease and hinder their spread. Similar to fungicides, they can be contact-based or systemic (Qi et al. 2023).
- Biopesticides: This category includes naturally occurring substances, beneficial microorganisms, and certain minerals utilized for plant disease management.

Examples include copper and sulfur compounds, which possess fungicidal and bactericidal properties. Biopesticides are considered environmentally friendly and are increasingly utilized in organic farming practices (Li et al. 2021).

- Nematicides: These chemicals are aimed at controlling nematodes, microscopic worms that attack plant roots, leading to substantial damage. Nematicides can be either chemical-based or derived from natural sources, and they are employed to safeguard plants from nematode infections (Ebone et al. 2019).
- Plant resistance inducers: These chemicals do not directly target pathogens but function by enhancing the plant's innate defense mechanisms. They stimulate the plant to produce more of its natural protective substances (Sandroni et al. 2020).

11.6.2.2 Resistance of Plant Diseases to Pesticides

Bacteria can reproduce with a generation time as short as 20 min. Fungi, on the other hand, have widely varying generation times, with some fungi capable of producing spores and reproducing within just a few days to weeks. Moreover, certain fungi exhibit both asexual and sexual reproduction. During sexual reproduction, genes are combined, significantly increasing the likelihood of developing resistance (Kema et al. 2018). Certain plant pathogens exhibit a broad spectrum of hosts, facilitating the development of resistance (Wei Li et al. 2020). Examples include *Botrytis*, certain bacteria such as *Erwinia*, several types of powdery mildews, and pathogens like *Rhizoctonia* and *Pythium*, which induce root rot (Bhattacharjee and Dey 2014). Conversely, there are pathogens with a narrower host range, such as certain leaf spots and Fusarium wilt (Sampaio et al. 2020).

Using an incorrect product for a disease often leads to ineffective control, which can be mistaken for resistance. Employing the appropriate product at the proper rate and interval is crucial to avoid exerting pressure on the fungus or bacterium, which could lead to resistance development (Hawkins et al. 2019). It is widely recognized that administering doses lower than lethal levels of a product selectively targets the weakest or most susceptible individuals, gradually increasing the proportion of the population resistant to the treatment until it becomes predominant (Hawkins et al. 2019).

Frequent application of a product within a program exposes fungal or bacterial pathogens to a single mode of action, leading to the development of resistance. Tank mixing or alternating products with the same mode of action achieves the same result (Farhaoui et al. 2023a).

11.6.3 Biological Control Approaches

In plant pathology, biological control encompasses the utilization of microbial antagonists to suppress diseases and the deployment of host-specific pathogens to manage weed populations. The concept of biological control has evolved with modern biotechnological advancements. The U.S. National Research Council's definition includes the use of natural or genetically modified organisms, genes, or gene products to mitigate the impact of harmful organisms and promote beneficial ones like crops, beneficial insects, and microorganisms. However, this definition has sparked significant debate within the scientific community, with many finding it overly broad (Steele 2020).

Researchers have extensively investigated the mechanisms underlying biological control, aiming to understand the diverse interactions between organisms (Köhl et al. 2019). These mechanisms encompass various strategies. Direct antagonism occurs through physical contact or selective targeting of the pathogen. Indirect antagonisms result from activities that don't directly detect or target the pathogen. Some biocontrol microorganisms stimulate the host plant's defense pathways. Hyperparasites, including fungal parasites like *Coniothyrium minitans*, target pathogen structures or living tissues (Sundheim and Tronsmo 2023). Additionally, microorganisms produce lytic enzymes and metabolic byproducts that inhibit pathogen development. Notably, lytic enzymes hydrolyze compounds such as chitin, proteins, cellulose, hemicellulose, and DNA, while substances like hydrogen cyanide (HCN) disrupt aerobic microorganisms' vital pathways (Farhaoui et al. 2023a).

11.6.4 Integrated Pest Management

The objective of pest management is to effectively, economically, and safely control pests. Integrated Pest Management (IPM) offers a comprehensive approach to managing various pests, including insects, weeds, plant diseases, slugs, birds, and mammal pests. With IPM, the goal is to reduce pest populations below damaging levels rather than eliminate all pests (Bueno et al. 2021). Originally developed for agricultural pests, IPM has been successfully applied to landscape, forestry, structural, and home and garden pests since the 1980s (Barzman et al. 2015). By integrating various pest control methods, IPM helps minimize reliance on chemical pesticides, resulting in cost savings and reduced environmental risks (Bueno et al. 2021). The main stages of integrated control are:

11.6.4.1 Prevention

Preventive measures are proactively implemented to prevent organisms from becoming problematic. They involve changes in the management of crops, buildings, or other sites to avert pest issues. This approach is more cost-effective and provides better long-term results than reactive strategies (Singh et al. 2023). Preventive measures include using disease-resistant plants and maintaining sanitation (Constantine et al. 2020).

11.6.4.2 Identification

Accurate identification of pests and beneficial organisms is crucial for effective management. Factors like extreme weather or nutrient deficiencies can cause damage similar to diseases. Correct identification ensures the selection of appropriate treatments. Understanding the biology and life cycle of pests or diseases helps target treatments more effectively (Odhano et al. 2016).

11.6.4.3 Monitoring

Continuous monitoring of pest and beneficial organism populations is essential. It helps determine the need for treatments, their optimal timing, and location (DuPont et al. 2021). Early detection through monitoring makes it easier to resolve emerging issues (Parkash et al. 2023).

11.6.4.4 Treatments

Various control methods are used, such as cultural (crop rotation, resistant varieties), mechanical (traps, barriers), biological (natural enemies), behavioral (pheromone traps), or chemical, often in combination, while minimizing environmental impact (Kaur and Kaur 2020; Farhaoui et al. 2022; Rajashekhar et al. 2024).

11.6.4.5 Evaluation

The effectiveness of pest management plans is assessed to evaluate their success and identify areas for improvement (Vreysen et al. 2007).

11.7 Challenges and Future Directions

A comprehensive understanding of how plant pathogens and their hosts acclimate to their surroundings is essential for advancing rapid methods for identifying, treating, and monitoring plant pathogens. In agriculture, a significant challenge lies in developing sophisticated plant disease models that integrate more nuanced climate projections. Numerous investigations have underscored that one of the most significant predictors of the extent of the impact of climate change is the adaptability of pathogen populations and their hosts (Bebber 2015). Nevertheless, incorporating fitness parameters into disease prediction models has been challenging due to the limited understanding of predicting phytopathogen and plant fitness across diverse environments. Systems biology offers a framework to address this gap by leveraging the wealth of molecular insights accumulated over decades to predict the fitness and evolution of both plant pathogens and plants.Plant defense against numerous pathogens constitutes a complex and interdependent phenomenon with multiple layers of defense, all mediated by rapidly coevolving effectors during plant-pathogen interactions. Achieving durable resistance in crop plants against these harmful pathogens necessitates a comprehensive understanding of the molecular mechanisms underlying pathogen-host interactions and the identification of diverse genetic resistance sources. Although numerous R genes have been identified in wild plants and used to breed cultivars resistant to specific pathogens (Naveed et al. 2020), resistance was overcome because the majority of these R genes exhibit highly specific recognition for their pathogen counterparts (avr genes), which show considerable diversity not only between different species of pathogens but also among diverse strains and races within the same species (Rodewald and Trognitz 2013). Consequently, relying solely on R gene-mediated resistance proves inadequate in providing broad-spectrum and long-lasting protection. Besides the R gene-effector model, current research suggests that multiple RXLR effectors secreted by certain oomycetes and pathogenic fungi target host S factors, presenting new opportunities for developing resistant crops against these phytopathogens. CRISPR/Casgenome editing emerges as a promising approach for removing negative immune regulators from host plants. Measures to apply these new findings are underway and showing encouraging outcomes. For instance, disabling 6 S genes in potato cultivars has led to improved resistance against *Phytophthora infestans* (Sun et al. 2016). In this age characterized by advanced high-throughput technology, investigations into pathogen-plant interactions extensively employ various "omics"methodologies, including metabolomics, proteomics, genomics, effectomics, and transcriptomics. These approaches aim to uncover novel components of plant defense mechanisms. For instance, through comparative transcriptome analyses of wild tomato resistant to *P. infestans*, researchers identified transcription factors like SpWRKY3 (Cui et al. 2017), as well as a long non-coding RNA (lncRNA16397) in conjunction with the glutaredoxin (SpGRX) gene (Naveed et al. 2020). The introduction of these non-race-specific genes into plants resulted in providing broad-spectrum resistance against the pathogen *P. infestans* (Cui et al. 2018).

One significant challenge in researching host-microbe interactions is distinguishing between pathogen and host metabolites. Addressing this challenge, Pang et al. (2018) employed labeling techniques in *Arabidopsis* stomata cells inoculated with *Pseudomonas syringae*. By labeling pathogen metabolites, they successfully differentiated them from host metabolites using targeted metabolite detection. This method facilitated the identification of bacterial organic and amino acids that might have otherwise been misidentified as plant compounds. Thus, their study revealed the pathogen's influence on the primary metabolism and cellular signaling of the infected plant, impacting stomatal movement (Pang et al. 2018). Another hurdle in studying pathogen-host interactions is identifying unknown metabolites due to the absence of specific databases for phytopathogen or plant metabolites, unlike in yeast (Jewison et al. 2012) or human (Wishart et al. 2018) investigations. The intricate interplay between host plants and pathogens, coupled with the complexity of secondary metabolism, poses a significant challenge in annotating involved metabolites. Thus, characterizing specific metabolites often involves labor-intensive processes like isolation, purification, analysis by NMR, or prediction-based synthesis. Many research institutes maintain private metabolite libraries tailored to their research interests (Johnson and Lange 2015). Sharing spectral information on secondary metabolites among research groups could foster a comprehensive plant-pathogen metabolomics database. The increasing use of metabolomics to detect and identify pathogenic compounds synthesized during infection contributes to a deeper understanding of plant-pathogen interactions. This knowledge is crucial for developing novel strategies aimed at enhancing plant resistance to biotic stress and efficiently managing plant pathogens in the field. It also facilitates the implementation of integrated disease management strategies, including cultural practices and preventive treatments, to mitigate yield losses caused by pathogens.

11.8 Conclusion

Human mobility, recent expansions in international trade, and global environmental changes have contributed to an increase in emerging plant diseases caused primarily by bacteria, fungi, and oomycetes, which pose a threat to food security. Plant-pathogen interactions are intricate and multifaceted, with extensive research conducted over the past decade aimed at unraveling their underlying mechanisms. Within these interactions, two primary nodes are evident: PTI and ETI. PTI involves the process detection and neutralization of conserved MAMP or PAMP signatures, while ETI relies on plant resistance genes to provide immunity. Conversely, certain phytopathogens may secrete powerful effectors to evade plant defenses, facilitating further infection and colonization. Understanding the diverse genetic, physiological, and molecular processes implicated, alongside continued exploration of pathosystems, holds promise for crop improvement and protection. This review aims to comprehend these investigations, shedding light on the mechanisms of plant-pathogen interactions and aiding in the interpretation of various plant resistance pathways. Ultimately, this knowledge will inform the reasoned management of phytopathogens, contributing to food security and sustainable agricultural development in the future.

Acknowldegements This research was financially supported by the Phytopathology Unit in the Department of Plant Protection and Environment at the Ecole Nationale d'Agriculture de Meknès, Morocco.

References

Abdul Malik NA, Kumar IS, Nadarajah K (2020) Elicitor and receptor molecules: orchestrators of plant defense and immunity. Int J Mol Sci 21:963–963

Abebe W (2021) Wheat leaf rust disease management: a review. J Plant Pathol Microbiol 12:1–8
Akbar MU, Aqeel M, Shah MS, Jeelani G, Iqbal N, Latif A, Elnour RO, Hashem M, Alzoubi OM, Habeeb T (2023) Molecular regulation of antioxidants and secondary metabolites act in conjunction to defend plants against pathogenic infection. South Afr J Bot 161:247–257
Akhtar SS, Mekureyaw MF, Pandey C, Roitsch T (2020) Role of cytokinins for interactions of plants with microbial pathogens and pest insects. Front Plant Sci 10:1777. https://doi.org/10.3389/fpls.2019.01777
Aljawasim BD, Samtani JB, Rahman M (2023) New insights in the detection and management of Anthracnose diseases in strawberries. Plan Theory 12:3704. https://doi.org/10.3390/plants12213704
Alkan N, Friedlander G, Ment D, Prusky D, Fluhr R (2015) Simultaneous transcriptome analysis of C olletotrichum gloeosporioides and tomato fruit pathosystem reveals novel fungal pathogenicity and fruit defense strategies. New Phytol 205:801–815
Almási A, Harsányi A, Gáborjányi R (2001) Photosynthetic alterations of virus infected plants. Acta Phytopathol Entomol Hung 36:15–29
Alotaibi KD, Alharbi HA, Yaish MW, Ahmed I, Alharbi SA, Alotaibi F, Kuzyakov Y (2023) Date palm cultivation: a review of soil and environmental conditions and future challenges. Land Degrad Dev 34:2431–2444. https://doi.org/10.1002/ldr.4619
Andret-Link P, Laporte C, Valat L, Ritzenthaler C, Demangeat G, Vigne E, Laval V, Pfeiffer P, Stussi-Garaud C, Fuchs M (2004) Grapevine fanleaf virus: still a major threat to the grapevine industry. J Plant Pathol 86:183–195
Atay M (2023) The importance of crop rotation in control of plant diseases. 4. International Hasankeyf Scientific Research and Innovation Congress, Turkey
Baldrich P, Rutter BD, Karimi HZ, Podicheti R, Meyers BC, Innes RW (2019) Plant extracellular vesicles contain diverse small RNA species and are enriched in 10-to 17-nucleotide "tiny" RNAs. Plant Cell 31:315–324
Barnum CR, Endelman BJ, Shih PM (2021) Utilizing plant synthetic biology to improve human health and wellness. Front Plant Sci 12:691462
Barzman M, Bàrberi P, Birch ANE, Boonekamp P, Dachbrodt-Saaydeh S, Graf B, Hommel B, Jensen JE, Kiss J, Kudsk P (2015) Eight principles of integrated pest management. Agron Sustain Dev 35:1199–1215
Baweja P, Kumar S, Kumar G (2020) Fertilizers and pesticides: their impact on soil health and environment. In: Soil health, pp 265–285
Baysal-Gurel F, Gardener BM, Miller SA (2012) Soil borne disease management in organic vegetable production. Org Agric
Baysal-Gurel F, Liyanapathiranage P, Mullican J (2018) Biofumigation: opportunities and challenges for control of soilborne diseases in nursery production. Plant Health Prog 19:332–337. https://doi.org/10.1094/PHP-08-18-0049-RV
Bebber DP (2015) Range-expanding pests and pathogens in a warming world. Annu Rev Phytopathol 53:335–356. https://doi.org/10.1146/annurev-phyto-080614-120207
Bhattacharjee R, Dey U (2014) An overview of fungal and bacterial biopesticides to control plant pathogens/diseases. Afr J Microbiol Res 8:1749–1762
Bowers GR Jr, Goodman RM (1979) Soybean mosaic virus: infection of soybean seed parts and seed transmission. Phytopathology 69:569–572
Bueno AF, Panizzi AR, Hunt TE, Dourado PM, Pitta RM, Gonçalves J (2021) Challenges for adoption of integrated pest management (IPM): the soybean example. Neotrop Entomol 50:5–20. https://doi.org/10.1007/s13744-020-00792-9
Campos MD, Patanita M, Varanda C, Materatski P, do Rosário Félix M (2021) Plant-pathogen interaction. Biology 10:444
Card SD, Pearson MN, Clover GRG (2007) Plant pathogens transmitted by pollen. Australas Plant Pathol 36:455–461

Cardarelli M, Woo SL, Rouphael Y, Colla G (2022) Seed treatments with microorganisms can have a biostimulant effect by influencing germination and seedling growth of crops. Plan Theory 11:259–259

Chauhan RS, Singh H, Singh N, Chauhan P, Yadav NK, Kumar A (2023) Epidemiology and Management of Downy Mildew of Isabgol: a review. Asian J Agric Ext Econ Sociol 41:741–750

Chlebek D, Grebtsova V, Piński A, Żur-Pińska J, Hupert-Kocurek K (2022) Genetic determinants of antagonistic interactions and the response of new endophytic strain Serratia quinivorans KP32 to fungal phytopathogens. Int J Mol Sci 23:15561

Chowdhary VA, Tank JG (2023) Biomolecules regulating defense mechanism in plants. Proc Natl Acad Sci India Sect B Biol Sci 93:17–25. https://doi.org/10.1007/s40011-022-01387-7

Constantine KL, Kansiime MK, Mugambi I, Nunda W, Chacha D, Rware H, Makale F, Mulema J, Lamontagne-Godwin J, Williams F, Edgington S, Day R (2020) Why don't smallholder farmers in Kenya use more biopesticides? Pest Manag Sci 76:3615–3625. https://doi.org/10.1002/ps.5896

Cui J, Luan Y, Jiang N, Bao H, Meng J (2017) Comparative transcriptome analysis between resistant and susceptible tomato allows the identification of lnc RNA 16397 conferring resistance to *Phytophthora infestans* by co-expressing glutaredoxin. Plant J 89:577–589. https://doi.org/10.1111/tpj.13408

Cui J, Xu P, Meng J, Li J, Jiang N, Luan Y (2018) Transcriptome signatures of tomato leaf induced by *Phytophthora infestans* and functional identification of transcription factor SpWRKY3. Theor Appl Genet 131:787–800. https://doi.org/10.1007/s00122-017-3035-9

Dai P, Jiang Y, Liang X, Gleason ML, Zhang R, Sun G (2020) Trichothecium roseum enters 'Fuji' apple cores through Stylar fissures. Plant Dis 104:1060–1068. https://doi.org/10.1094/PDIS-07-19-1559-RE

Dalio RJD, Paschoal D, Arena GD, Magalhães DM, Oliveira TS, Merfa MV, Maximo HJ, Machado MA (2021) Hypersensitive response: from NLR pathogen recognition to cell death response. Ann Appl Biol 178:268–280. https://doi.org/10.1111/aab.12657

Dang SZ, Zhang HJ, Li YZ (2024) Germination and invasion of *Paraphoma radicina* on roots of a susceptible and a resistant alfalfa cultivar (*Medicago sativa*). Phytopathology 114:102–110. https://doi.org/10.1094/PHYTO-11-22-0437-R

De Bruyne L, Höfte M, De Vleesschauwer D (2014) Connecting growth and defense: the emerging roles of brassinosteroids and gibberellins in plant innate immunity. Mol Plant 7:943–959

Deliopoulos T, Kettlewell PS, Hare MC (2010) Fungal disease suppression by inorganic salts: a review. Crop Prot 29:1059–1075

Denancé N, Sánchez-Vallet A, Goffner D, Molina A (2013) Disease resistance or growth: the role of plant hormones in balancing immune responses and fitness costs. Front Plant Sci 4:44526

Dent D (2000) Host plant resistance. In: Insect pest management. CABI Publishing, pp 123–179

Dixon GR, Tilston EL (2010) Soil-borne pathogens and their interactions with the soil environment. In: Soil microbiology and sustainable crop production, pp 197–271

DuPont ST, Strohm C, Nottingham L, Rendon D (2021) Evaluation of an integrated pest management program for central Washington pear orchards. Biol Control 152:104390

Ebone LA, Kovaleski M, Deuner CC (2019) Nematicides: history, mode, and mechanism action. Plant Sci Today 6:91–97

Elik A, Yanik DK, Istanbullu Y, Guzelsoy NA, Yavuz A, Gogus F (2019) Strategies to reduce post-harvest losses for fruits and vegetables. Strategies 5:29–39

Erdel E, Şimşek U, Kesimci TG (2023) Effects of fungi on soil organic carbon and soil enzyme activity under agricultural and pasture land of Eastern Türkiye. Sustain For 15:1765

Fang Y, Ramasamy RP (2015) Current and prospective methods for plant disease detection. Biosensors 5:537–561

Farhaoui A, Adadi A, Tahiri A, El Alami N, Khayi S, Mentag R, Ezrari S, Nabil R, Mokrini F, Belabess Z, Lahlali R (2022) Biocontrol potential of plant growth-promoting rhizobacteria (PGPR) against *Sclerotiorum rolfsii* diseases on sugar beet (*Beta vulgaris* L.). Physiol Mol Plant Pathol 101829. https://doi.org/10.1016/j.pmpp.2022.101829

Farhaoui A, El Alami N, Khadiri M, Ezrari S, Radouane N, Baala M, Tahiri A, Lahlali R (2023a) Biological control of diseases caused by *Rhizoctonia solani* AG-2-2 in sugar beet (*Beta vulgaris* L.) using plant growth-promoting rhizobacteria (PGPR). Physiol Mol Plant Pathol 124:101966

Farhaoui A, Tahiri A, Khadiri M, El Alami N, Lahlali R (2023b) Fungal root rots of sugar beets: a review of common causal agents and management strategies. Gesunde Pflanz 75:1411–1440

Farooq T, Adeel M, He Z, Umar M, Shakoor N, Da Silva W, Elmer W, White JC, Rui Y (2021) Nanotechnology and plant viruses: an emerging disease management approach for resistant pathogens. ACS Nano 15:6030–6037. https://doi.org/10.1021/acsnano.0c10910

Fones HN, Preston GM (2013) Trade-offs between metal hyperaccumulation and induced disease resistance in metal hyperaccumulator plants. Plant Pathol 62:63–71

Foresto E, Carezzano ME, Giordano W, Bogino P (2023) Ascochyta blight in chickpea: an update. J Fungi 9:203

Geeta, Mishra R (2018) Fungal and bacterial biotrophy and necrotrophy. Mol Asp Plant-Pathog Interact, pp 21–42

Hamel D, Rozman V, Liška A (2020) Storage of cereals in warehouses with or without pesticides. Insects 11:846–846

Hawkins NJ, Bass C, Dixon A, Neve P (2019) The evolutionary origins of pesticide resistance. Biol Rev 94:135–155

Hirooka T, Ishii H (2013) Chemical control of plant diseases. J Gen Plant Pathol 79:390–401

Holland LA, Travadon R, Lawrence DP, Nouri MT, Trouillas F (2021) Evaluation of pruning wound protection products for the management of Almond Canker diseases in California. Plant Dis 105:3368–3375. https://doi.org/10.1094/PDIS-11-20-2371-RE

Hu T, Chitnis N, Monos D, Dinh A (2021) Next-generation sequencing technologies: an overview. Hum Immunol 82:801–811

Jaiswal AK, Graber ER, Elad Y, Frenkel O (2019) Biochar as a management tool for soilborne diseases affecting early stage nursery seedling production. Crop Prot 120:34–42

Jan R, Asaf S, Numan M, Lubna, Kim K-M (2021) Plant secondary metabolite biosynthesis and transcriptional regulation in response to biotic and abiotic stress conditions. Agron 11:968. https://doi.org/10.3390/agronomy11050968

Jewison T, Knox C, Neveu V, Djoumbou Y, Guo AC, Lee J, Liu P, Mandal R, Krishnamurthy R, Sinelnikov I (2012) YMDB: the yeast metabolome database. Nucleic Acids Res 40:D815–D820

Johnson SR, Lange BM (2015) Open-access metabolomics databases for natural product research: present capabilities and future potential. Front Bioeng Biotechnol 3:22

Jones N (2013) Planetary disasters: it could happen one night. Nature 493:154–156

Kashyap A, Planas-Marquès M, Capellades M, Valls M, Coll NS (2021) Blocking intruders: inducible physico-chemical barriers against plant vascular wilt pathogens. J Exp Bot 72:184–198

Katan J (2017) Diseases caused by soilborne pathogens: biology, management and challenges. J Plant Pathol 99:305–315

Kaur T, Kaur M (2020) Integrated pest management: a paradigm for modern age. IntechOpen

Kaur B, Bhatia D, Mavi GS (2021) Eighty years of gene-for-gene relationship and its applications in identification and utilization of R genes. J Genet 100:50–50

Kaur S, Samota MK, Choudhary M, Choudhary M, Pandey AK, Sharma A, Thakur J (2022) How do plants defend themselves against pathogens-biochemical mechanisms and genetic interventions. Physiol Mol Biol Plants 28:485–504. https://doi.org/10.1007/s12298-022-01146-y

Keane PJ (2012) Horizontal or generalized resistance to pathogens in plants. Plant Pathol 13:337–342

Kema GHJ, Mirzadi Gohari A, Aouini L, Gibriel HAY, Ware SB, van Den Bosch F, Manning-Smith R, Alonso-Chavez V, Helps J, Ben M'Barek S (2018) Stress and sexual reproduction affect the dynamics of the wheat pathogen effector AvrStb6 and strobilurin resistance. Nat Genet 50:375–380

Khajuria YP, Akhoon BA, Kaul S, Dhar MK (2023) Avirulence (Avr) genes in fungal pathogen *Venturia inaequalis*, a causal agent of scab disease on apple trees. Physiol Mol Plant Pathol 127:102101

Khan AU (2021) Insect pests management in the agriculture plants and crops. In: Insect pests, diseases and weeds management, pp 1–50
Khan AH, Hassan M, Khan MN (2020) Conventional plant breeding program for disease resistance. In: Ul Haq I, Ijaz S (eds) Plant disease management strategies for sustainable agriculture through traditional and modern approaches. Springer, Cham, pp 27–51
Khan I, Bai Y, Zha L, Ullah N, Ullah H, Shah SRH, Sun H, Zhang C (2021) Mechanism of the gut microbiota colonization resistance and enteric pathogen infection. Front Cell Infect Microbiol 11:716299
Köhl J, Kolnaar R, Ravensberg WJ (2019) Mode of action of microbial biological control agents against plant diseases: relevance beyond efficacy. Front Plant Sci 10:454982–454982
Krishnamoorthy KK, Subramaniam KK (2023) Water resource management for alleviating crop diseases in semi-arid regions. In: Pande CB, Kumar M, Kushwaha NL (eds) Surface and groundwater resources development and management in semi-arid region. Springer, Cham, pp 191–206
Krivitsky V, Granot E, Avidor Y, Borberg E, Voegele RT, Patolsky F (2021) Rapid collection and Aptamer-based sensitive electrochemical detection of soybean rust fungi airborne urediniospores. ACS Sens 6:1187–1198. https://doi.org/10.1021/acssensors.0c02452
Kumar P, Murali V (2023) A review on soil and phytomicrobiome for plant disease management. Int J Environ Clim Change 13:2890–2904
Kumar Y, Yadav BC (2020) Plant-parasitic nematodes: Nature's most successful plant parasite. Int J Res Rev 7:379–386
Kumar K, Pal S, Devi YK (2020) Morpho-physical characteristic in tomato imparting resistance to sucking pest whitefly, *Bemisia tabaci* (*Gennadius*) Aleyrodidae: Hemiptera: a review. J Emerg Technol Innov 7:1512–1522
Kumara UA, Cooray P, Ambanpola N, Thiruchchelvan N (2022) Plant-pathogen interaction: mechanisms and evolution. In: Trends of applied microbiology for sustainable economy. Elsevier, pp 655–687
Li W, Deng Y, Ning Y, He Z, Wang G-L (2020) Exploiting broad-spectrum disease resistance in crops: from molecular dissection to breeding. Annu Rev Plant Biol 71:575–603. https://doi.org/10.1146/annurev-arplant-010720-022215
Li S, Yang B, Tan G-Y, Ouyang L-M, Qiu S, Wang W, Xiang W, Zhang L (2021) Polyketide pesticides from actinomycetes. Curr Opin Biotechnol 69:299–307
Lin F, Chhapekar SS, Vieira CC, Da Silva MP, Rojas A, Lee D, Liu N, Pardo EM, Lee Y-C, Dong Z, Pinheiro JB, Ploper LD, Rupe J, Chen P, Wang D, Nguyen HT (2022) Breeding for disease resistance in soybean: a global perspective. Theor Appl Genet 135:3773–3872. https://doi.org/10.1007/s00122-022-04101-3
Lukan T, Coll A (2022) Intertwined roles of reactive oxygen species and salicylic acid signaling are crucial for the plant response to biotic stress. Int J Mol Sci 23:5568
Maheswary T, Nurul AA, Fauzi MB (2021) The insights of microbes' roles in wound healing: a comprehensive review. Pharmaceutics 13:981
Mastouri F, Björkman T, Harman GE (2010) Seed treatment with Trichoderma harzianum alleviates biotic, abiotic, and physiological stresses in germinating seeds and seedlings. Phytopathology 100:1213–1221
Mihajlović M, Rekanović E, Hrustić J, Grahovac M, Tanović B (2017) Methods for management of soilborne plant pathogens. Pestic Fitomedicina 32:9–24
Mittler R, Zandalinas SI, Fichman Y, Van Breusegem F (2022) Reactive oxygen species signalling in plant stress responses. Nat Rev Mol Cell Biol 23:663–679
Mohamed HI, El-Shazly HH, Badr A (2020) Role of salicylic acid in biotic and abiotic stress tolerance in plants. In: Lone R, Shuab R, Kamili AN (eds) Plant Phenolics in sustainable agriculture. Springer, Singapore, pp 533–554
Mourou M, Raimondo ML, Lops F, Carlucci A (2023) Brassicaceae fungi and chromista diseases: molecular detection and host–plant interaction. Plan Theory 12:1033

Narayan H, Srivasatava P, Bhatt SC, Joshi D, Soni R (2022) Plant pathogenesis and disease control. In: Plant protection: from chemicals to biologicals, p 95

Naveed ZA, Wei X, Chen J, Mubeen H, Ali GS (2020) The PTI to ETI continuum in *Phytophthora*-plant interactions. Front Plant Sci 11:593905

Nazarov PA, Baleev DN, Ivanova MI, Sokolova LM, Karakozova MV (2020) Infectious plant diseases: etiology, current status, problems and prospects in plant protection. Acta Nat 12:46

Negut I, Grumezescu V, Grumezescu AM (2018) Treatment strategies for infected wounds. Molecules 23:2392

Ngou BPM, Ahn H-K, Ding P, Jones JDG (2021) Mutual potentiation of plant immunity by cell-surface and intracellular receptors. Nature 592:110–115

NurulNahar E, Adam P, Mazidah M, Roslan I, Rafii YM (2020) Rice blast disease in Malaysia: options for its control. J Trop Agric Fd Sci 48(1):11–23

Odhano SA, Bojoi R, Armando E, Homrich G, Flores Filho AF, Popescu M, Dorrell DG (2016) Identification of three-phase IPM machine parameters using torque tests. IEEE Trans Ind Appl 53:1883–1891

Pandey AK, Kumar A, Mbeyagala EK, Barbetti MJ, Basandrai A, Basandrai D, Nair RM, Lamichhane JR (2023) Anthracnose resistance in legumes for cropping system diversification. Crit Rev Plant Sci 42:177–216. https://doi.org/10.1080/07352689.2023.2228122

Pang Q, Kong W, Guan Q, Chen S (2018) Metabolomics of early stage plant cell–microbe interaction using stable isotope labeling. Front Plant Sci 9:343150

Panigrahi GK, Satapathy KB (2020) Sacrificed surveillance process favours plant defense: a review. Plant Arch 09725210:20

Panno S, Davino S, Caruso AG, Bertacca S, Crnogorac A, Mandić A, Noris E, Matić S (2021) A review of the most common and economically important diseases that undermine the cultivation of tomato crop in the mediterranean basin. Agronomy 11:2188

Panth M, Hassler SC, Baysal-Gurel F (2020) Methods for management of soilborne diseases in crop production. Agriculture 10:16

Parkash J, Rajat AK, Jeevan BG (2023) Integrated pest management: a comprehensive overview. Adv Entomol 3:125–138

Pathak J, Ahmed H, Kumari N, Pandey A, Rajneesh SRP (2020) Role of calcium and potassium in amelioration of environmental stress in plants. In: Roychoudhury A, Tripathi DK (eds) Protective chemical agents in the amelioration of plant abiotic stress, 1st edn. Wiley, pp 535–562

Patnaik BB, Baliarsingh S, Sarkar A, Hameed ASS, Lee YS, Jo YH, Han YS, Mohanty J (2024) The role of pattern recognition receptors in crustacean innate immunity. Rev Aquac 16:190–233. https://doi.org/10.1111/raq.12829

Pech J-C, Latché A, Bouzayen M (2010) Ethylene biosynthesis. In: Plant hormones: biosynthesis, signal transduction, action! Springer, pp 115–136

Qi P-Y, Zhang T-H, Wang N, Feng Y-M, Zeng D, Shao W-B, Meng J, Liu L-W, Jin L-H, Zhang H, Zhou X, Yang S (2023) Natural products-based botanical bactericides discovery: novel abietic acid derivatives as anti-virulence agents for plant disease management. J Agric Food Chem 71:5463–5475. https://doi.org/10.1021/acs.jafc.2c08392

Rajapaksha L, Gunathilake D, Pathirana SM, Fernando T (2021) Reducing post-harvest losses in fruits and vegetables for ensuring food security—case of Sri Lanka. MOJ Food Process Technol 9:7–16

Rajashekhar M, Rajashekar B, Reddy TP, Manikyanahalli Chandrashekara K, Vanisree K, Ramakrishna K, Sunitha V, Shaila O, Sathyanarayana E, Shahanaz (2024) Evaluation of farmers friendly IPM modules for the management of fall armyworm, *Spodoptera frugiperda* (JE Smith) in maize in the hot semiarid region of India. Sci Rep 14:7118

Ramos-Cruz D, Troyee AN, Becker C (2021) Epigenetics in plant organismic interactions. Curr Opin Plant Biol 61:102060. https://doi.org/10.1016/j.pbi.2021.102060

Reid MS (1995) Ethylene in plant growth, development, and senescence. In: Plant hormones: physiology, biochemistry and molecular biology. Springer, pp 486–508

Rhouma A, Hajji-Hedfi L, Kouadri MEA, Atallaoui K, Okon OG, Khrieba MI (2023) Verticillium wilt of olive and its control caused by the hemibiotrophic soil-borne fungus *Verticillium dahliae*. Microb Biosyst 8:25–36

Rodewald J, Trognitz B (2013) S *olanum* resistance genes against P *hytophthora infestans* and their corresponding avirulence genes. Mol Plant Pathol 14:740–757. https://doi.org/10.1111/mpp.12036

Sampaio AM, de Araujo S, Rubiales D, Vaz Patto MC (2020) Fusarium wilt management in legume crops. Agronomy 10:1073–1073

Sandroni M, Liljeroth E, Mulugeta T, Alexandersson E (2020) Plant resistance inducers (PRIs): perspectives for future disease management in the field. CABI Rev 15:1–10

Savary S, Willocquet L, Pethybridge SJ, Esker P, McRoberts N, Nelson A (2019) The global burden of pathogens and pests on major food crops. Nat Ecol Evol 3:430–439

Schmey T, Tominello-Ramirez CS, Brune C, Stam R (2024) Alternaria diseases on potato and tomato. Mol Plant Pathol 25(3):e13435. https://doi.org/10.1111/mpp.13435. PMID: 38476108; PMCID: PMC10933620

Sekhwal MK, Li P, Lam I, Wang X, Cloutier S, You FM (2015) Disease resistance gene analogs (RGAs) in plants. Int J Mol Sci 16:19248–19290

Selvakumar G, Shagol CC, Kang Y, Chung BN, Han SG, Sa TM (2018) Arbuscular mycorrhizal fungi spore propagation using single spore as starter inoculum and a plant host. J Appl Microbiol 124:1556–1565

Shafique HA, Sultana V, Ehteshamul-Haque S, Athar M (2016) Management of soil-borne diseases of organic vegetables. J Plant Prot Res 56:221–230

Shah KK, Modi B, Pandey HP, Subedi A, Aryal G, Pandey M, Shrestha J (2021) Diversified crop rotation: an approach for sustainable agriculture production. Adv Agric 2021:1–9

Shekhawat K, Fröhlich K, García-Ramírez GX, Trapp MA, Hirt H (2022) Ethylene: a master regulator of plant–microbe interactions under abiotic stresses. Cells 12:31

Shi X, Zhang Z, Zhang C, Zhou X, Zhang D, Liu Y (2021) The molecular mechanism of efficient transmission of plant viruses in variable virus–vector–plant interactions. Hortic Plant J 7:501–508

Shi T-T, Li G-H, Zhao P-J (2023) Appressoria—small but incredibly powerful structures in plant–pathogen interactions. Int J Mol Sci 24:2141

Silva CJ, Adaskaveg JA, Mesquida-Pesci SD, Ortega-Salazar IB, Pattathil S, Zhang L, Hahn MG, Van Kan JA, Cantu D, Powell AL (2023) Botrytis cinerea infection accelerates ripening and cell wall disassembly to promote disease in tomato fruit. Plant Physiol 191:575–590

Singh D, Verma N, Varma A (2008) The fungal transmitted viruses. In: Varma A (eds) Mycorrhiza. Springer, Berlin, Heidelberg. https://doi.org/10.1007/978-3-540-78826-3_24

Steele C (2020) An analysis of U.S. federal environmental legislation in the nineteenth, twentieth and beginning twenty-first centuries, with emphasis on presidential party and political majorities in congress. J Legis Stud 26:295–313. https://doi.org/10.1080/13572334.2020.1726634

Sun L, Zhang J (2020) Regulatory role of receptor-like cytoplasmic kinases in early immune signaling events in plants. FEMS Microbiol Rev 44:845–856

Sun K, Wolters A-MA, Vossen JH, Rouwet ME, Loonen AEHM, Jacobsen E, Visser RGF, Bai Y (2016) Silencing of six susceptibility genes results in potato late blight resistance. Transgenic Res 25:731–742. https://doi.org/10.1007/s11248-016-9964-2

Sundheim L, Tronsmo A (2023) Hyperparasites in biological control. In: Biocontrol of plant diseases. CRC Press, pp 53–70

Tahat MM, Alananbeh KM, Othman YA, Leskovar DI (2020) Soil health and sustainable agriculture. Sustain For 12:4859

Tanner F, Tonn S, De Wit J, Van Den Ackerveken G, Berger B, Plett D (2022) Sensor-based phenotyping of above-ground plant-pathogen interactions. Plant Methods 18:35. https://doi.org/10.1186/s13007-022-00853-7

Thordal-Christensen H (2020) A holistic view on plant effector-triggered immunity presented as an iceberg model. Cell Mol Life Sci 77:3963–3976. https://doi.org/10.1007/s00018-020-03515-w

Tian D, Yang F, Niu Y, Lin Y, Chen Z, Li G, Luo Q, Wang F, Wang M (2020) Loss function of SL (sekiguchi lesion) in the rice cultivar Minghui 86 leads to enhanced resistance to (hemi) biotrophic pathogens. BMC Plant Biol 20:507. https://doi.org/10.1186/s12870-020-02724-6

Tiku AR (2020) Antimicrobial compounds (phytoanticipins and phytoalexins) and their role in plant defense. In: Co-evolution of secondary metabolites. Springer, Berlin, pp 845–868

Timilsina S, Potnis N, Newberry EA, Liyanapathiranage P, Iruegas-Bocardo F, White FF, Goss EM, Jones JB (2020) *Xanthomonas* diversity, virulence and plant–pathogen interactions. Nat Rev Microbiol 18:415–427

Toffolatti SL, Maddalena G, Passera A, Casati P, Bianco PA, Quaglino F (2021) Role of terpenes in plant defense to biotic stress. In: Biocontrol agents and secondary metabolites. Elsevier, pp 401–417

Tong T, Li Q, Jiang W, Chen G, Xue D, Deng F, Zeng F, Chen Z-H (2021) Molecular evolution of calcium signaling and transport in plant adaptation to abiotic stress. Int J Mol Sci 22:12308

Toruño TY, Stergiopoulos I, Coaker G (2016) Plant-pathogen effectors: cellular probes interfering with plant defenses in spatial and temporal manners. Annu Rev Phytopathol 54:419–441

Tripathi AN, Tiwari SK, Behera TK (2022) Postharvest diseases of vegetable crops and their management. In: Postharvest technology-recent advances, new perspectives and applications. IntechOpen

Van Dijk M, Morley T, Rau ML, Saghai Y (2021) A meta-analysis of projected global food demand and population at risk of hunger for the period 2010–2050. Nat Food 2:494–501

Venbrux M, Crauwels S, Rediers H (2023) Current and emerging trends in techniques for plant pathogen detection. Front Plant Sci 14:1120968. https://doi.org/10.3389/fpls.2023.1120968

Verma RC, Amulothu D, Suhas PD, Nain M, Kumar D, Saikanth DRK, Pandey SK (2023) A comprehensive overview of plant pathology understanding disease mechanisms and control. Int J Plant Soil Sci 35:1207–1214

Vreysen MJB, Robinson AS, Hendrichs J, Kenmore P (2007) Area-wide integrated pest management (AW-IPM): principles, practice and prospects. In: Area-wide control insect pests. From research to field implementation. Springer, Dordrecht, pp 3–33

Wang Y, Li X, Fan B, Zhu C, Chen Z (2021) Regulation and function of defense-related callose deposition in plants. Int J Mol Sci 22:2393

Wang Y, Pruitt RN, Nuernberger T, Wang Y (2022) Evasion of plant immunity by microbial pathogens. Nat Rev Microbiol 20:449–464

Welke SE (2005) The effect of compost extract on the yield of strawberries and the severity of *Botrytis cinerea*. J Sustain Agric 25:57–68. https://doi.org/10.1300/J064v25n01_06

Welsh MJ, Turner JA, Epanchin-Niell RS, Monge JJ, Soliman T, Robinson AP, Kean JM, Phillips C, Stringer LD, Vereijssen J, Liebhold AM, Kompas T, Ormsby M, Brockerhoff EG (2021) Approaches for estimating benefits and costs of interventions in plant biosecurity across invasion phases. Ecol Appl 31:e02319. https://doi.org/10.1002/eap.2319

Wishart DS, Feunang YD, Marcu A, Guo AC, Liang K, Vázquez-Fresno R, Sajed T, Johnson D, Li C, Karu N (2018) HMDB 4.0: the human metabolome database for 2018. Nucleic Acids Res 46:D608–D617

Wyatt TT, Wösten HAB, Dijksterhuis J (2013) Fungal spores for dispersion in space and time. Adv Appl Microbiol 85:43–91

Yang Y, Wei F, Li Z, Han F, Guan H, Huang J (2023) Improper crop rotation may enrich soil-borne pathogens of *Panax notoginseng*. J Phytopathol 171:567–576. https://doi.org/10.1111/jph.13212

Yu M, Wang Y-C, Huang C-J, Ma L-S, Lai E-M (2021) Agrobacterium tumefaciens deploys a versatile antibacterial strategy to increase its competitiveness. J Bacteriol 203. https://doi.org/10.1128/JB.00490-20

Zanini AA, Di Feo L, Luna DF, Paccioretti P, Collavino A, Rodriguez MS (2021) Cassava common mosaic virus infection causes alterations in chloroplast ultrastructure, function, and carbohydrate metabolism of cassava plants. Plant Pathol 70:195–205

Zhang C, Wang W, Xue M, Liu Z, Zhang Q, Hou J, Xing M, Wang R, Liu T (2021) The combination of a biocontrol agent *Trichoderma asperellum* SC012 and hymexazol reduces the effective fungicide dose to control fusarium wilt in cowpea. J Fungi 7:685. https://doi.org/10.3390/jof7090685

Zheng Y, Yanful EK, Bassi AS (2005) A review of plastic waste biodegradation. Crit Rev Biotechnol 25:243–250. https://doi.org/10.1080/07388550500346359

Zimmerman AE, Howard-Varona C, Needham DM, John SG, Worden AZ, Sullivan MB, Waldbauer JR, Coleman ML (2020) Metabolic and biogeochemical consequences of viral infection in aquatic ecosystems. Nat Rev Microbiol 18:21–34

Zubler AV, Yoon J-Y (2020) Proximal methods for plant stress detection using optical sensors and machine learning. Biosensors 10:193

Chapter 12
Exploring Microbial Consortia for Enhanced Crop Protection and Plant Growth Promotion in Agriculture

Shiying Qu, Óscar Santamaría, Jorge Martín-García, and Jorge Poveda

Abstract Nowadays, the combination of population growth, climate change and increasing food demands has created an urgent requirement for agricultural systems to transition towards a more eco-friendly and sustainable approach. Using microorganisms has been extensively researched as an effective strategy to address this challenge. A variety of microorganisms are used either alone or in combination to form consortia, including bacteria-bacteria, fungus-fungus and bacteria-fungus. The employment of these consortia showcases remarkable potential as both biostimulants and biocontrol agents. This chapter examines how microbial consortia operate in crops and provides recent studies that demonstrate the effectiveness of microbial consortia in promoting plant growth and development, improving crop yield, and enhancing tolerance to environmental and biological stresses. Nevertheless, further research is still needed to thoroughly comprehend the mechanisms of action of specific microbial communities in various plant species. Besides, there is a requirement for a more extensive technological exploration at the production level to optimize the utilization of microbial consortia in agricultural products.

Keywords Biostimulant · Biological control · Sustainable agriculture · Yield increase · Bio-alternative products

S. Qu · Ó. Santamaría · J. Martín-García · J. Poveda (✉)
Recognised Research Group AGROBIOTECH, UIC-370 (JCyL), Department of Plant Production and Forest Resources, Higher Technical School of Agricultural Engineering of Palencia, University Institute for Research in Sustainable Forest Management (iuFOR), University of Valladolid, Palencia, Spain
e-mail: jorge.poveda@uva.es

O. O. Babalola, A. S. Ayangbenro (eds.), *Microbial Allies*,
https://doi.org/10.1007/978-3-031-90530-8_12

12.1 Introduction

The world's population is expected to reach 9.7 billion by 2050 (ONU 2019). Rapid population growth is placing higher demands on agricultural systems, needing more arable land or more productivity to meet human requirements for food. However, climate change and extreme weather events may cause water scarcity, soil degradation and biosphere destruction which will likely hinder further agricultural challenges (Cramer et al. 2001). As a result, these pressures have led to agricultural systems relying on synthetic chemical farming products to enhance plant growth, eradicate pests and diseases, and thus rapidly boost crop yields (Santos et al. 2012).

Along with increased crop yields, the overuse of fertilizers may increase greenhouse gas emissions that contribute to climate change (Saroha et al. 2023). At the same time, the excessive use of synthetic chemicals has led to problems, such as soil acidification, eutrophication, leaching of nutrients into water bodies and pesticide residues, which have seriously harmed the ecological environment and human health (Khanna and Gupta 2018; Ullah et al. 2019). In more serious cases, agrochemicals can cause acute or chronic poisoning in humans, with skin issues, neurological or immune system impairment, and harm to the liver and kidneys (Wong and Hagen 2022). To avoid or limit those adverse aspects, agricultural systems need to be transformed to meet the growing demand for food in more sustainable ways (Jhariya et al. 2019). Within those ways, the application of new alternatives to agrochemicals, such as microorganisms, has received growing attention and increased research.

In an effort to improve agricultural productivity and overall quality, we have embarked on a quest to promote environmentally sustainable agriculture and explore viable alternatives to the prevailing dominance of chemical-based crop production. This chapter begins by examining the mechanisms used by bacteria and fungi to protect and enhance crop growth and yield. This is followed by a compilation of the various forms of microbial consortia found in agricultural systems. The specific applications of bacterial consortia, fungal consortia and bacterial-fungal consortia in agriculture are then detailed. Finally, the functional properties of microbial consortia and the complex interactions that take place between plants and microorganisms are illustrated.

12.2 Bacteria as Agents of Crop Protection and Plant Growth Promotion

Common names for plant rhizosphere bacteria are: plant-growth-promoting bacteria (PGPB), plant-growth-promoting rhizobacteria (PGPR) or plant-growth-promoting microbes (PGPM) (Zhao et al. 2024). They are capable of acting as biofertilizers by improving plant access to nutrients and producing phytohormones that promote plant growth and development. In addition, PGPB have an exceptional ability to protect plants under various abiotic stresses, such as limited water availability, elevated salinity and heavy metal contamination (Gupta et al. 2022; Patel et al. 2023). Furthermore, PGPB show remarkable efficacy against biotic stresses acting as biocontrol agents, protecting plants from the invasion of harmful insects and pathogens (Poria et al. 2022; Santhosh et al. 2023).

Bacteria play a vital role in plant nitrogen fixation, converting atmospheric nitrogen into bioavailable N forms (e.g., ammonium), through a process, known as biological nitrogen fixation (BNF), mediated by nitrogenase enzymes, which are able to break the chemical triple bonds in the nitrogen molecule and producing ammonia (NH_3), which is converted to ammonium (NH_4^+) for protein synthesis in plants (Shameem et al. 2023). Consequently, nitrogen-fixing bacteria can improve nitrogen utilization, plant growth and development, and ultimately leading to increased crop yields. *Burkholderia* spp. and *Methylobacterium* spp. have been reported to be very efficient in the BNF process in plants such as *Zea mays*, *Glycine max*, *Saccharum officinarum*, *Fragaria vesca* or *Oryza sativa*, among others (Rana et al. 2022; Torres Vera et al. 2024). *Rhizobium* is also a bacterium that establishes a symbiotic nitrogen fixation (SNF) with legumes. These bacteria exhibit a profound and intricate relationship with leguminous plants, which represents a distinctive configuration of the root structure that offers an optimal place for the process of nitrogen fixation (Saranraj et al. 2023). *Pseudomonas* sp. is another widely distributed nitrogen-fixing bacterium (e.g. *P. stutzeri* A1501, *P. stutzeri* DSM4166 or *P. szotifigens* 6HT33bT). They synthesize nitrogenase enzymes through genes in nitrogen-fixing islands, assisting plants in better nitrogen fixation (Sanow et al. 2023). This nitrogen fixation is a sustainable and eco-friendly alternative to traditional nitrogen fertilizers.

Bacterial biofertilizers (e.g. *Methylobacterium* spp.) can also perform their functions by making phosphorus more accessible and available to plants (Nysanth et al. 2023). These biofertilizers comprise phosphate-solubilizing bacteria (PSB) that can mobilize insoluble inorganic phosphates and mineralize organic phosphorus-containing compounds (Bandeppa et al. 2022). In some studies, inoculation with PSB increased the concentration of phosphorus into soil and improved soil enzyme activity, resulting in a significant increase in grain yield in PSB-treated wheat crops (Dasila et al. 2023). Studies have also shown that the application of PSB strains isolated from different sources, such as agroforestry systems and halophytes, can lead to significant increases in the growth and yield of wheat and strawberry crops (Valle-Romero et al. 2023; Paganin et al. 2023). PGB are increasingly being employed as a natural fertilizer because of its environmental safety, low cost and effectiveness (Timofeeva et al. 2022).

There are also potassium-solubilizing bacteria (KSB) that convert insoluble potassium into a soluble form that plants can take up. This fact can be done by secreting organic acids into soil that can solubilize potassium-containing minerals (Jain et al. 2022). Different bacteria genera, such as *Chromobacterium azotobium*, *Mesorhizobium*, *Agrobacterium*, *Bacillus*, *Burkholderia*, *Microbacterium*, *Myroides*, *Pseudomonas* and *Pantoea*, have been identified as potassium solubilizing agents (Yadav 2022). Bacteria are also known for their ability to secrete organic acids, inorganic acids or chelating agents that solubilize micronutrients in soil, that can be easily obtained and absorbed by plants, thereby improving its general growth (Rajkumar and Kurinjimalar 2021). This is the case of certain rhizobacteria that secrete chelating agents (e.g. catechols, hydroxyamic acid and pyrrolizin) which favor the uptake of iron and zinc from the soil by plants (Ahmad et al. 2023). Another works have stated that certain *Bacillus*, which secretes gluconic acid and its derivatives capable of solubilizing insoluble and precipitated zinc compounds, provides a key aid to plant growth and health (Gorain et al. 2022). Likewise, it is known that bacteria such as *Methylobacterium* spp. are able to produce siderophores enhancing the iron absorption (Nysanth et al. 2023). Overall, bacterial biofertilizers could provide a way to improve crop yields, reducing the amount of traditional fertilizers used and consequently the environmental hazards they may produce.

In addition to their ability to act as biofertilizers, bacteria (e.g. *Azospirillum*, *Bacillus*, *Methylobacterium*, *Paenibacillus* or *Pseudomonas*) have been found to produce phytohormones that may promote plant growth, such as auxins (IAAs), cytokinins (CKs), gibberellins (GAs), abscisic acid (ABA), and ethylene (ET) (Poveda and González-Andrés 2021; Nysanth et al. 2023). It has been reported that a specific PGPR strain, *Pseudomonas putida* GAP-P45, can modify endogenous phytohormone levels in *Arabidopsis thaliana*, being able to increase the levels of IAAs and CKs in branches and roots, while decreasing the stress hormone ABA in the whole plant, thus helping to alleviate the plant's water stress (Ghosh et al. 2019). Certain bacteria (e.g., *Agrobacterium*, *Burkholderia*, and *Pseudomonas*) have been found to produce 1-aminocyclopropane-1-carboxylate deaminase (ACCd), compound that reduces ET production by decomposing its precursors (Etesami et al. 2020). Keeping ET at a low level in plant roots has been identified as a factor that may allow plants to growth better under stressful situations (high salinity, drought, etc.) (Rabbi et al. 2018). Likewise, *Enterobacter aerogenes* G3 may produce a large amount of ACCd, helping rice plants to grow better in harsh environments, such as drought (Zhang et al. 2020). Moreover, high degrees of ET can also negatively affect crop nodulation, reducing the nitrogen-fixing capacity of legumes and inducing defense response in nodules (Berrabah et al. 2018). Therefore, bacteria strains with ACC deaminase activity within the rhizomes, not only may favor plants under stressful conditions by reducing ET levels in the plant, but also may enhance nodulation in legumes.

Plant growth can also be promoted by a soil remediation approach in polluted soils. In this sense, bacterial strains, such as *Bacillus hexacanthus*, *B. spizenii*, *Pantoea dispera* and *Pseudomonas aeruginosa*, isolated from contaminated soil have shown the ability to degrade organic pollutants, like insecticides (methoxystrobin), and clean up damaged soil environments (Omeiri et al. 2022). Vergani et al. (2019) found three strains of *Rhodococcus erythropolis* bacteria able to break down a harmful chemical called polychlorinated biphenyls (PCB) and to increase the weight of *Arabidopsis* roots. Moreover, selecting the appropriate region for screening the strain holds significant importance, given the apparent correlation between strain development and pollution levels in the surrounding environment (Dourado et al. 2012). Dourado et al. (2012) suggested that particular bacterial strains, such as

Methylobacterium, sourced from petroleum-contaminated areas, demonstrate heightened resistance to heavy metals such as cadmium, lead, and arsenic. These bacterial strains play a crucial role in soil remediation by immobilizing harmful heavy metals, thereby potentially bolstering bioremediation initiatives in polluted regions.

As biocontrol agents (BCAs), bacteria (e.g. *Bacillus*, *Streptomyces*, *Pseudomonas* or *Serratia*) can produce different antibiotics/antifungals to fight against harmful bacteria or fungi. Additionally, they may prompt the activation of the plant's defense mechanism while obstructing the communication between harmful bacteria (Lee et al. 2023). As an example of that, *Streptomyces* spp. induces systemic resistance (ISR) in plants by producing lipopolysaccharides (LPS), iron carriers, biosurfactants, volatile organic compounds (VOCs) and other substances that enhance their defenses against a wide range of harmful organisms (Ebrahimi-Zarandi et al. 2022). Plant pathogens use quorum sensing (QS) signals for communication and subsequent interaction to infect the plant. In this sense, numerous studies have demonstrated that the enzyme lactonase AiiA, produced by *Bacillus subtilis*, has the ability to interfere with QS signals, thereby reducing the risk of plant disease (Lyng and Kovács 2023). Bacteria can also generate secondary metabolites (such as VOCs), via glycolysis, lipolysis, protein hydrolysis, and sulfur metabolism, with the capacity of hindering fungi, insects, and pathogens, while also contributing to the induction of plant tolerance and activation of plant defense mechanisms (Poveda 2021a). For instance, the actinomycetes *Streptomyces setonii*, which produces VOCs such as 2-ethyl-5-methylpyrazine and dimethyl disulfide, prevents the growth and the spore production of the fungal pathogen *Ceratocystis fimbriata* (Gong et al. 2022).

12.3 Fungi as Agents of Crop Protection and Plant Growth Promotion

Those fungi that boost host plant growth by producing phytohormones, increasing nutrient utilization and inducing systemic resistance to plant diseases are referred as plant growth-promoting fungi (PGPF) (Argumedo-Delira et al. 2022). They have been shown to enhance root growth, promote seed germination, increase chlorophyll concentration, and raise overall crop yields (Adedayo and Babalola 2023; García-Latorre et al. 2023a). PGPF can also protect plants from biotic and abiotic stresses by producing defense-related enzymes and metabolites that can inhibit or destroy pathogens (Rodrigo et al. 2021; Sharma et al. 2023b). PGPF in agricultural research have focused on *Aspergillus flavus*, *Actinomucor elegans*, arbuscular mycorrhizal fungi (AMF), *Gliocladium virens*, *Penicillium digitatum*, *Podospora bulbillosa*, *Sarocladium terricola*, *Trichoderma* spp. and *Xylariaceae* sp. (Adedayo and Babalola 2023; Giehl et al. 2023). Therefore in general, PGPF have a great potential for application in a sustainable agricultural development.

Fungi can improve plant uptake of nutrients required for growth through various mechanisms. AMF enhance the nutrient uptake area of the root system of the plants they establish a symbiotic relationship with, thereby augmenting the efficiency of nutrient absorption by plants (Tanwir et al. 2023; Zhou et al. 2024a). These symbiotic fungi also release phosphatases and organic acids that enhance the solubility of phosphorus in the soil thereby increasing its uptake by the symbiotic plant (Leroy 2023). AMF also facilitate the absorption of various essential nutrients by plant, such as the assimilation of organic nitrogen and sulfur compounds. The mycelial hyphae have the capacity to effectively raise these essential elements within the soil matrix, subsequently facilitating their transportation to various tissues of plant (Cozzolino et al. 2021). In other respects, endophytic fungi living within plant tissues can produce substances that may promote root growth and solubilize nutrients, as well as secrete hydrolytic enzymes to enhance nutrient solubility and facilitate nutrient uptake by the plant (Poveda et al. 2021; García-Latorre et al. 2023b). In addition, fungi can form beneficial interactions with other soil organisms (e.g., bacteria and mycorrhizal fungi) to enhance nutrient uptake by plants (Bandyopadhyay et al. 2022). All these mechanisms help plants to improve nutrient uptake and could be used to reduce dependence on chemical fertilizers, thus making fungi one of the most promising options for effective biofertilizers.

Fungi can also increase tolerance to abiotic stresses (e.g. salinity, drought, reactive oxygen species (ROS), etc.) by inducing stress resistance mechanisms in plants. Endophytic fungi, such as *Periconia macrospinosa*, *Neocamarosporium goegapense* or *N. chichastianum*, have been found to enhance the tolerance of barley plants to salinity and drought stresses by elevating chlorophyll concentration, proline content, and antioxidant enzyme activity (Hosseyni Moghaddam et al. 2022). In addition, endophytic fungi can modulate local and systemic mechanisms in plants, thereby increasing antioxidants secretion and scavenging of ROS (Inamuddin et al. 2022). It has also been suggested that AMF may establish a reciprocal relationship with the host plant through the exopolysaccharides they can produce, improving the growth parameters of maize during water shortage (Tiepo et al. 2024). Therefore, the use of fungal endophytes could be a practical and valuable method to improve plant tolerance to abiotic stresses.

Similarly to bacteria, fungi can be applied as BCAs in agricultural systems to indirectly activate local and systemic defense responses controlled by phytohormones, such as salicylic acid (SA), jasmonic acid (JA) and ET, to better face attacks by pathogens, insect pests or plant parasitic nematodes (Poveda et al. 2020; Poveda and Baptista 2021). As part of the defense mechanism, different secondary metabolites (e.g. terpenes, phenolic compounds) are also produced, which directly attack pathogens and protect the plant against them (Poveda and Baptista 2021; Rodrigo et al. 2021). For example, *Trichoderma* species produce and use secondary metabolites with insecticidal effects, such as adhesins and lytic enzymes (chitinases, proteases, lipases), to fight against the insect's defenses and colonize its body (Poveda 2021b). In addition, fungi have shown beneficial antibacterial properties. For example, *Trichoderma* helps tomato plants to prevent bacterial spot caused by *Xanthomonas euvesicatoria*, and has the ability to induce resistance against the infection of *Pseudomonas syringae* pv. lachrymans in cucumber plants (Abdullah et al. 2021). Also, the endophyte *Alternaria leptinellae* has been found to protect plants against *P. syringae* pv. tomato in tomato plants through the secondary metabolites it produces. These findings provide new strategies for enhancing plant disease resistance and improving crop protection.

12.4 Microbial Consortia

Microbial consortia (MC) is a group of two or more microorganisms (bacteria, fungi, etc.) that are able to work together, co-operate with each other, and maximize each other's effectiveness, resulting in a synergistic effect (Díaz-Urbano et al. 2023). They have been widely researched and applied in multiple areas, such as bioremediation, agriculture, and pollution reduction. Within agriculture, these consortia have been found to have applications in different areas, such as improving soil health and crop yields, controlling soil-borne diseases, and producing valuable metabolites, which can reduce the need for chemical products in modern agriculture and thus contribute to sustainable development (Poveda and Eugui 2022; Nunes et al. 2024) (Fig. 12.1.).

Bernstein and Carlson (2012) classified microbial consortia into four types, namely natural microbial consortia (NMC), artificial microbial consortia (AMC), synthetic microbial consortia (SMC), and cross-kingdom consortia (CKC). According to Massot et al. (2022), the CKC is considered a minor derivative of the NMC, AMC and SMC. Therefore, microbial consortia can be divided into three

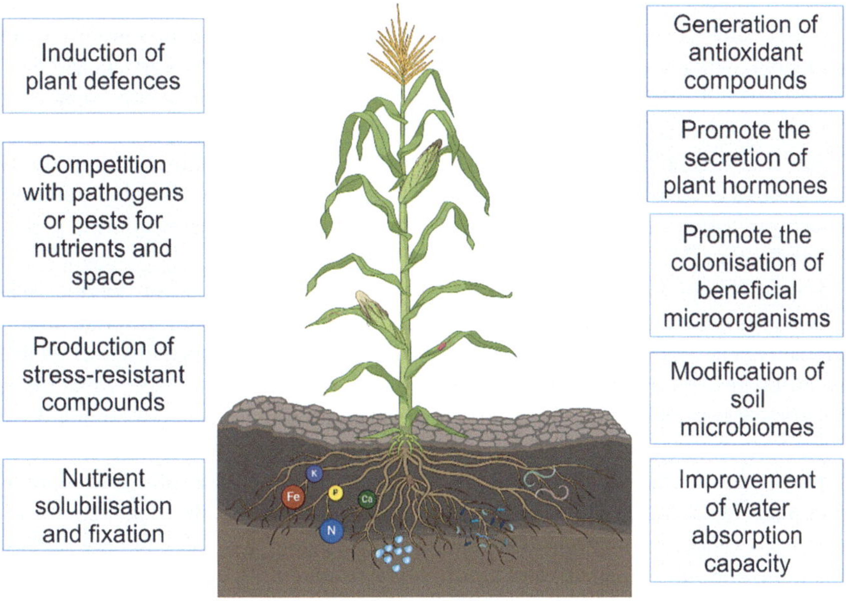

Fig. 12.1 Effect of co-inoculation with bacterial-fungus consortia in different cultures

types: NMC, AMC and SMC. NMC is composed of multiple microorganisms from the same environment that have developed synergistic interactions on their own. AMC, in contrast, is a confluence of microorganisms carefully selected by scientists from different habitats, with the types adjusted to meet different needs. SMC is a structurally well-defined group of microorganisms that are selected from different carefully selected microbial groups, communities that match specific needs to fulfill specific biological functions (Massot et al. 2022; Zhou et al. 2024b).

In order to apply the microbial consortia at a field scale is required the previous development of a bioproduct with the appropriated formulation which maintains the effectiveness of the beneficial micro-organisms under different environmental conditions and allows an easy storage, handling and distribution into the field. To achieve this, it is imperative to thoroughly evaluate and take into account the effectiveness, viability, and overall practicality of these formulations in order to optimize their potential impact and benefits on the crops (Singh and Kumar 2024). Factors, such as the way of using the microbial consortium, soil type, plant type, native microbiome or environmental conditions may affect the specifications of the formulation (Bhattacharjee et al. 2023).

It is possible to broadly classify microbial formulations into two distinct types, namely solid and liquid, based on the carrier medium employed (Bhattacharjee et al. 2023). Solid formulations are predominantly employed for consortia aimed to enhance crop yields by facilitating the uptake of nutrients from the soil environment. These formulations typically comprise of microbial bioinoculants, as well as additives, such as osmoprotectants, sticking agents, nutrients and solid carriers (Verma et al. 2022). Solid dosage products can be categorized into two groups, namely granular and powdery, depending on the measurement of their particle diameter (Stamenković et al. 2018). There are many kinds of commonly used carrier media, which are roughly divided into two categories: organic carriers and inorganic carriers, being the most commonly used ones peat, clay, wheat bran, sugarcane bagasse, chitosan, chitin and alginate (Balla et al. 2022; Bhattacharjee et al. 2023). Regardless the medium or type employed, the objective is to maintain the viability and biological activity of microbial solid formulations for an extended duration, ensuring a gradual and regulated impact, while minimizing expenses during practical application (Vassilev et al. 2020).

Liquid formulations often use aqueous, oil-based, water-in-oil or polymer suspensions as media to accommodate the different characteristics of different strains (Mahanty et al. 2017). In addition, there are certain cell protectors and additives that are different from those found in solid formulations, which also serve to facilitate the storage, transport and application of these liquid bioproducts (Brahmaprakash and Sahu 2012). Liquid formulations possess a broader spectrum of uses and can be employed on both seeds and plants in a manner equivalent to solid formulations (Sharma et al. 2023a). Liquid formulations possess several notable benefits in comparison to solid formulations. These advantages encompass safeguarding against contamination, an extended shelf life ranging from 1 to 2 years, reduced dosage requirements, increased tolerance to higher temperatures (up to 45 °C), improved environmental friendliness, and heightened efficacy (Mahanty et al. 2017; Riaz et al. 2020). However, due to the lack of protection of the aqueous carrier media toward microbial inoculants, the survival rate cannot be guaranteed. Therefore, the update and improvement of the formulation technology have to be continued (Mahanty et al. 2017). Several studies have been conducted to increase the survival rate of microorganisms in liquid formulations. As stated in the review by Mącik et al. (2020), the incorporation of additives, such as sucrose, arabic gum, glycerol, and polyvinylpyrrolidone has improved the adhesion between seeds and liquid microbial formulations. Consequently, this has resulted in an improved survival of microbial consortia in field applications.

12.4.1 *Bacteria-Bacteria Consortia for Agricultural Applications*

In addition to the utilization of single bacterial bioinoculants, in recent times, there has been an increase in investigation focused on the co-inoculation with multiple bacterial species, as it displays various advantages: (i) boosting plant growth (Díaz et al. 2023; Gustab et al. 2024), (ii) alleviating the impacts of abiotic stresses (Gao et al. 2023; Ercole et al. 2023), (iii) enhancing soil health (Waseem et al. 2024; Zeng et al. 2024), and (iv) augmenting crop disease resistance (Srinivasan and Mathivanan 2009). There are multiple examples of synergistic interactions between different bacteria demonstrating their ability to improve crop growth and increase agricultural impact (Table 12.1). In experiments on barley crops, Kaur et al. (2022) showed that a combined application of *Erwinia* sp. EU-B2SNL1 (N-fixer), *Chryseobacterium arthrosphaerae* EU-LWNA-37 (P-solubiliser) and *Pseudomonas gessardii* EU-MRK-19 (K-solubiliser) was more effective compared to the use of bacterial strains alone. It was evidenced by increased barley root/branch length, biomass, chlorophyll, carotenoids, phenolics, flavonoids, and soluble sugar content (Kaur et al. 2022). However, the complexity in the relationship between microbial inoculants produces also a lack of consistency in the results obtained, especially when compared those of the greenhouse and those of the field. Therefore, when opting to use bacteria consortia, not only is important to consider the compatibility between the strains used, but also the rest of the environmental conditions in order to get as much efficacy as possible.

In addition to a growth promotion, the use of bacterial consortia can help crops to cope with abiotic stresses, such as salt, drought and heavy metals (HMs). Several bacterial groups have the potential to produce secondary compounds, like plant hormones, or trigger plant defense mechanisms, to minimize the harmful impacts caused by those stresses. Shilev et al. (2020) have provided evidences that the co-inoculation with *Pseudomonas putida, P. fluorescens, P. synxantha and Bacillus thuringiensis* induced the metabolization of stress-related chemicals (ACCs) by the treated plants, thereby diminishing the accumulation of HMs in spinach (Shilev et al. 2020). Dixit et al. (2023) demonstrated that a group of bacteria, including *Bacillus safensis* NBRI 12 M, *B. subtilis* NBRI 28B and *B. subtilis* NBRI 33 N, has

Table 12.1 Effect of co-inoculation with bacteria-bacteria consortia in different cultures

Bacteria-bateria consortia	Experiment	Crop	Effects	Mode of Action	Reference
Bacillus licheniformis (MML2501), *Bacillus* sp. (MML2551) and *Pseudomonas aeruginosa* (MML2212)	In field	Sunflower	Control of plant diseases (Sunflower necrosis virus disease (SNVD)) and growth promotion	Resistance induction	Srinivasan and Mathivanan (2009)
Pseudomonas putida (REN5) and *Pseudomonas fluorescens* (REN1)	*In vitro* and greenhouse	Rice	Growth promotion	Improvements in nitrogen uptake	Etesami and Alikhani (2016)
Bradyrhizobium and *Actinomadura* spp. or *Paenibacillus graminis*	In greenhouse	Cowpea	Mitigating salt stress	Improvement of antioxidant enzyme activity and antioxidant defense system	Santos et al. (2018)
Azospirillum brasilense (Ab-V5), *Burkholderia cepacia* (CCMA 0056) and *Enterobacter cloacae* (CCMA 1285)	In greenhouse	Strawberry	Growth promotion	Promote nitrogen-fixation and production of phytohormones (such as IAA)	De Andrade et al. (2019)
Rhodococcus sp. (AW3) and *Pseudomonas* sp. (AW5)	In pots	Soybean	Growth promotion	Unidentified	Wevar Oller et al. (2020)
Pseudomonas sp. (AW4) and *Pseudomonas* sp. (AW6)					
Massilia sp. (Pn2), *Sphingobium* sp. (RS1), *Sphingobium* sp. (RS2), *Diaphorobacter* sp. (Phe15), *Mycobacterium* sp. (Pyr3) and *Mycolicibacterium* sp. (Pyr9)	In greenhouse	Water spinach, Chinese cabbage and pakchoi	Reduction of harmful chemicals in soil and plants	Degradation of polycyclic aromatic hydrocarbon (PAH)	Zhang et al. (2021a)
Streptomyces rochei (IT20) and *S. vinaceusdrappus* (SS14)	In greenhouse	Bell pepper	Control of plant diseases, increased crop yield	Reduced nitrate content, improvement of defense gene expression and antioxidant enzyme activity	Abbasi et al. (2022)
Piriformospora indica (DSM 11,827 and *Azorhizobium caulinodans* (B81176)	In greenhouse	Tomato	Promoting growth and mitigating saline stress	Increase superoxide dismutase, catalase, peroxidase activities	Xu et al. (2022)

Kosakonia radicincitans (CABV2) and *Streptomyces corchorusii* (CASL5)	In greenhouse	Tomato	Reduction of heavy metal (aluminium) stress, promoting growth	Improvement of antioxidant enzyme activity	Silambarasan et al. (2022)
Enterobacter sp. (Z1) and *Klebsiella* sp. (Z2)	In pots	Soybean	Growth promotion	Improvement of nitrogen fixation capacity，flavonoid levels and expression of genes related to flavonoid synthesis	Zhang et al. (2023)
Azospirillum argentinense (Az39) *and P. fluorescens* (A506)	*In vitro* and greenhouse	Lettuce	Growth promotion	Unidentified	Díaz et al. (2023)
Bradyrhizobium japonicum and *Bacillus subtilis*	In greenhouse	Bambara groundnut	Mitigating drought stress	Improving the growth, photosynthetic performance, and carbon assimilation	Udeh et al. (2023)
Sinorhizobium meliloti and *Enterobacter ludwigii*	In pot and field	Alfalfa	Mitigating salt stress，increased crop yield	Increased expression of genes associated with nodulation nitrogen fixation and antioxidant enzyme activity	Gao et al. (2023)
Bacillus velezensis and *Stenotrophomonas maltophilia*	In greenhouse	Maize	Mitigating salt stress	Unidentified	Ercole et al. (2023)
Rhodopseudomonas palustris and *B. subtilis*	In pot	Rice	Increased crop yield	Unidentified	Huang et al. (2024)
Bacillus licheniformis and *Pseudomonas aeruginosa*	In greenhouse	Rice	Reduction of heavy metal (Cd, As) accumulation in plants	Mitigation of oxidative stress	Zeng et al. (2024)
Bradyrhizobium spp. and *Azospirillum brasilense*	In field	Soybean	Increased crop yield	Improvements in nitrogen uptake	Prando et al. (2024)
Chryseobacterium lathyri, Bacillus cereus, Lysinibacillus fusiformis and *Pseudomonas protegens*	In greenhouse	*Brassica* crops	Growth promotion	Unidentified	Gustab et al. (2024)

the ability to help *Arabidopsis* plants combat the negative effects of salt stress. In this context, the utilization of bacterial consortia inoculants emerges as an efficient and environmentally-friendly solution to face the negative impact of the abiotic stresses.

Bacterial co-inoculant formulations have also been found to produce an increased disease resistance in the treated plants. For example, Kim et al. (2008) conducted a field experiment on chilli pepper crop, where the application of bio-formulated products comprised of three chitinolytic bacteria, i.e., *Serratia plymuthica*, *Chromobacterium* sp. and *Lysobacter enzymogenes*, provided an excellent disease control against the pathogenic oomycete *Phytophthora capsici*. The mechanisms involved in such a biocontrol were attributed to their ability to enzymatically degrade the cellular structures of the pathogenic oomycete, reducing its viability and virulence, and to the promotion of pre-existing microorganisms in soil capable of combating this plant disease (Kim et al. 2008). The dosage and conditions of utilization of bacterial formulations are very important. In the experiment of Kim et al. (2008), the application of a diluted solution of the bioproduct was still effective when used in solar-sterilised greenhouses and on a rotationally-cropped farmland. It is thus evident that the development of microbial formulations for multi-bacterial consortia and their commercial application in modern agriculture may be a new strategy to protect commercial crops from diseases.

Soil quality can be adversely affected by the presence of HMs or organic compounds, which may produce a suboptimal plant development. A commonly used strategy to cope with this problem is the use of bacterial consortia to reduce the concentration of harmful substances in the soil (Zhang et al. 2021a). As reported by Waseem et al. (2024), inoculation of sunflower plants with *Stutzerimonas stutzeri* and *Pseudomonas sundara* enhanced the heavy metals uptake (Cd, Cr and Pb) by sunflower. A similar investigation was also performed by Zhang et al. (2021b), where employing a mixture of six endophytic bacterial strains as inoculants, the content of polycyclic aromatic hydrocarbons (PAHs) in soil was significantly reduced due to the increase of genes responsible for PAHs metabolism (*NidA*, *bphalf* and *phe*).

12.4.2 Fungus-Fungus Consortia for Agricultural Applications

The beneficial effects of individual fungi on plants have sparked the researchers' interest in studying the fungi as a community. The objective of the combination can be either to obtain a synergistic effect on the same goal or to achieve multiple effects on several goals at the same time (Nunes et al. 2024). When the plant-fungus-fungus interactions have been explored, many positive effects have been found that are beneficial to agricultural production and sustainable development (Díaz-Urbano et al. 2023). As summarized in Table 12.2, fungi consortia not only can act as

Table 12.2 Effect of co-inoculation with fungus-fungus consortia in different cultures

Fungus-fungus consortia	Experiment	Crop	Effects	Mode of Action	Reference
Glomus mosseae, *Acaulospora laevis*, *Trichoderma viride* and *Pseudomonas fluorescens* (MTCC No. 103)	In greenhouse	Sunflower	Growth promotion	Increased MF root colonization, improvement of nutrient acquisition (especially P) and growth hormone content	Yadav et al. (2015)
Metarhizium brunneum (KVL 04–57), *M. flavoviride* (KVL 14–112), *M. robertsii* (KVL 12–35) and *Clonostachys rosea* (IK726)	In greenhouse	Wheat	Control of insect's pest (*Tenebrio molitor* and *Fusarium culmorum*)	Unidentified	Keyser et al. (2016)
Trichoderma harzianum (CECT 2413) and AMF	In greenhouse	Rapeseed, Arabidopsis	Increased siliques yield	Increased *T. harzianum*-root colonization and allows colonization of cruciferous plants by AMF	Poveda et al. (2019)
T. harzianum (T8), *Gigaspora margarita*, *Glomus hoi* and *Scutelospora gigantea*	In virto and greenhouse	Bean	Growth promotion, control of plant diseases (*Fusarium. solani*)	Increased P solubilization, against pathogen mycelial	Eke et al. (2019)
Funneliformis monosporus, *Rhizoglomus clarum*, *Dentiscutata nigra* and *Acaulospora laevis*	In greenhouse	Onion	Growth promotion	Increased MF root colonization	Metwally and Al-Amri (2020)
Trichoderma gamsii (A5MH) and *Trichoderma harzianum* (Tr906)	In greenhouse	Wheat	Growth promotion, control of plant diseases (*Fusarium pseudograminearum*)	Decreased wheat pathogenic	Stummer et al. (2022)
T. harzianum (MZ025966) and AMF	In greenhouse	Tomato	Growth promotion, control of plant diseases (*Meloidogyne javanica*)	Induction of systemic resistance, and increased enzymatic activity	Nafady et al. (2022)
Metarhizium brunneum (KVL 16–36), *M. robertsii* (KVL 12–35), *B. bassiana* (KVL 13–39) and *Funneliformis mosseae* (BEG12)	In greenhouse	Tomato	Growth promotion, control of plant diseases (*Botrytis cinerea*)	Systemic shifts in plant defenses	Zitlalpopoca-Hernandez et al. (2022)

(continued)

Table 12.2 (continued)

Fungus-fungus consortia	Experiment	Crop	Effects	Mode of Action	Reference
T. koningiopsis NBRI-PR5 (MTCC 25372) and *T. asperellum* NBRI-K14 (MTCC 25373)	In pot	Rice	Growth promotion, salinity tolerance	Improve catalase activity and proline content	Anshu et al. (2022)
Beauveria bassiana, *Funneliformis mosseae*, *Funneliformis constrictum*, *Gigaspora margarita* and *Rhizophagus irregularis*	In greenhouse	Cotton	Control of insect's pest (cotton leafworm, *Spodoptera littoralis*)	Improve enzymes activity and proline content	Metwally et al. (2022)

biostimulants for plant growth promotion, increased crop yields and enhancement of tolerance towards abiotic stresses, but also serve as BCAs against biotic stresses.

In recent years, the use of fungi and fungal co-inoculation has been one of the most effective and sustainable strategy for crop growth promotion, as well as crop yield and quality enhancement. Although a number of studies have not identified the specific mechanism of action of the microorganisms, growth parameters (e.g. germination rate) were significantly enhanced in different crops compared to inoculation with a particular fungus alone. For example, the co-inoculation with *T. harzianum*, *Beauveria bassiana* and *Metarhizium anisopliae* produced in eggplant crops a lower requirement of days to grow and a longer shoot and root lengths (Upamanya et al. 2020). Also, the co-inoculation of the fungal strains *T. atroviride* and *T. citrinoviride* into pakchoi resulted in increased levels of seed germination and growth of the radicle compared to the values obtained by single inoculation (Chen et al. 2021). One of the reasons for the better crop growth and increased yield may be the synergistic effect between fungi that promotes the colonization of plant roots by beneficial fungi. For example, *Trichoderma viride* promotes AMF colonization in onion plants resulting in significant increase in plant height, root length, bulb and neck diameter parameters (Metwally and Al-Amri 2020). Growth promotion and yield increase can also be achieved by increasing nutrient uptake and utilization of various nutrients by plants. For example, Liu et al. (2021) found that the inoculation of *Epulorhiza repens* and *Umbelopsis nana* resulted in an increase in the levels of nutrients, such as phosphorus, potassium, calcium, iron and zinc, in *Cymbidium hybridum* plants.

In the case of adverse conditions, the application of multiple fungal consortia can largely mitigate their negative effects of plant stress. One of the mechanisms involved, as in the case of crop growth promotion, is the promotion of a plant colonization by beneficial fungi. An example of that was the co-inoculation of *Rhizoglomus intraradices* and *Mucor* sp. in *Lactuca serriola* crop growth in areas contaminated with HMs (Ważny et al. 2018). Likewise, co-inoculation of wheat with AMF, *Penicillium funiculosum* and *Fusarium oxysporum* improved crop growth under saline conditions (Elgharably and Nafady 2021). The better growth caused by the fungi consortia under environmental stress can also be explained by the increased tolerance of the plant in the face of this adversity due to the increased antioxidant activity produced by the consortia, how was described by Khalvandi et al. (2021) after the co-inoculation with AMF and EF in peppermint growing under salt stress conditions.

The third big realm for the application of fungal consortia in agriculture is the plant protection against diverse pests and diseases. Although, the study conducted by Cruz et al. (2006) did not reveal a precise mode of action, it showed that a combination of three strains of *B. bassiana* was very effective in the control of the coffee berry borer, causing a significant increase in its mortality rate. Similar works have been undertaken against the pest *Plutella xylostella* (Soth et al. 2022) or *Duponchelia fovealis* (Stuart et al. 2022). *B. bassiana* is able to generate a variety of toxins, which are secondary metabolites such as beauvericin, bassianin, bassianolide, beauverolides, tenellin, oosporein, and oxalic acid. So, the fungal *B. bassiana* consortium could entail different mechanisms of action, even the same toxin could exert different mechanisms of actions according to the host (Wang et al. 2021).

12.4.3 Bacteria-Fungus Consortia for Agricultural Applications

As indicated in Table 12.3, different bacteria-fungus strains have been used in combination as biostimulants or BCAs in the context of a sustainable agriculture. However, several reports have provided evidences of antagonism between fungi and bacteria (Erjaee et al. 2019). Consequently, when studying bacterial and fungal consortia as bioinoculants it is imperative to take into account the compatibility between microorganisms (Poveda and Eugui 2022).

The promotion of crop growth and development through bacteria-fungus consortia exhibits a diverse range of mechanisms. These alliances can, for instance, enhance enzyme activity by synergizing fungi to contribute to nutrient cycling (El-Katatny 2010). Once again, by means of synergistic interactions, the PGPB strains are able to facilitate the dissolution of insoluble nutrients, such as phosphate, and facilitate the absorption of nutrients (Devi et al. 2022). Moreover, co-culture may induce the synthesis of more secondary compounds (such as fatty acids or amino acids) or plant hormones (such as IAA) that directly enhance the growth and maturation of crops (Karuppiah et al. 2019a, b). However, there are still studies in which the enhancement of growth or physiological parameters occurs without a clear presentation of the underlying mechanism (Firdu et al. 2020; Neelipally et al. 2020). Of greater significance, bacterial-fungal consortia have demonstrated the ability to effectively increase crop yields (El-Katatny and Idres 2014; Kumar et al. 2015; Ajay et al. 2015), thereby establishing a foundation for achieving high rates of sustainable agriculture.

The use of bacterial-fungal consortia has made it possible for various crops to cope better with abiotic stresses and to respond more quickly to unsuitable habitats. A study by Singh et al. (2020) demonstrated that co-inoculation of *T. harzianum* and *Achromobacter xylosoxidans* was able to reduce the damage suffered by basil in cold environments, due to the production of the ethylene precursor ACC. In another study, a bacterial-fungal consortium helped the maize plant better use ammonium in cold conditions. This helped the plant remove ROS and also led to a 76% increase in IAA levels in roots and a 33% increase in ABA levels in branches (Moradtalab et al. 2020).

In addition, bacterial-fungal consortia can act as BCAs to reduce the negative impact of disease infections through various mechanisms. The study by Karuppiah et al. (2019a) showed that *B. amyloliquefaciens* was able to promote the production of enzymes with antagonistic properties in *T. asperellum* (e.g. chitinase, glucoamylase), which can break down the cell walls of harmful fungi and hence prevent their

Table 12.3 Effect of co-inoculation with bacterial-fungus consortia in different cultures

Bacteria-fungus consortiA	Experiment	Crop	Effects	Mode of action	Reference
Pseudomonas fluorescens (Pf1) and *Beauveria bassiana* (AH1)	In greenhouse and field	Rice	Control of plant diseases (*Rhizoctonia solani*) and insect pest (*Cnaphalocrocis medinalis* and *Rhizoctonia solani*)	Resistance induction; enzyme accumulation, lipoxygenase and chitinase activity	Karthiba et al. (2010)
Azospirillum brasilense and *Trichoderma harzianum*	In greenhouse	Tomato	Promoting growth	Increase in activity of Chitinase, β-1,3-glucanase, carboxymethyl cellulase xylanase and polygalacturonase	El-Katatny (2010)
Trichoderma viride (TVA) and *P. fluorescens* (EBL 20-PF)	In greenhouse	Pepper	Control of plant diseases (*Pythium aphanidermatum*)	Activation of the defense mechanism	Muthukumar et al. (2011)
Trichoderma sp., *Bacillus* sp. and P. *fluorescens*	In greenhouse	Pea	Control of plant diseases (*Sclerotinia sclerotiorum*)	Resistance induction	Jain et al. (2012)
P. fluorescens (Pf1 e TDK1) and *B. bassiana* (B2)	In greenhouse	groundnut	Control of insect's pest (*Aproaerema modicella* and *Sclerotium rolfsii*)	Increase enzymatic activity and phenolic compounds	Senthilraja et al. (2013)
B. bassiana (B2) and *Bacillus subtilis* (EPC8)	In glasshouse and field	Tomato	Control of insect's pest (*Helicoverpa armigera* and *Fusarium oxysporum* f. sp. *lycopersici*)	Accumulation of defense enzymes (lipoxygenase, peroxidase and polyphenol oxidase)	Prabhukarthikeyan et al. (2014)
Azospirillum brasilense and T. harzianum	In greenhouse	Wheat, maize	Promoting growth, increased nutrient absorption	Improvements in nitrogen and phosphorus uptake	El-Katatny and Idres (2014)
Bacillus subtilis (OTPB1) and *T. harzianum* (OTPB3)	In greenhouse and field	Brinjal, beans, bitter gourd, bottle gourd, cabbage, chilli	Promoting growth, increased crop yield	Unidentified	Kumar et al. (2015)

(continued)

Table 12.3 (continued)

Bacteria-fungus consortiA	Experiment	Crop	Effects	Mode of action	Reference
P. fluorescens and *T. harizianum*	In greenhouse	Tomato	Promoting growth, increased crop yield	Unidentified	Ajay et al. (2015)
Azotobacter and *Trichoderma* sp.	In greenhouse	*Amaranthus spinosus*	Promoting growth	Improvement of IAA	Battini et al. (2015)
Azotobacter chroococcum W5 (MTCC 25045/ NAIMCC-B-00061) and *T. viride* (ITCC 2211)	In greenhouse	Chickpea	Promoting growth	Improvements in phosphorus uptake	Velmourougane et al. (2017)
P. fluorescens, Trichoderma spp. and AMF	In fields	Pepper	Increased crop yield	Unidentified	Duc (2017)
T. viride and P. fluorescens	In greenhouse	*Chrysanthemum indicum*	Promoting growth	Increased MF root colonization	Saini et al. (2019)
Brevibacterium halotolerans and *T. harizianum*	In greenhouse and field	Mint	Promoting growth, increased oil yield	Unidentified	Singh et al. (2019)
B. subtilis (znjdf1) and *Trichoderma harzianum* (znlkhc1)	In field	Potato	Control of plant diseases (*Streptomyces* spp.)	Improvement in the abundance of beneficial bacteria in rhizosphere	Wang et al. (2019)
Bacillus amyloliquefaciens and *T. asperellum*	*In vitro*	Wheat	Promoting growth	Unidentified	Karuppiah et al. (2019a)
B. amyloliquefaciens and *T. asperellum*	*In vitro*	Maize	Promoting growth	Unidentified	Karuppiah et al. (2019b)
T. harzianum, *B. bassiana* and *Metarhizium anisopliae*	In pot and field	Eggplant	Control of plant diseases (Phomopsis, Alternaria and Fusarium)	Growth promotion, suppressiveness increase and resistance induction	Upamanya et al. (2020)
B. cereus (UPM15) and *T. asperellum* (UPM16)	In greenhouse	Oil Palm	Promoting growth	Improvement of siderophores and IAA	Tuan Muhammad Syafiq et al. (2021)

B. subtilis and *Trichoderma sp.*	In greenhouse	Soybean	Promoting growth	Unidentified	Miftakhurrohmat and Sutarman (2021)
Funneliformis mosseae, *Glomus fasciculatum* and *Bacillus* sp.	In pot and field	Tomato	Control of plant diseases (*Fusarium oxysporum f.* sp. *Lycopersici*)	Production of lytic enzymes, and siderophore, antibiosis, and phosphate solubilization	Devi et al. (2022)
Bacillus nematocida (B16) and *Pochonia chlamydosporia* (ZK7)	In field	Tomato	Control of plant diseases (Meloidogyne incognita)	Production of VOCs, proteases and chitinases, efficient root colonization	Bo et al. (2022)
Bacillus sp. (CP4 e AHP3) and MF	In field	Wheat	Promoting growth	Increased MF colonization，improve antioxidant properties	Yadav (2022)
Piriformospora indica (DSM 11,827) *and Azorhizobium caulinodans* (B81176)	In greenhouse	Tomato	Promoting growth and mitigating salt stress	increase superoxide dismutase, catalase, peroxidase activities	Xu et al. (2022)
Bacillus licheniformis (MN718157) and *Aspergillus violaceofuscus* (MH220545)	In greenhouse	Tomato	Promoting growth and mitigating salt stress	Improve catalase activity and proline content	Muthuraja and Muthukumar (2022)
Glomus spp. and *Bacillus sp.*	In greenhouse	Corn	Mitigating drought stress	Downregulating pectin, affecting their methylation degree, and increasing hemicellulose content	Wilmowicz et al. (2022)
Glomus versiforme and *Bacillus methylotrophicus*	In greenhouse	Tobacco	Mitigating drought stress	Improve chlorophyll, carotenoids and photosynthesis	Begum et al. (2022)

attack to the wheat crop. The study also found that *B. amyloliquefaciens* co-cultured with *T. asperellum* produced a compound which ameliorated the disease caused by *Fusarium graminearum* to a large extent, which may be due to the ability of the compound to activate the expression of genes related to the suppression of fungal pathogens (such as *tvsp1/prb1*).

12.5 Conclusions and Future Perspectives

Worldwide investigation into diverse 'microbial consortia' suggests that their application in agriculture is becoming more prevalent. It is clear that the selection of diverse strains for mixtures, based on the relationship between the functions of microorganisms and their application to crop production, has a great potential to contribute to the progress of a sustainable agriculture. It can also have a positive effect on helping crops cope with salinity, drought tolerance and diseases. Improvements in these areas can hence contribute to the healthy development of agricultural systems. Thus, the development of microbial consortia can not only contribute to solving climate, food and population problems, but also to the environmental protection and the healthy development of ecosystems.

Although there is a growing body of literature on microbial consortia, much of it is currently based on *in vitro* or greenhouse experiments, with limited application in real agricultural settings. This chapter explores the implications and potential of microbial consortia in the context of sustainable agricultural development. However, the complex nature of the microbial consortium requires further investigation and discussion. The potential impact of pre-existing microorganisms in agricultural crops on the efficacy of microbial inoculant mixtures remains uncertain. As a result, most of the studies reviewed in this chapter only demonstrated an improvement in growth and/or physiological aspects, opening the door for further research into the specific mechanisms underlying these effects.

At a commercial production scale, the use of formulation carriers is important and directly impacts the efficacy of microbial inoculants. More attention should then be paid to formulation development and production processes to improve the cost effectiveness and practicality of formulated products. Overall, microbial consortia have been shown to provide suitable environmentally friendly solutions to various problems in current agriculture. However, there are still many issues that need to be refined and addressed, and more extensive research is needed to address the evolving agricultural systems and ecosystems. Although the use of microorganisms, alone or in consortia, has a great potential as plant growth promoters and BCAs, there is still a long road until they can completely substitute the synthetic chemicals (fertilizers or pesticides).

References

Abbasi S, Alipour Kafi S, Karimi E, Sadeghi A (2022) Streptomyces consortium improved quality attributes of bell pepper fruits, induced plant defense priming, and changed microbial communities of rhizosphere under commercial greenhouse conditions. Rhizosphere 23:100570. https://doi.org/10.1016/j.rhisph.2022.100570

Abdullah NS, Doni F, Mispan MS et al (2021) Harnessing *Trichoderma* in agriculture for productivity and sustainability. Agronomy 11:2559. https://doi.org/10.3390/agronomy11122559

Adedayo AA, Babalola OO (2023) Fungi that promote plant growth in the rhizosphere boost crop growth. JoF 9:239. https://doi.org/10.3390/jof9020239

Ahmad M, Hussain A, Dar A et al (2023) Combating iron and zinc malnutrition through mineral biofortification in maize through plant growth promoting *Bacillus* and *Paenibacillus* species. Front Plant Sci 13:1094551. https://doi.org/10.3389/fpls.2022.1094551

Ajay T, Ramji S, Manoj M (2015) Determination of compatibility of *Pseudomonas fluorescens* and *Trichoderma harizianum* grown on deoiled cakes of neem and jatropha for mass multiplication of P. *fluorescens* and T. *harizianum* in *vitro*. Afr J Agric Res 10:67–75. https://doi.org/10.5897/AJAR2014.8874

Anshu A, Agarwal P, Mishra K et al (2022) Synergistic action of *Trichoderma koningiopsis* and *T. asperellum* mitigates salt stress in paddy. Physiol Mol Biol Plants 28:987–1004. https://doi.org/10.1007/s12298-022-01192-6

Argumedo-Delira R, Gómez-Martínez MJ, Mora-Delgado J (2022) Plant growth promoting filamentous fungi and their application in the fertilization of pastures for animal consumption. Agronomy 12:3033. https://doi.org/10.3390/agronomy12123033

Balla A, Silini A, Cherif-Silini H et al (2022) Recent advances in encapsulation techniques of plant growth-promoting microorganisms and their prospects in the sustainable agriculture. Appl Sci 12:9020. https://doi.org/10.3390/app12189020

Bandeppa S, Kumar K, Latha PC et al (2022) Phosphate-solubilizing microbial inoculants for sustainable agriculture. In: Trends of applied microbiology for sustainable economy. Elsevier, pp 501–524

Bandyopadhyay P, Yadav BG, Kumar SG et al (2022) *Piriformospora indica* and *Azotobacter chroococcum* consortium facilitates higher Acquisition of N, P with improved carbon allocation and enhanced plant growth in *Oryza sativa*. J Fungi (Basel) 8:453. https://doi.org/10.3390/jof8050453

Battini K, Kasa P, Modugapalem H (2015) Isolation, screening, and molecular characterization of plant growth promoting rhizobacteria isolates of *Azotobacter* and *Trichoderma* and their beneficial activities. J Nat Sc Biol Med 6:360. https://doi.org/10.4103/0976-9668.160006

Begum N, Wang L, Ahmad H et al (2022) Co-inoculation of arbuscular mycorrhizal fungi and the plant growth-promoting rhizobacteria improve growth and photosynthesis in tobacco under drought stress by up-regulating antioxidant and mineral nutrition metabolism. Microb Ecol 83:971–988. https://doi.org/10.1007/s00248-021-01815-7

Bernstein HC, Carlson RP (2012) Microbial consortia engineering for cellular factories: in *vitro* to in Silico systems. Comput Struct Biotechnol J 3:e201210017. https://doi.org/10.5936/csbj.201210017

Berrabah F, Balliau T, Aït-Salem EH et al (2018) Control of the ethylene signaling pathway prevents plant defenses during intracellular accommodation of the rhizobia. New Phytol 219:310–323. https://doi.org/10.1111/nph.15142

Bhattacharjee A, Dubey S, Sharma S (2023) "Next-generation bioformulations" for plant growth promotion and stress mitigation: a promising approach for sustainable agriculture. J Plant Growth Regul 42:6741–6759. https://doi.org/10.1007/s00344-023-10996-z

Bo T, Kong C, Zou S et al (2022) *Bacillus nematocida* B16 enhanced the rhizosphere colonization of *Pochonia chlamydosporia* ZK7 and controlled the efficacy of the root-knot nematode meloidogyne incognita. Microorganisms 10:218. https://doi.org/10.3390/microorganisms10020218

Brahmaprakash GP, Sahu PK (2012) Biofertilizers for sustainability. J Indian Inst Sci 92(1):37–62

Chen D, Hou Q, Jia L, Sun K (2021) Combined use of two *Trichoderma* strains to promote growth of Pakchoi (*Brassica chinensis* L.). Agronomy 11:726. https://doi.org/10.3390/agronomy11040726

Cozzolino V, Monda H, Savy D et al (2021) Cooperation among phosphate-solubilizing bacteria, humic acids and arbuscular mycorrhizal fungi induces soil microbiome shifts and enhances plant nutrient uptake. Chem Biol Technol Agric 8:31. https://doi.org/10.1186/s40538-021-00230-x

Cramer W, Holten JI, Kaczmarek Z, Kundzewicz ZW, Parry ML (2001) Europe. In: McCarthy JJ, Canziani OF, Leary NA, Dokken DJ, White KS (eds) Climate Change 2001: Impacts, Adaptation, and Vulnerability. Contribution of Working Group II to the Third Assessment Report of the Intergovernmental Panel on Climate Change. Cambridge University Press, Cambridge, pp 641–692

Cruz LP, Gaitan AL, Gongora CE (2006) Exploiting the genetic diversity of *Beauveria bassiana* for improving the biological control of the coffee berry borer through the use of strain mixtures. Appl Microbiol Biotechnol 71:918–926. https://doi.org/10.1007/s00253-005-0218-0

Dasila H, Sah VK, Jaggi V et al (2023) Cold-tolerant phosphate-solubilizing *Pseudomonas* strains promote wheat growth and yield by improving soil phosphorous (P) nutrition status. Front Microbiol 14:1135693. https://doi.org/10.3389/fmicb.2023.1135693

De Andrade FM, De Assis PT, Souza TP et al (2019) Beneficial effects of inoculation of growth-promoting bacteria in strawberry. Microbiol Res 223–225:120–128. https://doi.org/10.1016/j.micres.2019.04.005

Devi NO, Tombisana Devi RK, Debbarma M et al (2022) Effect of endophytic *Bacillus* and arbuscular mycorrhiza fungi (AMF) against *Fusarium* wilt of tomato caused by *Fusarium oxysporum* f. sp. *lycopersici*. Egypt J Biol Pest Control 32:1. https://doi.org/10.1186/s41938-021-00499-y

Díaz PR, Merlo F, Carrozzi L et al (2023) Lettuce growth improvement by *Azospirillum argentinense* and *fluorescent pseudomonas* co-inoculation depends on strain compatibility. Appl Soil Ecol 189:104969. https://doi.org/10.1016/j.apsoil.2023.104969

Díaz-Urbano M, Goicoechea N, Velasco P, Poveda J (2023) Development of agricultural bio-inoculants based on mycorrhizal fungi and endophytic filamentous fungi: co-inoculants for improve plant-physiological responses in sustainable agriculture. Biol Control 182:105223. https://doi.org/10.1016/j.biocontrol.2023.105223

Dixit R, Bisht N, Misra S et al (2023) *Bacillus* consortia modulate transcriptional and metabolic machinery of *Arabidopsis* plants for salt tolerance. Curr Microbiol 80:77. https://doi.org/10.1007/s00284-023-03187-2

Dourado MN, Ferreira A, Araújo WL et al (2012) The diversity of Endophytic methylotrophic bacteria in an oil-contaminated and an oil-free mangrove ecosystem and their tolerance to heavy metals. Biotechnol Res Int 2012:1–8. https://doi.org/10.1155/2012/759865

Duc N (2017) Combined inoculation of arbuscular mycorrhizal fungi, *Pseudomonas fluorescens* and *Trichoderma* spp. for enhancing defense enzymes and yield of three pepper cultivars. Appl Ecol Environ Res 15:1815–1829. https://doi.org/10.15666/aeer/1503_18151829

Ebrahimi-Zarandi M, Saberi Riseh R, Tarkka MT (2022) Actinobacteria as effective biocontrol agents against plant pathogens, an overview on their role in eliciting plant defense. Microorganisms 10:1739. https://doi.org/10.3390/microorganisms10091739

Eke P, Wakam LN, Fokou PVT et al (2019) Improved nutrient status and Fusarium root rot mitigation with an inoculant of two biocontrol fungi in the common bean (*Phaseolus vulgaris* L.). Rhizosphere 12:100172. https://doi.org/10.1016/j.rhisph.2019.100172

Elgharably A, Nafady NA (2021) Inoculation with Arbuscular mycorrhizae, *Penicillium funiculosum* and *Fusarium oxysporum* enhanced wheat growth and nutrient uptake in the saline soil. Rhizosphere 18:100345. https://doi.org/10.1016/j.rhisph.2021.100345

El-Katatny MH (2010) Enzyme production and nitrogen fixation by free, immobilized and coimmobilized inoculants of *Trichoderma harzianum* and *Azospirillum brasilense* and their possible role in growth promotion of tomato. Food Technol Biotechnol 48(2):161–174

El-Katatny MH, Idres MM (2014) Effects of single and combined inoculations with *Azospirillum brasilense* and *Trichoderma harzianum* on seedling growth or yield parameters of wheat (*Triticum vulgaris L.*, Giza 168) and corn (*Zea mays L.*, hybrid 310). J Plant Nutr 37:1913–1936. https://doi.org/10.1080/01904167.2014.911322

Ercole TG, Kava VM, Aluizio R et al (2023) Co-inoculation of *bacillus velezensis* and *Stenotrophomonas maltophilia* strains improves growth and salinity tolerance in maize (Zea mays L.). Rhizosphere 27:100752. https://doi.org/10.1016/j.rhisph.2023.100752

Erjaee Z, Shekarforoush SS, Hosseinzadeh S (2019) Identification of endophytic bacteria in medicinal plants and their antifungal activities against food spoilage fungi. J Food Sci Technol 56:5262–5270. https://doi.org/10.1007/s13197-019-03995-0

Etesami H, Alikhani HA (2016) Co-inoculation with endophytic and rhizosphere bacteria allows reduced application rates of N-fertilizer for rice plant. Rhizosphere 2:5–12. https://doi.org/10.1016/j.rhisph.2016.09.003

Etesami H, Noori F, Ebadi A, Reiahi Samani N (2020) Alleviation of stress-induced ethylene-mediated negative impact on crop plants by bacterial ACC deaminase: perspectives and applications in stressed agriculture management. In: Yadav AN, Singh J, Rastegari AA, Yadav N (eds) Plant microbiomes for sustainable agriculture. Springer, Cham, pp 287–315

Firdu Z, Alemu T, Assefa F (2020) Field performance of *Trichoderma harzianum* AAUT14 and *Bacillus subtilis* AAUB95 on Faba Bean (*Vicia faba* L.) growth promotion and management of chocolate spot (Botrytis fabae Sard.). Int J Plant Soil Sci 32:35–45. https://doi.org/10.9734/ijpss/2020/v32i1230351

Gao H, Yang D, Yang L et al (2023) Co-inoculation with *Sinorhizobium meliloti* and *Enterobacter ludwigii* improves the yield, nodulation, and quality of alfalfa (*Medicago sativa* L.) under saline-alkali environments. Ind Crop Prod 199:116818. https://doi.org/10.1016/j.indcrop.2023.116818

García-Latorre C, Rodrigo S, Marin-Felix Y et al (2023a) Plant-growth promoting activity of three fungal endophytes isolated from plants living in dehesas and their effect on Lolium multiflorum. Sci Rep 13:7354. https://doi.org/10.1038/s41598-023-34036-8

García-Latorre C, Rodrigo S, Santamaría O (2023b) Potential of fungal endophytes isolated from pasture species in Spanish Dehesas to produce enzymes under salt conditions. Microorganisms 11:908. https://doi.org/10.3390/microorganisms11040908

Ghosh D, Gupta A, Mohapatra S (2019) Dynamics of endogenous hormone regulation in plants by phytohormone secreting rhizobacteria under water-stress. Symbiosis 77:265–278. https://doi.org/10.1007/s13199-018-00589-w

Giehl A, Dos Santos AA, Cadamuro RD et al (2023) Biochemical and biotechnological insights into fungus-plant interactions for enhanced sustainable agricultural and industrial processes. Plan Theory 12:2688. https://doi.org/10.3390/plants12142688

Gong Y, Liu J-Q, Xu M-J et al (2022) Antifungal volatile organic compounds from *Streptomyces setonii* WY228 control black spot disease of sweet potato. Appl Environ Microbiol 88:e02317-e02321. https://doi.org/10.1128/aem.02317-21

Gorain B, Paul S, Parihar M (2022) Role of soil microbes in micronutrient solubilization. In: New and future developments in microbial biotechnology and bioengineering. Elsevier, pp 131–150

Gupta A, Mishra R, Rai S et al (2022) Mechanistic insights of plant growth promoting bacteria mediated drought and salt stress tolerance in plants for sustainable agriculture. Int J Mol Sci 23:3741. https://doi.org/10.3390/ijms23073741

Gustab M, Ważny R, Jędrzejczyk RJ et al (2024) Beneficial impact of multi-bacterial inoculation on growth of selected *Brassicaceae* seedlings in a greenhouse culture. Sci Hortic 324:112575. https://doi.org/10.1016/j.scienta.2023.112575

Hosseyni Moghaddam MS, Safaie N, Rahimlou S, Hagh-Doust N (2022) Inducing tolerance to abiotic stress in *Hordeum vulgare* L. by halotolerant endophytic fungi associated with salt Lake plants. Front Microbiol 13:906365. https://doi.org/10.3389/fmicb.2022.906365

Huang K, Yin H, Zheng Q et al (2024) Microbial inoculation alters rhizoplane bacterial community and correlates with increased rice yield. Pedobiologia 104:150945. https://doi.org/10.1016/j.pedobi.2024.150945

Inamuddin ACO, Ahamed MI, Altalhi T (2022) Microbial biostimulants for sustainable agriculture and environmental bioremediation, 1st edn. CRC Press, Boca Raton

Jain A, Singh S, Kumar Sarma B, Bahadur Singh H (2012) Microbial consortium-mediated reprogramming of defence network in pea to enhance tolerance against *Sclerotinia sclerotiorum*: microbial consortium enhance defence network. J Appl Microbiol 112:537–550. https://doi.org/10.1111/j.1365-2672.2011.05220.x

Jain D, Saheewala H, Sanadhaya S et al (2022) Potassium solubilizing microorganisms as soil health engineers: an insight into molecular mechanism. In: Rhizosphere engineering. Elsevier, pp 199–214

Jhariya MK, Banerjee A, Meena RS, Yadav DK (eds) (2019) Sustainable agriculture, forest and environmental management. Springer, Singapore

Karthiba L, Saveetha K, Suresh S et al (2010) PGPR and entomopathogenic fungus bioformulation for the synchronous management of leafer pest and sheath blight disease of rice: bioformulation for the control of leafer and sheath blight of rice. Pest Manag Sci 66:555–564. https://doi.org/10.1002/ps.1907

Karuppiah V, Sun J, Li T et al (2019a) Co-cultivation of *Trichoderma asperellum* GDFS1009 and *bacillus amyloliquefaciens* 1841 causes differential gene expression and improvement in the wheat growth and biocontrol activity. Front Microbiol 10:1068. https://doi.org/10.3389/fmicb.2019.01068

Karuppiah V, Vallikkannu M, Li T, Chen J (2019b) Simultaneous and sequential based co-fermentations of *Trichoderma asperellum* GDFS1009 and *bacillus amyloliquefaciens* 1841: a strategy to enhance the gene expression and metabolites to improve the bio-control and plant growth promoting activity. Microb Cell Factories 18:185. https://doi.org/10.1186/s12934-019-1233-7

Kaur T, Devi R, Kumar S et al (2022) Microbial consortium with nitrogen fixing and mineral solubilizing attributes for growth of barley (*Hordeum vulgare* L.). Heliyon 8:e09326. https://doi.org/10.1016/j.heliyon.2022.e09326

Keyser CA, Jensen B, Meyling NV (2016) Dual effects of *Metarhizium* spp. and *Clonostachys rosea* against an insect and a seed-borne pathogen in wheat: dual effects of *Metarhizium* spp. and *Clonostachys rosea*. Pest Manag Sci 72:517–526. https://doi.org/10.1002/ps.4015

Khalvandi M, Amerian M, Pirdashti H, Keramati S (2021) Does co-inoculation of mycorrhiza and *Piriformospora indica* fungi enhance the efficiency of chlorophyll fluorescence and essential oil composition in peppermint under irrigation with saline water from the Caspian Sea? PLoS One 16:e0254076. https://doi.org/10.1371/journal.pone.0254076

Khanna R, Gupta S (2018) Agrochemicals as a potential cause of ground water pollution: a review. Int J Chem Stud 6(3):985–990

Kim YC, Jung H, Kim KY, Park SK (2008) An effective biocontrol bioformulation against Phytophthora blight of pepper using growth mixtures of combined chitinolytic bacteria under different field conditions. Eur J Plant Pathol 120:373–382. https://doi.org/10.1007/s10658-007-9227-4

Kumar SPM, Chowdappa P, Krishna V (2015) Development of seed coating formulation using consortium of *Bacillus subtilis* OTPB1 and *Trichoderma harzianum* OTPB3 for plant growth promotion and induction of systemic resistance in field and horticultural crops. Indian Phytopathol 68(1):25–31

Lee J, Kim S, Jung H et al (2023) Exploiting bacterial genera as biocontrol agents: mechanisms, interactions and applications in sustainable agriculture. J Plant Biol 66:485–498. https://doi.org/10.1007/s12374-023-09404-6

Leroy C (2023) Fungi in ant–plant interactions: a key to enhancing plant nutrient-acquisition strategies. New Phytol 238:1752–1754. https://doi.org/10.1111/nph.18830

Liu S, Liu M, Liao Q-G et al (2021) Effects of inoculated mycorrhizal fungi and non-mycorrhizal beneficial micro-organisms on plant traits, nutrient uptake and root-associated fungal community composition of the *Cymbidium hybridum* in greenhouse. J Appl Microbiol 131:413–424. https://doi.org/10.1111/jam.14967

Lyng M, Kovács ÁT (2023) Frenemies of the soil: *bacillus* and *pseudomonas* interspecies interactions. Trends Microbiol 31:845–857. https://doi.org/10.1016/j.tim.2023.02.003

Mącik M, Gryta A, Frąc M (2020) Biofertilizers in agriculture: an overview on concepts, strategies and effects on soil microorganisms. In: Advances in agronomy. Elsevier, pp 31–87

Mahanty T, Bhattacharjee S, Goswami M et al (2017) Biofertilizers: a potential approach for sustainable agriculture development. Environ Sci Pollut Res 24:3315–3335. https://doi.org/10.1007/s11356-016-8104-0

Massot F, Bernard N, Alvarez LMM et al (2022) Microbial associations for bioremediation. What does "microbial consortia" mean? Appl Microbiol Biotechnol 106:2283–2297. https://doi.org/10.1007/s00253-022-11864-8

Metwally RA, Al-Amri SM (2020) Individual and interactive role of *Trichoderma viride* and arbuscular mycorrhizal fungi on growth and pigment content of onion plants. Lett Appl Microbiol 70:79–86. https://doi.org/10.1111/lam.13246

Metwally RA, Azab HS, Al-Shannaf HM, Rabie GH (2022) Prospective of mycorrhiza and *Beauvaria bassiana* silica nanoparticles on *Gossypium hirsutum* L. plants as biocontrol agent against cotton leafworm, Spodoptera littoralis. BMC Plant Biol 22:409. https://doi.org/10.1186/s12870-022-03763-x

Miftakhurrohmat A, Sutarman (2021) The vegetative growth response of Detam soybean varieties towards *Bacillus subtilis* and *Trichoderma* sp. applications as bio-fertilizer. E3S Web Conf 232:03024. https://doi.org/10.1051/e3sconf/202123203024

Moradtalab N, Ahmed A, Geistlinger J et al (2020) Synergisms of microbial consortia, N forms, and micronutrients alleviate oxidative damage and stimulate hormonal cold stress adaptations in maize. Front Plant Sci 11:396. https://doi.org/10.3389/fpls.2020.00396

Muthukumar A, Eswaran A, Sangeetha G (2011) Induction of systemic resistance by mixtures of fungal and endophytic bacterial isolates against *Pythium aphanidermatum*. Acta Physiol Plant 33:1933–1944. https://doi.org/10.1007/s11738-011-0742-8

Muthuraja R, Muthukumar T (2022) Co-inoculation of halotolerant potassium solubilizing *Bacillus licheniformis* and *Aspergillus violaceofuscus* improves tomato growth and potassium uptake in different soil types under salinity. Chemosphere 294:133718. https://doi.org/10.1016/j.chemosphere.2022.133718

Nafady NA, Sultan R, El-Zawahry AM et al (2022) Effective and promising strategy in management of tomato root-knot nematodes by *Trichoderma harzianum* and arbuscular mycorrhizae. Agronomy 12:315. https://doi.org/10.3390/agronomy12020315

Neelipally RTKR, Anoruo AO, Nelson S (2020) Effect of co-inoculation of *Bradyrhizobium* and *Trichoderma* on growth, development, and yield of *Arachis hypogaea* L. (Peanut). Agronomy 10:1415. https://doi.org/10.3390/agronomy10091415

Nunes PSO, Lacerda-Junior GV, Mascarin GM et al (2024) Microbial consortia of biological products: do they have a future? Biol Control 188:105439. https://doi.org/10.1016/j.biocontrol.2024.105439

Nysanth NS, Rajan SA, Sivapriya SL, Anith KN (2023) Pink pigmented facultative methylotrophs (PPFMs): potential bioinoculants for sustainable crop production. J Pure Appl Microbiol 17:660–681. https://doi.org/10.22207/JPAM.17.2.17

Omeiri M, Khnayzer R, Yusef H et al (2022) *Bacillus* spp. isolated from soil in Lebanon can simultaneously degrade methomyl in contaminated soils and enhance plant growth. Biocatalysis and agricultural. Biotechnology 39:102280. https://doi.org/10.1016/j.bcab.2022.102280

ONU (2019) The Sustainable Development Goals Report 2019. United Nations. https://unstats.un.org/sdgs/report/2019/TheSustainable-Development-Goals-Report2019.pdf

Paganin P, Isca C, Tasso F et al (2023) A bacterial formula with native strains as alternative to chemical fertiliser for tomato crop. Plant Growth Regul 102(2):1–16. https://doi.org/10.1007/s10725-023-00993-3

Patel P, Sayyed RZ, Patel H (2023) PGPR: a sustainable agricultural mitigator for stressed agro-environments. In: Mawar R, Sayyed RZ, Sharma SK, Sattiraju KS (eds) Plant growth promoting microorganisms of arid region. Springer, Singapore, pp 303–318

Poria V, Dębiec-Andrzejewska K, Fiodor A et al (2022) Plant growth-promoting bacteria (PGPB) integrated phytotechnology: a sustainable approach for remediation of marginal lands. Front Plant Sci 13:999866. https://doi.org/10.3389/fpls.2022.999866

Poveda J (2021a) Beneficial effects of microbial volatile organic compounds (MVOCs) in plants. Appl Soil Ecol 168:104118. https://doi.org/10.1016/j.apsoil.2021.104118

Poveda J (2021b) Trichoderma as biocontrol agent against pests: new uses for a mycoparasite. Biol Control 159:104634. https://doi.org/10.1016/j.biocontrol.2021.104634

Poveda J, Baptista P (2021) Filamentous fungi as biocontrol agents in olive (Olea europaea L.) diseases: mycorrhizal and endophytic fungi. Crop Prot 146:105672. https://doi.org/10.1016/j.cropro.2021.105672

Poveda J, Eugui D (2022) Combined use of *Trichoderma* and beneficial bacteria (mainly *bacillus* and *pseudomonas*): development of microbial synergistic bio-inoculants in sustainable agriculture. Biol Control 176:105100. https://doi.org/10.1016/j.biocontrol.2022.105100

Poveda J, González-Andrés F (2021) *Bacillus* as a source of phytohormones for use in agriculture. Appl Microbiol Biotechnol 105:8629–8645. https://doi.org/10.1007/s00253-021-11492-8

Poveda J, Hermosa R, Monte E, Nicolás C (2019) *Trichoderma harzianum* favours the access of arbuscular mycorrhizal fungi to non-host Brassicaceae roots and increases plant productivity. Sci Rep 9:11650. https://doi.org/10.1038/s41598-019-48269-z

Poveda J, Abril-Urias P, Escobar C (2020) Biological control of plant-parasitic nematodes by filamentous fungi inducers of resistance: *Trichoderma,* mycorrhizal and endophytic fungi. Front Microbiol 11:992. https://doi.org/10.3389/fmicb.2020.00992

Poveda J, Eugui D, Abril-Urías P, Velasco P (2021) Endophytic fungi as direct plant growth promoters for sustainable agricultural production. Symbiosis 85:1–19. https://doi.org/10.1007/s13199-021-00789-x

Prabhukarthikeyan R, Saravanakumar D, Raguchander T (2014) Combination of endophytic *bacillus* and *Beauveria* for the management of *Fusarium* wilt and fruit borer in tomato: bioconsortia for the management of *Fusarium* wilt and fruit borer in tomato. Pest Manag Sci 70:1742–1750. https://doi.org/10.1002/ps.3719

Prando AM, Barbosa JZ, Oliveira ABD et al (2024) Benefits of soybean co-inoculation with *Bradyrhizobium* spp. and *Azospirillum brasilense*: large-scale validation with farmers in Brazil. Eur J Agron 155:127112. https://doi.org/10.1016/j.eja.2024.127112

Rabbi SMF, Tighe MK, Flavel RJ et al (2018) Plant roots redesign the rhizosphere to alter the three-dimensional physical architecture and water dynamics. New Phytol 219:542–550. https://doi.org/10.1111/nph.15213

Rajkumar R, Kurinjimalar C (2021) Microbes and plant mineral nutrition. In: Soni R, Suyal DC, Bhargava P, Goel R (eds) Microbiological activity for soil and plant health management. Springer, Singapore, pp 111–132

Rana KL, Kour D, Kaur T et al (2022) Endophytic nitrogen-fixing bacteria: untapped treasurer for agricultural sustainability. J App Biol Biotech 11:75–93. https://doi.org/10.7324/JABB.2023.110207

Riaz U, Mehdi SM, Iqbal S et al (2020) Bio-fertilizers: eco-friendly approach for plant and soil environment. In: Hakeem KR, Bhat RA, Qadri H (eds) Bioremediation and biotechnology. Springer, Cham, pp 189–213

Rodrigo S, García-Latorre C, Santamaria O (2021) Metabolites produced by fungi against fungal Phytopathogens: review, implementation and perspectives. Plants 11:81. https://doi.org/10.3390/plants11010081

Saini I, Yadav K et al (2019) Response of arbuscular mycorrhizal fungi along with *Trichoderma viride* and *Pseudomonas fluorescens* on the growth, biochemical attributes and vase life of *Chrysanthemum indicum*. J Environ Biol 40:183–191. https://doi.org/10.22438/jeb/40/2/MRN-848

Sanow S, Kuang W, Schaaf G et al (2023) Molecular mechanisms of *pseudomonas* -assisted plant nitrogen uptake: opportunities for modern agriculture. MPMI 36:536–548. https://doi.org/10.1094/MPMI-10-22-0223-CR

Santhosh CR, Nuthan BR, Mahadevakumar S et al (2023) Plant growth-promoting potential of endophytic bacteria for sustainable agriculture. In: Endophytic association: what, why and how. Elsevier, pp 457–486

Santos VB, Araújo ASF, Leite LFC et al (2012) Soil microbial biomass and organic matter fractions during transition from conventional to organic farming systems. Geoderma 170:227–231. https://doi.org/10.1016/j.geoderma.2011.11.007

Santos ADA, Silveira JAGD, Bonifacio A et al (2018) Antioxidant response of cowpea co-inoculated with plant growth-promoting bacteria under salt stress. Braz J Microbiol 49:513–521. https://doi.org/10.1016/j.bjm.2017.12.003

Saranraj P, Sayyed RZ, Sivasakthivelan P et al (2023) Symbiotic effectiveness of rhizobium strains in agriculture. In: Mawar R, Sayyed RZ, Sharma SK, Sattiraju KS (eds) Plant growth promoting microorganisms of arid region. Springer Nature Singapore, Singapore, pp 389–421

Saroha A, Kotiyal A, Thakur A (2023) Climate-resilient fertilizer management for crop production. In: Hasanuzzaman M (ed) Climate-resilient agriculture, vol 2. Springer, Cham, pp 61–79

Senthilraja G, Anand T, Kennedy JS et al (2013) Plant growth promoting rhizobacteria (PGPR) and entomopathogenic fungus bioformulation enhance the expression of defense enzymes and pathogenesis-related proteins in groundnut plants against leafminer insect and collar rot pathogen. Physiol Mol Plant Pathol 82:10–19. https://doi.org/10.1016/j.pmpp.2012.12.002

Shameem MR, Sonali JMI, Kumar PS et al (2023) Rhizobium mayense sp. Nov., an efficient plant growth-promoting nitrogen-fixing bacteria isolated from rhizosphere soil. Environ Res 220:115200. https://doi.org/10.1016/j.envres.2022.115200

Sharma P, Bano A, Singh SP, Tong YW (2023a) Microbial inoculants: recent progress in formulations and methods of application. In: Microbial inoculants. Elsevier, pp 1–28

Sharma S, Dhar MK, Kaul S (2023b) Antagonistic, plant growth promoting and extracellular hydrolytic enzyme activity of fungal endophytes of *Dioscorea bulbifera* L. Biocatal Agric Biotechnol 50:102694. https://doi.org/10.1016/j.bcab.2023.102694

Shilev S, Babrikova I, Babrikov T (2020) Consortium of plant growth-promoting bacteria improves spinach (*Spinacea oleracea* L.) growth under heavy metal stress conditions. J Chem Technol Biotechnol 95:932–939. https://doi.org/10.1002/jctb.6077

Silambarasan S, Logeswari P, Sivaramakrishnan R et al (2022) Amelioration of aluminum phytotoxicity in Solanum *lycopersicum* by co-inoculation of plant growth promoting *Kosakonia radicincitans* strain CABV2 and *Streptomyces corchorusii* strain CASL5. Sci Total Environ 832:154935. https://doi.org/10.1016/j.scitotenv.2022.154935

Singh V, Kumar B (2024) A review of agricultural microbial inoculants and their carriers in bioformulation. Rhizosphere 29:100843. https://doi.org/10.1016/j.rhisph.2023.100843

Singh S, Tripathi A, Maji D et al (2019) Evaluating the potential of combined inoculation of *Trichoderma harzianum* and *Brevibacterium halotolerans* for increased growth and oil yield in Mentha arvensis under greenhouse and field conditions. Ind Crop Prod 131:173–181. https://doi.org/10.1016/j.indcrop.2019.01.039

Singh S, Tripathi A, Chanotiya CS et al (2020) Cold stress alleviation using individual and combined inoculation of ACC deaminase producing microbes in *Ocimum sanctum*. Environ Sustain 3:289–301. https://doi.org/10.1007/s42398-020-00118-w

Soth S, Glare TR, Hampton JG et al (2022) Biological control of diamondback moth—increased efficacy with mixtures of beauveria fungi. Microorganisms 10:646. https://doi.org/10.3390/microorganisms10030646

Srinivasan K, Mathivanan N (2009) Biological control of sunflower necrosis virus disease with powder and liquid formulations of plant growth promoting microbial consortia under field conditions. Biol Control 51:395–402. https://doi.org/10.1016/j.biocontrol.2009.07.013

Stamenković S, Beškoski V, Karabegović I et al (2018) Microbial fertilizers: a comprehensive review of current findings and future perspectives. Span. J Agric Res 16:e09R01. https://doi.org/10.5424/sjar/2018161-12117

Stuart AKDC, Furuie JL, Cataldi TR et al (2022) Fungal consortium of two Beauveria bassiana strains increases their virulence, growth, and resistance to stress: a metabolomic approach. PLoS One 17:e0271460. https://doi.org/10.1371/journal.pone.0271460

Stummer BE, Zhang X, Yang H, Harvey PR (2022) Co-inoculation of *Trichoderma gamsii* A5MH and *Trichoderma harzianum* Tr906 in wheat suppresses in planta abundance of the crown rot pathogen Fusarium pseudograminearum and impacts the rhizosphere soil fungal microbiome. Biol Control 165:104809. https://doi.org/10.1016/j.biocontrol.2021.104809

Tanwir K, Abbas S, Shahid M et al (2023) Mycorrhizal symbiosis and nutrients uptake in plants. In: Singh VP, Siddiqui MH (eds) Plant ionomics, 1st edn. Wiley, pp 73–95

Tiepo AN, Fávaro MH, Amador TS, Tavares LF, Hertel MF, Calzavara AK, de Oliveira ALM, Oliveira HC, Dias-Pereira J, de Araújo HH, Bianchini E, Pimenta JA, Stolf-Moreira R (2024) Associative bacteria and arbuscular mycorrhizal fungus increase drought tolerance in maize (Zea mays L.) through morphoanatomical, physiological, and biochemical changes. Plants 13(12):1667. https://doi.org/10.3390/plants13121667

Timofeeva A, Galyamova M, Sedykh S (2022) Prospects for using phosphate-solubilizing microorganisms as natural fertilizers in agriculture. Plan Theory 11:2119. https://doi.org/10.3390/plants11162119

Torres Vera R, Bernabé García AJ, Carmona Álvarez FJ et al (2024) Application and effectiveness of *Methylobacterium symbioticum* as a biological inoculant in maize and strawberry crops. Folia Microbiol 69:121–131. https://doi.org/10.1007/s12223-023-01078-4

Tuan Muhammad Syafiq TH, Nusaibah SA, Rafii MY (2021) Effectiveness of bioinoculants *Bacillus cereus* and *Trichoderma asperellum* as oil palm seedlings growth promoters. J Trop Agric Sci 44:10.47836/pjtas.44.1.09

Udeh EL, Ngmenzuma TY, Maseko ST et al (2023) Gas exchange and carbon assimilation of Bambara groundnut landraces co-inoculated with *Bradyrhizobium japonicum* and *Bacillus subtilis* under simulated drought stress. Sci Afr 21:e01823. https://doi.org/10.1016/j.sciaf.2023.e01823

Ullah H, Santiago-Arenas R, Ferdous Z et al (2019) Improving water use efficiency, nitrogen use efficiency, and radiation use efficiency in field crops under drought stress: a review. In: Advances in agronomy. Elsevier, pp 109–157

Upamanya GK, Bhattacharyya A, Dutta P (2020) Consortia of entomo-pathogenic fungi and bio-control agents improve the agro-ecological conditions for brinjal cultivation of Assam. 3 Biotech 10:450. https://doi.org/10.1007/s13205-020-02439-3

Valle-Romero P, García-López JV, Redondo-Gómez S et al (2023) Biofertilization with PGP bacteria improve strawberry plant performance under sub-optimum phosphorus fertilization. Agronomy 13:335. https://doi.org/10.3390/agronomy13020335

Vassilev N, Vassileva M, Martos V et al (2020) Formulation of microbial inoculants by encapsulation in natural polysaccharides: focus on beneficial properties of carrier additives and derivatives. Front Plant Sci 11:270. https://doi.org/10.3389/fpls.2020.00270

Velmourougane K, Prasanna R, Singh S et al (2017) Modulating rhizosphere colonisation, plant growth, soil nutrient availability and plant defense enzyme activity through *Trichoderma viride-Azotobacter chroococcum* biofilm inoculation in chickpea. Plant Soil 421:157–174. https://doi.org/10.1007/s11104-017-3445-0

Vergani L, Mapelli F, Suman J et al (2019) Novel PCB-degrading Rhodococcus strains able to promote plant growth for assisted rhizoremediation of historically polluted soils. PLoS One 14:e0221253. https://doi.org/10.1371/journal.pone.0221253

Verma P, Shakya M, Swamy NK, Sandhu SS (2022) Microbial consortium. In: Microbial resource technologies for sustainable development. Elsevier, pp 23–46

Wang Z, Li Y, Zhuang L et al (2019) A rhizosphere-derived consortium of *Bacillus subtilis* and *Trichoderma harzianum* suppresses common scab of potato and increases yield. Comput Struct Biotechnol J 17:645–653. https://doi.org/10.1016/j.csbj.2019.05.003

Wang H, Peng H, Li W et al (2021) The toxins of *Beauveria bassiana* and the strategies to improve their virulence to insects. Front Microbiol 12:705343. https://doi.org/10.3389/fmicb.2021.705343

Waseem M, Khilji SA, Tariq S et al (2024) Phytoremediation of heavy metals from industrially contaminated soil using sunflower (*Helianthus annus* L.) by inoculation of two indigenous bacteria. Plant. Stress 11:100297. https://doi.org/10.1016/j.stress.2023.100297

Ważny R, Rozpądek P, Jędrzejczyk RJ, Śliwa M, Stojakowska A, Anielska T, Turnau K (2018) Does co-inoculation of Lactuca serriola with endophytic and arbuscular mycorrhizal fungi improve plant growth in a polluted environment? Mycorrhiza 28(3):235–246. https://doi.org/10.1007/s00572-018-0819-y

Wevar Oller AL, Regis S, Armendariz AL et al (2020) Improving soybean growth under arsenic stress by inoculation with native arsenic-resistant bacteria. Plant Physiol Biochem 155:85–92. https://doi.org/10.1016/j.plaphy.2020.07.015

Wilmowicz E, Kućko A, Bogati K et al (2022) *Glomus* sp. and *bacillus* sp. strains mitigate the adverse effects of drought on maize (*Zea mays* L.). Front. Plant Sci 13:958004. https://doi.org/10.3389/fpls.2022.958004

Wong E, Hagen J (2022) An evidence-based causal perspective of agrochemical pollution and its impact on health. Sci Insights 41:691–695. https://doi.org/10.15354/si.22.re086

Xu Z, Pehlivan N, Ghorbani A, Wu C (2022) Effects of *Azorhizobium caulinodans* and *Piriformospora indica* co-inoculation on growth and fruit quality of tomato (*Solanum lycopersicum* L.) under salt stress. Horticulturae 8:302. https://doi.org/10.3390/horticulturae8040302

Yadav AN (2022) Potassium-solubilizing microorganisms for agricultural sustainability. J App Biol Biotech 10:1–4. https://doi.org/10.7324/JABB.2022.105ed

Yadav A, Yadav K, Aggarwal A (2015) Impact of Arbuscular Mycorrhizal fungi with *Trichoderma viride* and *Pseudomonas fluorescens* on growth, yield and oil content in *Helianthus annuus* L. J Essent Oil Bear Plants 18:444–454. https://doi.org/10.1080/0972060X.2014.971066

Zeng R, Liu H, Hong Z et al (2024) Co-inoculation effects of *B. Licheniformis* and *P. Aeruginosa* on soil cd and as availability and rice accumulation. J Environ Manag 351:119739. https://doi.org/10.1016/j.jenvman.2023.119739

Zhang Y, Du H, Xu F et al (2020) Root-bacteria associations boost Rhizosheath formation in moderately dry soil through ethylene responses. Plant Physiol 183:780–792. https://doi.org/10.1104/pp.19.01020

Zhang G, Wang J, Zhao H et al (2021a) PAH degradation and gene abundance in soils and vegetables inoculated with PAH-degrading endophytic bacteria. Appl Soil Ecol 168:104193. https://doi.org/10.1016/j.apsoil.2021.104193

Zhang J, Cook J, Nearing JT et al (2021b) Harnessing the plant microbiome to promote the growth of agricultural crops. Microbiol Res 245:126690. https://doi.org/10.1016/j.micres.2020.126690

Zhang Y, Xu Q, Wang G, Shi K (2023) Mixed Enterobacter and Klebsiella bacteria enhance soybean biological nitrogen fixation ability when combined with rhizobia inoculation. Soil Biol Biochem 184:109100. https://doi.org/10.1016/j.soilbio.2023.109100

Zhao G, Zhu X, Zheng G et al (2024) Development of biofertilizers for sustainable agriculture over four decades (1980–2022). Geogr Sustain 5:19–28. https://doi.org/10.1016/j.geosus.2023.09.006

Zhou J, Gong J, Liu S et al (2024a) AM fungi reduce grass–legume competition by increasing nutrient access. Plant Soil 494:127–147. https://doi.org/10.1007/s11104-023-06261-7

Zhou S-P, Ke X, Jin L-Q et al (2024b) Sustainable management and valorization of biomass wastes using synthetic microbial consortia. Bioresour Technol 395:130391. https://doi.org/10.1016/j.biortech.2024.130391

Zitlalpopoca-Hernandez G, Pozo MJ, Hauser TP, Meyling NV (2022) Combined effects of root-associated entomopathogenic and mycorrhizal fungi on the foliar pathogen Botrytis cinerea in tomato. Biol Control 175:105034. https://doi.org/10.1016/j.biocontrol.2022.105034

Chapter 13
Scaling Up Microbial Solutions: Challenges and Triumphs in Agriculture

Samuel A. Fasusi, Modupe S. Ayilara, Abisola E. Ajakaye, Saheed A. Akinola, and Olubukola Oluranti Babalola

Abstract The application of microbial solutions in agriculture presents a viable approach to fight against food insecurity caused by different factors such as climate change. Therefore, this chapter gives information on identifying beneficial microbes that can increase crop yield, soil quality, and livestock production. It discusses the misconceptions around the use of microbial solutions for agricultural purposes, which reduces their acceptability. The general belief that microbes are solely harmful and pathogenic is erroneous. Fortunately, against these beliefs, microbial solutions have proved helpful in crop and livestock production. Different research has shown that probiotics in animals, and mycorrhizal fungi and rhizobia in crops increase crop yield and serve as biopesticides in plants. Other advantages and effects of microbial solutions in agriculture include increased soil fertility and minimal

---

S. A. Fasusi (✉)
Department of Crop Science and Biotechnology, College of Bioresource Science, Dankook University, Cheonan, South Korea

M. S. Ayilara
Food Security and Safety Focus Area, Faculty of Natural and Agricultural Sciences, North-West University, Mmabatho, South Africa

Department of Biotechnology, Environmental Pollution Science and Technology Limited, Ido-Ijesha, Nigeria

A. E. Ajakaye
Department of Public Health, University of Suffolk, Ipswich, UK

Pharmacy Department, Camp David Pharmacy and Stores, Abeokuta, Nigeria

S. A. Akinola
Food Security and Safety Focus Area, Faculty of Natural and Agricultural Sciences, North-West University, Mmabatho, South Africa

Department of Microbiology and Parasitology, School of Medicine and Pharmacy, College of Medicine and Health Sciences, University of Rwanda, Butare, Rwanda

O. O. Babalola (✉)
Food Security and Safety Focus Area, Faculty of Natural and Agricultural Sciences, North-West University, Mmabatho, South Africa
e-mail: olubukola.babalola@nwu.ac.za

O. O. Babalola, A. S. Ayangbenro (eds.), *Microbial Allies*,
https://doi.org/10.1007/978-3-031-90530-8_13

environmental pollution. However, several issues need to be addressed to improve the applicability of microbial solutions for agricultural purposes. Such issues border around formulation, quality, and specific applications to crops. Finally, the chapter underlines the need for research based on the compatibility of microbial solutions with specific environmental conditions, the application of new biotechnology technologies, and extensive education programs (awareness) to achieve the maximum outcome when utilising microbes. Future research should aim to develop environment-resistant and effective agricultural microbial solutions using technological improvements to ensure sustainable agricultural systems, especially in the face of climate change.

Keywords Biotechnology · Bioinoculant · Climate change · Food security · Microbes · Probiotics

13.1 Introduction

The severity of climate change is known to directly and indirectly impact both plant and animal physiology. Water scarcity, extremity in rainfall patterns, with elevated temperatures, and decreasing pollinators are observable effects of climate change (Myers et al. 2017). Farmers have been adversely affected by unstable weather patterns and extreme weather events in recent times. Most farmers, for instance, in developing nations practice rainfed agriculture, which is greatly threatened by climate change. With the severity of climate change, high cases of new insect pests and diseases have emerged, which constrain optimum crop production (Oladele and Adekunle 2021; Waheed et al. 2023).

The inadequacy of food production has led to different challenges in humans, including malnutrition and hunger. This is evident with the cases of about three million below age 5 children death yearly (Kumar et al. 2023). The incidence level of undernutrition predisposes the human population to morbidity and eventual mortality (Rohr et al. 2019). Different studies reveal that many infectious diseases occur in children and adults suffering from undernutrition compared to those who are well-nourished. This has called for the production of nutritious and sufficient food to feed the ever-increasing global population.

A complex relationship exists between climate change and agriculture, which has an overall effect on food security. A major challenge in recent times is minimising the risk effect of climate change on global food security. The vulnerability of sustainable food production to current climate change demands intentional activities to mitigate or adapt to the process of global warming through different innovations. Deliberate efforts have yielded a significant increase in global food production over the years through different technological advancements, including biotechnology and breeding improvement, mechanization of activities, and appropriate synthetic agrochemicals (Myers et al. 2017). The yield of crops has been reported to increase almost four times in recent years due to the extensive supply of fertiliser;

however, this incessant use of fertiliser has caused a sharp decrease in soil organic content (Sharma et al. 2022). The indiscriminate usage of other agrochemicals is linked to insect resistance or disease resistance, food contamination, environmental toxicity, and death of beneficial insects (parasitoids, predators) (Singh and Trivedi 2017).

The uncertainty of meeting food demand with present conventional agricultural practices warrants the immediate consideration of a complementary strategy. A present approach being considered is the use microbiome to attain food security. The use of microbes is found to have high potential in the optimum production of crops and animals. Therefore, microbes are explored as replacements or complements of agrochemicals. Although microbe solution is found beneficial, there is limited awareness. Hence, this chapter focuses on providing empirical evidence on the usefulness of beneficial microbes in agriculture. We also addressed different misconceptions that have limited the acceptance of microbial solutions in agriculture and provided future direction to scale up the usage of microbes.

13.2 Microbes as a Solution to Improve Agriculture

Different studies have focused on the structure and interactions of microbiomes with host plants or animals. Microbes exist in different locations in/on soil, plants, and animals. This and their ability to grow easily make them suitable for easy adaptation in agriculture and other processes. Several studies have been carried out to analyse the role of microbial-based solutions in agriculture (De Simone et al. 2021; Dukare et al. 2021; Yarzábal and Chica 2021). These include studies on using probiotics to improve livestock's health status. A notable example is a study by Al-Shawi et al. (2020), who reported that feeding probiotics to dairy cows increases milk production and quality through increased efficiency of the digestive system and immune response. Similarly, microbes like mycorrhizal fungi and rhizobia have enhanced soil productivity and plant growth (Ayilara et al. 2022). Microbes enhance nutrient cycling, organic matter decomposition, and soil structure (Akanmu et al. 2021; Ayilara and Babalola 2023). Studies have shown that mycorrhizal fungi colonise plant roots, help plants acquire phosphorus, a major crop nutrient, and increase the ability of plants to withstand stress (Adeleke et al. 2022a, b). In all, research outcomes have shown that microbial inoculants can foster a sustainable increase in agricultural productivity.

13.3 Microbes in Crop Production

With the alarming rate at which soil fertility is being depleted over the years due to human activities (tillage practices, overgrazing, continuous cropping, and bush burning) and the severity of climate change, food security isn't guaranteed (Fall et al. 2022). Globally, farmers have adopted monocropping and continuous cropping agriculture to meet the food demand of the world population. However, monocropping is linked to an imbalance in soil nutrients, an increase in soil-borne pathogens, a drastic change in the microbial community, and allelopathic autotoxicity (Wang et al. 2022). Incessant application of agrochemicals and continuous cropping damages the soil environment as it alters the soil's microbial community and physicochemical features (Wei et al. 2024). This disturbance of the microbial community directly causes a decline in the population and diversity of beneficial microbes, which is detrimental to plant growth. Continuous cropping in Punjab (India) resulted in heavy pesticide application which accounts for a diminishing population and diversity of nitrogen-fixing bacteria and microbes that solubilize phosphates (Buch et al., 2023). The overall effect of excessive application of agrochemicals includes acidification, soil pH imbalance, disruption of soil structure, soil salinisation, and compaction (soil degradation); this, therefore, limits soil porosity, hampers water percolation and aeration, and hinders plant root growth (Usharani et al. 2019; Wei et al. 2024). Continuous cropping leads to a constant decline in the organic matter component of cultivated soil, rendering soil infertile and depriving plants of nutrients. Different physical and chemical approaches have been tried to ameliorate the present state of soil degradation, but with limited results and complex processes (Bhattacharyya et al. 2015).

The significant contribution of soil microorganisms in maintaining the function and structure of the soil ecosystem is crucial in improving the present state of soil degradation. Soil microbes help to promote plant production through their plant growth promotion ability, pesticidal activity, stress alleviation, and nutrient acquisition (Fig. 13.1).

Soil microbes aid the decomposition of animal and plant residue in the soil, improving soil fertility and properties with symbiotic relationships through mycorrhizal structure formation with host plant roots (Wang et al. 2022). A typical example of microbes involved in the decomposition and replenishment of soil fertility is

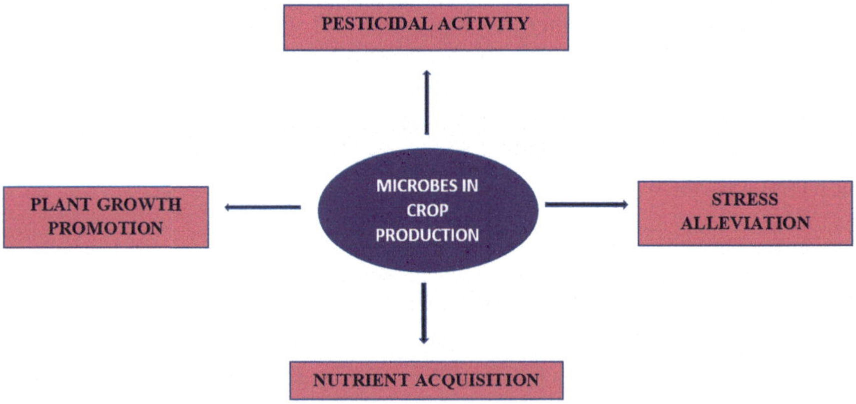

Fig. 13.1 Roles of microbes in crop production

the plant-growth-promoting rhizobacteria (PGPR) such as *Escherichia coli, Bacillus spp., Azotobacter spp.*, and arbuscular mycorrhizal fungi (AMF) (Ghosh et al. 2024). The mechanism adopted by PGPR in the decomposition process entails enzyme (ligninases and cellulases) production and solubilisation of potassium, and phosphate nutrients (Etesami and M Adi 2020; Singh et al. 2023). The use of microbes in improving soil state is found to be environmentally friendly, economical, and effective.

Presently, studies have focused on the inoculation of soil with microbes (Arbuscular Mycorrhizae Fungi (AMFs)) to enhance soil fertility (Dal Cortivo et al. 2018; Moreira et al. 2020). This involved the use of *Rhizoglomus irregulare* and *Rhizophagus irregularis* on maize fields (soil with elevated salt content) and wheat fields respectively helped in nutrient uptake improvement with high yield. This is an eco-friendly strategy to attain sustainable crop production. The use of microbial fertiliser to increase microbe diversity in the soil is presently a potent approach. Different countries the United States, China, France, Japan, Russia, the United Kingdom, and South Korea have actively adopted this approach (Wei et al. 2024). Mycorrhizae and rhizobia are the most important biofertilisers globally based on their prevalence. The growing adoption of the use of microbial fertiliser is not far-fetched from recent trends to achieve a greener ecology. The mechanism by which microbial fertiliser operates is the use of microbe life processes to make available absorbable nutrients (N, P, K) needed by plants to achieve sustainable production of crops (Liu et al. 2022b). The microbial community of soil is enhanced directly by the activities of these microbes, which include the production of organic acids, antimicrobial, and enzyme secretion. Furthermore, microbial fertilisers contribute to soil aggregate formation and ameliorate soil compaction (Alori et al. 2017). Microbial fertilisers foster organic matter decomposition which creates humic substances that help in the binding of soil particles into lumps or aggregates. Also, microbial secretions enhance the stability of these soil aggregates.

Also, studies have advanced the understanding of how microbial communities in soils contribute to the sustenance of the soil's overall health and nutrient quality. For instance, some soil bacteria and fungi play a role in decomposing the organic remains and releasing nutrients required by plants (Ayilara et al. 2020). Soil nutrient availability is achieved by soil bacteria and fungi via nitrogen fixation, mobilization of a few mineral content (Fe, P, and K), and decomposition of organic matter. The use of different fungal strains (*Glomus mosseae*, *Paraglomus occultum*, and *Rhizophagus irregularis*) has been explored in improving soil structure and fertility because they positively influence organic matter content, aggregate stability and formation (Rashid et al., 2016). In addition, microbes in the soil help to control disease-causing organisms in the soil and, therefore, prevent plant diseases. Various microbial strains have been reported to be useful in pest control. For example, *Bacillus thuringiensis* (Bt) produces a pesticide protein toxic to pests, which is an alternative to chemical control (Ayilara et al. 2024). Microbial-associated molecular patterns (MAMPs) are also produced by beneficial soil microbes to distinguish non-pathogens from pathogens to achieve a resistance response (Finkel et al. 2017). Microbial fertilisers enhance defense responses by producing important substances such as phenolic compounds, chitinase, antioxidant enzymes, and phytoalexins that promote plant resistance (Sood et al. 2021). Extracellular polysaccharides of important microbes (fungi and actinomycetes) help in the binding of soil particles to provide a solid soil aggregate (Pokharel et al. 2013). Examples of other roles played by microbes in plant growth are detailed in Table 13.1.

The success of such microbial biopesticides shows that these have a very encouraging potential to change pest control strategies and decrease the use of hazardous chemicals.

13.4 Microbes in Livestock Production

Probiotics, which is the administration of live microorganisms to confer health benefits to the host organism when provided in adequate quantity, are currently being explored in enhancing livestock productivity. Bacteria activities have been found to boost animals' gastrointestinal tract (GIT) functioning and immune system. Microorganisms clustered in animals' GIT systems are linked to the breakdown of ingesta, detoxification of compounds, immune development, short-chain fatty acid production, vitamin supplement provision, and exclusion of pathogens. Intestinal microflora significantly affects feedstuff digestion and metabolism in animals. This is possible because these microbial activities release short-chain fatty acids (SCFA), especially butyrate, acetate, and propionate, which are volatile FA.

The rumen makes up 20% of ruminant animals' total body mass due to microbial breakdown and fermentation activities on feedstuff consumed (Krehbiel 2014). Non-pathogenic microbes' interactions are important decomposing agents in the rumen of ruminant animals that ensure that fibrous plants of low quality are converted into needed energy with high-quality milk production and meat. This rumen microbiome is made of different microbes that coexist in a symbiotic relationship, some of which are pathogens and others that are beneficial and assist in ruminant animals' overall productivity (Cammack et al. 2018; Chen et al. 2020c). In the absence of required enzymes to aid the adequate digestion and fermentation process, microbes in the rumen actively assist ruminants to survive with general physiological activities. The microbes break down the fibrous plant materials that are consumed through the process of feeding, producing volatile fatty acids such as propionate, which act as essential energy for ruminant activities (Jami et al. 2013). This microbial digestion makes up about 70% of the accumulated total energy in ruminants. Some essential physiological processes in ruminants, such as immune system enhancement and rumen epithelium development, depend on the synthesised final products originating from microbial fermentation. Formic acids, vitamins, methane (CH4), and more bioactive molecules are examples of other final products. Inefficient microbial activities result in detrimental effects such as metabolic disorders (nitrate toxicity or acidosis) on ruminant animal's health and production (Perez et al. 2024). The state of the rumen environment greatly impacts the co-existing of microbes. A good balance in the rumen environment will support the

Table 13.1 Microbes and their role in plant production

Microbes	Effect	Crop	References
P. thivervalensis SC5S	Plant growth promotion: through phytohormones, nutrient uptake & microbe elimination	*Cucumis sativus*	Munshi et al. (2021)
Pseudomonas moraviensis, Bacillus halotolerans, Enterobacter hormaechei, and Pseudomonas frederiksbergensis	Stress tolerant through induced systemic resistance, phytohormone synthesis, phosphate solubilisation, Zn uptake, and modification of ROS.	*Ziziphus lotus*	Fahsi et al. (2021)
Xanthomonas oryzae, Rhizoctonia solani and *Fusarium verticillioides*	Plant growth promotion through induced system regulation, hormones, and nutrient uptake	Rice	Kumar et al. (2020)
CEDBE-1, CEDLE-6 and CEDLE-10	Biocontrol activity through metabolite antioxidant property	*Ceriops decandra*	Munshi et al. (2021)
Pseudomonas sp.	Biocontrol activity through induced systemic resistance & antibiotics	Cress and avocado	Scotti et al. (2020)
Chryseobacterium sp.	Biocontrol activity through antimicrobial action	Tomato	Tsolakidou et al. (2019)
Trichodema sp.	Plant growth promotion through IAA synthesis, microbe suppression & nutrient availability	Cowpea	Chagas et al. (2016) and Gómez-Muñoz et al. (2017)
Comamonadaceae and *Pseudomonadaceae*	Plant growth promotion through pathogen suppression, nitrogen fixation & induced systemic resistance	Maize	Beirinckx et al. (2020)
Rhizobium sp.	Nutrient acquisition through nitrogen fixation, polysaccharide secretion & hormone production	Legumes	Kalsoom et al. (2020)
Pseudomonas sp.	Plant growth promotion through microbe suppression, hormone production & nutrient availability	Maize	Gholami et al. (2019)
Rhizophagus intraradices	Plant growth promotion through enhanced water absorption, root development & hormone interaction	Tea	Sharma and Kayang (2017)
Sphingomonas sp.	Stress alleviation through improved root architecture and enhanced lateral root development	*Arabidopsis thaliana*	Luo et al. (2019)
Baciillus	Biocontrol activity through antimicrobial activities	Tomato	Tsolakidou et al. (2019)

(continued)

Table 13.1 (continued)

Microbes	Effect	Crop	References
Rhizobiaceae, Xanthomonadaceae, Burkholderiaceae, and *Enterobacteriaceae*	Plant growth promotion through phytohormone production, microbe suppression, and phosphate solubilisation	Sugarcane	Armanhi et al. (2018)
Acidobacteria, Proteobacteria, Actinobacteria, Bacteroidetes, Chloroflexi, Firmicutes and *Verrucomicrobia*	Plant growth promotion through nitrogen fixation, nutrient acquisition improvement	Rice	Zhang et al. (2019)
Pseudomonas stutzeri	Plant growth promotion through nitrogen fixation, phytohormone production, phosphate solubilisation & antimicrobial activities	Maize	Ke et al. (2019)
Bacillus mucilaginosus	Nutrient acquisition through potassium solubilisation & phytohormone production	Tomato	Liu et al. (2019)
Bacillus mucilaginosus	Plant growth promotion through biofertilizer potential, potassium solubilisation & production of bacillus mucilaginosus	Garlic	Mounir et al. (2020)
Pseudomonas aeruginosa	Plant growth promotion through phosphate solubilisation & antimicrobial activities	Sunflower	Tewari and Arora (2018)
Flavobacterium	Biocontrol activity through phosphate solubilisation, enhanced mechanism of plant defense & enzyme activities	Tomato	Kwak et al. (2018)

survival and growth of microbes with the end product production under microbes' symbiotic interaction. The rumen microbes include bacteriophages, bacteria, archaeaphages, fungi, and more. Bacteria are the major microbe in the rumen and exist under anaerobic conditions, while fungi make up about 10% of rumen microbes. Commonly found bacteria in the rumen include *Streptococcus, Ruminococcus, Prevotella, Ruminobacter, and Fibrobacter* (Perez et al. 2024).

A study on mice also establishes that intestinal microflora affects non-ruminant GIT function and metabolic processes (Misiukiewicz et al. 2021). In this study, mice with indigenous microbiota had about 42% more body mass or fat than mice nurtured in germless conditions (Lin and Li 2021). Further studies that introduced cecal microflora in germless mice recorded about 57% total body mass increase within the first 2 weeks regardless of any diet provided (Hoces et al. 2022).

Microbes, therefore, enhance a distinct nutritive profile as they successfully convert carbohydrate and protein components of feedstuff to acids. For instance, probiotics have been explored in cattle feedstuff (Iegorov et al. 2021). Although several people have challenged the use of probiotics, World Health Organization (WHO) considers the administration of live microorganisms in the feed beneficial if provided appropriately (Mara et al. 1989). Presently, there are selected and approved microbes in animal feed applications. Some examples of these are detailed in Table 13.2.

13.5 Misconception on Microbe Solution

Microbial applications in agriculture are generally viewed as complex, which causes reluctance and doubts from farmers and other people. Some common assumptions in microbes are that they are pathogenic and lead to diseases. This is due to the general view of microbes as pathogenic bacteria and viruses causing diseases in plants, animals and man. However, not all microbes are harmful; in fact, it is estimated that a high percentage of the microbes that exist in the world are useful in enhancing plant and animal growth, soil fertility, and other various benefits (Reid et al. 2003; Timmusk et al. 2017). This is the reason adequate awareness of the functional usefulness of these microbes is crucial to farmers and the public.

In addition, it is assumed that microbial products, when applied on fields, will yield instant and equal results as their chemical counterparts (Patil and Solanki 2016). It should therefore be noted that microbial solutions' efficacy depends on such factors as the type of the soil, climate conditions, kind of crops, and numbers and presence of other microorganisms. Hence, it is crucial to understand that microbial solutions are more likely to be designed to fit certain needs and may take some time before they can give optimum results.

Another myth about microbial solutions is the belief that microbial solutions are an inferior substitute for chemical inputs (Ayilara et al. 2023a). Some farmers and agricultural stakeholders believe that since chemical fertilisers and pesticides are fast in their action and produce visible results, they are in a better position than microbial products. This perception is partly due to constant marketing and earlier endorsement of chemical inputs that have satisfied farmers for many decades. However, microbial solutions have several advantages over chemical ones that are long-term in nature and relate to the enhancement of soil, reduction of negative impact on the physical environment, and development of rational farming practices (Thakur et al. 2020).

13.6 Challenges of Microbe Solution

There are a number of issues related to microbial solutions that impede their further growth and success. A fundamental problem is inconsistency in formulations and quality management (Ivase et al. 2021). The effectiveness of microbial products can be altered by factors, which include the strains used in producing the products, the

Table 13.2 Microbes and their role in livestock production

S/N	Genus	Beneficial microbe	Importance in livestock feed	References
1	*Aspergillus*	*Aspergillus niger*	Poultry source of enzymes (proteases) and bioactive compounds (gluconic & citric acid). Enzymes aid fiber digestibility through the ease of breaking physical barrier (lignin)	Sharma et al. (2022) and Fan et al. (2023)
		Aspergillus oryzae	Enhance milk production in lactating cows and heat stress tolerance; contain high crude protein	Guo et al. (2022) and Uwineza et al. (2023)
2	*Bacillus*	*Bacillus coagulans*		Sun et al. (2022), Sun et al. (2022) and
		Bacillus licheniformis	Improve broiler meat quality, & growth efficiency; suppress activities of *E. coli* in poultry	Yaderets et al. (2023)
		Bacillus pumilus	Decreases damage caused by oxidative stress linked to *Salmonella enteritidis* infection in poultry. Enhance growth and immune in goat	Liu et al. (2023a) and Zhang et al. (2023)
		Bacillus subtilis	Enhance growth piglet growth; improve broiler meat quality, nutrient usage & growth efficiency	EFSA Panel on Additives and Products or Substances used in Animal Feed (FEEDAP) et al. (2017)
3	*Bacteroides*	*Bacteroides amylophilus*	Enhance the breakdown of protein into ammonia in the course of fermentation	Xu et al. (2021) and Bain and Rahadi (2022)
		Bifidobacterium animalis	Enhance antioxidant activities and growth of piglets	Pang et al. (2022)
		Bifidobacterium bifidum		
		Bifidobacterium infantis		
		Bifidobacterium longum		
		Bifidobacterium thermophilum		

(continued)

Table 13.2 (continued)

S/N	Genus	Beneficial microbe	Importance in livestock feed	References
5	*Enterococcus*	*Enterococcus cremoris*		
		Enterococcus diacetylactis		
		Enterococcus faecium	Enhance growth, immune, and antimicrobial activity in the intestinal environment	Liu et al. (2023b)
		Enterococcus intermedius		
		Enterococcus lactis	Lower microbe population in the intestinal tract (ileum and duodenum)	Bernbom et al. (2006)
6	*Kluyveromyces*	*Kluyveromyces marxianus*	Enhance antimicrobial property against activities of pathogenic bacteria Enhance immune and efficient feed utilisation in broiler	Díaz-Vergara et al. (2017) and Wang et al. (2022)
7	*Lactobacillus*	*Lactobacillus acidophilus*		
		Lactobacillus brevis		
		Lactobacillus buchneri (cattle only)		
		Lactobacillus bulgaricus		
		Lactobacillus casei		
		Lactobacillus cellobiosus		
		Lactobacillus curvatus		
		Lactobacillus delbrueckii	Enhance suckling piglets' immune and improve antioxidant profile	Chen et al. (2020a)
8	*Pediococcus*	*Pediococcus acidilacticii*	Boost fish immune against infectious pathogens (Aspergillus)	EFSA Panel on Additives and Products or Substances used in Animal Feed (FEEDAP) et al. (2023) and Eissa et al. (2023)
		Pediococcus pentosaceus	Enhance Holstein calves' growth performance, limit respiratory tract infections	Otunba et al. (2021) and Liu et al. (2022a)

(continued)

Table 13.2 (continued)

S/N	Genus	Beneficial microbe	Importance in livestock feed	References
9	*Propionibacterium*	*Propionibacterium freudenreichii*	Synthesis of folic acid, Vit. B12, trehalose, acetate, and propionate; health-enhancing additive; synthesis of bacteriocins against survival of pathogenic microbes	Piwowarek et al. (2018) and Chen et al. (2020b)
		Propionibacterium shermanii	Synthesis of acetate, propionate, and health-enhancing additive;	Piwowarek et al. (2018)

method of microbial production, and the storage procedure (Batista and Singh 2021). Maintaining the standard and quality of the microbial formulations is very important for producing reliable and quality results. Also, to ensure the safety and effective use of microbial products, better and more effective regulatory measures are required to supervise the quality of microbial products. Prescribing regulatory standards and quality assurance measures can go a long way in combating such challenges and creating a connection to microbial solutions.

In addition, microbial solutions do not have the same efficiency in different environmental conditions regarding usage and results obtained (Ayilara et al. 2023b). Some of the antecedent conditions that may affect the efficacy of microbial products include the type and quality of soil environment, climate, and populations of other microorganisms (Ojuederie et al. 2021). For example, the microbial inoculant that is efficient in one area may not be efficient in another due to variations in the nature of soils and climatic conditions. The volatility causes a lot of difficulty in finding out general microbial solutions that can be implemented in different cases. More emphasis should be placed on developing formulation and availing farmers' inwardness on how to augment microbial products depending on their region.

The presence of pathogenic microbes (*Escherichia coli*, Salmonella, *Clostridium perfringens*, and Campylobacter) in livestock poses a threat to food safety and health of animals (Olson et al. 2023). They are responsible for colibacillosis with observable diarrhea (He et al. 2022). They are mostly transmitted by contaminated feed or food products or through the animal-to-human route of transfer. Therefore, it is important to curtain these pathogenic microbes in ruminant animals through the adoption of hygienic practices, adequate vaccinations, and regular veterinary attention especially in serious cases.

Another issue connected with microbial solutions is the economic factors. The perceived risk associated with the change from conventional chemical inputs is the main drawback that affects farmers' decisions on microbial solutions (Fenibo et al. 2022). For example, smallholder farmers will not be interested in purchasing microbial solutions (a new technology) if the payback is not guaranteed. Hence, there is a need to offer financial incentives and subsidies and utilise demonstration projects proving that microbial options are profitable and effective in the long term.

Additionally, the provision of affordable products to the farmers and good technical information on microbial products will increase the adoption of such products.

Lack of awareness also poses problems to the use of microbial solutions. Most farmers and stakeholders in the agricultural sector have little or no information concerning the advantages of microbial products and their correct application. The cultural belief that microbial solutions are harmful is overrated. Hence, farmers are required to have extensive education and awareness about the benefits of microbial products and how to apply them. Also. research organisations, extension services, and industry members should partner in the promotion of knowledge and the development of confidence in microbial applications.

13.7 Conclusion and Future Studies

Microbial solutions have proven to be a promising approach towards improving production in agriculture. Specifically, they bring about advantages such as increased soil fertility, enhanced yields of crops, and the well-being and efficiency of livestock, as well as preventing and minimising negative environmental impacts. However, the misconception and challenges surrounding the implementation of microbial solutions into agriculture have reduced their efficiency, acceptability and adoption. Therefore, it is vital to break through the current barriers and enhance research and development to promote the application of microbial solutions in agriculture.

Further research should be directed towards identifying microbial products specific to the different regions since environmental conditions are a factor that affects the effectiveness of microbial solutions. Such studies should involve the assessment of microbial species that are potentially compatible with specific soil types and climates. Furthermore, research should be intensified on the combination of different microbial consortiums with other conservation agriculture practices, such as minimum tillage, crop rotation, and use of organic inputs to achieve maximum results in the application of microbial solutions for agricultural productivity.

Biotechnology and genomics provide novel opportunities to improve microbial solutions. Subsequent research might use these technologies to design and enhance microorganisms to promote food security. For instance, genetic engineering can be applied to enhance the efficiency of microbes in nitrogen fixation, phosphorus solubilisation and as biocontrol agents against plant diseases.

References

Adeleke BS, Akinola SA, Adedayo AA, Glick BR, Babalola OO (2022a) Synergistic relationship of endophyte-nanomaterials to alleviate abiotic stress in plants. Front Environ Sci 10:1015897

Adeleke BS, Ayilara MS, Akinola SA, Babalola OO (2022b) Biocontrol mechanisms of endophytic fungi. Egypt J Biol Pest Control 32(1):1–17
Akanmu AO, Babalola OO, Venturi V, Ayilara MS, Adeleke BS, Amoo AE, Sobowale AA, Fadiji AE, Glick BR (2021) Plant disease management: leveraging on the plant-microbe-soil interface in the biorational use of organic amendments. Front Plant Sci 12:700507
Alori ET, Dare MO, Babalola OO (2017) Microbial inoculants for soil quality and plant health. Sustain Agric Rev 2:281–307
Al-Shawi SG, Dang DS, Yousif AY, Al-Younis ZK, Najm TA, Matarneh SK (2020) The potential use of probiotics to improve animal health, efficiency, and meat quality: a review. Agriculture 10(10):452
Armanhi JSL, De Souza RSC, Damasceno N d B, De Araujo LM, Imperial J, Arruda P (2018) A community-based culture collection for targeting novel plant growth-promoting bacteria from the sugarcane microbiome. Front Plant Sci 8:2191
Ayilara MS, Babalola OO (2023) Bioremediation of environmental wastes: the role of microorganisms. Front Agron 5:1183691
Ayilara MS, Olanrewaju OS, Babalola OO, Odeyemi O (2020) Waste management through composting: challenges and potentials. Sustain For 12(11):4456
Ayilara MS, Abberton M, Oyatomi OA, Odeyemi O, Babalola OO (2022) Potentials of underutilized legumes in food security. Front Soil Sci 2:1020193
Ayilara MS, Adeleke BS, Akinola SA, Fayose CA, Adeyemi UT, Gbadegesin LA, Omole RK, Johnson RM, Uthman QO, Babalola OO (2023a) Biopesticides as a promising alternative to synthetic pesticides: a case for microbial pesticides, phytopesticides, and nanobiopesticides. Front Microbiol 14:1040901
Ayilara MS, Adeleke BS, Babalola OO (2023b) Bioprospecting and challenges of plant microbiome research for sustainable agriculture, a review on soybean endophytic bacteria. Microb Ecol 85(3):1113–1135
Ayilara MS, Akinola SA, Adeleke BS, Gbadegesin LA, Adejumo GD, Glick BR, Babalola OO (2024) Biopesticides versus synthetic pesticides usage in Africa. In: Microbiome-based decontamination of environmental pollutants, vol 1. Elsevier, pp 417–450
Bain A, Rahadi S (2022) Effect of micronutrient supplementation to reduce heavy metal toxicity in rations from local feed ingredients in vitro using rumen fluid of ettawa crossbreeds. In: International conference on Improving Tropical Animal Production for Food Security (ITAPS 2021). Atlantis Press, pp 320–325
Batista BD, Singh BK (2021) Realities and hopes in the application of microbial tools in agriculture. Microb Biotechnol 14(4):1258–1268
Beirinckx S, Viaene T, Haegeman A, Debode J, Amery F, Vandenabeele S, Nelissen H, Inzé D, Tito R, Raes J (2020) Tapping into the maize root microbiome to identify bacteria that promote growth under chilling conditions. Microbiome 8:1–13
Bernbom N, Licht TR, Brogren C-H, Jelle B, Johansen AH, Badiola I, Vogensen FK, Nørrung B (2006) Effects of Lactococcus lactis on composition of intestinal microbiota: role of nisin. Appl Environ Microbiol 72(1):239–244
Bhattacharyya R, Ghosh BN, Mishra PK, Mandal B, Rao CS, Sarkar D, Das K, Anil KS, Lalitha M, Hati KM (2015) Soil degradation in India: challenges and potential solutions. Sustain For 7(4):3528–3570
Buch K, Saikanth DRK, Singh BV, Mallick B, Pandey SK, Prabhavathi N, Satapathy SN (2023) Impact of agrochemicals on beneficial microorganisms and human health. Int J Environ Clim Change 13(10):1135–45. https://doi.org/10.9734/ijecc/2023/v13i102763
Cammack KM, Austin KJ, Lamberson WR, Conant GC, Cunningham HC (2018) Ruminnat nutrition symposium: tiny but mighty: the role of the rumen microbes in livestock production. J Anim Sci 96(2):752–770
Chagas LFB, De Castro HG, Colonia BSO, De Carvalho Filho MR, Miller LDO, Chagas AFJ (2016) Efficiency of Trichoderma spp. as a growth promoter of cowpea (Vigna unguicu-

lata) and analysis of phosphate solubilization and indole acetic acid synthesis. Rev Bras Bot 39(2):437–445
Chen F, Wang H, Chen J, Liu Y, Wen W, Li Y, Huang X (2020a) Lactobacillus delbrueckii ameliorates intestinal integrity and antioxidant ability in weaned piglets after a lipopolysaccharide challenge. Oxidative Med Cell Longev 1:6028606
Chen J, Harstad OM, McAllister T, Dörsch P, Holo H (2020b) Propionic acid bacteria enhance ruminal feed degradation and reduce methane production in vitro. Acta Agric Scandinavica, Sect A Anim Sci 69(3):169–175
Chen Y, Wang Y, Wang W, Zhang Z, Si X, Cao Z, Li S, Yang H (2020c) Beneficial effect of Rhodopseudomonas palustris on in vitro rumen digestion and fermentation. Benefic Microbes 11(1):91–100
Dal Cortivo C, Barion G, Ferrari M, Visioli G, Dramis L, Panozzo A, Vamerali T (2018) Effects of field inoculation with VAM and bacteria consortia on root growth and nutrients uptake in common wheat. Sustain For 10(9):3286
De Simone N, Capozzi V, Amodio ML, Colelli G, Spano G, Russo P (2021) Microbial-based biocontrol solutions for fruits and vegetables: recent insight, patents, and innovative trends. Recent Pat Food Nutr Agric 12(1):3–18
Díaz-Vergara L, Pereyra CM, Montenegro M, Pena GA, Aminahuel CA, Cavaglieri LR (2017) Encapsulated whey–native yeast Kluyveromyces marxianus as a feed additive for animal production. Food Addit Contam Part A Chem Anal Control Expo Risk Assess 34(5):750–759
Dukare A, Paul S, Kumar R, Sharma V (2021) Microbial-based inoculants in sustainable agriculture: current perspectives and future prospects. Biofertilizers 1:167–181
EFSA Panel on Additives and Products or Substances used in Animal Feed (FEEDAP), Rychen G, Aquilina G, Azimonti G, Bampidis V, Bastos MDL, Bories G, Chesson A, Cocconcelli PS, Flachowsky G (2017) Safety and efficacy of Bacillus subtilis PB 6 (Bacillus subtilis ATCC PTA-6737) as a feed additive for sows. EFSA J 15(5):e04855
EFSA Panel on Additives and Products or Substances used in Animal Feed (FEEDAP), Bampidis V, Azimonti G, de Lourdes Bastos M, Christensen H, Durjava M, Dusemund B, Kouba M, López-Alonso M, López Puente S (2023) Safety and efficacy of a feed additive consisting of Pediococcus acidilactici CNCM I-4622 for all insect species (Danstar ferment AG). EFSA J 21(12):e8468
Eissa ME, Alaryani FS, Elbahnaswy S, Khattab MS, Elfeky A, AbouelFadl KY, Eissa E-SH, Ahmed RA, Van Doan H, El-Haroun E (2023) Dietary inclusion of Pediococcus acidilactici probiotic promoted the growth indices, hemato-biochemical indices, enzymatic profile, intestinal and liver histomorphology, and resistance of Nile Tilapia against Aspergillus flavus. Anim Feed Sci Technol 306:115814
Etesami H, Adl SM (2020) Plant growth-promoting rhizobacteria (PGPR) and their action mechanisms in availability of nutrients to plants. In: Phyto-Microbiome in Stress Regulation. Singapore: Springer. p. 147–203. https://doi.org/10.1007/978-981-15-2576-6_9
Fahsi N, Mahdi I, Mesfioui A, Biskri L, Allaoui A (2021) Plant growth-promoting Rhizobacteria isolated from the jujube (Ziziphus lotus) plant enhance wheat growth, Zn uptake, and heavy metal tolerance. Agriculture 11(4):316
Fall AF, Nakabonge G, Ssekandi J, Founoune-Mboup H, Apori SO, Ndiaye A, Badji A, Ngom K (2022) Roles of arbuscular mycorrhizal fungi on soil fertility: contribution in the improvement of physical, chemical, and biological properties of the soil. Front Fungal Biol 3:723892
Fan Z, Chen T, Cai G, Huang X, Zhong S, Li X, Zhang E (2023) Effect of aspergillus Niger fermentation on the metabolites in corn stalks. Fermentation 9(1):50
Fenibo EO, Ijoma GN, Matambo T (2022) Biopesticides in sustainable agriculture: current status and future prospects. In: New and future development in biopesticide research: biotechnological exploration, vol 2. Springer, pp 1–53
Finkel OM, Castrillo G, Paredes SH, González IS, Dangl JL (2017) Understanding and exploiting plant beneficial microbes. Curr Opin Plant Biol 38:155–163

Gholami A, Shahsavani S, Nezarat S (2019) The effect of plant growth promoting rhizobacteria (PGPR) on germination, seedling growth and yield of maize. Int J Agric Biosyst Eng 3(1):9–14
Ghosh A, Acharya R, Shaw S, Gangopadhyay D (2024) Plant growth-promoting rhizobacteria (PGPR): a potential alternative tool for sustainable agriculture. In: Updates on rhizobacteria. London: IntechOpen. https://doi.org/10.5772/intechopen.1004252
Gómez-Muñoz B, Pittroff SM, de Neergaard A, Jensen LS, Nicolaisen MH, Magid J (2017) Penicillium bilaii effects on maize growth and P uptake from soil and localized sewage sludge in a rhizobox experiment. Biol Fertil Soils 53:23–35
Guo L, Zhang D, Du R, Li F, Li F, Ran T (2022) Supplementation of Aspergillus oryzae culture improved the feed dry matter digestibility and the energy supply of total volatile fatty acid concentrations in the rumen of Hu sheep. Front Nutr 9:847156
He L, Wang C, Simujide H, Aricha H, Zhang J, Liu B, et al (2022) Effects of pathogenic Escherichia coli infection on the flora composition, function, and content of short-chain fatty acids in calf feces. Anim 12(8):959. https://doi.org/10.3390/ani12080959
Hoces D, Lan J, Sun W, Geiser T, Stäubli ML, Cappio Barazzone E, Arnoldini M, Challa TD, Klug M, Kellenberger A (2022) Metabolic reconstitution of germ-free mice by a gnotobiotic microbiota varies over the circadian cycle. PLoS Biol 20(9):e3001743
Iegorov B, Kananykhina O, Turpurova T (2021) Probiotic feed additives in fattening of agricultural animals. Grain Prod Mix Fodder's 21(4):25–31
Ivase T, Nyakuma BB, Otitolaiye VO, Utume LN, Ayoosu MI, Jagun ZT, Oladokun O, Dodo YA (2021) Standardization, quality control, and bio-enhancement of botanical insecticides: a review. J Environ Agric Energy 2:104–111
Jami E, Israel A, Kotser A, Mizrahi I (2013) Exploring the bovine rumen bacterial community from birth to adulthood. ISME J 7(6):1069–1079
Kalsoom M, Rehman FU, Shafique T, Junaid S, Khalid N, Adnan M, Zafar I, Tariq MA, Raza M, Zahra A (2020) Biological importance of microbes in agriculture, food and pharmaceutical industry: a review. Innovare J Life Sci 8(6):1–4
Ke X, Feng S, Wang J, Lu W, Zhang W, Chen M, Lin M (2019) Effect of inoculation with nitrogen-fixing bacterium Pseudomonas stutzeri A1501 on maize plant growth and the microbiome indigenous to the rhizosphere. Syst Appl Microbiol 42(2):248–260
Krehbiel C (2014) Invited review: applied nutrition of ruminants: fermentation and digestive physiology. Professional Anim Sci 30(2):129–139
Kumar V, Jain L, Jain SK, Chaturvedi S, Kaushal P (2020) Bacterial endophytes of rice (Oryza sativa L.) and their potential for plant growth promotion and antagonistic activities. S Afr J Bot 134:50–63
Kumar A, Kerketta A, Dewali S, Sharma N, Kumari Panda A, Singh Bisht S (2023) Tackling hidden hunger: understanding micronutrient deficiency and effective mitigation strategies. In: Emerging solutions in sustainable food and nutrition security. Springer, pp 305–319
Kwak M-J, Kong HG, Choi K, Kwon S-K, Song JY, Lee J, Lee PA, Choi SY, Seo M, Lee HJ (2018) Rhizosphere microbiome structure alters to enable wilt resistance in tomato. Nat Biotechnol 36(11):1100–1109
Lin X, Li H (2021) Obesity: epidemiology, pathophysiology, and therapeutics. Front Endocrinol 12:706978
Liu D, Yan R, Fu Y, Wang X, Zhang J, Xiang W (2019) Antifungal, plant growth-promoting, and genomic properties of an endophytic actinobacterium Streptomyces sp. NEAU-S7GS2. Front Microbiol 10:2077
Liu B, Wang C, Huasai S, Han A, Zhang J, He L, Aorigele C (2022a) Compound probiotics improve the diarrhea rate and intestinal microbiota of newborn calves. Animals 12(3):322
Liu H, Li S, Qiang R, Lu E, Li C, Zhang J, Gao Q (2022b) Response of soil microbial community structure to phosphate fertilizer reduction and combinations of microbial fertilizer. Front Environ Sci 10:899727
Liu Y, Li Z, Li H, Wan S, Tang S (2023a) Bacillus pumilus TS1 alleviates salmonella Enteritidis-induced intestinal injury in broilers. BMC Vet Res 19(1):41

Liu Z-L, Chen Y-J, Meng Q-L, Zhang X, Wang X-L (2023b) Progress in the application of Enterococcus faecium in animal husbandry. Front Cell Infect Microbiol 13:1168189
Luo Y, Wang F, Huang Y, Zhou M, Gao J, Yan T, Sheng H, An L (2019) Sphingomonas sp. Cra20 increases plant growth rate and alters rhizosphere microbial community structure of Arabidopsis thaliana under drought stress. Front Microbiol 10:1221
Mara DD, Cairncross S, Organization, W. H (1989) Guidelines for the safe use of wastewater and excreta in agriculture and aquaculture: measures for public health protection. World Health Organization
Misiukiewicz A, Gao M, Filipiak W, Cieslak A, Patra A, Szumacher-Strabel M (2021) Methanogens and methane production in the digestive systems of nonruminant farm animals. Animal 15(1):100060
Moreira H, Pereira SI, Vega A, Castro PM, Marques AP (2020) Synergistic effects of arbuscular mycorrhizal fungi and plant growth-promoting bacteria benefit maize growth under increasing soil salinity. J Environ Manag 257:109982
Mounir AM, Osman YM, Khalil O (2020) Impact of potassium solubilizing bacteria on growth and yield of garlic. Plant Arch 20(2):8374–8388
Munshi M, Sohrab MH, Begum MN, Rony SR, Karim MA, Afroz F, Hasan MN (2021) Evaluation of bioactivity and phytochemical screening of endophytic fungi isolated from Ceriops decandra (Griff.) W. Theob, a mangrove plant in Bangladesh. Clin Phytoscience 7:1–10
Myers SS, Smith MR, Guth S, Golden CD, Vaitla B, Mueller ND, Dangour AD, Huybers P (2017) Climate change and global food systems: potential impacts on food security and undernutrition. Annu Rev Public Health 38(1):259–277
Ojuederie OB, Chukwuneme CF, Samuel O, Olanrewaju MA, Adegboyega TT, Babalola OO (2021) Contribution of microbial inoculants in sustainable maintenance of human health, including test methods and evaluation of safety of microbial pesticide microorganisms. In: Biopesticides: botanicals and microorganisms for improving agriculture and human health, vol 2. Logos Verlag Berlin GmbH, Berlin, pp 207–240
Oladele O, Adekunle O (2021) Management of root-knot and other phytonematodes infecting okra with soil amendments. Sci Agric Bohem 52:67–75
Olson EG, Grenda T, Ghosh A, Ricke SC (2023) Microbial pathogen contamination of animal feed. In: Present Knowledge in Food Safety. Academic Press, pp 378–393. https://doi.org/10.1016/B978-0-12-819470-6.00023-8
Otunba AA, Osuntoki AA, Olukoya DK, Babalola BA (2021) Genomic, biochemical and microbial evaluation of probiotic potentials of bacterial isolates from fermented sorghum products. Heliyon 7(12):e08536
Pang J, Liu Y, Kang L, Ye H, Zang J, Wang J, Han D (2022) Bifidobacterium animalis promotes the growth of weaning piglets by improving intestinal development, enhancing antioxidant capacity, and modulating gut microbiota. Appl Environ Microbiol 88(22):e01296–e01222
Patil HJ, Solanki MK (2016) Microbial inoculant: modern era of fertilizers and pesticides. In: Microbial inoculants in sustainable agricultural productivity: vol. 1: research perspectives, vol 2. Springer, pp 319–343
Perez HG, Stevenson CK, Lourenco JM, Callaway TR (2024) Understanding rumen microbiology: an overview. Encyclopedia 4(1):148–157
Piwowarek K, Lipińska E, Hać-Szymańczuk E, Kieliszek M, Ścibisz I (2018) Propionibacterium spp.—source of propionic acid, vitamin B12, and other metabolites important for the industry. Appl Microbiol Biotechnol 102:515–538
Pokharel AK, Jannoura R, Heitkamp F, Kleikamp B, Wachendorf C, Dyckmans J, Ludwig B, Joergensen RG (2013) Development of aggregates after application of maize residues in the presence of mycorrhizal and non-mycorrhizal pea plants. Geoderma 202:38–44
Rashid MI, Mujawar LH, Shahzad T, Almeelbi T, Ismail IM, Oves M (2016) Bacteria and fungi can contribute to nutrients bioavailability and aggregate formation in degraded soils. Microbiol Res 183:26–41. https://doi.org/10.1016/j.micres.2015.11.007

Reid G, Jass J, Sebulsky MT, McCormick JK (2003) Potential uses of probiotics in clinical practice. Clin Microbiol Rev 16(4):658–672

Rohr JR, Barrett CB, Civitello DJ, Craft ME, Delius B, DeLeo GA, Hudson PJ, Jouanard N, Nguyen KH, Ostfeld RS (2019) Emerging human infectious diseases and the links to global food production. Nat Sustain 2(6):445–456

Scotti R, Mitchell AL, Pane C, Finn RD, Zaccardelli M (2020) Microbiota characterization of agricultural green waste-based suppressive composts using omics and classic approaches. Agriculture 10(3):61

Sharma D, Kayang H (2017) Effects of arbuscular mycorrhizal fungi (AMF) on Camellia sinensis (L.) O. Kuntze under greenhouse conditions. J Exp Biol 5:235–241

Sharma MK, White DL, Singh AK, Liu H, Tan Z, Peng X, Kim WK (2022) Effect of dietary supplementation of probiotic Aspergillus niger on performance and cecal microbiota in Hy-Line W-36 laying Hens. Animals 12(18):2406

Singh BK, Trivedi P (2017) Microbiome and the future for food and nutrient security. Microb Biotechnol 10(1):50

Singh A, Yadav VK, Chundawat RS, Soltane R, Awwad NS, Ibrahium HA, Yadav KK, Vicas SI (2023) Enhancing plant growth promoting rhizobacterial activities through consortium exposure: a review. Front. Bioeng. Biotechnol 11:1099999. https://doi.org/10.3389/fbioe.2023.1099999

Sood M, Kapoor D, Kumar V, Kalia N, Bhardwaj R, Sidhu GP, Sharma A (2021) Mechanisms of plant defense under pathogen stress: a review. Curr Protein Peptide Sci 22(5):376–395

Sun T, Miao H, Zhang C, Wang Y, Liu S, Jiao P, Li W, Li Y, Huang Z (2022) Effect of dietary Bacillus coagulans on the performance and intestinal microbiota of weaned piglets. Animal 16(7):100561

Tewari S, Arora NK (2018) Role of salicylic acid from Pseudomonas aeruginosa PF23 EPS+ in growth promotion of sunflower in saline soils infested with phytopathogen Macrophomina phaseolina. Environ Sustain 1:49–59

Thakur N, Kaur S, Tomar P, Thakur S, Yadav AN (2020) Microbial biopesticides: current status and advancement for sustainable agriculture and environment. In: New and future developments in microbial biotechnology and bioengineering. Elsevier, pp 243–282

Timmusk S, Behers L, Muthoni J, Muraya A, Aronsson A-C (2017) Perspectives and challenges of microbial application for crop improvement. Front Plant Sci 8:49

Tsolakidou M-D, Stringlis IA, Fanega-Sleziak N, Papageorgiou S, Tsalakou A, Pantelides IS (2019) Rhizosphere-enriched microbes as a pool to design synthetic communities for reproducible beneficial outputs. FEMS Microbiol Ecol 95(10):fiz138

Usharani K, Roopashree K, Naik D (2019) Role of soil physical, chemical and biological properties for soil health improvement and sustainable agriculture. J Pharmacogn Phytochem 8(5):1256–1267

Uwineza C, Bouzarjomehr M, Parchami M, Sar T, Taherzadeh MJ, Mahboubi A (2023) Evaluation of in vitro digestibility of Aspergillus oryzae fungal biomass grown on organic residue derived-VFAs as a promising ruminant feed supplement. J Anim Sci Biotechnol 14(1):120

Waheed A, Haxim Y, Islam W, Ahmad M, Muhammad M, Alqahtani FM, Hashem M, Salih H, Zhang D (2023) Climate change reshaping plant-fungal interaction. Environ Res 238:117282

Wang G, Ren Y, Bai X, Su Y, Han J (2022) Contributions of beneficial microorganisms in soil remediation and quality improvement of medicinal plants. Plan Theory 11(23):3200

Wei X, Xie B, Wan C, Song R, Zhong W, Xin S, Song K (2024) Enhancing soil health and plant growth through microbial fertilizers: mechanisms, benefits, and sustainable agricultural practices. Agronomy 14(3):609

Xu Q, Qiao Q, Gao Y, Hou J, Hu M, Du Y, Zhao K, Li X (2021) Gut microbiota and their role in health and metabolic disease of dairy cow. Front Nutr 8:701511

Yaderets V, Karpova N, Glagoleva E, Shibaeva A, Dzhavakhiya V (2023) Bacillus subtilis RBT-7/32 and bacillus licheniformis RBT-11/17 as new promising strains for use in probiotic Feed Additives. Microorganisms 11(11):2729

Yarzábal LA, Chica EJ (2021) Microbial-based technologies for improving smallholder agriculture in the Ecuadorian Andes: current situation, challenges, and prospects. Front Sustain Food Syst 5:617444

Zhang J, Liu Y-X, Zhang N, Hu B, Jin T, Xu H, Qin Y, Yan P, Zhang X, Guo X (2019) NRT1.1B is associated with root microbiota composition and nitrogen use in field-grown rice. Nat Biotechnol 37(6):676–684

Zhang N, Ding J, Li J, Wang L, Wei Y (2023) Effects of Bacillus pumilus on growth performance, serum indexes and muscle development-related gene expression of weanling Jintang black goat. Benefic Microbes 1(aop):1–9

Part IV
Innovations in Microbial Applications and Technologies

Chapter 14
Food Safety in a Changing World: Challenges and Innovations

Afolake Atinuke Olanbiwoninu, Theresa Abimbola Awotundun, Bukola Margaret Popoola, Racheal Oluwayemisi Fashogbon, and Kolawole Banwo

Abstract Food safety assurance is paramount to prevent foodborne illness and safeguard public health. Over time, various challenges have been encountered in ensuring food safety brought about by globalisation, climate change, and urbanisation. However, some key innovations have been employed to overcome these food safety challenges. This chapter explores the current challenges and innovations in achieving food safety. Food trade globalisation has made food supply chains longer and more complicated, making it difficult to trace the origin of food and identify points of contamination. Climate change is affecting farming lands by causing floods, droughts, and extreme heat waves leading to adverse effects on crops including compromised food integrity and the creation of suitable conditions for the proliferation of pathogenic microorganisms and toxins production. The innovative strategies utilised to preserve food safety standards and overcome food safety challenges are also discussed in this chapter. They include high-pressure technology, cold plasma systems, nanotechnology, and 3D printing. Additional techniques include the use of big data and blockchain for food traceability, the use of cost-effective and efficient automation and robotics for food quality as well as uniformity of products and employing the One Health approach to identifying, preventing, responding to threats caused by unsafe foods with an ultimate goal of safeguarding human health, animal health and environment.

Keywords Food safety · Food quality · Climate change · Big data · Blockchains · One health

A. A. Olanbiwoninu (✉) · T. A. Awotundun · B. M. Popoola · R. O. Fashogbon
Department of Microbiology and Biotechnology, Faculty of Natural Sciences, Ajayi Crowther University, Oyo, Oyo State, Nigeria
e-mail: aa.olanbiwoninu@acu.edu.ng

K. Banwo
Department of Microbiology, University of Ibadan, Ibadan, Nigeria

O. O. Babalola, A. S. Ayangbenro (eds.), *Microbial Allies*,
https://doi.org/10.1007/978-3-031-90530-8_14

14.1 Introduction

Food has always been and remains a basic human need. Various foods with different nutrients are the main building material for multiple tissues, organs and cells of a living body. The quality and fullness of our lives depend on more than 50% of the food we eat. Food quality is the totality of the food characteristics that can satisfy human food needs under normal conditions of use. With the increase in world population, there is an escalating demand for secure, nourishing and sustainable sources of food. This awareness emphasizes the fundamental significance of ensuring that provisions are made for food safety within the entire value chain from farming to consumption. Food safety involves various aspects including prevention of foodborne illnesses; management of contaminants and assurance regarding quality and nutritional content (Odeyemi et al. 2019; Ehuwa et al. 2021; Ahmad et al. 2023). The primary goals of food safety are to protect consumers from potential health hazards and to ensure that food products meet established quality standards. World Health Organization [WHO] (2022), reported that foodborne illnesses affect millions of people each year, leading to significant morbidity and mortality worldwide. Specifically, an estimated 600 million people worldwide—nearly one in ten people—fall ill after consuming contaminated food, and 420,000 of them die annually, costing 33 million healthy life expectancy (DALYs). This affects children, the elderly, pregnant women, and individuals with compromised immune systems (Njoagwuani et al. 2023). Ensuring food safety helps to protect these classes of individuals and reduce the overall burden of disease. When food safety is prioritised, healthier, secure, and sustainable foods can be created for the present and future generations.

In the past, concerns about food safety were concentrated at the downstream end especially with respect to processing, packaging and distribution activities. However, recent changes have shown that issues around food safety need to be tackled right from the beginning of the production cycle—crop production (Dudeja and Singh 2017; Melotto et al. 2020). Food safety begins on the farm before proceeding to the entire food chain. A resilient food system cannot exist without certainty concerning crop safety. Crop safety encompasses issues around hygiene, pesticide use and natural contaminants, including microorganisms, mycotoxins and heavy metals (Galagarza et al. 2021). Failure to apply the highest food safety standards can have a severe impact on the crops, which can spread to the whole of the food supply chain, even in product sectors or countries not associated with the failures. A reputation for safe and healthy products is hard won but can all too easily be lost and can take considerable time and effort to recover.

Food safety is important not only to public health protection but also to economic stability and social trust in food systems. Food safety is also very relevant to attaining quite a number of the SDGs as set by the United Nations within the context of global health and sustainable development. These include SDG 2: Zero hunger—end hunger, achieve food security and improved nutrition, and promote sustainable agriculture; SDG 3: Good health and well-being for all people at all ages; and SDG

12: Responsible consumption and production to ensure safe and sustainable food production practices.

The global food system faces current transformations and changes driven by interrelated factors, including globalisation, climate change, and rising urbanisation (Morgan and Fanzo 2020; Elechi et al. 2022). Such changes transform the way food is grown, harvested, processed, distributed, and consumed. They bring about opportunities and challenges to food safety and security simultaneously. Globalisation has interlinked the food supply chains in the world, thereby increasing access to diverse types of foods but bringing new risks into the picture (Davis et al. 2021). Climate change events affect agricultural productivity and food safety and thus require adaptive and resilient food systems (Duchenne-Moutien and Neetoo 2021). Urbanisation is having its impacts on patterns of consumption and how food is distributed, requiring creative ways to keep food safe within urban environments (FAO, IFAD, UNICEF, WFP and WHO 2023). Noting how these dynamics function is relevant to the design of strategies that would adapt and mitigate these changes' risks, at the same time opening opportunities for food systems to be sustainable and equitable. Ensuring food safety nowadays has become an increasingly complex challenge in view of the rapid processes of global change underway. The agriculture and food sector needs to respond to novel threats emerging with increasing interconnectivity in today's world and unprecedented changes in the environment while remaining vigilant in applying innovative technologies that safeguard consumer health. The present chapter is aimed at reviewing the challenges and recent innovations in food safety in our changing world focused majorly on foods of plant origin, with the key sub-theme of globalization, climate change, emerging technologies, shifting consumer demands as food safety challenges, and innovations in food processing and storage technologies, big data and blockchain for traceability, automation and robotics, and the One Health approach (Fig. 14.1).

14.2 Challenges to Food Safety

Safe food sources promote national economies, trade, and tourism, as well as sustainable development, food and nutrition security, and tourism. Some of the factors that challenges food safety from farm to fork are;

14.2.1 Globalisation of Food Trade

Globalisation can generally be defined as the process that integrates national and regional markets into one global market (Gabriela 2018). This has been identified as the key driver of consumption homogenization and hence the food behaviour of consumers. Globalization has dramatically changed the food system worldwide, enabling a strongly connected global web of food production, distribution, and

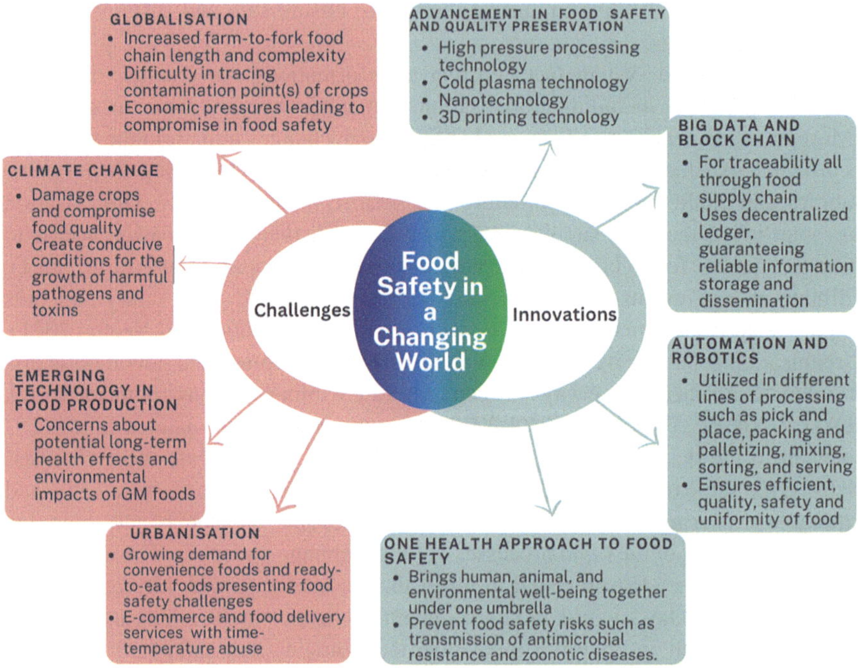

Fig. 14.1 Schematic illustration of food safety in a changing world: challenges and innovations

consumption (Caron et al. 2018; Elechi et al. 2022). This has introduced several advantages, including increased food availability, diversity, and raising economic opportunities. However, the farm-to-fork food chains have increased in length and complexity and require a new strategic approach to the identification of emerging and unknown food-borne threats such as potential toxins, allergens and pathogens throughout the food chain. The accidental introduction of pathogens and pests in new geographic areas has huge potential to disrupt food systems and public health (Davis et al. 2021). Globalisation has also created strong incentive for food fraud.

Economic factors may, in this case, come out as being very critical to food safety. Economic pressures can lead to cost-cutting measures that compromise food safety, such as inadequate sanitation, insufficient training, and the use of unsafe additives (Soria-Lopez et al. 2022). Developing countries, especially those with deficient facilities and support, may not be able to meet the international food safety standards (Oloo et al. 2018). Majority of farmers in developing countries still practice the conventional farming practices which increase the risk of producing contaminated food products. In such countries, where growing conditions are frequently unhygienic and protective measures are lacking, microbial contamination of food may start at the farm (Aworh 2021). Specifically, the use of untreated manure as fertilizer can be a source of food contamination as they may harbor large populations of pathogenic microorganisms including *Bacillus anthracis, Brucella* spp., *Clostridium perfringes, Escherichia coli* O157:H7, *Klebsiella* spp., *Listeria monocytogenes*, *Mycobacterium* spp., *Salmonella* spp., and *Yersinia enterocolitica*, harbouring antibiotic resistance genes (Dudeja and Singh 2017). Another way by which crops can get contaminated on the farm include the use of contaminated/unclean water for irrigation (Srivastav et al. 2024). Additionally, certain harvesting techniques can make crops more susceptible to microbial contamination. These techniques frequently include slamming fruit-bearing tree branches to the ground or striking them with sticks or other similar objects, which can cause harvest injuries that encourage contamination (Aworh 2021). Cultivation and harvest of crops under these circumstances can result in the export of substandard or contaminated food products, which not only poses health risks but also affects trade relationships and market access. The challenge of globalization is to balance the benefits in terms of economic growth with establishing sound food safety measures. Economic growth should not be at the expense of public health. Investments in infrastructure, capacity building of farmers through relevant trainings, and support is imperative.

14.2.1.1 Difficulty in Tracing Food Origin and Identifying Contamination Points

The intricate nature of global food supply chains poses significant challenges in tracing the origin of contaminated products and identifying specific points of contamination which starts from planting of crop produce and follow through the cultivation process, monitor the use of pesticide and transportation to market. This difficulty can lead to delayed responses to foodborne illness outbreaks and ineffective recall procedures (Ahmad et al. 2023). Implementing robust traceability systems is crucial for rapid identification and containment of food safety issues. Rigorous standards and practices across all stages of the supply chain combine with effective traceability systems are essential for managing. Examples of standards that need to be put in place to minimize the risk of food contamination are:

(a) Use of good agricultural practice
   - Use approved pesticides and herbicides
   - Practice proper irrigation and crop management
   - Maintain clean and sanitary facilities
(b) Develop and implement food safety plans by identifying the associated risks and hazards to set up a defense against pathogenic microbes before they constitute any risk
(c) Maintain good hygiene practices using;
   - HACCP principles
   - Establish and follow strict cleaning procedures of crop and plant produce
   - Maintain cleanliness in all areas of the food establishment
   - Properly sanitise all utensils, equipment, and food contact surfaces
(d) Follow proper cooking methods

14.2.1.2 Need for Robust International Food Safety Standards and Regulations

The globalisation of food trade necessitates the development and enforcement of harmonised international food safety standards and regulations. International collaboration and the establishment of global standards are crucial for ensuring food safety in a globalised world. International food safety standards are developed in large part by organizations like the Codex Alimentarius Commission, the Food and Agriculture Organization (FAO), and the World Health Organization (WHO) (Wu et al. 2021; Ahmad et al. 2023). While groups like the Codex Alimentarius Commission strive to create international food standards (Codex Alimentarius Commission 2020), significant disparities still exist between countries' regulatory frameworks and enforcement capabilities. Different countries have varying capacities to enforce food safety regulations, influenced by factors such as economic development, infrastructure, and governance. Developing countries, in particular, may struggle to align with international standards due to limited resources and expertise. Moreover, even in countries with robust food safety systems, inconsistencies in regulatory frameworks and enforcement can create gaps. A report by the Food and Agriculture Organization (FAO) highlights the significance of increasing developing nations' capabilities to improve their ability to adhere to international food safety standards. This includes improving laboratory infrastructure, training personnel, and strengthening regulatory oversight.

14.2.2 Climate Change

Climate change is becoming more widely recognized as a serious problem that needs immediate attention. Climate change is the continuous modifications to the statistical characteristics of the climatic system which includes the weather, such as adjustments to the average weather (i.e., extreme weather occurrences) that have an impact on atmospheric carbon dioxide, precipitation, and global temperatures (Wu et al. 2016). The primary cause of climate change is greenhouse gas (GHG) emissions, which also influence other climate parameters including temperature, relative humidity, precipitation, and UV radiation, all of which have an impact on crop output (Wu et al. 2016). Climate change is an important driver of emerging risks, plant, animal and human health, nutritional quality and threatening food and feed safety. This affects various aspects of food production, processing, and distribution with the potential to exacerbate the frequency and severity of certain foodborne diseases (Duchenne-Moutien and Neetoo 2021).

14.2.2.1 Impact on Food Production

Climate change is altering agricultural landscapes worldwide, with profound implications for food safety. Extreme weather events, such as floods, droughts, and heat waves, can damage crops, compromise food quality, and create conditions conducive to the growth of harmful foodborne microorganisms and toxins. There are more risks and obstacles involved in successfully cultivating crops due to the reduction of arable land and the anticipated increase in human migration linked to climate change. Food security is thus in jeopardy, famine may increase in areas like sub-Saharan Africa and many other regions, and food safety can be greatly impaired, putting human health at stake (Duchenne-Moutien and Neetoo 2021). Therefore, climate change adversely affects all the five pillars of food security: availability, access, utilization, stability and safety (IPCC 2019).

Lower crop yields are a result of both the high temperature rise and the altered global weather patterns. This is mostly because drought, heat waves, and flooding increase the shortage of water. Climate change has a negative impact on crop yields, lowers the nutritional value of foods, and increases the frequency of extreme weather events. These effects are interwoven with those of food security. Food production, availability, accessibility, and safety are all at risk due to these changes, especially in low- and middle-income nations.

Microorganisms, particularly bacteria and fungi, are essential to the maintenance of soil health, the cycling of nutrients (carbon and nitrogen), and the production and consumption of greenhouse gases (methane and carbon dioxide) (EPA 2023). In addition to improving soil structure and suppressing dangerous pathogens, they increase the availability of vital nutrients for plants. The balance of ecosystems may be impacted by the way that climate change affects microbial communities, changing their roles (Affoh et al. 2022; IMF 2022; Wang et al. 2022; Malkawi and Kapiel 2024). Harnessing the potential of microbes to mitigate the challenges posed by climate change to food safety is essential and this can be achieved by innovative agricultural practices, such as the use of microbial-based fertilizers and bioremediation techniques, to enhance crop resilience and productivity in the face of changing climate conditions (Chen et al. 2023).

The extensive use of agrochemicals in contemporary agriculture has created serious problems for the sustainability of agriculture. Undoubtedly, these pesticides have helped boost crop yields, but they have also put food supply and safety at risk. Animals, humans, beneficial soil organisms, and the total nutritional worth of crops are all negatively impacted by the careless use of pesticides and fertilizers. To make matters worse, the overuse of pesticides has resulted in the emergence of infections and pests that are resistant to them. Furthermore, a significant decrease in crop diversity has been brought about by current agricultural methods, increasing agriculture's susceptibility to disease infestations, insect outbreaks, and environmental challenges including reduced water availability, high temperatures, and increased salinity. A thorough approach is required to solve these complex difficulties in an efficient manner. This strategy should include developing climate-resilient plans, implementing soil conservation measures, and adopting sustainable agriculture practices. By adhering to these guidelines, we can work toward achieving both long-term ecological sustainability and food security (Brzozowski and Mazourek 2018; Damalas and Eleftherohorinos 2011; Dhanaraju et al. 2022).

Protecting plants from environmental and biological pressures is necessary to ensure continuous and strong crop production. Crops with increased genetic variability are more resilient to environmental stresses, more productive, better able to fight off diseases, and more biodiversity-promoting. Developing crop varieties with high production and resistance to pests, diseases, and unfavorable environmental factors including drought, high temperatures, and salinity of the soil is essential to achieving long-term agricultural sustainability. The identification of highly performing and well-adapted organisms has been the main focus of traditional approaches. Environmental stressors operate as selection forces, affecting the acclimatization, adaption, and evolutionary processes of species. They include biotic factors such as parasites, diseases, and pests, as well as abiotic ones such as chemicals, temperature, light, and other climatic variables (Bailey-Serres et al. 2019; Cheng and Cheng 2015; Dong and Ronald 2019; Das et al. 2023) (Table 14.1).

Developing sustainable and resilient food production systems requires an understanding of and management of the relationships between microorganisms, food security, climate change, and food safety. The detrimental effects of climate change can be lessened, and future generations can be guaranteed a secure and sustainable food supply, by encouraging microbial diversity and improving microbial activities (Aguilar-Paredes et al. 2023).

14.2.3 Emerging Technologies in Food Production

As the food industry grapples with the challenges of globalisation and climate change, it is increasingly turning to innovative technologies to enhance food safety and production efficiency. However, these emerging technologies also bring new considerations and potential risks that must be carefully evaluated.

14.2.3.1 Potential Benefits and Risks Associated with New Technologies

Genetically Modified Organisms (GMOs) have the potential to enhance crop yields, improve nutritional content, and increase resistance to pests and diseases (Brookes and Barfoot 2020). A meta-analysis by Klümper and Qaim (2014) found that GM technology has reduced pesticide use by 37% and increased crop yields by 22% on average. An example is the Bt-corn, which has a gene from the soil bacterium *Bacillus thuringensis* (Zhang et al. 2016). This gene helps produce sophisticated pesticides by instructing maize to produce a protein that kills a variety of insects and pests (Yali 2022). A study found that between 1996 and 2018, farmers using GMO plants cut back on pesticide use by 775.4 million kilograms (8.3%) (Hirschi 2020). Reduced use of pesticides in plants may also mean less harm to the environment and

Table 14.1 Impacts of environmental stresses influenced by climate change on foodborne pathogen behaviour and implications for food safety

Environmental stressors	Effect of climate on environmental factors involved in the growth, survival, and pathogenicity of foodborne pathogens	References
Temperature		
Increase	Increased occurrence of and ultimately exposure of plants to pests and pathogenic microbes Detection of new mycotoxin-producing fungal species in maize	Singh et al. (2023) and Moretti et al. (2018)
Decrease	Increased contamination of crops by norovirus, hepatitis A and other foodborne virus	EFSA (2014)
Precipitation and humidity		
Increase in precipitation	Promote the development and spread of plant diseases, diminishing the effectiveness of chemical and biological control measures Increased risk of splash dispersal and aerosolized pathogens infecting crops	Lim et al. (2023) and Singh et al. (2023)
Decrease in precipitation and humidity	Increased mycotoxin contamination by xerophilic fungi in crops at preharvest stage Increase the susceptibility of plants to drought stress and other pathogens	Moretti et al. (2018) and Maurya et al. (2022)
Extreme weather events		
Storms	Inflict physical wounds on crops, serving as entry points for pathogens	Hong et al. (2021)
Heat waves	Induce heat and oxidative stress in plants, compromising their photosynthesis and respiration and rendering them more susceptible to pathogens	Rivero et al. (2022) and Tanveer et al. (2023)
Hail	Causes damage to protective structures like the cuticle and epidermis can increase plant susceptibility to diseases	Khadiri et al. (2023)

fewer health risks for individuals who consume it, thus enhancing the populace food safety concerns. Additionally, genetically engineered soybean, cotton, and maize types yielded 20%, 15%, and 7% higher yields than non-genetically modified varieties, according to Hirschi (2020). However, concerns remain about potential long-term environmental impacts, necessitating ongoing research and rigorous safety assessments (Shen et al. 2022).

14.2.4 Urbanisation- Consumer Demands and Habits

Changing consumer preferences and lifestyles are reshaping the food industry, presenting both opportunities and challenges for food safety. Understanding and addressing these evolving demands is crucial for maintaining public health and consumer trust.

14.2.4.1 Growing Demand for Convenience Foods and Ready-to-Eat Meals

The rise of busy lifestyles has fueled a growing demand for convenience foods and ready-to-eat meals. While these products offer timesaving benefits to consumers, they also present unique food safety challenges. A study by Bao et al. (2017) found that ready-to-eat foods were responsible for a significant proportion of foodborne illness outbreaks, highlighting the need for stringent safety measures in their production and handling.

To address these risks, food manufacturers must implement robust food safety management systems, such as Hazard Analysis and Critical Control Points (HACCP), throughout the production process. Additionally, innovative packaging technologies, such as modified atmosphere packaging and active packaging systems, can help extend shelf life and maintain food safety.

The rapid growth of e-commerce and food delivery services has transformed the way many consumers access food. While this trend offers convenience, it also introduces new food safety risks, particularly related to temperature control during transportation.

A study by Li et al. (2020) found that food delivered through online platforms was at higher risk of time-temperature abuse compared to traditional foodservice channels. To mitigate these risks, food delivery companies must invest in proper insulation and temperature monitoring systems, while regulatory bodies need to develop and enforce guidelines specific to this emerging sector.

14.2.4.2 Importance of Consumer Education on Safe Food Handling Practices

As the food landscape becomes more complex, educating consumers about safe food handling practices is increasingly important. A survey by the International Food Information Council Foundation (2019) revealed that many consumers lack basic knowledge about food safety, with only 67% of respondents aware of the importance of hand washing in preventing foodborne illnesses.

Consumers awareness can be increased to ensure safety and security of foods by conducting effective education program, advertisement through the news media to disseminate information regarding food safety. Other areas of creating consumer awareness are through government agencies, scientific journals, health care professionals, social media platforms for younger generations and health-focused websites among others.

14.3 Innovations in Food Safety

14.3.1 Advancements in Food Processing and Preservation of Quality

Food processing technologies are of great importance due to the increasing demand for food, which is projected to expand with the advance in the world population (Fellows 2016; United Nations 2019; van Dijk et al. 2021; Hassoun et al. 2023). Food processing techniques improve the nutritional constituents, preservation, food safety potentials, and food security (Sachs et al. 2019; Hassoun et al. 2023). Food processing comprises stages of conversion of raw or fresh agricultural produce into partially finished or finished products, enabling food products to consumers all year round (Teng et al. 2021; Banwo et al. 2021, 2022; Qian et al. 2022). This process has complex challenges: food safety, plant-based foods, quality control, and food adulteration, a significant aspect of food security (Augusto 2020; Hassoun et al. 2023).

Food safety is a major challenge in food processing due to the concerns related to outbreaks of food-borne disease, especially with contamination of pathogenic microorganisms such as *Salmonella* and *Campylobacter* species, *Listeria monocytogenes* and *Escherichia coli* (Centers for Diseases Control and Prevention 2022). In addition to food-borne illnesses, heavy metals, mycotoxins, and pesticide residue contamination are getting severe due to environmental pollution. These contaminants have been observed in routine food testing and have led to constant recalls (Falade et al. 2018; Banwo et al. 2023; Hassoun et al. 2023).

Quality control is an essential component of food properties and processing conditions, which is vital to the monitoring of compliance-defined criteria (Ali and Hashim 2021). However, there are limitations related to the continuous evaluation of food qualities and processing in modern lines of production. This process entails the regular acquisition of representative samples and the interpretation of the results obtained. These procedures are carried out by employing protocols available in the requisite laboratories with trained and skilled laboratory technologists who will obtain constant and reproducible results. These activities must support the objectives of the study and assist in the improvement of the quality control systems (Di Rosa et al. 2017; Hassoun et al. 2023).

Food adulteration is a vital concept of food security defined as the process in which the quality of food is reduced by replacing food components, the addition of non-authenticated substances or removal of vital components from food for the sake of earning profits or due to other incidental reasons (Posudin et al. 2015). Additional way of food adulteration is food altering, whereby food is tampered with by the deliberate addition of compounds and materials to inflict harm to consumers and induce food outbreaks (FDA 2018). This food alteration is usually not frequent but has been indicated in some recent reports (European Commission 2020; Hassoun et al. 2023). The increase in lucidity, availability, safety, and fixity of data from food production can drastically diminish the menace of food fraud (Antonucci et al. 2019; Hassoun et al. 2023).

Some key innovations are employed in preserving the quality and safety of foods. These are the use of high-pressure processing technology, cold plasma technology, and nanotechnology.

High-pressure processing is a cold, non-thermal pasteurization technique that employs the use of liquid, usually water, as a channel to generate the desired pressure at a temperature (0–90 °C). This process involves sealing a food product in the final packaging, followed by submersion in water at a cold or room temperature within an enclosed vessel (Pérez-Lamela et al. 2021). This procedure successfully inactivates or kills pathogens and microorganisms by inactivating their cellular functions, thereby enhancing the safety and extension of the shelf life of food (Pérez-Lamela et al. 2021; Nie et al. 2022).

Cold plasma is a technology that is employed to decontaminate food surfaces and packaging, thereby reducing the use of chemical preservatives. The mechanism by which cold plasma works is the impact of reactive species, which causes damage by ultraviolet radiation to cellular components and the breakage of DNA strands. It has gained recognition due to the quality attributes it contributes to food processing during treatment and storage, in addition to the extension of the shelf life of food based on the inactivation of microorganisms and enzymes. This technique ensures the safety potential of food (Sruthi et al. 2022).

Nanotechnology is utilized as an advanced food packaging system, which enhances food safety, flavour, and texture. Nanomaterials have received enormous awareness in recent times because of their safety and capacity to preserve important nutrients. These nanomaterials can enhance the properties of food packaging and assist in the detection of contaminants (Suvarna et al. 2022; Rodrigues et al. 2024).

14.3.2 Big Data and Blockchain for Food Traceability

Production of food with consistent quality and safety is the bedrock of a successful food industry. The supply chain system is dependent on the processing of raw materials into the final product (Ellahi et al. 2023). There are several challenges that the food supply chains face, which are related to food quality, food safety, and food traceability (Ahmad and Bailey 2021; Vistro et al. 2022; Ellahi et al. 2023). Consumer demand for an all-year-round supply of food products has placed immense obligation on businesses to provide thorough details about characteristics such as food safety standards, food originality, and food traceability throughout the supply chain (Rejeb and Rejeb 2020). Food traceability organizations raise concerns over patented innovative concepts due to insufficient technology implementation and dependence on paper-based processes (Wang et al. 2019; Ellahi et al. 2023). Blockchain technology can be utilized in the food sector to enhance food quality product recalls, track the inventory, transparency, and consumer trust (Ahmad and Bailey 2021). This technology runs as a decentralized ledger, guaranteeing reliable information storage and dissemination (Iansiti and Lakhani 2017; Nofer et al. 2017; Ellahi et al. 2023). Food entrepreneurs are utilizing blockchain technology to enhance the safety and traceability of food, which is apparent in the cooperation among big departmental chains of stores such as Walmart, IBM, and Tsinghua University in China (Kshetri 2019; Ahmad and Bailey 2021; Ellahi et al. 2023). This technology has been utilized in the food value chain system of SmartRice in Nigeria, where the dynamics of food safety, food quality, and food traceability of the overall supply chain were enhanced (Akazue et al. 2023). Therefore, blockchain technology is essential in the food processing industry and will further enhance the food supply chain concerning automation, sustainability, safety, tracking, and traceability.

14.3.3 Role of Automation and Robotics

Food safety is a vital concern that must ensure the hygienic quality standard of food and beverages to minimize or eliminate the contamination of germs leading to compromise of the food (Iqbal et al. 2017; Hassoun et al. 2023). An advancement in technology has made this possible through the use of robotics and automated systems in the food processing industry (Zohaib et al. 2014; Iqbal et al. 2017).

Food manufacturing and processing industries have resolved to the utilization of cost-effective and efficient automation systems for increased productivity as compared to manual technologies (Buckenhüskes and Oppenhäuser 2014; Iqbal et al. 2017). The employment of robots is vital in the different lines of processing such as pick and place, packing and palletizing, mixing, sorting, and serving (Iqbal et al. 2017; Dzedzickis et al. 2022; Wang et al. 2022). These robots are designed to work as humans in their activities and functions, thus making them more flexible than automated machines. They are comprised of an arm, wrist, and hands that perform the tasks (Iqbal et al. 2017; Dzedzickis et al. 2022). The utilization of robotics in the food and beverage industry possesses beneficial abilities that ensure efficient quality and uniformity, leading to consistency in the product's value. In addition, there is a minimal injury to personnel due to monotonous programs, thus enhancing work efficiency with a reduction in time and marginal wastage. These advantages improve the industry's competitive advantage (Bader and Rahimifard 2020; Chen and Yu 2022). However, there are downsides to the use of robotics, such as the high cost of purchase and maintenance when implementing robotics in the food processing industry (Wang et al. 2022).

14.3.4 One Health Approach to Food Safety

Food safety is a multifaceted issue that needs the involvement of several collaborative efforts to make it succeed (Forghani et al. 2023). One health approach is imperative in collective determinations to alleviate it by assisting in reducing the occurrence of foodborne illness and improving public health integrated systems. One health concept brings human, animal, and environmental well-being together under one umbrella. This concept is essential to food safety because of the transmission of diseases from animals to humans, known as zoonoses (Kniel et al. 2018). The most important issue with this concept is the problem of antimicrobial resistance (AMR) in both humans and animals; more especially the disease occurrence associated with *Salmonella* and *Campylobacter* species (King 2013; Kniel et al. 2018). This has led to World Health Organization (WHO) guidelines on the unified investigation of AMR in food-borne microorganisms and pathogens related to animals that produce foods, foods, and humans (Fleming-Dutra et al. 2016). This situation is threatening in low-middle-income countries (LMICs), where there is indiscriminate use of antibiotics and limited regulations on the purchase of over-the-counter medications without prescription (McGettigan et al. 2017; Kniel et al. 2018). The transmission of AMR to the environment is usually through the presence of bacteria in the waste products of unhealthy animals and man. This is very critical because wildlife constitutes a risk due to its extensive gamut of pathogenic diseases (Wielinga and Schlundt 2014). The strategies employed to alleviate AMR in the food and livestock chain systems using the One health principle are the utilization of antimicrobials for the prevention of diseases must be validated where the presence of the disease is indicated or may occur, use of antimicrobials as prophylaxis must be discouraged and the use as growth promoters must be urgently phased-out (WHO 2000; Wielinga and Schlundt 2014).

Other approaches are at the significant stages of the supply chain, such as from farm to fork. Since zoonoses start from animals before transmission to humans, an efficient intervention may be achieved at the source, which is the farm. To reduce the risk of zoonoses related to foodborne, the health of the animal must be of utmost importance and kept at the barest minimum of disease infestation. This is very important because a good population in LMICs is reliant on livestock animals to survive and for consumption (Wielinga and Schlundt 2014). Strategies employed to mitigate risks of zoonotic foodborne-related diseases must start with surveillance of livestock in the farm by a veterinarian and at the food production stage by a food safety specialist, microbiologists, and scientists must be carried out to ensure safe, wholesome, and nutritious foods. Keeping to all these guidelines will lower the risks of zoonotic foodborne diseases from farm-to-fork and supply chain systems (Wielinga and Schlundt 2014; Kniel et al. 2018).

The fostering of collaboration among the various sectors, such as public health specialists, agriculturists, food microbiologists and scientists, veterinarians, and food safety specialists in the One Health approach enhances the ability to detect, prevent, and respond to food safety hazards, ultimately protecting human health, animal health, and the environment.

14.4 Conclusion

Facing the reality of the rapidly evolving landscape of the global food system, ensuring food safety at all levels presents numerous challenges and opportunities for innovation in the food system. The primary challenges identified include the intricacies introduced by globalisation, the deep impacts of climate change, and the rapid urbanisation affecting our food system from farm to fork. These factors contribute to the increasing difficulty of ensuring safe food supply chains, as they introduce new vectors for contamination and complicate traceability and accountability. Innovations in food safety are crucial for addressing some of these challenges. Advances in technology, such as rapid detection methods for pathogens, blockchain for enhanced traceability, and smart packaging, are paving the way for more robust food safety measures. In addition, integrated strategies that include surveillance, risk assessment, and education are essential for building resilient food systems.

Continuous improvement and adaptation in food safety practices are imperative. The dynamic nature of the global food system means that food safety protocols must evolve in response to emerging threats and technological advancements. Regularly updating standards, conducting research, and implementing new technologies are necessary steps to keep pace with the changing landscape.

Furthermore, the importance of international cooperation and public-private partnerships in sustaining the global food system cannot be overlooked. There is an urgent call for collaborative efforts across borders and sectors to harmonise food safety standards and respond to global food safety incidents. Public-private partnerships can also leverage the strengths of both sectors to drive innovation, improve compliance, and enhance overall food safety.

By adopting the One Health approach and promoting collaboration among public health, agriculture, and all stakeholders, could help in addressing the interconnected challenges of food safety.

References

Affoh R, Zheng H, Dangui K, Dissani BM (2022) The impact of climate variability and change on food security in sub- saharan africa: perspective from panel data analysis. Sustain For 14(2):759. https://doi.org/10.3390/su14020759

Aguilar-Paredes A, Valdés G, Araneda N, Valdebenito E, Hansen F, Nuti M (2023) Microbial community in the composting process and its positive impact on the soil biota in sustainable agriculture. Agronomy 13(2):542. https://doi.org/10.3390/agronomy13020542

Ahmad A, Bailey K (2021) Blockchain in food traceability: a systematic literature review. 32nd Irish signals and systems conference (ISSC). 978-1-6654-3429-4/20/IEEE. https://doi.org/10.1109/ISSC52156.2021.9467848

Ahmad SR, Munawar H, Saima H, Siddique F (2023) Introductory chapter: food safety. IntechOpen. https://doi.org/10.5772/intechopen.113289

Akazue MI, Yoro RE, Malasowe BO, Nwankwo O, Ojugo AA (2023) Improved services traceability and management of a food value chain using block-chain network: a case of Nigeria. Indonesian J Electr Eng Comp Sci 29:1623–1633

Ali MM, Hashim N (2021) Non-destructive methods for detection of food quality. In: Bhat R (ed) Future foods: global trends, opportunities, and sustainability challenges, pp 645–667. https://doi.org/10.1016/B978-0-323-91001-9.00003-7

Antonucci F, Figorilli S, Costa C, Pallottino F, Raso L, Menesatti P (2019) A review of blockchain applications in the agri-food sector. J Sci Food Agric 99(14):6129–6138. https://doi.org/10.1002/jsfa.9912

Augusto PED (2020) Challenges, trends and opportunities in food processing. Curr Opin Food Sci 35:72–78. https://doi.org/10.1016/j.cofs.2020.03.005

Aworh OC (2021) Food safety issues in fresh produce supply chain with particular reference to sub-Saharan Africa. Food Control 123:107737. https://doi.org/10.1016/j.foodcont.2020.107737

Bader F, Rahimifard S (2020) A methodology for the selection of industrial robots in food handling. Innov Food Sci Emerg Technol 64:102379. https://doi.org/10.1016/J.IFSET.2020.102379

Bailey-Serres J, Parker JE, Ainsworth EA, Oldroyd GED, Schroeder JI (2019) Genetic strategies for improving crop yields. Nature 575(7781):109–118. https://doi.org/10.1038/s41586-019-1679-0

Banwo K, Olojede AO, Adesulu-Dahunsi AT, Verma DK, Thakur M, Tripathy S, Singh S, Patel AR, Gupta AK, Aguilar CN, Utama GL (2021) Functional importance of bioactive compounds of foods with potential health benefits: a review on recent trends. Food Biosci 43:101320. https://doi.org/10.1016/j.fbio.2021.101320

Banwo K, Ogunremi OR, Adesulu-Dahunsi AT (2022) Sustainable use of food-grade microorganisms in traditional fermented food production. In: Verma DK, Patel AR, Billoria S, Kaushik G, Kaur M (eds) Microbial biotechnology in food processing and health: advances, challenges and potential. Apple Academic Press, CRC Taylor and Francis Group, LLC. ISBN: 978-1-77463-728-9, pp 237–261

Banwo K, Adesina T, Aribisala O, Falade TDO (2023) Detoxification of aflatoxins in fermented cereal gruel (*Ogi*) by probiotic lactic acid bacteria and yeasts with differences in amino acid profiles. Toxins 15:210. https://doi.org/10.3390/toxins15030210

Bao M, Brauer A, Datta AR, Gao Y (2017) Burden of listeriosis attributable to ready-to-eat foods in the United States, 1998–2016. Foodborne Pathog Dis 14(11):651–657

Brookes G, Barfoot P (2020) Environmental impacts of genetically modified (GM) crop use 1996–2018: impacts on pesticide use and carbon emissions. GM Crops Food 11(4):215–241. https://doi.org/10.1080/21645698.2020.1773198

Brzozowski L, Mazourek M (2018) A sustainable agricultural future relies on the transition to organic agroecological pest management. Sustain For 10(6):2023. https://doi.org/10.3390/su10062023

Buckenhüskes HJ, Oppenhäuser G (2014) DLG-Trendmonitor: roboter in der Lebensmittel- und Getränkeindustrie, DLG trend report: 'robots in the food and beverage industry'. DLG Lebensmittel 9(6):16–17

Caron P, Ferrero y de Loma-Osorio G, Nabarro D et al (2018) Food systems for sustainable development: proposals for a profound four-part transformation. Agron Sustain Dev 38:41. https://doi.org/10.1007/s13593-018-0519-11

Centers for Disease Control and Prevention (2022) Foodborne outbreaks. Retrieved July 7, 2024, from List of Multistate Foodborne Outbreak Notices website: https://www.cdc.gov/foodsafety/outbreaks/lists/outbreaks-list.htmla

Chen TC, Yu SY (2022) The review of food safety inspection system based on artificial intelligence, image processing, and robotic. Food Sci Technol 42. https://doi.org/10.1590/fst.35421

Chen W, Modi D, Picot A (2023) Soil and phytomicrobiome for plant disease suppression and management under climate change: a review. Plan Theory 12(14):2736. https://doi.org/10.3390/plants12142736

Cheng F, Cheng Z (2015) Research progress on the use of plant allelopathy in agriculture and the physiological and ecological mechanisms of allelopathy. Frontiers. Plant Sci 6. https://doi.org/10.3389/fpls.2015.01020

Codex Alimentarius Commission (2020) Hazard analysis and critical control point (HACCP) system and guidelines for its application. Codex Alimentarius

Damalas CA, Eleftherohorinos IG (2011) Pesticide exposure, safety issues, and risk assessment indicators. Int J Environ Res Public Health 8(5):1402–1419

Das A, Chakraborty S, Nath H, Pathak BB (2023) Impact of climate change induced pest and disease attack on crop production. In: Dayal P, Ranjan S, Sow S, Pyare R, Kumar A, Kumar A (eds) Climate smart approaches towards sustainable crop production. Vital Biotech Publication, pp 251–272

Davis KF, Downs S, Gephart JA (2021) Towards food supply chain resilience to environmental shocks. Nat Food 2:54–65. https://doi.org/10.1038/s43016-020-00196-3

Dhanaraju M, Chenniappan P, Ramalingam K, Pazhanivelan S, Kaliaperumal R (2022) Smart farming: internet of things (IoT)- based sustainable agriculture. Agriculture 12(10):1745

Di Rosa AR, Leone F, Cheli F, Chiofalo V (2017) Fusion of electronic nose, electronic tongue and computer vision for animal source food authentication and quality assessment—a review. J Food Eng 210:62–75. https://doi.org/10.1016/j.jfoodeng.2017.04.024

Dong OX, Ronald PC (2019) Genetic engineering for disease resistance in plants: recent progress and future perspectives. Plant Physiol 180(1):26–38. https://doi.org/10.1104/pp.18.01224

Duchenne-Moutien RA, Neetoo H (2021) Climate change and emerging food safety issues: a review. J Food Prot 84(11):1884–1897., ISSN 0362-028X. https://doi.org/10.4315/JFP-21-141

Dudeja P, Singh A (2017) Food-safety issues related to plant foods at farms. In: Gupta RK, Dudeja P, Minhas AS (eds) Food safety in 21st century. Academic, pp 179–191. https://doi.org/10.1016/B978-0-12-801773-9.00013-3

Dzedzickis A, Subačiūtė-Žemaitienė J, Šutinys E, Samukaitė-Bubnienė U, Bučinskas V (2022) Advanced applications of industrial robotics: new trends and possibilities. Appl Sci 12(1):135. https://doi.org/10.3390/app12010135

Ehuwa O, Jaiswal AK, Jaiswal S (2021) Salmonella, food safety and food handling practices. Food 10(5):907

Elechi JOG, Nwiyi IU, Adamu CS (2022) Global food system transformation for resilience. IntechOpen. https://doi.org/10.5772/intechopen.102749

Ellahi RM, Wood LC, Bekhit AE-DA (2023) Blockchain-based frameworks for food traceability: a systematic review. Food Secur 12:3026. https://doi.org/10.3390/foods12163026

EPA (2023) Climate change impacts on agriculture and food supply. Retrieved Nov 22. https://www.epa.gov/climateimpacts/climate-change-impacts-agriculture-and-food-supply

European Commission (2020) The rapid alert system for food and feed annual report 2019. https://op.europa.eu/en/publication-detail/-/publication/2c5c7729-0c31-11ebbc07-01aa75ed71a1/language-en/format-PDF/source-174742448. Accessed 10 July 2024

European Food Safety Authority (EFSA) (2014) Panel on Biological Hazards. Scientific opinion on the risk posed by pathogens in food of non-animal origin. Part 2 (*Salmonella* and Norovirus in berries). EFSA J 12:95

Falade TDO, Chrysanthopoulos PK, Hodson MP, Sultanbawa Y, Fletcher M, Darnell R, Korie S, Fox G (2018) Metabolites identified during varied doses of aspergillus species in *Zea mays* grains, and their correlation with aflatoxin levels. Toxins 10:187

FAO, IFAD, UNICEF, WFP, WHO (2023) The State of Food Security and Nutrition in the World 2023. In: Urbanization, agrifood systems transformation and healthy diets across the rural–urban continuum. FAO, Rome. https://doi.org/10.4060/cc3017en

FDA (2018) Food tampering: an extra ounce of caution. Retrieved July 10, 2024, from https://www.fda.gov/food/buy-store-serve-safe-food/food-tampering-extraounce-caution

Fellows P (2016) Food processing technology: principles and practice, 4th edn. Woodhead Publishing

Fleming-Dutra KE, Hersh AL, Shapiro DJ, Bartoces M, Enns EA, File TM Jr, Finkelstein JA, Gerber JS, Hyun DY, Linder JA, Lynfield R, Margolis DJ, May LS, Merenstein D, Metlay JP, Newland JG, Piccirillo JF, Roberts RM, Sanchez GV, Suda KJ, Thomas A, Woo TM, Zetts RM, Hicks LA (2016) Prevalence of inappropriate antibiotic prescriptions among US ambulatory care visits, 2010–2011. JAMA 315:1864–1873

Forghani F, Sreenivasa MY, Jeon B (2023) Editorial: one health approach to improve food safety. Front Microbiol 14:1269425. https://doi.org/10.3389/fmicb.2023.1269425

Gabriela H (2018) The impact of globalisation on the food behaviour of consumers. CBU Int Conf Proc 6:170. https://doi.org/10.12955/cbup.v6.1151

Galagarza OA, Ramirez-Hernandez A, Oliver HF, Álvarez Rodríguez MV, Valdez Ortiz MC, Pachari Vera E, Cereceda Y, Diaz-Valencia YK, Deering AJ (2021) Occurrence of chemical contaminants in Peruvian produce: a food-safety perspective. Food Secur 10(7):1461. https://doi.org/10.3390/foods10071461

Hassoun A, Jagtap S, Trollman H, Garcia-Garcia G, Abdullah NA, Goksen G, Bader F, Ozogul F, Barba FJ, Croptova J, Munekata PES, Lorenzo JM (2023) Food processing 4.0: current and future developments spurred by the fourth industrial revolution. Food Control 145:109507

Hirschi KD (2020) Genetically modified plants: nutritious, sustainable, yet underrated. J Nutr 150:2628–2634

Hong, C. X., Hansen, M. A., Bush, E. A., Day, E. R., Del-Pozo, A., & Derr, J. F. (2021). 2021 home grounds and animals PMG-home ornamentals

Iansiti M, Lakhani KR (2017) The truth about blockchain. Harvard Bus Rev 5:118–127

IMF (2022) Climate change and chronic food insecurity in sub- Saharan Africa. International Monetary Fund, WP/22/16. https://www.imf.org/en/Publications/Departmental-Papers-Policy-Papers/Issues/2022/09/13/Climate-Change-and-Chronic-Food-Insecurity-in-Sub-Saharan-Africa-522211

International Food Information Council Foundation (2019) Food and Health Survey

IPCC (2019) Summary for policymakers. In: Shukla PR, Skea J, Buendia EC, Masson-Delmotte V, Pörtner H-O, Roberts DC et al (eds) Climate change and land: an IPCC special report on climate change, desertification, land degradation, sustainable land management, food security, and greenhouse gas fluxes in terrestrial ecosystems. Chapter 5: Food security. IPCC

Iqbal J, Khan ZH, Khalid A (2017) Prospects of robotics in food industry. Food Sci Technol 37(2):159–165. https://doi.org/10.1590/1678-457X.14616

Khadiri M, Boubaker H, Askarne L, Ezrari S, Radouane N, Farhaoui A et al (2023) *Bacillus cereus* B8W8 an effective bacterial antagonist against major postharvest fungal pathogens of fruit. Postharvest Biol Technol 200:112315. https://doi.org/10.1016/j.postharvbio.2023.112315

King LJ (2013) Combating the triple threat: the need for a one health approach. Microbiol Spectr 1:OH-00122012. https://doi.org/10.1128/microbiolspec.OH-0012-2012

Klümper W, Qaim M (2014) A meta-analysis of the impacts of genetically modified crops. PLoS One 9(11):e111629

Kniel KE, Kumar D, Thakur S (2018) Understanding the complexities of food safety using a "one health" approach. Microbiol Spectr 6(1):PFS-0021-2017. https://doi.org/10.1128/microbiolspec.PFS-0021-2017

Kshetri N (2019) Blockchain and the economics of food safety. IT Professionals 21:63–66

Li C, Mirosa M, Bremer P (2020) Review of online food delivery platforms and their impacts on sustainability. Sustain For 12(14):1–11

Lim JA, Yaacob JS, Mohd Rasli SRA, Eyahmalay JE, El Enshasy HA, Zakaria MRS (2023) Mitigating the repercussions of climate change on diseases affecting important crop commodities in Southeast Asia, for food security and environmental sustainability—a review. Front Sustain Food Syst 6:1030540. https://doi.org/10.3389/fsufs.2022.1030540

Malkawi HI, Kapiel TYS (2024) Microbial biotechnology: a key tool for addressing climate change and food insecurity. Eur J Biol Biotechnol 5(2):1–15. https://doi.org/10.24018/ejbio.2024.5.2.503

Maurya MK, Yadav VK, Singh SP, Jatoth R, Singh HK, Singh D (2022) Impact of climate change on diseases of crops and their management—a review. J Agric Sci Technol B 1. https://doi.org/10.17265/2161-6264/2022.01.001
McGettigan P, Roderick P, Kadam A, Pollock AM (2017) Access, watch, and reserve antibiotics in India: challenges for WHO stewardship. Lancet Global Health 5:e1075–e1076
Melotto M, Brandl MT, Jacob C, Jay-Russell MT, Micallef SA, Warburton ML, Van Deynze A (2020) Breeding crops for enhanced food safety. Front Plant Sci 11:428. https://doi.org/10.3389/fpls.2020.00428
Moretti A, Pascale M, Logrieco AF (2018) Mycotoxin risks under a climate change scenario in Europe. Trends Food Sci Technol 84:38–40
Morgan AE, Fanzo J (2020) Nutrition transition and climate risks in Nigeria: moving towards food systems policy coherence. Curr Environ Health Rep 7:392–403. https://doi.org/10.1007/s40572-020-00292-3
Nie X, Zhang R, Cheng L, Zhu W, Li S, Chen X (2022) Mechanisms underlying the deterioration of fish quality after harvest and methods of preservation. Food Control 135:Article 108805. https://doi.org/10.1016/J.FOODCONT.2021.108805
Njoagwuani EI, Onyeaka H, Mazi IM, Akegbe H, Oladunjoye IO, Ochulor CE, Omotosho AD, Odeyemi OA, Nwaiwu O, Tamasiga P (2023) Food safety in vulnerable populations: a perspective on the challenges and solutions. FASEB J 37(5):e22872. https://doi.org/10.1096/fj.202201713R
Nofer M, Gomber P, Hinz O, Schiereck D (2017) Blockchain. Bus Inform Syst Eng 59:183–187
Odeyemi OA, Sani NA, Obadina AO, Saba CKS, Bamidele FA, Abughoush M, Ashghar A, Dongmo FFD, Macer D, Aberoumand A (2019) Food safety knowledge, attitudes and practices among consumers in developing countries: an international survey. Food Res Int 116:1386–1139
Oloo B, Daisy L, Oniang'o R (2018) Food safety legislation in some developing countries. Intech. https://doi.org/10.5772/intechopen.75587
Pérez-Lamela C, Franco I, Falque E (2021) Impact of high-pressure processing on antioxidant activity during storage of fruits and fruit products: a review. Molecules 26(17):5265. https://doi.org/10.3390/MOLECULES26175265
Posudin YI, Peiris K, Kays SJ (2015) Non-destructive detection of food adulteration to guarantee human health and safety. Ukr Food J 4:207–353
Qian J, Dai B, Wang B, Zha Y, Song Q (2022) Traceability in food processing: problems, methods, and performance evaluations—a review. Crit Rev Food Sci Nutr 62(3):679–692. https://doi.org/10.1080/10408398.2020.1825925
Rejeb A, Rejeb K (2020) Blockchain and supply chain sustainability. Logforum 16:363–372
Rivero RM, Mittler R, Blumwald E, Zandalinas SI (2022) Developing climate-resilient crops: improving plant tolerance to stress combination. Plant J 109(2):373–389. https://doi.org/10.1111/tpj.15483
Rodrigues S, da Silva JMO, Banwo K, Lima CMG, Aldarhami A, Guine RPF, Verruck S, Pagnossa J (2024) Nanoencapsulation of natural products and their role in the preservation and control of contaminations in the food industry. Food Sci Technol (Campinas) 44:1–16. https://doi.org/10.5327/fst.00222
Sachs JD, Schmidt-Traub G, Mazzucato M, Messner D, Nakicenovic N, Rockstr¨om, J. (2019) Six transformations to achieve the sustainable development goals. Nat Sustain 2(9):805–814. https://doi.org/10.1038/s41893-019-0352-9
Shen C, Yin X, Jiao B, Jia P, Zhang X, Cheng X, Ren J, Lan H, Hou W, Fang M, Li X, Fei Y, Robinson N, Liu J (2022) Evaluation of adverse effects/events of genetically modified food consumption: a systematic review of animal and human studies. Environ Sci Eur 34:8. https://doi.org/10.1186/s12302-021-00578-9
Singh BK, Delgado-Baquerizo M, Egidi E, Guirado E, Leach JE, Liu H, Trivedi P (2023) Climate change impacts on plant pathogens, food security and paths forward. Nat Rev Microbiol 21(10):640–656. https://doi.org/10.1038/s41579-023-00900-7

Soria-Lopez A, Garcia-Perez P, Carpena M, Garcia-Oliveira P, Otero P, Fraga-Corral M, Cao H, Prieto MA, Simal-Gandara J (2022) Challenges for future food systems: from the green revolution to food supply chains with a special focus on sustainability. Food Front 4(1):9–20. https://doi.org/10.1002/fft2.173

Srivastav AL, Patel N, Rani L et al (2024) Sustainable options for fertilizer management in agriculture to prevent water contamination: a review. Environ Dev Sustain 26:8303–8327. https://doi.org/10.1007/s10668-023-03117-z

Sruthi NU, Josna K, Pandiselvam R, Kothakota A, Gavahian M, Mousavi Khaneghah A (2022) Impacts of cold plasma treatment on physicochemical, functional, bioactive, textural, and sensory attributes of food: a comprehensive review. Food Chem 368:Article 130809. https://doi.org/10.1016/J.FOODCHEM.2021.130809

Suvarna V, Nair A, Mallya R, Khan T, Omri A (2022) Antimicrobial nanomaterials for food packages. Antibiotics (Basel) 11(6):72

Tanveer M, Mahmood A, Sarfraz B, Zia MA, Javaid MM, Bibi S et al (2023) Mechanism and approaches to enhancing heat stress tolerance in crop plants. In: Climate-resilient agriculture, Vol 2: Agro-Biotechnological advancement for crop production. Springer, Cham, pp 499–520. https://doi.org/10.1007/978-3-031-37428-9_22

Teng X, Zhang M, Mujumdar AS (2021) Potential application of laser technology in food processing. Trends Food Sci Technol 118:711–722. https://doi.org/10.1016/J.TIFS.2021.10.031

United Nations (2019) World population prospects 2019—Highlights. In Department of economic and social Affairs. World population prospects 2019. Retrieved from http://www.ncbi.nlm.nih.gov/pubmed/12283219

van Dijk M, Morley T, Rau ML, Saghai Y (2021) A meta-analysis of projected global food demand and population at risk of hunger for the period 2010–2050. Nat Food 2(7):494–501. https://doi.org/10.1038/s43016-021-00322-9

Vistro DM, Rehman AU, Farooq MS, Khalid F (2022) Role of blockchain technology in agriculture supply chain: a systematic literature review. In: Proceedings of the 2022 IEEE 2nd Mysore sub section international conference (MysuruCon), Mysuru, India, 16–17 October 2022. IEEE, Piscataway, NJ, USA, pp 1–8

Wang Y, Han JH, Beynon-Davies P (2019) Understanding blockchain technology for future supply chains: a systematic literature review and research agenda. Supply Chain Manag Int J 24:62–84

Wang Z, Hirai S, Kawamura S (2022) Challenges and opportunities in robotic food handling: a review. Front Robot AI 8:433. https://doi.org/10.3389/FROBT.2021.789107/BIBTEX

Wielinga PR, Schlundt J (2014) One health and food safety. In: Confronting emerging zoonoses: the one health paradigm. Springer, pp 213–232. https://doi.org/10.1007/978-4-431-55120-1_10

World Health Organization [WHO] (2000) WHO global principles for the containment of antimicrobial resistance in animals intended for food. In: Report of a WHO consultation with the participation of the Food and Agriculture of the United Nations and the Office International des Epizooties, 5–9 June, Geneva, Switzerland. WHO/CDS/CSR/APH/2000.4. http://apps.who.int/iris/bitstream/10665/68931/1/WHO_CDS_CSR_APH_2000.4.pdfE/

World Health Organization [WHO] (2022). Food safety. https://www.who.int/news-room/fact-sheets/detail/food-safety

Wu X, Lu Y, Zhou S, Chen L, Xu B (2016) Impact of climate change on human infectious diseases: empirical evidence and human adaptation. Environ Int 86:14–23

Wu D, Elliott C, Wu Y (2021) Food safety strategies: the one health approach to global challenges and China's actions. China CDC Weekly 3(24):507–513. https://doi.org/10.46234/ccdcw2021.131

Yali W (2022) Application of genetically modified organism (GMO) crop technology and its implications in modern agriculture. Int J Agric Sci Food Technol 8(1):14.20

Zhang C, Wohlhueter R, Zhang H (2016) Genetically modified foods: a critical review of their promise and problems. Food Sci Human Wellness 5:116–123

Zohaib M, Pasha SM, Javaid N, Salaam A, Iqbal J (2014) An improved algorithm for collision avoidance in environments having U and H shaped obstacles. Stud Inform Control 23(1):97–106

Chapter 15
Exploring Metabolomics for Unraveling Microbial Contributions to Enhancing Plant Health

Dedat Prismantoro, Sulistya Ika Akbari, Rizky Riscahya Pratama Syamsuri, Nia Rossiana, Mia Miranti, Muhamad Shakirin Mispan, and Febri Doni

Abstract The interaction between plants and beneficial microbes is crucial for enhancing plant productivity, with a specific focus on strengthening plant health and productivity. Metabolomics emerges as a novel approach currently under exploration to confer plant health and fitness, particularly against biotic stress responses. It presents a promising avenue for identifying various metabolites produced to improve stress tolerance and stimulate plant defense mechanisms. However, exploring the metabolomics approach to elucidate microbial contributions to enhancing plant resilience against diseases leaves many questions unanswered. Metabolomics can be utilized to elucidate complex interactions between plants and microbes, emphasizing microbial contributions to enhance plant health. Metabolomics analyses, including gas chromatography-mass spectrometry (GC-MS), liquid chromatography-mass spectrometry (LC-MS), high-performance liquid chromatography (HPLC), and nuclear magnetic resonance (NMR), among others, can be employed to unravel complex plant metabolic and signaling pathways. Understanding these metabolites and pathways is key to exploiting the diverse potential of microbes for stress mitigation and improved plant health. Therefore, metabolomics analysis

D. Prismantoro · S. I. Akbari · N. Rossiana · M. Miranti
Department of Biology, Faculty of Mathematics and Natural Sciences, Universitas Padjadjaran, Jatinangor, West Java, Indonesia

R. R. P. Syamsuri
Doctorate Program in Biotechnology, Graduate School, Universitas Padjadjaran, Bandung, West Java, Indonesia

M. S. Mispan
Institute of Biological Sciences, Faculty of Science, Universiti Malaya, Kuala Lumpur, Malaysia

F. Doni (✉)
Department of Biology, Faculty of Mathematics and Natural Sciences, Universitas Padjadjaran, Jatinangor, West Java, Indonesia

Department of Global Development, Cornell University, Ithaca, USA
e-mail: febri@unpad.ac.id; fd263@cornell.edu

O. O. Babalola, A. S. Ayangbenro (eds.), *Microbial Allies*,
https://doi.org/10.1007/978-3-031-90530-8_15

offers the opportunity to detect and identify microbial metabolites while profiling changes in plant metabolomes in response to microbial interactions. This chapter also examines future opportunities and challenges associated with metabolomics in plant-microbe studies, thereby paving the way for the exploration of future prospects in this field.

Keywords Metabolomics · Microbial metabolites · Microbiome · Plant-microbe interactions · Stress tolerance

15.1 Introduction

A major challenge of the twenty-first century is climate change, which significantly impacts plant health and fitness (Eastburn et al. 2011; Carvajal-Yepes et al. 2019; Jeger et al. 2021). Changes in climate can modulate plant susceptibility to pathogens, impacting both plant health and production (Dixon 2012; Romero et al. 2022). Its effects can have significant global impacts, affecting economic losses, food security, and the environment (Waheed et al. 2023). Research conducted over the past 150 years consistently demonstrates that plant-microbe interactions effectively enhance growth, defense, productivity, and health of plants (Doni et al. 2014, 2023; Amoo et al. 2023). Beneficial microorganisms are important for preserving plant health and fitness by bolstering resistance to environmental stresses, providing disease resistance, and facilitating nutrient availability and absorption (Olanrewaju et al. 2017; Doni et al. 2023; Akbari et al. 2024). Thus, a deep understanding of beneficial plant-microbe interactions is essential for uncovering microbial mechanisms that can provide plant fitness and health to achieve sustainability in agricultural production.

Multiple microbial mechanisms promote plant health by increasing resistance to phytopathogens, alleviating abiotic stresses such as pollution, nutrient deficiency, drought, and salinity, enhancing nutrient uptake efficiency, boosting photosynthetic rates, maintaining optimal internal cellular environments for crucial metabolic processes, and producing diverse bioactive compounds (enzymes, antibiotics, metabolites, phytohormones, etc.) that are toxic to phytopathogens (Babalola 2010; Sansinenea 2019; Harman et al. 2021; Ortiz and Sansinenea 2022). Numerous research studies have shown that metabolite compounds produced by microbes are vital not only for growth but also for protecting against pathogens, aiding in adapting to challenging environmental conditions, and modulating interactions to enhance mutual benefits in plant defense (Pang et al. 2021; Anjali et al. 2023; Bano et al. 2023). Thus, harnessing the benefits of microbial metabolites represents one of the most promising and sustainable approaches for advancing crop enhancement in the future.

Plant growth-promoting bacteria (PGPR) have emerged as beneficial microbes, contributing to plant health through diverse mechanisms by generating various secondary metabolites (Mishra and Arora 2018; Adeleke et al. 2021). For instance,

Pseudomonas spp. produce phenazines (PHZ), 2,4-diacetylphloroglucinol (DAPG), pyocyanin (PYO), pyoluteorin (PLT), pyoverdine (PVDI), cyclic lipopeptides (CLPs), pyrrolnitrin (PRN), and volatile organic compounds (VOCs) such as hydrogen cyanide (HCN), which possess antifungal, antibacterial, antiviral, and antinematodal properties (Spago et al. 2014; Deveau et al. 2016; Shahid et al. 2021b). *Bacillus* spp., synthesize a variety of potential bioactive compounds, comprising volatile and non-volatile compounds such as bacitracin, petrobactin, rhizocticin, bacilysin, and bacillibactin (Shahid et al. 2021a; Dimkić et al. 2022). These metabolites exhibit antibacterial and antifungal activities, effectively inhibiting plant diseases (Ruparelia et al. 2022). *Rhizobium* spp. are reported to produce a diverse array of antibiotics, including bacteriocins and trifolitoxin, which hold significant potential for inhibiting phytopathogens (Hirsch et al. 1980; Triplett and Barta 1987; Das et al. 2017). Additionally, these bacteria produce HCN and siderophores, as well as lytic enzymes such as chitinases, cellulases, β-1,3-glucanases, proteases, and pectinases (Arora et al. 2001; Arfaoui et al. 2006; Chandra et al. 2007). These compounds enhance their ability to efficiently manage plant diseases (Singh et al. 2019). *Azotobacter* spp. and *Azospirillum* spp. also produce metabolites such as siderophores, antimicrobial substances, and toxins that can suppress a wide range of phytopathogens (Moretti et al. 2020; Sumbul et al. 2020; Deepa et al. 2022). Another potential PGPR is *Serratia* spp., which produces prodigiosin, a compound that exhibits antimicrobial activities (Alijani et al. 2022).

Plant growth-promoting fungi (PGPF) provide protection by producing direct antimicrobial compounds and by stimulating the plant to produce its own defense metabolites (Jogaiah et al. 2018; Manganyi et al. 2019; Fadiji and Babalola 2020b; Techaoei et al. 2020). Indeed, *Trichoderma* spp. produce numerous potential secondary metabolites, such as antibacterial peptides, chlorotrichomycin, gelatinomycin, phytoalexins, trichomycin, and VOCs like aromatic compounds, butenolactones, polyketides, and terpenoids (Vinale et al. 2014; Maruyama et al. 2020). These compounds function as antimicrobial agents against plant pathogens (Yao et al. 2023). *Penicillium* spp. produces metabolites with antibacterial, antifungal, and antioxidant properties, including alkaloids, phenolic compounds, and triterpenoids, as well as anti-insect compounds such as 15-deoxyoxalicine B and decaturins A–B (Nicoletti et al. 2023; Tamariz-Angeles et al. 2023). The effectiveness of microbial metabolites in suppressing and inhibiting plant pathogens, as well as enhancing plant resistance responses, has been reported in previous studies (e.g., Borriss et al. 2019; Yao et al. 2023). Hence, novel approaches are necessary to investigate their underlying mechanisms.

One of the key ways in which secondary metabolites can enhance plant health is by directly combating phytopathogens with antimicrobial activity while also profoundly influencing plant metabolisms (Vinale et al. 2014; Nicoletti et al. 2023). These compounds act as signaling molecules, elicitors, or growth regulators, affecting various physiological processes and pathways related to plant defense mechanisms (Ortíz-Castro et al. 2009; Erb and Kliebenstein 2020). For instance, peptaibols, a metabolite from *Trichoderma*, can activate plant defense responses, leading to the production of defense compounds (Khan et al. 2020). They also inhibit enzymes for cell wall synthesis, thereby enhancing plant resistance to pathogens (Mukhopadhyay and Kumar 2020). Due to the extremely complex processes within microbial metabolites that can contribute to plant health (i.e., phytochemical analysis, biochemical changes, biosynthetic pathways), platforms capable of analyzing them to reveal their mechanism and function are needed (Sumner et al. 2003; Allwood et al. 2008; Jorge et al. 2016).

Metabolomics, an advanced technology with diverse research applications, enables the profiling of metabolites by identifying various compounds, including those produced by plants under stress conditions, molecules involved in stress signal transduction, and compounds critical for plant acclimation responses (Shulaev et al. 2008; Ye et al. 2022; van der Velden and Jansen 2023). Metabolomics represents a rapidly advancing approach capable of revealing various metabolic pathways and elucidating the mechanisms underlying plant-microbe interactions (Ye et al. 2022; Abdullah et al. 2024). Presently, microbial metabolomics is extensively used to investigate the roles and interactions of beneficial microbes, particularly in tomato (Mazzei et al. 2016; Coppola et al. 2019; Ji et al. 2021), chili (Thumanu et al. 2017; Kashyap et al. 2022), onion (Abdelrahman et al. 2016), rice (Valette et al. 2020), wheat (Mironenka et al. 2021), and maize (Saravanakumar et al. 2016). This approach has become a powerful technique for deciphering the complex metabolic processes in plant-microbe interactions, offering crucial insights into improving plant health and agricultural productivity (Kataoka et al. 2021; Olanrewaju et al. 2024).

Metabolomics is essential for comprehending the intricate interactions between microbes and host plants, enabling the identification and characterization of secondary metabolite compounds produced by beneficial microbes (Gupta et al. 2022; Olanrewaju et al. 2024). Metabolomics employs various approaches or analytical techniques, such as gas chromatography–mass spectrometry (GC-MS), liquid chromatography–mass spectrometry (LC-MS), high-performance liquid chromatography (HPLC), fourier transform infrared spectroscopy (FT-IR), and nuclear magnetic resonance (NMR) (González-Peña et al. 2015; Adeniji et al. 2020). Each technique has specific advantages and applications, particularly in the analysis of specific metabolites (Pezzatti et al. 2020; Gao et al. 2021; Manickam et al. 2023).

Recently, metabolomics profiling has significantly contributed to unraveling key mechanisms underlying plant-microbe interactions and has opened avenues for the discovery of beneficial microbial metabolites that enhance crop resilience and yield (Abdullah et al. 2024). However, metabolomics studies face complex challenges, resulting in limited information on the microbial metabolite contributions to plant health (Doni et al. 2022a). This chapter explores the current role of metabolomics in elucidating the processes involved in interactions between plants and microbes, as well as the influence of beneficial microbial metabolites on crop production. Additionally, it assesses how metabolomics studies can elucidate the underlying mechanisms behind the enhancement of plant health by microbes. The integration of metabolomics into plant-microbe interactions represents a promising and sustainable method to enhance plant resilience and productivity.

15.2 Beneficial Microbes for Enhancing Plant Health

Plant-microbe interactions significantly contribute to improving plant health through various mechanisms, such as directly inhibiting and suppressing pathogen growth and modulating plant defense responses (Berg 2009; Abdullah et al. 2021; Harman et al. 2021). Previous studies have shown that various beneficial microbes associated with plants can significantly enhance stress resilience, boost disease resistance, facilitate nutrient availability and uptake, and improve plant immunity without causing any negative impacts on plants or their environment (Trivedi et al. 2020; Doni et al. 2022b; Ortiz and Sansinenea 2022). To understand their mechanism of action, it is crucial to first investigate the interaction between plants and beneficial microbes in the surrounding environment, facilitated by the successful colonization of the host plant by these microbes (Pieterse et al. 2014; Santos et al. 2014; Doni et al. 2019).

Plant colonization by beneficial microbes for defense enhancement involves complex processes include (1) plants release root exudates into the soil to attract specific beneficial microbes, (2) beneficial microbes possess chemotactic abilities, which enable them to detect and migrate towards plant root exudates, (3) microbes establish symbiotic relationships with plants by recognizing and adhering to root surfaces through specific signaling molecules secreted by plant roots, (4) microbes form root biofilms to ensure persistence and protection, comprising complex communities encased in extracellular polymeric substances (EPS), (5) microbes communicate with plants using signaling molecules (phytohormones and elicitors), thereby triggering defense mechanisms that enhance pathogen resistance, and (6) as colonization progresses, mutualistic interactions develop, providing reciprocal benefits as microbes aid in nutrient acquisition, while plants offer a carbon source and a protected niche for microbial growth (Li et al. 2017; Durán et al. 2018; Olanrewaju et al. 2019; Knights et al. 2021; Sun et al. 2021; Ahlawat et al. 2024).

The induction of plant defense against various pathogens involves a wide array of highly complex and dynamic mechanisms, mediated both directly and indirectly (Jamalizadeh et al. 2011; Olanrewaju et al. 2017; Tariq et al. 2020; Roca-Couso et al. 2021; Bonaterra et al. 2022; Akbari et al. 2023). Antibiosis is a crucial mechanism through which beneficial bacterial and fungal antagonists enhance plant health by synthesizing antimicrobial metabolites that directly inhibit and disrupt the growth of phytopathogens, thus safeguarding plants from diseases (Köhl et al. 2019; Lahlali et al. 2022; Ayaz et al. 2023).

In addition, beneficial microbes effectively compete for nutrients within the host environment (Janisiewicz et al. 2000; Liu et al. 2013; Oszust et al. 2020). Those with higher metabolic efficiency gain an advantage in resource acquisition while simultaneously occupying or colonizing a niche or space through biofilm formation (Kaur et al. 2010; Li et al. 2011; Klein and Kupper 2018). This process limits pathogen attachment and growth on the host plants (Speranza et al. 2020; Heredia-Ponce et al. 2021). Lastly, hyperparasitism acts as a direct mechanism in which beneficial microbes invade and eliminate the mycelium, spores, and dormant structures of fungal pathogens, along with the cells of bacterial pathogens (Parratt and Laine 2016; Köhl et al. 2019). By directly attacking and eliminating pathogens, hyperparasitism effectively suppresses disease development and enhances plant health and fitness (Maciag et al. 2023).

Another equally important mechanism involves beneficial microbes enhancing plant health and disease resistance through various indirect pathways (Messa 2021; Yu et al. 2022). One of these mechanisms, induced systemic resistance (ISR) and systemic acquired resistance (SAR), involves plants developing resistance against pathogens (Pieterse et al. 2014). SAR is typically induced by pathogenic microbes, while ISR is triggered by beneficial microbes (Kamle et al. 2020; Yu et al. 2022). Related to ISR, microbes stimulate the plants' innate immune system, initiating underlying signaling pathways that are dependent on jasmonic acid (JA) and ethylene (ET) signaling that could be harnessed for plant protection strategy (Niu et al. 2011; Samaras et al. 2021).

Plants activate a salicylate-dependent signaling cascade that promotes the systemic development of broad-spectrum and long-lasting disease resistance, effectively combating various plant pathogens (Ilham et al. 2019). SAR, induced by pathogen exposure or elicitors, enhances plant resistance to future infections by accumulating pathogenesis-related proteins, activating defense genes, and reinforcing cell walls through salicylic acid (SA) signaling (Yuan et al. 2019). This provides lasting protection against specific pathogens (Nadarajah 2016; Kamle et al. 2020).

However, among all effective defense mechanisms, the suppression of phytopathogen growth through the production of phytohormones, antimicrobial compounds (including antibacterial and antifungal agents), secondary metabolites (such as enzymes, peptides, and VOCs), toxins, and biosurfactants stands out as one of the most critical strategies (Srivastava et al. 2016; Fadiji and Babalola 2020a; Raheel et al. 2022; Egamberdieva et al. 2023). Moreover, antibiotics and toxins produced by beneficial microbes have been recognized as pivotal factors in promoting plant health (Mendoza et al. 2015; Ghorai et al. 2021; Velmurugan et al. 2023).

As reported in numerous studies, the production of antimicrobial secondary metabolites by beneficial microorganisms involves the production and release of substances can directly inhibit growth or activities of various phytopathogens (Arseneault and Filion 2017; Narayanan and Glick 2022). Some secondary metabolites, such as surfactin, a group of lipopeptides synthesized by *B. subtilis*, have the ability to disrupt the integrity of the cell membrane in pathogens, causing leakage of cellular contents and ultimately resulting in cell death (Arseneault and Filion 2017; Liu et al. 2023; Rosier et al. 2023). Furthermore, metabolites within this category demonstrate activities such as antiviral, antifungal, antibacterial, biofilm formation, and induction of plant resistance (Meena and Kanwar 2015; Malviya et al. 2020).

Secondary metabolites can also interfere with enzymatic activity essential for pathogen survival, as evidenced by the antibiotic bacillomycin produced by *B. amyloliquefaciens* targeting enzymes involved in cell wall synthesis in fungal pathogens such as *Fusarium graminearum* (Ramarathnam et al. 2007; Yuan et al. 2012; Gu et al. 2017). Understanding the biosynthetic process, metabolic pathways, and functions, as well as screening and identifying the potential of secondary metabolites, offers a promising approach for managing plant diseases, enhancing plant fitness, and supporting sustainable agricultural methods. The impact of secondary metabolites produced by beneficial microbes on plant health is summarized in Table 15.1.

15.3 The Roles of Microbial Metabolites in Improving Plant Health

Many beneficial microbes function as biocontrol agents, synthesizing and releasing specific metabolites with antimicrobial properties, thus inhibiting the growth or metabolic activity of pathogens (Palyzová et al. 2019; Bonaterra et al. 2022; Ayaz et al. 2023). Previous studies have demonstrated that antibiotics, siderophores, and VOCs produced by beneficial microbes serve as natural defense mechanisms against various plant pathogens (Poveda 2021; Roca and Matilla 2023; Deb and Tatung 2024). These metabolites can directly inhibit the growth and survival of pathogens by targeting specific cellular processes or structures, such as disrupting their reproduction and chelating essential nutrients like iron (Arseneault and Filion 2017; Singh et al. 2022). For instance, *T. pseudokoningii* produces peptaiboles, a class of antibiotics that exhibit antimicrobial activity against plant pathogens (Shi et al. 2012).

ISR in plants facilitated by microbial metabolites activates plant defense mechanisms throughout the plant, thereby significantly enhancing resistance (Choudhary et al. 2007). Other metabolites act as antioxidants, scavenging reactive oxygen species (ROS) to reduce oxidative stress, which is crucial for maintaining cellular balance, especially under environmental stress (Das and Roychoudhury 2014; Zandi and Schnug 2022). Microbial phytohormones, such as indole-3-acetic acid (IAA), ACC deaminase, cytokinins, gibberellins, and siderophores, enhance plant growth and development by stimulating root elongation, improving nutrient uptake, and boosting photosynthesis (Koza et al. 2022; Orozco-Mosqueda et al. 2023). Metabolites produced by microbes play a crucial role in modulating hormonal balance, exerting a significant influence on the process of seed germination (Khoso et al. 2024). Additionally, microbial metabolites can induce the synthesis of phytoalexins, thus fortifying the plants' immune response against pathogens and enhancing their overall fitness (Ahuja et al. 2012).

15.3.1 Antibiotics and Antibacterial

Plant-associated bacteria-derived antibiotics exhibit significant potential as protective agents against phytopathogens (Roca and Matilla 2023). The identification of new antibiotics is crucial, given that only 3% of the bacterial genetic potential for producing bioactive compounds has been experimentally investigated (Gavriilidou et al. 2022). However, antibiotics have been extensively used in controlling plant

Table 15.1 Examples of metabolites produced by various microorganisms and their effect on plant health

Beneficial microbes	Metabolites	Class of metabolites	Mechanism of actions	Diseases/Pathogens	Observed effect on plants	References
B. subtilis, *B. amyloliquefaciens*, and *B. velezensis*	Bacillomycin D	Cyclic lipopeptides	Disrupt the pathogen's cell membrane by increasing the expression of genes related to ergosterol synthesis and oxidative stress	Anthracnos (*Colletotrichum gloeosporioide*)	Mango (*Mangifera indica* Linn.)	Jin et al. (2020) and Tran et al. (2022)
B. amyloliquefaciens BUZ-14	Iturin A	Cyclic lipopeptides	Antifungal activity and induced systemic resistance (ISR)	Brown rot (*Monilinia fructicola*)	Peach (*Prunus persica*)	Calvo et al. (2019)
B. subtilis M4	Fengycin	Cyclic lipopeptides	Providing plant protection through its antifungal compounds against various pathogens and participating in ISR	Damping-off (*Pythium ultimum*)	Bean (*Phaseolus vulgaris*)	Ongena et al. (2005)
Bacillus NH-100 and NH-217	Surfactin A	Cyclic lipopeptides	Antifungal activity and ISR	Bakanae (*F. moniliforme*)	Rice (*Oryza sativa*)	Sarwar et al. (2018)
Fluorescent pseudomonads (FPs)	Phenazine	Cyclic lipopeptides	Antifungal activity	Sheath blight (*Rhizoctonia solani*)	Rice (*O. sativa*)	Chin-A-Woeng et al. (2003) and Karmegham et al. (2020)
Bacillus spp. and *Pseudomonas* spp.	Dimethyl disulfide (DMDS)	VOCs	Antifungal activity, disruption of cell membrane, and ISR	Sclerotinia rot (*Sclerotinia minor*)	Tomato (*Solanum lycopersicum*)	Tyagi et al. (2020)
T. longibrachiatum	Trichogin GA IV	Peptaibols	Antimicrobial activity against gram-positive bacteria	Black rot (*Xanthomonas campestris* pv. *campestris*)	*Brassica* spp.	Caracciolo et al. (2023)

diseases since the 1950s due to their effectiveness in inhibiting pathogenic bacteria (Islam et al. 2024). Therefore, beneficial plant-associated bacteria are a promising source for discovering novel antibiotics that can improve plant health (Castronovo et al. 2020; Roca and Matilla 2023).

One of the most effective mechanisms of antibiotics in combating plant pathogens is by actively preventing the growth and proliferation of phytopathogens through disruption of essential cellular processes (Raaijmakers and Mazzola 2012). For instance, zwittermicin A and kanosamine produced by *B. cereus* UW85 enhance its capacity to suppress specific plant diseases and directly target soilborne phytopathogens such as *Phytophthora medicaginis* (Milner et al. 1996). In addition, specific antibiotics that target bacterial cell wall synthesis, such as lipopeptides, bacteriocins, and biosurfactants, induce cell lysis and subsequent death (Bonaterra et al. 2022). This study employed HPLC to reveal that lipopeptides such as surfactin, produced by *B. amyloliquefaciens* KPS46, exhibit antibacterial activity by inhibiting the pathogen *X. axonopodis* pv. *glycines*, which causes bacterial pustule disease in soybeans (Preecha et al. 2010).

15.3.2 Antifungal

Numerous studies have shown that using microbial antifungals provides a natural and sustainable substitute to chemical fungicides for improving plant health (Sellitto et al. 2021; Fenta and Mekonnen 2024). Certain microbial antifungal agents have the capacity to stimulate plant defense responses, consequently augmenting resistance against fungal pathogens (Akram et al. 2023). Various fungal species produce a diverse range of substances that enable plants to withstand various stresses, including protecting them from diverse pathogens (Ayaz et al. 2023).

For instance, *Trichoderma* spp. and *Gliocladium* spp. are renowned for being prolific producers of numerous antifungal compounds, including gliotoxin, gliovirin, heptelidic acid, valinotrocin, viridin, and viridiol (El-Saadony et al. 2022). Furthermore, Bacillomycin D, a primary metabolite compound isolated from *B. velezensis* HN-2, exhibits stronger antifungal activity against *C. gloeosporioides* compared to conventional fungicides (Jin et al. 2020). *P. protegens* secretes three major antifungal compounds with biocontrol activity: 2,4-diacetylphloroglucinol (2,4-DAPG), pyoluteorin (PLT), and pyrrolnitrin (PRN), which are effective against *Heterobasidion* spp. (Pellicciaro et al. 2022). Furthermore, *T. asperellum* also produces 6-Pentyl-α-Pyrone to inhibit *Magnaporthiopsis maydis*, the causative agent of maize late wilt disease (Degani et al. 2021). Various studies have identified antifungal compounds from diverse sources using a range of metabolomic techniques such as GC-MS, LC-MS, and HPLC (Wang et al. 2012).

15.3.3 Volatile Organic Compounds (VOCs)

The secondary metabolites exhibit VOC properties effective in combating various plant diseases (Srikamwang et al. 2023). VOCs produced by *Bacillus* spp. demonstrated the capability to promote growth and trigger ISR in *Arabidopsis* (Ryu et al. 2003, 2004; Lee et al. 2012). Substantial evidence from a multitude of studies highlights the potent inhibitory effects of these metabolites in controlling diverse plant pathogens, underscoring their potential as effective antimicrobial agents (Košćak et al. 2023). These metabolites exhibit the ability to selectively target specific pathogens or exert a broad-spectrum effect (Schulz-Bohm et al. 2017). In addition, VOCs can directly and indirectly increase plant biomass, enhance disease resistance, and improve tolerance to environmental stresses (Srikamwang et al. 2023).

The diverse advantages offered by these metabolites have sparked extensive research endeavors focused on their identification and profiling using metabolomics (Márquez et al. 2022; Mhlongo et al. 2022; Almeida et al. 2023). Certain VOCs, such as 2,3-butanediol, acetoin, and short-chain alcohols, produced by the PGPR strain *Paenibacillus polymyxa* E681, have been demonstrated to enhance plant growth and development and trigger ISR against *P. syringae* (Lee et al. 2012). Furthermore, *B. amyloliquefaciens* produces 2,5-Dimethyl pyrazine, 2-Dodecanone, and 2-Tetradecanone, which exhibit antifungal activity against *C. gloeosporioides* and *Fusarium* spp. (Guevara-Avendaño et al. 2019). Therefore, numerous studies indicate that beneficial microbes are capable of enhancing plant health by combating fungi, bacteria, oomycetes, and nematodes while also serving as elicitors of plant immunity through the modulation of signaling pathways such as SA and JA (Bitas et al. 2013; Schalchli et al. 2016; Poveda 2021).

In order to harness the potential of metabolites generated by beneficial microbes, which serve as biostimulants, biofertilizers, and biocontrol agents, a more comprehensive understanding of metabolomics is essential. This includes acquiring deeper insights into the metabolic changes that take place in both plants and microbes during their interaction, identifying the specific metabolites involved in signaling between plants and microbes, and elucidating the metabolic pathways that are either activated or suppressed in this process.

15.4 Metabolomics in Plant-Microbe Interactions Research

The relationship between beneficial microbes and their host plants provides several advantages for plant health, which are being investigated using a variety of techniques (Gamalero et al. 2022). Metabolomics has become a critical approach that enables in-depth exploration of intricate metabolic processes and their products, as well as the elucidation of secondary metabolic pathways, relying on the analysis of precision instrumentation (Liu et al. 2021; Shen et al. 2023). In the past decade, this burgeoning field has garnered widespread attention (Fig. 15.1). However, one of these omics methods remains underutilized compared to genomic and

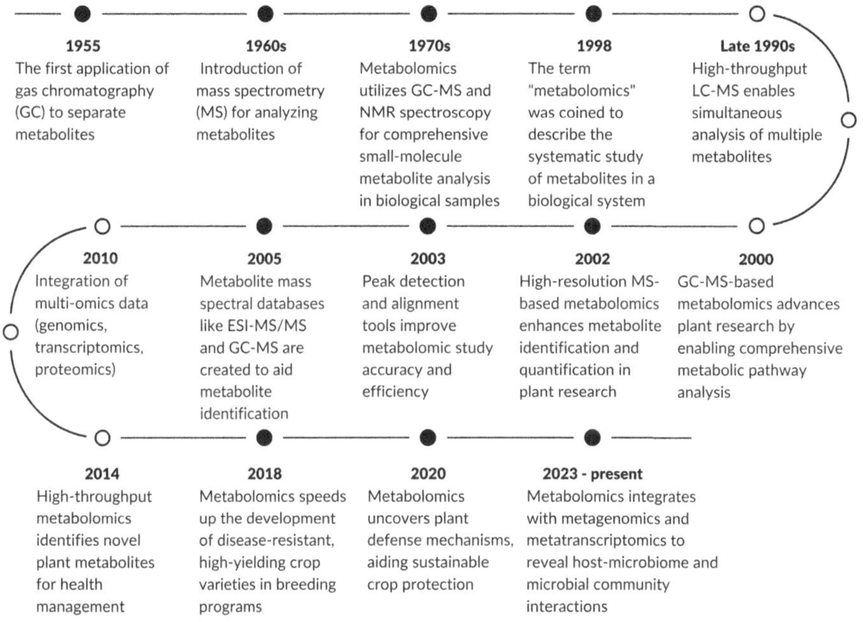

Fig. 15.1 Timeline and milestone of metabolomics research

transcriptomic approaches in studies of plant-microbe interactions (Gupta et al. 2022). Yet, various metabolomics studies indicate that with its ability to provide comprehensive coverage of metabolites, it offers a powerful tool for studying plant-microbe interactions by elucidating the biochemical basis of symbiosis, plant tolerance to stresses, and mechanisms of resistance and susceptibility (Manickam et al. 2023).

Metabolomics encompasses the documentation, identification, and quantification of diverse metabolites within biological systems—such as cells, tissues, or organisms—utilizing advanced analytical platforms (Marchev et al. 2021; Akyol et al. 2023). These advanced analytical tools offer high sensitivity, resolution, and accuracy, enabling researchers to detect and quantify a broad spectrum of metabolites found in microorganisms participating in interactions with plants (Gupta et al. 2022; Shen et al. 2023). GC/MS is a powerful analytical technique that is highly effective for separating and identifying volatile and semi-volatile compounds often associated with interactions between plants and microbes, while LC/MS excels in analyzing a diverse array of polar and non-polar metabolites crucial for comprehending the intricate biochemical communication between plants and microbes (Fiehn 2016). Additionally, NMR spectroscopy offers valuable structural insights into metabolites involved in these interactions without the need for extensive sample preparation, further enhancing our understanding of their roles (Emwas 2015).

Metabolomics studies are essential for tackling plant health challenges caused by climate change and for elucidating the comprehensive interactions between plants and microbes, which offer numerous benefits to plant health (Kellogg and Kang 2020;

Table 15.2 Microbial metabolite analysis utilizing metabolomics techniques and its impact on plant health

Microbial species	Metabolites	Effects in plant species	Metabolomics techniques	References
B. amyloliquefaciens FZB42	Fengycin	Inhibiting disease growth and preventing infection in seeds	HP-LC and MALDI-TOF-MS	Hanif et al. (2019)
B. velezensis EM-1	Iturin A	Inhibiting *Ralstonia solanacearum* growth and stimulating plant resistance by enhancing CAT and PPO activities in tobacco	LC-MS/MS and GC-MS/MS	Sui et al. (2022)
T. virens T23	Gliotoxin	Inhibiting *Sclerotium rolfsii* growth in peanut seedlings	HP-LC	Hua et al. (2021)
T. virens HZA14	Gliotoxin	Reducing pathogen development and controlling chili pepper blight	HP-LC	Tomah et al. (2020)
B. amyloliquefaciens	Acetoin and 2,3-butanediol	Inducing stomatal closure in *A. thaliana* and *Nicotiana benthamiana* to block pathogen entry	GC-MS	Wu et al. (2018)
Pseudomonas spp., *Serratia* spp., and *Burkholderia* spp.	Pyrrolnitrin (PRN)	Providing comprehensive protection to plants against soil and seed-borne diseases	HP-LC, FT-IR, NMR, LC-MS, and GC-MS	Pawar et al. (2019)

Manickam et al. 2023). Identifying bioactive compounds or secondary metabolites associated with disease resistance, stress response, growth promotion, and enhancing plant productivity across various environmental conditions is crucial (Olanrewaju et al. 2024). Indeed, metabolomics is considered a valuable approach for exploring the discovery of metabolites, as it is believed to be able to elucidate plant-pathogen interactions and the exploration of bioactive compounds, as demonstrated by the findings presented in Table 15.2.

15.5 Metabolomics Analytical Techniques Involved in Plant-Microbe Interactions

Metabolomics enables the identification of metabolic changes within an organism, delineating the pathways governing these changes, and qualitatively and quantitatively detecting active metabolites (Kumar et al. 2017; Abadie et al. 2022; Olanrewaju et al. 2024). A primary approach in metabolomics research for investigating the

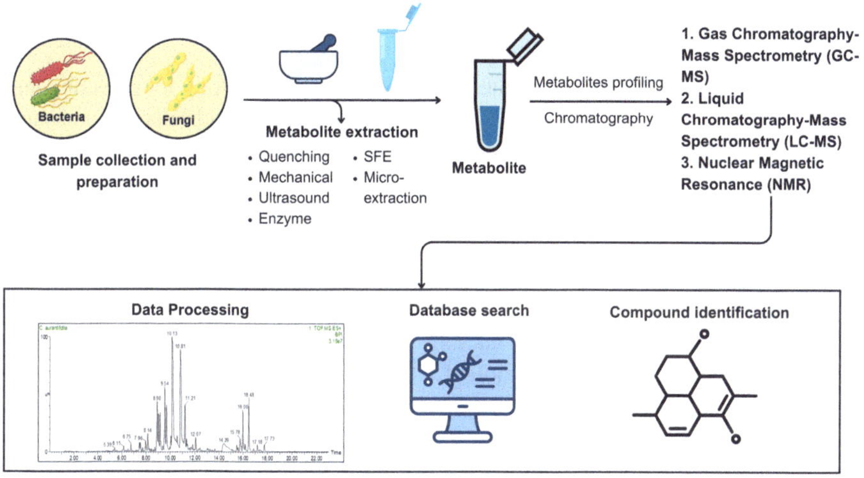

Fig. 15.2 Schematic representation of the metabolomics study illustrates the workflow, including sample collection, extraction, analysis, and data interpretation. From (Farag et al. 2022; Sehim et al. 2023), the journals do not require permissions to use the materials

functional roles of specific metabolites in plant-microbe interactions involves using high-throughput technologies such as GC-MS, LC-MS, and HPLC methods, with a schematic workflow depicted in Fig. 15.2;. Metabolomics is a highly dynamic and rapidly advancing field, continually evolving with new developments and insights (Kumar et al. 2017; Perez de Souza et al. 2019; Manickam et al. 2023).

15.5.1 GC-MS

GC-MS metabolomics, also known as metabolic profiling, has become a prevalent technique widely employed to analyze various metabolites involved in plant-microbe interactions and biocontrol strategies (Hill and Roessner 2013; Ajilogba and Babalola 2019). Its extensive application underscores its pivotal role in elucidating the complex metabolic dynamics underlying these biological processes (Lakhdari et al. 2023). Identified compounds are listed with their retention times, mass spectra, and chemical names, aiding in precise identification (Gomathi et al. 2015). Retention times, crucial for identification, measure how long compounds take to travel through the GC column (Chevalier and Sommerer 2011; Zhang et al. 2016). Mass spectra show the mass-to-charge ratio (m/z) of ions during fragmentation, helping in structural elucidation (Chevalier and Sommerer 2011; Zhang et al. 2016). Peak areas in chromatograms reflect compound abundance, enabling quantitative analysis for concentration determination (Fiehn 2016). Both qualitative and quantitative analyses are performed in GC-MS, with qualitative analysis identifying compounds and quantitative analysis determining concentrations through peak areas or calibration curves (Koek et al. 2011).

15.5.2 LC-MS

LC-MS, a robust analytical technique widely used in microbiology for detecting and quantifying microbial metabolites, allows for highly sensitive and selective measurement of metabolites, providing comprehensive details regarding the metabolic changes occurring during these interactions (Xiao et al. 2012). This methodology enables precise quantification of metabolites with exceptional sensitivity and selectivity, thus offering extensive insights into the metabolic changes that underlie these interactions (Halket et al. 2005). LC-MS exhibits exceptional sensitivity, enabling the detection of metabolites at low concentrations, and can investigate complex biological phenomena such as plant-microbe interactions (Bauermeister et al. 2016, 2022; Shahid et al. 2018). Furthermore, LC-MS exhibits flexibility by accommodating the analysis of various metabolites, including polar and non-polar compounds, thereby supporting comprehensive metabolomics studies (Metz et al. 2007). Its rapid analytical speed also allows for efficient processing of large sample volumes, enhancing its effectiveness across various research applications (Chen and Kim 2013).

15.5.3 HP-LC

This analytical method enables the identification and measurement of diverse compounds, including secondary metabolites produced by plants or microbes, as well as changes in metabolite profiles under different experimental conditions (Kumar 2017). HPLC results may indicate the presence or absence of specific compounds known to be involved in biocontrol mechanisms or microbial interactions with plants (Ali 2022). Additionally, HPLC can provide understanding into the dynamics of metabolic pathways and the influence of environmental factors on metabolite production (Kumar 2017; Wang et al. 2020). Metabolite profiling through HPLC enables the separation and quantification of metabolites crucial in plant-microbe interactions, encompassing phytohormones, secondary metabolites, and signaling molecules (Chen et al. 2012; Long et al. 2014). Additionally, HPLC facilitates the identification of specific compounds produced by beneficial microbes that participate in biocontrol or pathogenesis (Sarwar et al. 2018). It also aids in the quantification of plant defense compounds like phenolics, flavonoids, and alkaloids, which are integral to plant defense mechanisms (Mizzi et al. 2020). Moreover, HPLC serves to monitor changes in metabolite levels over time or in response to treatments, thereby offering insights into the metabolic pathways (Kumar 2017).

15.6 Future Perspectives and Research Opportunities

Metabolomics, ideal for unraveling complex rhizosphere interactions, provides insights into plant-microbe dynamics by identifying pivotal metabolites crucial for plant defense and microbial activities (Gupta et al. 2022; Manickam et al. 2023).

However, microbial metabolomics aimed at improving plant health encounters several challenges, including the complex composition of the microbiome within a plant, the dynamic production of metabolites, and the need for advanced analytical methodologies (Diwan et al. 2022; Gupta et al. 2022). The diversity of the plant microbiome poses challenges in identifying and characterizing all interacting microbes and their produced metabolites (Chen et al. 2021; Diwan et al. 2022). Furthermore, metabolite production can fluctuate due to environmental factors, complicating efforts to obtain a complete metabolomics profile (Bayona et al. 2020; Kim and Kang 2021). Additionally, the expertise and advanced data analysis tools needed for interpreting results add further complexity to metabolomics analyses. Therefore, integrating metabolomics with quantitative genetics, transcriptomics, and genetic manipulation has proven effective in enhancing crop traits and tackling global agricultural challenges (Razzaq et al. 2019; Doni et al. 2022a; Manickam et al. 2023).

All promising research on microbial metabolomics for plant health offers opportunities to explore new metabolites synthesized by beneficial microbes and their contributions to improving plant health and stress resilience (Olanrewaju et al. 2024). Metabolomics offers a holistic perspective on both the metabolic pathways engaged in plant-microbe interactions and the fundamental mechanisms governing host-microbe relationships. Continuous studies are crucial to address various challenges, including identifying the origins of analyzed metabolites, unraveling the complexity of metabolism within interacting organisms, and integrating metabolomics data with complementary omics methodologies. With technological advancements and analytical methods improving, metabolomics is anticipated to significantly contribute to understanding microbial metabolites, thereby enhancing plant health as an innovative and sustainable solution for the future.

15.7 Conclusion

Metabolomics has become a potent tool for exploring the intricacies of microbial metabolism and comprehending its multifaceted roles in promoting plant health. Metabolomics stands as a growing novel approach that can unveil the diverse potential of microbial metabolites crucial in enhancing plant health, aided by highly advanced analytical techniques. This tool can be utilized to devise sustainable and environmentally friendly approaches to crop management, including employing biocontrol agents and optimizing microbial populations within the rhizosphere. Metabolomics aids in enhancing the understanding of microbial metabolites and pathways. Upon addressing all of the previously mentioned challenges, the field of metabolomics can advance towards broader applications across various studies and disciplines. Therefore, integrating metabolomics data analysis with other multi-omics approaches, such as genomics, transcriptomics, and proteomics, provides a more comprehensive understanding of plant-microbe interactions.

Acknowledgments This work was funded by Universitas Padjadjaran through Riset Keunggulan Keilmuan Unpad (RKKU), grant number 950/UN6.3.1/PT.00/2025 awarded to Febri Doni.

Conflicts of Interest The authors declare no conflict of interest.

References

Abadie C, Lalande J, Tcherkez G (2022) Exact mass GC-MS analysis: protocol, database, advantages and application to plant metabolic profiling. Plant Cell Environ 45:3171–3183

Abdelrahman M, Abdel-Motaal F, El-Sayed M, Jogaiah S, Shigyo M, Ito S, Tran L-SP (2016) Dissection of *Trichoderma longibrachiatum*-induced defense in onion (*Allium cepa* L.) against *Fusarium oxysporum* f. sp. *cepa* by target metabolite profiling. Plant Sci 246:128–138

Abdullah NS, Doni F, Mispan MS, Saiman MZ, Yusuf YM, Oke MA, Suhaimi NSM (2021) Harnessing *Trichoderma* in agriculture for productivity and sustainability. Agronomy 11:2559

Abdullah NS, Doni F, Awal MA, Mispan MS, Saiman MZ, Mohd-Yusuf Y, Suhaimi NS (2024) Multi-omics tools for understanding *Trichoderma*-plant symbiosis: biotechnological developments and future directions. Symbiosis. https://doi.org/10.1007/s13199-024-00996-2

Adeleke BS, Babalola OO, Glick BR (2021) Plant growth-promoting root-colonizing bacterial endophytes. Rhizosphere 20:100433

Adeniji AA, Babalola OO, Loots DT (2020) Metabolomic applications for understanding complex tripartite plant-microbes interactions: strategies and perspectives. Biotechnol Rep 25:e00425

Ahlawat OP, Yadav D, Walia N, Kashyap PL, Sharma P, Tiwari R (2024) Root exudates and their significance in abiotic stress amelioration in plants: a review. J Plant Growth Regul. https://doi.org/10.1007/s00344-024-11237-7

Ahuja I, Kissen R, Bones AM (2012) Phytoalexins in defense against pathogens. Trends Plant Sci 17:73–90

Ajilogba CF, Babalola OO (2019) GC–MS analysis of volatile organic compounds from Bambara groundnut rhizobacteria and their antibacterial properties. World J Microbiol Biotechnol 35:83

Akbari SI, Abdullah NS, Permadi N, Rossiana N, Suhaimi NSM, Uphoff N, Doni F (2023) *Trichoderma*-based bioinoculant: a potential tool for sustainable rice cultivation. In: Bioinoculants: biological option for mitigating global climate change. Springer, Singapore, pp 239–264

Akbari SI, Prismantoro D, Permadi N, Rossiana N, Miranti M, Mispan MS, Mohamed Z, Doni F (2024) Bioprospecting the roles of *Trichoderma* in alleviating plants' drought tolerance: principles, mechanisms of action, and prospects. Microbiol Res 283:127665

Akram S, Ahmed A, He P, He P, Liu Y, Wu Y, Munir S, He Y (2023) Uniting the role of endophytic fungi against plant pathogens and their interaction. J Fungi 9:72

Akyol S, Ashrafi N, Yilmaz A, Turkoglu O, Graham SF (2023) Metabolomics: an emerging "omics" platform for systems biology and its implications for Huntington disease research. Meta 13:1203

Ali AH (2022) High-performance liquid chromatography (HPLC): a review. Ann Adv Chem 6:10–20

Alijani Z, Amini J, Karimi K, Pertot I (2022) Characterization of the mechanism of action of *Serratia rubidaea* Mar61-01 against *Botrytis cinerea* in strawberries. Plan Theory 12:154

Allwood JW, Ellis DI, Goodacre R (2008) Metabolomic technologies and their application to the study of plants and plant–host interactions. Physiol Plant 132:117–135

Almeida OAC, de Araujo NO, Mulato ATN, Persinoti GF, Sforça ML, Calderan-Rodrigues MJ, de Oliveira JVC (2023) Bacterial volatile organic compounds (VOCs) promote growth and induce metabolic changes in rice. Front Plant Sci 13:1056082

Amoo AE, Olanrewaju OS, Babalola OO, Ajilogba CF, Chukwuneme CF, Ojuederie OB, Omomowo OI (2023) The functionality of plant-microbe interactions in disease suppression. J King Saud Univ Sci 35:102893

Anjali KS, Korra T, Thakur R, Arutselvan R, Kashyap AS, Nehela Y, Chaplygin V, Minkina T, Keswani C (2023) Role of plant secondary metabolites in defence and transcriptional regulation in response to biotic stress. Plant Stress 8:100154

Arfaoui A, Sifi B, Boudabous A, El Hadrami I, Cherif M (2006) Identification of *Rhizobium* isolates possessing antagonistic activity against *Fusarium oxysporum* f.sp *ciceris*, the causal agent of *Fusarium wilt* of chickpea. J Plant Pathol 88:67–75

Arora N, Kang S, Maheshwari D (2001) Isolation of siderophore producing strains of *Rhizobium meliloti* and their biocontrol potential against *Macrophomina phaseolina* that causes charcoal rot of groundnut. Curr Sci 81:673–677

Arseneault T, Filion M (2017) Biocontrol through antibiosis: exploring the role played by subinhibitory concentrations of antibiotics in soil and their impact on plant pathogens. Can J Plant Pathol 39:267–274

Ayaz M, Li C-H, Ali Q et al (2023) Bacterial and fungal biocontrol agents for plant disease protection: journey from lab to field, current status, challenges, and global perspectives. Molecules 28:6735

Babalola OO (2010) Beneficial bacteria of agricultural importance. Biotechnol Lett 32:1559–1570

Bano A, Qadri TA, Mahnoor KN (2023) Bioactive metabolites of plants and microbes and their role in agricultural sustainability and mitigation of plant stress. S Afr J Bot 159:98–109

Bauermeister A, Zucchi TD, Moraes LAB (2016) Mass spectrometric approaches for the identification of anthracycline analogs produced by actinobacteria. J Mass Spectrom 51:437–445

Bauermeister A, Mannochio-Russo H, Costa-Lotufo LV, Jarmusch AK, Dorrestein PC (2022) Mass spectrometry-based metabolomics in microbiome investigations. Nat Rev Microbiol 20:143–160

Bayona LM, van Leeuwen G, Erol Ö, Swierts T, van der Ent E, de Voogd NJ, Choi YH (2020) Influence of geographical location on the metabolic production of giant barrel sponges (*Xestospongia* spp.) revealed by metabolomics tools. ACS Omega 5:12398–12408

Berg G (2009) Plant–microbe interactions promoting plant growth and health: perspectives for controlled use of microorganisms in agriculture. Appl Microbiol Biotechnol 84:11–18

Bitas V, Kim H-S, Bennett JW, Kang S (2013) Sniffing on microbes: diverse roles of microbial volatile organic compounds in plant health. Mol Plant Microbe 26:835–843

Bonaterra A, Badosa E, Daranas N, Francés J, Roselló G, Montesinos E (2022) Bacteria as biological control agents of plant diseases. Microorganisms 10:1759

Borriss R, Wu H, Gao X (2019) Secondary metabolites of the plant growth promoting model rhizobacterium *Bacillus velezensis* FZB42 are involved in direct suppression of plant pathogens and in stimulation of plant-induced systemic resistance. In: Secondary metabolites of plant growth promoting Rhizomicroorganisms. Springer, Singapore, pp 147–168

Calvo H, Mendiara I, Arias E, Blanco D, Venturini ME (2019) The role of iturin A from *B. amyloliquefaciens* BUZ-14 in the inhibition of the most common postharvest fruit rots. Food Microbiol 82:62–69

Caracciolo R, Sella L, De Zotti M, Bolzonello A, Armellin M, Trainotti L, Favaron F, Tundo S (2023) Efficacy of *Trichoderma longibrachiatum* Trichogin GA IV peptaibol analogs against the black rot pathogen *Xanthomonas campestris* pv. *campestris* and other phytopathogenic bacteria. Microorganisms 11:480

Carvajal-Yepes M, Cardwell K, Nelson A et al (2019) A global surveillance system for crop diseases. Science 364:1237–1239

Castronovo LM, Calonico C, Ascrizzi R, Del Duca S, Delfino V, Chioccioli S, Vassallo A, Strozza I, De Leo M, Biffi S, Bacci G, Bogani P, Maggini V, Mengoni A, Pistelli L, Lo Nostro A, Firenzuoli F, Fani R (2020) The cultivable bacterial microbiota associated to the medicinal plant *Origanum vulgare* L.: from antibiotic resistance to growth-inhibitory properties. Front Microbiol 11:862

Chandra S, Choure K, Dubey RC, Maheshwari DK (2007) Rhizosphere competent *Mesorhizobium loti* MP6 induces root hair curling, inhibits *Sclerotinia sclerotiorum* and enhances growth of Indian mustard (*Brassica campestris*). Braz J Microbiol 38:124–130

Chen C, Kim S (2013) LC-MS-based metabolomics of xenobiotic-induced toxicities. Comput Struct Biotechnol J 4:e201301008

Chen Q, Li Y, Chen Z (2012) Separation, identification, and quantification of active constituents in Fructus Psoraleae by high-performance liquid chromatography with UV, ion trap mass spectrometry, and electrochemical detection. J Pharm Anal 2:143–151

Chen L, Schwier M, Krumbach J, Kopriva S, Jacoby RP (2021) Metabolomics in plant-microbe interactions in the roots. In: Advances in botanical research. Elsevier, pp 133–161

Chevalier F, Sommerer N (2011) Analytical methods | mass spectrometric methods. In: Encyclopedia of dairy sciences. Elsevier, pp 198–205

Chin-A-Woeng TFC, Bloemberg GV, Lugtenberg BJJ (2003) Phenazines and their role in biocontrol by *Pseudomonas* bacteria. New Phytol 157:503–523

Choudhary DK, Prakash A, Johri BN (2007) Induced systemic resistance (ISR) in plants: mechanism of action. Indian J Microbiol 47:289–297

Coppola M, Diretto G, Digilio MC, Woo SL, Giuliano G, Molisso D, Pennacchio F, Lorito M, Rao R (2019) Transcriptome and metabolome reprogramming in tomato plants by *Trichoderma harzianum* strain T22 primes and enhances defense responses against aphids. Front Physiol 10:745

Das K, Roychoudhury A (2014) Reactive oxygen species (ROS) and response of antioxidants as ROS-scavengers during environmental stress in plants. Front Environ Sci 2:53

Das K, Prasanna R, Saxena AK (2017) Rhizobia: a potential biocontrol agent for soilborne fungal pathogens. Folia Microbiol 62:425–435

Deb CR, Tatung M (2024) Siderophore producing bacteria as biocontrol agent against phytopathogens for a better environment: a review. S Afr J Bot 165:153–162

Deepa N, Chennappa G, Deepthi BNV, Kumari, Naik MK, Ramesha KP, Amaresh YS et al (2022) Antifungal potential of *Azotobacter* species and its metabolites against *Fusarium verticillioides* and biodegradation of fumonisin. J Appl Microbiol 133:2430–2444

Degani O, Khatib S, Becher P, Gordani A, Harris R (2021) *Trichoderma asperellum* secreted 6-Pentyl-α-Pyrone to control magnaporthiopsis maydis, the maize late wilt disease agent. Biology (Basel) 10:897

Deveau A, Gross H, Palin B, Mehnaz S, Schnepf M, Leblond P, Dorrestein PC, Aigle B (2016) Role of secondary metabolites in the interaction between *Pseudomonas fluorescens* and soil microorganisms under iron-limited conditions. FEMS Microbiol Ecol 92:fiw107

Dimkić I, Janakiev T, Petrović M, Degrassi G, Fira D (2022) Plant-associated *Bacillus* and *Pseudomonas* antimicrobial activities in plant disease suppression via biological control mechanisms—a review. Physiol Mol Plant Pathol 117:101754

Diwan D, Rashid MM, Vaishnav A (2022) Current understanding of plant-microbe interaction through the lenses of multi-omics approaches and their benefits in sustainable agriculture. Microbiol Res 265:127180

Dixon GR (2012) Climate change—impact on crop growth and food production, and plant pathogens. Can J Plant Pathol 34:362–379

Doni F, Isahak A, Che Mohd Zain CR, Wan Yusoff WM (2014) Physiological and growth response of rice plants (*Oryza sativa* L.) to *Trichoderma* spp. inoculants. AMB Express 4:45

Doni F, Fathurrahman F, Mispan MS, Suhaimi NSM, Yusoff WMW, Uphoff N (2019) Transcriptomic profiling of rice seedlings inoculated with the symbiotic fungus *Trichoderma asperellum* SL2. J Plant Growth Regul 38:1507–1515

Doni F, Miranti M, Mispan MS, Mohamed Z, Uphoff N (2022a) Multi-omics approaches for deciphering the microbial modulation of plants' genetic potentials: what's known and what's next? Rhizosphere 24:100613

Doni F, Suhaimi NSM, Mispan MS, Fathurrahman F, Marzuki BM, Kusmoro J, Uphoff N (2022b) Microbial contributions for rice production: from conventional crop management to the use of 'omics' technologies. Int J Mol Sci 23:737

Doni F, Safitri R, Suhaimi NSM, Miranti M, Rossiana N, Mispan MS, Anhar A, Uphoff N (2023) Evaluating the underlying physiological and molecular mechanisms in the system of rice intensification performance with *Trichoderma*-rice plant symbiosis as a model system. Front Plant Sci 14:1214213

Durán P, Thiergart T, Garrido-Oter R, Agler M, Kemen E, Schulze-Lefert P, Hacquard S (2018) Microbial interkingdom interactions in roots promote arabidopsis survival. Cell 175:973–983.e14

Eastburn DM, McElrone AJ, Bilgin DD (2011) Influence of atmospheric and climatic change on plant–pathogen interactions. Plant Pathol 60:54–69

Egamberdieva D, Eshboev F, Shukurov O, Alaylar B, Arora NK (2023) Bacterial bioprotectants: biocontrol traits and induced resistance to phytopathogens. Microbiol Res 14:689–703

El-Saadony MT, Saad AM, Soliman SM et al (2022) Plant growth-promoting microorganisms as biocontrol agents of plant diseases: mechanisms, challenges and future perspectives. Front Plant Sci 13:923880

Emwas A-HM (2015) The strengths and weaknesses of nmr spectroscopy and mass spectrometry with particular focus on metabolomics research. Methods Mol Biol 1277:161–193

Erb M, Kliebenstein DJ (2020) Plant secondary metabolites as defenses, regulators, and primary metabolites: the blurred functional trichotomy. Plant Physiol 184:39–52

Fadiji AE, Babalola OO (2020a) Elucidating mechanisms of endophytes used in plant protection and other bioactivities with multifunctional prospects. Front Bioeng Biotechnol 8:467

Fadiji AE, Babalola OO (2020b) Exploring the potentialities of beneficial endophytes for improved plant growth. Saudi J Biol Sci 27:3622–3633

Farag MA, Zayed A, Sallam IE, Abdelwareth A, Wessjohann LA (2022) Metabolomics-based approach for coffee beverage improvement in the context of processing, brewing methods, and quality attributes. Food Secur 11:864

Fenta L, Mekonnen H (2024) Microbial biofungicides as a substitute for chemical fungicides in the control of phytopathogens: current perspectives and research directions. Scientifica (Cairo) 2024:1–12

Fiehn O (2016) Metabolomics by gas chromatography–mass spectrometry: combined targeted and untargeted profiling. Curr Protoc Mol Biol. https://doi.org/10.1002/0471142727.mb3004s114

Gamalero E, Bona E, Glick BR (2022) Current techniques to study beneficial plant-microbe interactions. Microorganisms 10:1380

Gao Y, Hou L, Gao J, Li D, Tian Z, Fan B, Wang F, Li S (2021) Metabolomics approaches for the comprehensive evaluation of fermented foods: a review. Food Secur 10:2294

Gavriilidou A, Kautsar SA, Zaburannyi N, Krug D, Müller R, Medema MH, Ziemert N (2022) Compendium of specialized metabolite biosynthetic diversity encoded in bacterial genomes. Nat Microbiol 7:726–735

Ghorai AK, Patsa R, Jash S, Dutta S (2021) Microbial secondary metabolites and their role in stress management of plants. In: Biocontrol agents and secondary metabolites. Elsevier, pp 283–319

Gomathi D, Kalaiselvi M, Ravikumar G, Devaki K, Uma C (2015) GC-MS analysis of bioactive compounds from the whole plant ethanolic extract of *Evolvulus alsinoides* (L.) L. J Food Sci Tech 52:1212–1217

González-Peña D, Dudzik D, Colina-Coca C, de Ancos B, García A, Barbas C, Sánchez-Moreno C (2015) Evaluation of onion as a functional ingredient in the prevention of metabolic impairments associated to diet-induced hypercholesterolaemia using a multiplatform approach based on LC-MS, CE-MS and GC-MS. J Funct Foods 19:363–375

Gu Q, Yang Y, Yuan Q, Shi G, Wu L, Lou Z, Huo R, Wu H, Borriss R, Gao X (2017) Bacillomycin D produced by *bacillus amyloliquefaciens* is involved in the antagonistic interaction with the plant-pathogenic fungus *Fusarium graminearum*. Appl Environ Microbiol. https://doi.org/10.1128/AEM.01075-17

Guevara-Avendaño E, Bejarano-Bolívar AA, Kiel-Martínez A-L, Ramírez-Vázquez M, Méndez-Bravo A, von Wobeser EA, Sánchez-Rangel D, Guerrero-Analco JA, Eskalen A, Reverchon F (2019) Avocado rhizobacteria emit volatile organic compounds with antifungal activity against

Fusarium solani, *Fusarium* sp. associated with kuroshio shot hole borer, and *Colletotrichum gloeosporioides*. Microbiol Res 219:74–83

Gupta S, Schillaci M, Roessner U (2022) Metabolomics as an emerging tool to study plant–microbe interactions. Emerg Top Life Sci 6:175–183

Halket JM, Waterman D, Przyborowska AM, Patel RKP, Fraser PD, Bramley PM (2005) Chemical derivatization and mass spectral libraries in metabolic profiling by GC/MS and LC/MS/MS. J Exp Bot 56:219–243

Hanif A, Zhang F, Li P et al (2019) Fengycin produced by *Bacillus amyloliquefaciens* FZB42 inhibits *Fusarium graminearum* growth and mycotoxins biosynthesis. Toxins (Basel) 11:295

Harman G, Khadka R, Doni F, Uphoff N (2021) Benefits to plant health and productivity from enhancing plant microbial symbionts. Front Plant Sci 11:610065

Heredia-Ponce Z, Gutiérrez-Barranquero JA, Purtschert-Montenegro G, Eberl L, de Vicente A, Cazorla FM (2021) Role of extracellular matrix components in the formation of biofilms and their contribution to the biocontrol activity of *Pseudomonas chlororaphis PCL1606*. Environ Microbiol 23:2086–2101

Hill CB, Roessner U (2013) Metabolic profiling of plants by GC–MS. In: The handbook of plant metabolomics. Wiley, pp 1–23

Hirsch PR, Van Montagu M, Johnston AWB, Brewin NJ, Schell J (1980) Physical identification of bacteriocinogenic, nodulation and other plasmids in strains of *Rhizobium leguminosarum*. Microbiology 120:403–412

Hua L, Zeng H, He L, Jiang Q, Ye P, Liu Y, Sun X, Zhang M (2021) Gliotoxin is an important secondary metabolite involved in suppression of *Sclerotium rolfsii* of *Trichoderma virens* T23. Phytopathology 111:1720–1725

Ilham B, Noureddine C, Philippe G, Mohammed EG, Brahim E, Sophie A, Martine N, Muriel M (2019) Induced systemic resistance (ISR) in *Arabidopsis thaliana* by *Bacillus amyloliquefaciens* and *Trichoderma harzianum* used as seed treatments. Agriculture 9:166

Islam T, Haque MA, Barai HR, Istiaq A, Kim J-J (2024) Antibiotic resistance in plant pathogenic bacteria: recent data and environmental impact of unchecked use and the potential of biocontrol agents as an eco-friendly alternative. Plan Theory 13:1135

Jamalizadeh M, Etebarian HR, Aminian H, Alizadeh A (2011) A review of mechanisms of action of biological control organisms against post-harvest fruit spoilage. EPPO Bull 41:65–71

Janisiewicz WJ, Tworkoski TJ, Sharer C (2000) Characterizing the mechanism of biological control of postharvest diseases on fruits with a simple method to study competition for nutrients. Phytopathology 90:1196–1200

Jeger M, Beresford R, Bock C, Brown N, Fox A, Newton A, Vicent A, Xu X, Yuen J (2021) Global challenges facing plant pathology: multidisciplinary approaches to meet the food security and environmental challenges in the mid-twenty-first century. CABI Agric Biosci 2:20

Ji C, Zhang M, Kong Z, Chen X, Wang X, Ding W, Lai H, Guo Q (2021) Genomic analysis reveals potential mechanisms underlying promotion of tomato plant growth and antagonism of soilborne pathogens by *Bacillus amyloliquefaciens* Ba13. Microbiol Spectr 9:e01615–e01621

Jin P, Wang H, Tan Z, Xuan Z, Dahar GY, Li QX, Miao W, Liu W (2020) Antifungal mechanism of bacillomycin D from *Bacillus velezensis* HN-2 against *Colletotrichum gloeosporioides* Penz. Pestic Biochem Physiol 163:102–107

Jogaiah S, Abdelrahman M, Tran LP, Ito S (2018) Different mechanisms of *Trichoderma virens*-mediated resistance in tomato against *Fusarium* wilt involve the jasmonic and salicylic acid pathways. Mol Plant Pathol 19:870–882

Jorge TF, Rodrigues JA, Caldana C, Schmidt R, van Dongen JT, Thomas-Oates J, António C (2016) Mass spectrometry-based plant metabolomics: metabolite responses to abiotic stress. Mass Spectrom Rev 35:620–649

Kamle M, Borah R, Bora H, Jaiswal AK, Singh RK, Kumar P (2020) Systemic acquired resistance (SAR) and induced systemic resistance (ISR): role and mechanism of action against phytopathogens. In: Fungal biotechnology and bioengineering. Springer, Cham, pp 457–470

Karmegham N, Vellasamy S, Natesan B, Sharma MP, Al Farraj DA, Elshikh MS (2020) Characterization of antifungal metabolite phenazine from rice rhizosphere fluorescent pseudomonads (FPs) and their effect on sheath blight of rice. Saudi J Biol Sci 27:3313–3326

Kashyap AS, Manzar N, Nebapure SM, Rajawat MVS, Deo MM, Singh JP, Kesharwani AK, Singh RP, Dubey SC, Singh D (2022) Unraveling microbial volatile elicitors using a transparent methodology for induction of systemic resistance and regulation of antioxidant genes at expression levels in chili against bacterial wilt disease. Antioxidants 11:404

Kataoka R, Akashi M, Taniguchi T, Kinose Y, Yaprak A, Turgay O (2021) Metabolomics analyses reveal metabolites affected by plant growth-promoting endophytic bacteria in roots of the halophyte *Mesembryanthemum crystallinum*. Int J Mol Sci 22:11813

Kaur R, Kaur J, Singh R (2010) Nonpathogenic *Fusarium* as a biological control agent. Plant Pathol J 9:79–91

Kellogg J, Kang S (2020) Metabolomics, an essential tool in exploring and harnessing microbial chemical ecology. Phytobiomes J 4:195–210

Khan RAA, Najeeb S, Hussain S, Xie B, Li Y (2020) Bioactive secondary metabolites from *Trichoderma* spp. against phytopathogenic fungi. Microorganisms 8:817

Khoso MA, Wagan S, Alam I, Hussain A, Ali Q, Saha S, Poudel TR, Manghwar H, Liu F (2024) Impact of plant growth-promoting rhizobacteria (PGPR) on plant nutrition and root characteristics: current perspective. Plant Stress 11:100341

Kim HM, Kang JS (2021) Metabolomic studies for the evaluation of toxicity induced by environmental toxicants on model organisms. Meta 11:485

Klein MN, Kupper KC (2018) Biofilm production by *Aureobasidium pullulans* improves biocontrol against sour rot in citrus. Food Microbiol 69:1–10

Knights HE, Jorrin B, Haskett TL, Poole PS (2021) Deciphering bacterial mechanisms of root colonization. Environ Microbiol Rep 13:428–444

Koek MM, Jellema RH, van der Greef J, Tas AC, Hankemeier T (2011) Quantitative metabolomics based on gas chromatography mass spectrometry: status and perspectives. Metabolomics 7:307–328

Köhl J, Kolnaar R, Ravensberg WJ (2019) Mode of action of microbial biological control agents against plant diseases: relevance beyond efficacy. Front Plant Sci 10:845

Koščak L, Lamovšek J, Đermić E, Prgomet I, Godena S (2023) Microbial and plant-based compounds as alternatives for the control of phytopathogenic bacteria. Horticulturae 9:1124

Koza N, Adedayo A, Babalola O, Kappo A (2022) Microorganisms in plant growth and development: roles in abiotic stress tolerance and secondary metabolites secretion. Microorganisms 10:1528

Kumar BR (2017) Application of HPLC and ESI-MS techniques in the analysis of phenolic acids and flavonoids from green leafy vegetables (GLVs). J Pharm Anal 7:349–364

Kumar R, Bohra A, Pandey AK, Pandey MK, Kumar A (2017) Metabolomics for plant improvement: status and prospects. Front Plant Sci 8:1302

Lahlali R, Ezrari S, Radouane N, Kenfaoui J, Esmaeel Q, El Hamss H, Belabess Z, Barka EA (2022) Biological control of plant pathogens: a global perspective. Microorganisms 10:596

Lakhdari W, Benyahia I, Bouhenna MM et al (2023) Exploration and evaluation of secondary metabolites from *Trichoderma harzianum*: GC-MS analysis, phytochemical profiling, antifungal and antioxidant activity assessment. Molecules 28:5025

Lee B, Farag MA, Park HB, Kloepper JW, Lee SH, Ryu C-M (2012) Induced resistance by a long-chain bacterial volatile: elicitation of plant systemic defense by a C13 volatile produced by *Paenibacillus polymyxa*. PLoS One 7:e48744

Li R, Zhang H, Liu W, Zheng X (2011) Biocontrol of postharvest gray and blue mold decay of apples with *Rhodotorula mucilaginosa* and possible mechanisms of action. Int J Food Microbiol 146:151–156

Li S, Xu C, Wang J, Guo B, Yang L, Chen J, Ding W (2017) Cinnamic, myristic and fumaric acids in tobacco root exudates induce the infection of plants by *Ralstonia solanacearum*. Plant Soil 412:381–395

Liu P, Luo L, Long C (2013) Characterization of competition for nutrients in the biocontrol of *Penicillium italicum* by *Kloeckera apiculata*. Biol Control 67:157–162

Liu R, Bao Z-X, Zhao P-J, Li G-H (2021) Advances in the study of metabolomics and metabolites in some species interactions. Molecules 26:3311

Liu L, Jin X, Lu X, Guo L, Lu P, Yu H, Lv B (2023) Mechanisms of surfactin from B*acillus subtilis* SF1 against *Fusarium foetens*: a novel pathogen inducing potato wilt. J Fungi (Basel) 9:367

Long Y, Zhang W, Wang F, Chen Z (2014) Simultaneous determination of three curcuminoids in *Curcuma longa* L. by high performance liquid chromatography coupled with electrochemical detection. J Pharm Anal 4:325–330

Maciag T, Kozieł E, Rusin P, Otulak-Kozieł K, Jafra S, Czajkowski R (2023) Microbial consortia for plant protection against diseases: more than the sum of its parts. Int J Mol Sci 24:12227

Malviya D, Sahu PK, Singh UB et al (2020) Lesson from ecotoxicity: revisiting the microbial lipopeptides for the management of emerging diseases for crop protection. Int J Environ Res Public Health 17:1434

Manganyi MC, Regnier T, Tchatchouang C-DK, Bezuidenhout CC, Ateba CN (2019) Antibacterial activity of endophytic fungi isolated from *Sceletium tortuosum* L. (Kougoed). Ann Microbiol 69:659–663

Manickam S, Rajagopalan VR, Kambale R, Rajasekaran R, Kanagarajan S, Muthurajan R (2023) Plant metabolomics: current initiatives and future prospects. Curr Issues Mol Biol 45:8894–8906

Marchev AS, Vasileva LV, Amirova KM, Savova MS, Balcheva-Sivenova ZP, Georgiev MI (2021) Metabolomics and health: from nutritional crops and plant-based pharmaceuticals to profiling of human biofluids. Cell Mol Life Sci 78:6487–6503

Márquez DGP, Munaro D, Uarrota VG (2022) Metabolomics as a tool to study volatile organic compounds produced by plant growth-promoting rhizobacteria. In: Secondary metabolites and volatiles of PGPR in plant-growth promotion. Springer, pp 21–44

Maruyama CR, Bilesky-José N, de Lima R, Fraceto LF (2020) Encapsulation of *Trichoderma harzianum* preserves enzymatic activity and enhances the potential for biological control. Front Bioeng Biotechnol 8:225

Mazzei P, Vinale F, Woo SL, Pascale A, Lorito M, Piccolo A (2016) Metabolomics by proton high-resolution magic-angle-spinning nuclear magnetic resonance of tomato plants treated with two secondary metabolites isolated from *Trichoderma*. J Agric Food Chem 64:3538–3545

Meena KR, Kanwar SS (2015) Lipopeptides as the antifungal and antibacterial agents: applications in food safety and therapeutics. Biomed Res Int 2015:1–9

Mendoza JLH, Pérez MIS, Prieto JMG, Velásquez JDQ, Olivares JGG, Langarica HRG (2015) Antibiosis of *Trichoderma* spp strains native to northeastern Mexico against the pathogenic fungus *Macrophomina phaseolina*. Braz J Microbiol 46:1093–1101

Messa VR (2021) Biocontrol by induced systemic resistance using plant growth promoting rhizobacteria. Rhizosphere 17:100323

Metz TO, Zhang Q, Page JS, Shen Y, Callister SJ, Jacobs JM, Smith RD (2007) Future of liquid chromatography–mass spectrometry in metabolic profiling and metabolomic studies for biomarker discovery. Biomark Med 1:159–185

Mhlongo MI, Piater LA, Dubery IA (2022) Profiling of volatile organic compounds from four plant growth-promoting rhizobacteria by SPME–GC–MS: a metabolomics study. Meta 12:763

Milner JL, Silo-Suh L, Lee JC, He H, Clardy J, Handelsman J (1996) Production of kanosamine by *Bacillus cereus* UW85. Appl Environ Microbiol 62:3061–3065

Mironenka J, Różalska S, Bernat P (2021) Potential of *Trichoderma harzianum* and its metabolites to protect wheat seedlings against *Fusarium culmorum* and 2,4-D. Int J Mol Sci 22:13058

Mishra J, Arora NK (2018) Secondary metabolites of fluorescent pseudomonads in biocontrol of phytopathogens for sustainable agriculture. Appl Soil Ecol 125:35–45

Mizzi L, Chatzitzika C, Gatt R, Valdramidis V (2020) HPLC analysis of phenolic compounds and flavonoids with overlapping peaks. Food Technol Biotechnol 58:1–12

Moretti LG, Crusciol CAC, Bossolani JW, Momesso L, Garcia A, Kuramae EE et al (2020) Bacterial consortium and microbial metabolites increase grain quality and soybean yield. J Soil Sci Plant Nutr 20:1923–1934

Mukhopadhyay R, Kumar D (2020) *Trichoderma*: a beneficial antifungal agent and insights into its mechanism of biocontrol potential. Egypt J Biol Pest Control 30:133

Nadarajah KK (2016) Induced systemic resistance in rice. In: Microbial-mediated induced systemic resistance in plants. Springer, Singapore, pp 103–124

Narayanan Z, Glick BR (2022) Secondary metabolites produced by plant growth-promoting bacterial endophytes. Microorganisms 10:2008

Nicoletti R, Andolfi A, Becchimanzi A, Salvatore MM (2023) Anti-insect properties of *Penicillium* secondary metabolites. Microorganisms 11:1302

Niu D-D, Liu H-X, Jiang C-H, Wang Y-P, Wang Q-Y, Jin H-L, Guo J-H (2011) The plant growth–promoting rhizobacterium *Bacillus cereus* AR156 induces systemic resistance in *Arabidopsis thaliana* by simultaneously activating salicylate- and jasmonate/ethylene-dependent signaling pathways. Mol Plant Microbe 24:533–542

Olanrewaju OS, Glick BR, Babalola OO (2017) Mechanisms of action of plant growth promoting bacteria. World J Microbiol Biotechnol 33:197

Olanrewaju OS, Ayangbenro AS, Glick BR, Babalola OO (2019) Plant health: feedback effect of root exudates-rhizobiome interactions. Appl Microbiol Biotechnol 103:1155–1166

Olanrewaju OS, Glick BR, Babalola OO (2024) Metabolomics-guided utilization of beneficial microbes for climate-resilient crops. Curr Opin Chem Biol 79:102427

Ongena M, Jacques P, Touré Y, Destain J, Jabrane A, Thonart P (2005) Involvement of fengycin-type lipopeptides in the multifaceted biocontrol potential of *Bacillus subtilis*. Appl Microbiol Biotechnol 69:29–38

Orozco-Mosqueda MDC, Santoyo G, Glick BR (2023) Recent advances in the bacterial phytohormone modulation of plant growth. Plan Theory 12:606

Ortiz A, Sansinenea E (2022) The role of beneficial microorganisms in soil quality and plant health. Sustain For 14:5358

Ortíz-Castro R, Contreras-Cornejo HA, Macías-Rodríguez L, López-Bucio J (2009) The role of microbial signals in plant growth and development. Plant Signal Behav 4:701–712

Oszust K, Cybulska J, Frąc M (2020) How do *Trichoderma* genus fungi win a nutritional competition battle against soft fruit pathogens? A report on niche overlap nutritional potentiates. Int J Mol Sci 21:4235

Palyzová A, Svobodová K, Sokolová L, Novák J, Novotný Č (2019) Metabolic profiling of *Fusarium oxysporum* f. sp. *conglutinans* race 2 in dual cultures with biocontrol agents *Bacillus amyloliquefaciens*, *Pseudomonas aeruginosa*, and *Trichoderma harzianum*. Folia Microbiol (Praha) 64:779–787

Pang Z, Chen J, Wang T, Gao C, Li Z, Guo L, Xu J, Cheng Y (2021) Linking plant secondary metabolites and plant microbiomes: a review. Front Plant Sci 12:621276

Parratt SR, Laine A-L (2016) The role of hyperparasitism in microbial pathogen ecology and evolution. ISME J 10:1815–1822

Pawar S, Chaudhari A, Prabha R, Shukla R, Singh DP (2019) Microbial Pyrrolnitrin: natural metabolite with immense practical utility. Biomol Ther 9:443

Pellicciaro M, Padoan E, Lione G, Celi L, Gonthier P (2022) Pyoluteorin produced by the biocontrol agent *Pseudomonas protegens* is involved in the inhibition of *Heterobasidion* species present in Europe. Pathogens 11:391

Perez de Souza L, Alseekh S, Naake T, Fernie A (2019) Mass spectrometry-based untargeted plant metabolomics. Curr Protoc Plant Biol 4:e20100

Pezzatti J, Boccard J, Codesido S, Gagnebin Y, Joshi A, Picard D, González-Ruiz V, Rudaz S (2020) Implementation of liquid chromatography–high resolution mass spectrometry methods for untargeted metabolomic analyses of biological samples: a tutorial. Anal Chim Acta 1105:28–44

Pieterse CMJ, Zamioudis C, Berendsen RL, Weller DM, Van Wees SCM, Bakker PAHM (2014) Induced systemic resistance by beneficial microbes. Annu Rev Phytopathol 52:347–375

Poveda J (2021) Beneficial effects of microbial volatile organic compounds (MVOCs) in plants. Appl Soil Ecol 168:104118

Preecha C, Sadowsky MJ, Prathuangwong S (2010) Lipopeptide surfactin produced by *Bacillus amyloliquefaciens* KPS46 is required for biocontrol efficacy against *Xanthomonas axonopodis* pv. *glycines*. Agric Nat Resour 44:84–99

Raaijmakers JM, Mazzola M (2012) Diversity and natural functions of antibiotics produced by beneficial and plant pathogenic bacteria. Annu Rev Phytopathol 50:403–424

Raheel M, Aatif HM, Ali S, Shakeel Q, Ahmad A, Shaheen MR (2022) Biological control activity of biosurfactant against plant pathogens. In: Applications of biosurfactant in agriculture. Elsevier, pp 17–28

Ramarathnam R, Bo S, Chen Y, Fernando WGD, Xuewen G, de Kievit T (2007) Molecular and biochemical detection of fengycin- and bacillomycin D-producing *Bacillus* spp., antagonistic to fungal pathogens of canola and wheat. Can J Microbiol 53:901–911

Razzaq A, Sadia B, Raza A, Khalid Hameed M, Saleem F (2019) Metabolomics: a way forward for crop improvement. Meta 9:303

Roca A, Matilla MA (2023) Microbial antibiotics take the lead in the fight against plant pathogens. Microb Biotechnol 16:28–33

Roca-Couso R, Flores-Félix JD, Rivas R (2021) Mechanisms of action of microbial biocontrol agents against *Botrytis cinerea*. J Fungi 7:1045

Romero F, Cazzato S, Walder F, Vogelgsang S, Bender SF, van der Heijden MGA (2022) Humidity and high temperature are important for predicting fungal disease outbreaks worldwide. New Phytol 234:1553–1556

Rosier A, Pomerleau M, Beauregard PB, Samac DA, Bais HP (2023) Surfactin and Spo0A-dependent antagonism by *Bacillus subtilis* strain UD1022 against *Medicago sativa* phytopathogens. Plan Theory 12:1007

Ruparelia J, Rabari A, Mitra D, Panneerselvam P, Das-mohapatra PK, Jha CK (2022) Efficient applications of bacterial secondary metabolites for management of biotic stress in plants. Plant Stress 6:100125

Ryu C-M, Farag MA, Hu C-H, Reddy MS, Wei H-X, Paré PW, Kloepper JW (2003) Bacterial volatiles promote growth in *Arabidopsis*. Proc Natl Acad Sci 100:4927–4932

Ryu C-M, Farag MA, Hu C-H, Reddy MS, Kloepper JW, Paré PW (2004) Bacterial volatiles induce systemic resistance in *Arabidopsis*. Plant Physiol 134:1017–1026

Samaras A, Roumeliotis E, Ntasiou P, Karaoglanidis G (2021) *Bacillus subtilis* MBI600 promotes growth of tomato plants and induces systemic resistance contributing to the control of soil-borne pathogens. Plan Theory 10:1113

Sansinenea E (2019) *Bacillus* spp.: as plant growth-promoting bacteria. In: Secondary metabolites of plant growth promoting rhizomicroorganisms. Springer, Singapore, pp 225–237

Santos F, Peñaflor MFGV, Paré PW, Sanches PA, Kamiya AC, Tonelli M, Nardi C, Bento JMS (2014) A novel interaction between plant-beneficial rhizobacteria and roots: colonization induces corn resistance against the root herbivore *Diabrotica speciosa*. PLoS One 9:e113280

Saravanakumar K, Fan L, Fu K, Yu C, Wang M, Xia H, Sun J, Li Y, Chen J (2016) Cellulase from *Trichoderma harzianum* interacts with roots and triggers induced systemic resistance to foliar disease in maize. Sci Rep 6:35543

Sarwar A, Hassan MN, Imran M, Iqbal M, Majeed S, Brader G, Sessitsch A, Hafeez FY (2018) Biocontrol activity of surfactin A purified from *Bacillus* NH-100 and NH-217 against rice bakanae disease. Microbiol Res 209:1–13

Schalchli H, Tortella GR, Rubilar O, Parra L, Hormazabal E, Quiroz A (2016) Fungal volatiles: an environmentally friendly tool to control pathogenic microorganisms in plants. Crit Rev Biotechnol 36:144–152

Schulz-Bohm K, Martín-Sánchez L, Garbeva P (2017) Microbial volatiles: small molecules with an important role in intra- and inter-kingdom interactions. Front Microbiol 8:248

Sehim AE, Amin BH, Yosri M, Salama HM, Alkhalifah DH, Alwaili MA, Abd Elghaffar RY (2023) GC-MS analysis, antibacterial, and anticancer activities of *Hibiscus sabdariffa* L. methanolic extract: in vitro and in silico studies. Microorganisms 11:1601

Sellitto VM, Zara S, Fracchetti F, Capozzi V, Nardi T (2021) Microbial biocontrol as an alternative to synthetic fungicides: boundaries between pre- and postharvest applications on vegetables and fruits. Fermentation 7:60

Shahid I, Rizwan M, Mehnaz S (2018) Identification and quantification of secondary metabolites by LC-MS from plant-associated *Pseudomonas aurantiaca* and *Pseudomonas chlororaphis*. Bio Protoc 8:e2702

Shahid I, Han J, Hanooq S, Malik KA, Borchers CH, Mehnaz S (2021a) Profiling of metabolites of *bacillus* spp. and their application in sustainable plant growth promotion and biocontrol. Front Sustain Food Syst 5:605195

Shahid I, Han J, Hardie D, Baig DN, Malik KA, Borchers CH, Mehnaz S (2021b) Profiling of antimicrobial metabolites of plant growth promoting *Pseudomonas* spp. isolated from different plant hosts. 3 Biotech 11:48

Shen S, Zhan C, Yang C, Fernie AR, Luo J (2023) Metabolomics-centered mining of plant metabolic diversity and function: past decade and future perspectives. Mol Plant 16:43–63

Shi M, Chen L, Wang X-W, Zhang T, Zhao P-B, Song X-Y, Sun C-Y, Chen X-L, Zhou B-C, Zhang Y-Z (2012) Antimicrobial peptaibols from *Trichoderma pseudokoningii* induce programmed cell death in plant fungal pathogens. Microbiology 158:166–175

Shulaev V, Cortes D, Miller G, Mittler R (2008) Metabolomics for plant stress response. Physiol Plant 132:199–208

Singh P, Rajput RS, Ram RM, Singh HB (2019) A deeper insight into the symbiotic mechanism of *Rhizobium* spp. from the perspective of secondary metabolism. In: Secondary metabolites of plant growth promoting rhizomicroorganisms. Springer, Singapore, pp 265–291

Singh P, Chauhan PK, Upadhyay SK, Singh RK, Dwivedi P, Wang J, Jain D, Jiang M (2022) Mechanistic insights and potential use of siderophores producing microbes in rhizosphere for mitigation of stress in plants grown in degraded land. Front Microbiol 13:898979

Spago FR, Ishii Mauro CS, Oliveira AG et al (2014) *Pseudomonas aeruginosa* produces secondary metabolites that have biological activity against plant pathogenic *Xanthomonas species*. Crop Prot 62:46–54

Speranza B, Liso A, Russo V, Corbo MR (2020) Evaluation of the potential of biofilm formation of *Bifidobacterium longum* subsp. *infantis* and *Lactobacillus reuteri* as competitive biocontrol agents against pathogenic and food spoilage bacteria. Microorganisms 8:177

Srikamwang C, Onsa NE, Sunanta P, Sangta J, Chanway CP, Thanakkasaranee S, Sommano SR (2023) Role of microbial volatile organic compounds in promoting plant growth and disease resistance in horticultural production. Plant Signal Behav 18:e2227440

Srivastava S, Bist V, Srivastava S, Singh PC, Trivedi PK, Asif MH, Chauhan PS, Nautiyal CS (2016) Unraveling aspects of *Bacillus amyloliquefaciens* mediated enhanced production of rice under biotic stress of *Rhizoctonia solani*. Front Plant Sci 7:587

Sui X, Han X, Cao J, Li Y, Yuan Y, Gou J, Zheng Y, Meng C, Zhang C (2022) Biocontrol potential of *Bacillus velezensis* EM-1 associated with suppressive rhizosphere soil microbes against tobacco bacterial wilt. Front Microbiol 13:940156

Sumbul A, Ansari RA, Rizvi R, Mahmood I (2020) Azotobacter: a potential bio-fertilizer for soil and plant health management. Saudi J Biol Sci 27:3634–3640

Sumner LW, Mendes P, Dixon RA (2003) Plant metabolomics: large-scale phytochemistry in the functional genomics era. Phytochemistry 62:817–836

Sun H, Jiang S, Jiang C, Wu C, Gao M, Wang Q (2021) A review of root exudates and rhizosphere microbiome for crop production. Environ Sci Pollut R 28:54497–54510

Tamariz-Angeles C, Olivera-Gonzales P, Santillán-Torres M, Briceño-Luna V, Silva-Villafana A, Villena GK (2023) Diverse biological activities and secondary metabolites profile of *Penicillium brevicompactum* HE19ct isolated from the high-Andean medicinal plant *Perezia coerulescens*. Fungal Biol 127:1439–1450

Tariq M, Khan A, Asif M, Khan F, Ansari T, Shariq M, Siddiqui MA (2020) Biological control: a sustainable and practical approach for plant disease management. Acta Agric Scand B Soil Plant Sci 70:507–524

Techaoei S, Jirayuthcharoenkul C, Jarmkom K, Dumrongphuttidecha T, Khobjai W (2020) Chemical evaluation and antibacterial activity of novel bioactive compounds from endophytic fungi in *Nelumbo nucifera*. Saudi J Biol Sci 27:2883–2889

Thumanu K, Wongchalee D, Sompong M, Phansak P, Le Thanh T, Namanusart W, Vechklang K, Kaewnum S, Buensanteai N (2017) Synchrotron-based FTIR microspectroscopy of chili resistance induced by *Bacillus subtilis* strain D604 against anthracnose disease. J Plant Interact 12:255–263

Tomah AA, Abd Alamer IS, Li B, Zhang J-Z (2020) A new species of *Trichoderma* and gliotoxin role: a new observation in enhancing biocontrol potential of *T. virens* against *Phytophthora capsici* on chili pepper. Biol Control 145:104261

Tran C, Cock IE, Chen X, Feng Y (2022) Antimicrobial *Bacillus*: metabolites and their mode of action. Antibiotics 11:88

Triplett EW, Barta TM (1987) Trifolitoxin production and nodulation are necessary for the expression of superior nodulation competitiveness by *Rhizobium leguminosarum* bv. *trifolii* strain T24 on clover. Plant Physiol 85:335–342

Trivedi P, Leach JE, Tringe SG, Sa T, Singh BK (2020) Plant–microbiome interactions: from community assembly to plant health. Nat Rev Microbiol 18:607–621

Tyagi S, Lee K-J, Shukla P, Chae J-C (2020) Dimethyl disulfide exerts antifungal activity against *Sclerotinia minor* by damaging its membrane and induces systemic resistance in host plants. Sci Rep 10:6547

Valette M, Rey M, Gerin F, Comte G, Wisniewski-Dyé F (2020) A common metabolomic signature is observed upon inoculation of rice roots with various rhizobacteria. J Integr Plant Biol 62:228–246

van der Velden PMM, Jansen RS (2023) Microbial metabolomics: an overview of applications. In: Metabolomics. Springer, Cham, pp 165–208

Velmurugan S, Ashajyothi M, Charishma K et al (2023) Enhancing defense against rice blast disease: unveiling the role of leaf endophytic firmicutes in antifungal antibiosis and induced systemic resistance. Microb Pathog 184:106326

Vinale F, Sivasithamparam K, Ghisalberti EL et al (2014) *Trichoderma* secondary metabolites active on plants and fungal pathogens. Open Mycol J 8:127–139

Waheed A, Haxim Y, Islam W, Ahmad M, Muhammad M, Alqahtani FM, Hashem M, Salih H, Zhang D (2023) Climate change reshaping plant-fungal interaction. Environ Res 238:117282

Wang H, Yan Y, Wang J, Zhang H, Qi W (2012) Production and characterization of antifungal compounds produced by *Lactobacillus plantarum* IMAU10014. PLoS One 7:e29452

Wang XX, Yu P, Li J (2020) High-throughput metabolomics for identification of metabolic pathways and deciphering the effect mechanism of dioscin on rectal cancer from cell metabolic profiles coupled with chemometrics analysis. Front Pharmacol 11:68

Wu L, Li X, Ma L, Borriss R, Wu Z, Gao X (2018) Acetoin and 2,3-butanediol from *Bacillus amyloliquefaciens* induce stomatal closure in *Arabidopsis thaliana* and *Nicotiana benthamiana*. J Exp Bot 69:5625–5635

Xiao JF, Zhou B, Ressom HW (2012) Metabolite identification and quantitation in LC-MS/MS-based metabolomics. TrAC Trends Anal Chem 32:1–14

Yao X, Guo H, Zhang K, Zhao M, Ruan J, Chen J (2023) *Trichoderma* and its role in biological control of plant fungal and nematode disease. Front Microbiol 14:1160551

Ye D, Li X, Shen J, Xia X (2022) Microbial metabolomics: from novel technologies to diversified applications. TrAC Trends Anal Chem 148:116540

Yu Y, Gui Y, Li Z, Jiang C, Guo J, Niu D (2022) Induced systemic resistance for improving plant immunity by beneficial microbes. Plan Theory 11:386

Yuan J, Li B, Zhang N, Waseem R, Shen Q, Huang Q (2012) Production of Bacillomycin- and Macrolactin-Type antibiotics by *Bacillus amyloliquefaciens* NJN-6 for suppressing soilborne plant pathogens. J Agric Food Chem 60:2976–2981

Yuan M, Huang Y, Ge W, Jia Z, Song S, Zhang L, Huang Y (2019) Involvement of jasmonic acid, ethylene and salicylic acid signaling pathways behind the systemic resistance induced by *Trichoderma longibrachiatum* H9 in cucumber. BMC Genomics 20:144

Zandi P, Schnug E (2022) Reactive oxygen species, antioxidant responses and implications from a microbial modulation perspective. Biology (Basel) 11:15

Zhang Z, Hu X, Li P (2016) Chromatography: combined chromatography and mass spectrometry. In: Encyclopedia of food and health. Elsevier, pp 79–84

Chapter 16
Each to Their Own: X-Omics Understanding of Genotype Specificity to Develop Smart Bioinoculant Formulations

Francesca Vaccaro, Camilla Fagorzi, Iacopo Passeri, and Alessio Mengoni

Abstract Mutualistic interactions between microbes and plants play a crucial rolein plant development, physiology, and stress resistance. The taxonomic and functional characteristics of the plant microbiome are being increasingly understood through various omics techniques (e.g. metagenomics, transcriptomics, metabolomics), including genome-wide association studies and quantitative trait loci mapping also. This emerging knowledge has the potential to revolutionize agricultural practices and breeding programs for novel crops. However, several studies suggest that the desired phenotype in the host plant is often the result of genotype-specific interactions between the partners. The rhizobium-legume symbiosis model has been extensively studied using a range of omics methods, from comparative genomics to transcriptomics, to uncover the genetic basis of these genotype-specific interactions and, in turn, support the development of customized bioinoculants for selected plant varieties.

Keywords Genotypic interaction · Rhizobium · Nitrogen fixation · Synthetic communities · Systems biology · Microbiome

16.1 Comparative Genomics and Epigenomics of Plant Symbiotic Bacteria Matters

The microbiome has been defined as a sort of "second organism", composed of thousands of microbial species that cohabit with their host, such as a plant or an animal. In fact, plants, like animals, are colonized by a surprising number of

F. Vaccaro (✉) · C. Fagorzi · I. Passeri · A. Mengoni
Department of Biology, University of Florence, Florence, Italy
e-mail: francesca.vaccaro@unifi.it

O. O. Babalola, A. S. Ayangbenro (eds.), *Microbial Allies*,
https://doi.org/10.1007/978-3-031-90530-8_16

microorganisms that can far exceed the number of plant cells that make up the plant itself (Hassani et al. 2018). The microorganisms of the microbiome (or microbiota, a term used to refer to the list of species present, regardless of their functions) are found both on the surface of the plant (for example on the leaves and roots) and within the tissues, where they colonize the vessels (xylem and phloem), the intercellular spaces, but sometimes also the inside of plant cells. The level of root colonization is perhaps the most relevant; this is where the majority of microorganisms and the greatest functional and taxonomic diversity (of species) reside. Indeed, the root of plants has been compared, in terms of its associated microbial community, to the intestine of animal species. Consequently, the root system presents some of the most relevant aspects at the microbial level and is the primary focus of studies on interactions between microorganisms and plants.

The plant microbiome has been studied both for its adaptability to environmental conditions and for its positive impact on the growth and health of host plants. In particular, for the microbiome of the rhizosphere (the soil layer surrounding the root) has been extensively studied, with a focus on the arbuscular mycorrhizal fungi (AMF) and the so-called plant growth-promoting bacteria (PGPB), which include both commensals and symbionts. PGPB are microorganisms closely and commonly associated with plants, which, independently of the environmental characteristics, are able to improve their growth by producing phytohormones, mobilizing nutrients present in the soil, providing protection from various stresses (including pathogens) and fixing atmospheric nitrogen (Mendes et al. 2013). Plant domestication and conventional agricultural practices have resulted in a loss of biodiversity in the plant microbiome, negatively impacting the ability of plants to protect and assimilate nutrients (Raaijmakers and Kiers 2022). Bioinoculants are designed to introduce novel microbes into the plant microbiome, aiding in the restoration of biodiversity and consequently enhancing resilience and yield (Ke et al. 2021). Within these many members of the plant microbiome, some are recognized to be key for particular crops and constitute model cases in the study and application of bioinoculants (de Souza et al. 2020; French et al. 2021; Sessitsch et al. 2019; Bellabarba et al. 2019). In particular, the mutualistic relationships between leguminous plants and nitrogen-fixing symbionts, i.e. rhizobia, are among the most relevant and studied models (Checcucci and Marchetti 2020).

This chapter summarizes the insights gained from studying rhizobia, including the significance of their genomic differences and the predictive power of integrating multiple omics data for developing effective bioinoculants.

16.1.1 The Genomic Landscape of Plant Symbiotic Bacteria

The genomic composition of rhizobia significantly affects their ability to form symbiotic relationships with host plants. Comparative genomics enables researchers to examine the genomes of various bacterial strains, pinpointing key genes involved in plant-microbe interactions. A thorough understanding of the genetic diversity among symbiotic bacteria is essential for unravelling the mechanisms that drive host-specific interactions. Rhizobial species demonstrate a remarkable diversity in both phenotype and genome, with individual strains exhibiting significantly varying symbiotic behaviors influenced by host plant genotypes and soil conditions. Furthermore, these species typically possess an open pangenome, characterized by a significant accessory genome containing thousands of genes (Checcucci et al. 2019). Plant symbiotic bacteria usually harbour large genomes (up to nearly 9 Mbp in size), often multipartite, viz. composed by multiple large replicons, as the chromosome, chromids and megaplasmids (Fig. 16.1). These replicons could also be relevant and specialized toward the colonization of different environments, as the soil, the plant rhizosphere and the symbiotic structures of leguminous plants. For example, studies employing comparative genomics, genome manipulation, and genome-scale metabolic model reconstruction have demonstrated that in the *Medicago* nitrogen-fixing symbiont, *Sinorhizobium meliloti*, the three primary genomic replicons (chromosome, approximately 3.6 Mbp in size, the chromid, approximately 1.6 Mbp, and the megaplasmid, approximately 1.3 Mbp) are associated with soil fitness (the chromosome), rhizosphere colonization (the chromid), and nitrogen-fixing symbiosis (the megaplasmid) (diCenzo and Finan 2015; diCenzo et al. 2014, 2016). Moreover, when comparing strains of *S. meliloti* with different symbiotic performances, the sequence and structure of the megaplasmid was shown to be highly different, possibly explaining at least some of the symbiotic

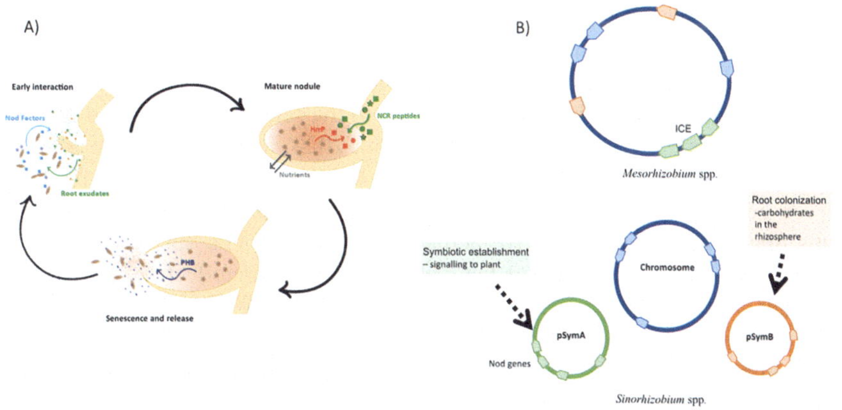

Fig. 16.1 Brief overview of main rhizobia biological features. (**a**) Schematic diagram of step in rhizobia-host plant symbiosis. Starting from an early interaction with the perception of root exudates and the release of Nod factors by rhizobia, there is the formation of an indeterminate root nodule, where control of bacteroid differentiation is modulated by the plant via NCR peptides and bacteria can control via HrrP peptidase. In the root nodule, exchange of nutrients between bacteroids and plants occurs. During senescence of the nodule, bacteria are released from the nodule. Here, the accumulation of polyhydroxybutyrate (PHB) can help rhizobia to survive after nodule dehiscence. (**b**) Genomes of rhizobia. Examples from the single replicon genome of *Mesorhizobium* spp. and the multipreplicon genome structure of *Sinorhizobium* spp. are shown. Integrative and conjugative element (ICE) is reported in green. Partially redrawn from (Cangioli et al. 2022a)

differences (Galardini et al. 2013). Phylogeographic analyses done on a large panel of strains of *S. meliloti* indicated that, in spite of a strong geographical structuring related to the chromosome and the chromid, the symbiotic megaplasmid showed a lack of phylogeographic structuring, indicating high level of horizontal gene transfer and recombination on this element (Riley et al. 2023). A similar pattern of uncoupled phylogenomic diversity has also been observed in other rhizobia, such as as *Rhizobium leguminosarum* and *Mesorhizobium* spp. (Kumar et al. 2015; Jorrin and Imperial 2015). For this latter in particular, genetic determinants of symbiosis are located on a genomic island, known as ICE (Integrative Conjugative Element) which mobilizes among strains, giving rise to novel strains, as was shown after the introduction of nitrogen-fixing inoculants of *Mesorhizobium* in New Zealand, where ICEs from introduced strains mobilized to indigenous strains, lacking ICEs (Colombi et al. 2023). Still concerning the plasticity of the symbiotic part of rhizobial genome, interesting results emerged from strains with multipartite genome structure, as *S. meliloti*. An experiment of genomic manipulation, where the megaplasmid from one strain was mobilized to another strain indicated that such a swapping of the megaplasmid produced a significant difference in symbiotic performance (Checcucci et al. 2018). However, in this experiment the symbiotic performance did not mirror that of the megaplasmid donor strains but was different from that of both parental strains (donor and acceptor strain), indicating that epistatic interactions among genes residing on the three replicons of *S. meliloti* contribute to the overall symbiotic phenotype. Figure 16.2 summarizes the main knowledge derived from comparative genomics and genome-scale manipulation of rhizobia.

As mentioned at the beginning of this paragraph, the formulation of new bioinoculants must rely also on comparative genomics studies of both natural isolates and SynComs, with the aim to identify key genes involved in plant-microbe interactions and PGP traits. Such studies will bring to the selection of the most effective strains from a strain collection (Agrahari et al. 2020) and consequently boost the formulation of consortia or SynComs. For example, recent studies identified conserved genetic determinants, involved in chemotaxis, motility, and secretion systems, that are consistently active across different microbial species during root colonization, paving the way for the development of more effective bioinoculants (Vannier et al. 2023). Moreover, over the past few years, the emergence of gene editing methods, such as CRISPR-Cas9-based tools, enabled the study of the molecular mechanisms of symbiotic nitrogen fixation in both (model and non-model) legume hosts and their microbial partners, thus contributing to the advancing of bioinoculant technology. This technique offers a considerable potential for the refining and the optimization of rhizobial strains to boost nitrogen fixation (Jain et al. 2023). With regard to this point, possible targets could be *nod* gene clusters to strengthen host interactions, as well as *nif* gene clusters to enhance nitrogen-fixing efficiency (Wang et al. 2021) . Recent studies showed that in the nitrogen-fixing bacterium *Azorhizobium caulinodans,* CRISPR-driven gene editing was able to enhance biofilm formation and chemotaxis toward its legume host (Wang et al. 2020). While the release of genetically modified microorganisms into the environment remains restricted in many countries, molecular screening of native strain culture collections

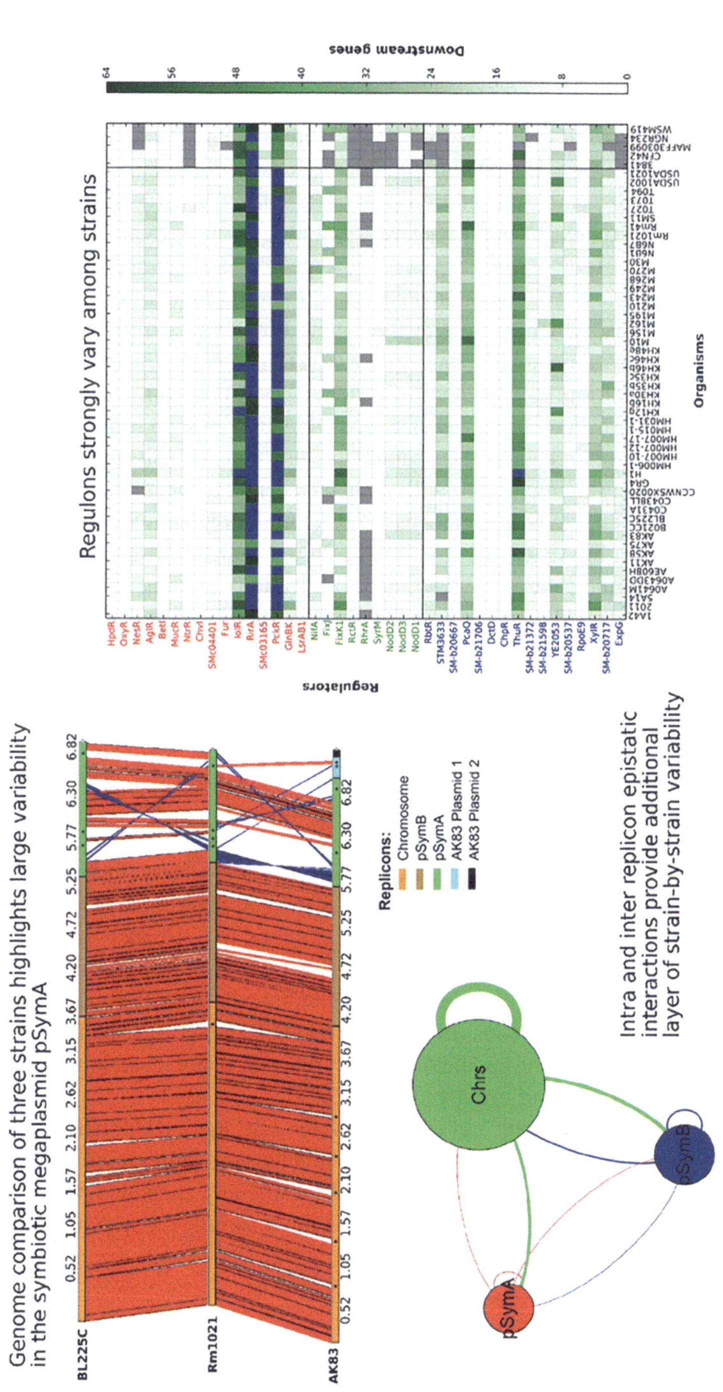

Fig. 16.2 Lessons learned from rhizobia comparative genomics in terms of plant symbiotic performances

offers a viable alternative. While these studies are still in the proof-of-concept stage, they are illuminating potential avenues for a more sustainable and productive agricultural system.Beyond genomics, epigenetic factors are increasingly attracting the attention of researchers, as they may significantly influence plant-microbe interactions. Epigenomic studies demonstrate how environmental signals shape the epigenetic landscape of both plants and bacteria, thereby affecting the establishment and maintenance of symbiotic relationships. The interplay between genetics and epigenetics in the context of plant-bacteria symbiosis could provide a comprehensive understanding of the factors shaping genotype-specific interactions (discussed below). Recent studies have shown that strains of rhizobia have differences in DNA methylation patterns, which can affect their gene flow and then the possibility to exchange symbiotic elements (diCenzo et al. 2022). Indeed, in *S. meliloti*, the presence of different restriction-modification systems was shown to impact the level of gene transfer *in vitro* (Ferri et al. 2010). Since DNA methylation has shown to also affect gene expression in bacteria, we cannot exclude that such an epigenetic modification may impact the interaction with plants. Recent advancements in computational tools have enabled the analysis of bacterial epigenomic features, with applications in rhizobia (Riccardi et al. 2023). In the near future, we anticipate the ability to elucidate the role of epigenomic modifications in plant-microbe interactions, facilitating targeted manipulation and optimization of bioinoculants and their formulation. Formulating bioinoculants is a crucial step that requires ensuring the viability of strains and their optimal application to the crop (Chaudhary and Shukla 2018). A key factor is desiccation resistance, particularly when strains are applied to seeds as a sprayed liquid suspension or powder (Berninger et al. 2018). Epigenomic studies may uncover potential molecular mechanisms that can enhance strain viability and shelf life of bioinoculant formulations.

The insights from comparative genomics and epigenomics reported above can strongly contribute to the development of effective bioinoculants. In fact, it is now possible to use genomic information to identify genetic determinants in the rhizobial genome that are responsible for different symbiotic performances. Researchers can then use this information to either screen rhizobial collections for the presence of such genetic determinants or genetically or genomically modify rhizobial strains to incorporate these determinants. In other words, we can leverage the large pangenome of rhizobial species to specifically extract relevant genomic resources and use them to select or breed novel improved strains.

By identifying key genomic and epigenomic features associated with successful symbiosis, researchers can formulate bioinoculants that enhance plant-microbe interactions, thereby promoting plant growth and resilience. The integration of these findings into bioinoculant formulations represents a significant step towards sustainable and efficient agricultural practices.

16.2 Genotype × Genotype Interactions in the Rhizomicrobiome Disclose a Panoply of Genes Related to Plant-Microbe Dialogue and Microbe-Microbe Dialogue

The rhizomicrobiome, as previously discussed, serves as a pivotal mediator of plant-microbe interactions and encompasses a significant portion of the plant microbiome‘s diversity. Such a plant microbiome in large part is recruited from the soil microbiome, however evidence is accumulating on the importance of the seed microbiome for a vertical transmission of microbes (from parents to offspring) (Berg and Raaijmakers 2018; Shade et al. 2017), which can have profound importance and impact on crop biofortification through the production of spontaneously bioinoculated seeds. Regardless of the plant microbiome‘s origin, it's evident that not all soil microbes can interact with plants. The genome manipulation experiment referred above, where reduction of the genome of the rhizobium *S. meliloti* to the sole chromosome was performed, sorted out a strain with identical fitness on soil, but lacking the ability to colonize the plant rhizosphere. To extend this evidence, which indicate that not all soil bacteria can associate with plant, needing a set of genetic determinants allowing interaction with plant, comparative genomic analyses highlighted the presence of several set of genetic and metabolic functions which are overrepresented in plant-associated bacteria (Levy et al. 2018a, b). Such genetic determinants can be considered the conditions necessary and sufficient to establish interaction with plants. Recent studies have also examined plant genomes, identifying loci that facilitate the recruitment of specific microbiomes (Deng et al. 2021; Escudero-Martinez et al. 2022; Oyserman et al. 2022).

Population analyses of rhizobia have revealed that within a single species, diverse strains exist, each with varying capacities to colonize plants and exhibit different symbiotic phenotypes (e.g., symbiotic quality) (Klinger et al. 2016; Batstone et al. 2022). The quality of symbiosis is influenced by both the symbiont and host plant genotypes, demonstrating a quantitative trait where genotype-by-genotype interactions are pivotal (Batstone et al. 2020, 2022; Fagorzi et al. 2021). Understanding the molecular mechanisms underlying genotype-by-genotype interactions within the rhizomicrobiome can offer valuable insights for developing elite bioinoculants tailored to specific plant genotypes (e.g., varieties). Additionally, this knowledge can contribute to our understanding of the complex plant-microbe interactions that inform microbiome-focused breeding programs aimed at re-wilding the plant microbiome (Raaijmakers and Kiers 2022).

Several research have in fact revealed that different plant genotypes exhibit specific responses to microbial communities in the rhizosphere (Wagner 2021). For instance, in wheat and potato different varieties recruit different rhizosphere microbiome in relation to both the variety, the site of cultivation and the specific fertilization regime (Cangioli et al. 2022b, c), highlighting the importance of host genotype x environment interaction in plant-microbiome association. These responses are governed by a complex interplay of genetic and environmental factors.

Genome-wide association studies (GWAS) have been performed on the legume-rhizobium symbiosis, aiming to decipher the quantitative genetic determinants of modulation in symbiotic quality (including symbiotic nodule formation, nodule colonization and plant growth). These studies (Batstone et al. 2022; Epstein et al. 2018; Bellabarba et al. 2021) identified a plethora of possible candidate genes linked to symbiotic quality, which could be further confirmed and exploited for genetic amelioration of bioinoculant rhizobia. For instance, concerning competitiveness for nodule colonization, several candidate loci residing on the symbiotic megaplasmid of *S. meliloti* were found, suggesting that, again, creation of hybrid strains mobilizing the symbiotic megaplasmid can increase the quality of the overall symbiotic phenotype. Furthermore, to highlight the numerous genetic factors influencing the quantitative and genotype-dependent modulation of symbiotic quality in model rhizobial strains, transcriptomic studies have revealed a complex network of genes differentially expressed in response to both rhizobial and host plant genotypes (Fagorzi et al. 2021; Cangioli et al. 2021; Burghardt et al. 2017). For instance, a study comparing three *S. meliloti* strains and three *Medicago sativa* (alfalfa) varieties found that 29% of all differentially expressed genes following incubation with plant root exudates could be statistically attributed to the interaction between strain identity and plant variety (Table 16.1). Similar indication from transcriptomic studies on an involvement of a large set of genes on the interaction taking place in the rhizomicrobiome is also coming from other studies investigating rhizobial-fungal interaction. In a recent study it was demonstrated that genotype-specific molecular communication between rhizobia and *Trichoderma* species can occur, leading to changes in the gene expression of a significant portion of the rhizobial genomes (up to 23.4% of total genes) and resulting in synergistic, neutral, or antagonistic interactions on host plant growth promotion (Vaccaro et al. 2024). Such genetic dissection of fungal-bacteria interactions could strongly help to identify the best combinations of strains for developing a consortium of a SynCom used as bioinoculants.

All these genomic, transcriptomic and metagenomic studies, coupled with powerful statistical analysis of data (e.g. GWAS or QTL mapping) identified genes

Table 16.1 Transcriptomic signatures of genotype x genotype interaction account for many genes

Variable	Number of genes over total *S. meliloti* genes	Percentage of genes
Strain	2028	25.0
Plant variety	87	1.0
Strain x plant variety	2301	29.0
Strain and plant variety	1201	15.0

Results of transcriptomic analysis on *S. meliloti* strains treated with root exudates of three alfalfa varieties. The number of genes and the percentage over total genes whose differential expression is statistically linked to rhizobial strain identity and alfalfa variety are reported (from Fagorzi et al. 2021) related to carbohydrate metabolism, type III secretion system, phytohormones production and phosphorus solubilization (de Souza et al. 2019). Moreover, genes related with the interaction with plant immune system and plant antimicrobial peptides are relevant (Cao et al. 2017). Indeed, nodule cysteine-rich peptides (NCR peptides) produced by the plant in the symbiotic organs are responsible for many of the metabolic and morphological changes observed for symbiotic rhizobial cells (bacteroids) (Nicoud et al. 2021). A host range restriction peptidase (HrrP) was shown to allow modulating *S. meliloti* host range toward different *Medicago* species by specific proteolysis of host-derived NCR peptides (Price et al. 2015). Additionally, different rhizobial secretion systems (type III, type IV and type VI) influence rhizobial host specificity and the number of nodules, possibly modulating the transport of effector proteins into host cells (Nelson and Sadowsky 2015). Finally, other genes include that encoding 1-aminocyclopropane-1-carboxylate (ACC) deaminase, which can affect rhizosphere colonization and modulation of plant ethylene production (Checcucci et al. 2017) and several transport functions (for a more detailed review see (Cangioli et al. 2022a)).

The identification of genotype-by-genotype interactions within the rhizomicrobiome presents exciting opportunities for developing sustainable agricultural practices. Leveraging our understanding of plant and microbial genetics, researchers can develop strategies to enhance plant-microbe interactions, ultimately leading to increased crop yields, nutrient uptake, and resilience to environmental challenges. While omics studies have identified a vast array of genes, suggesting that many genes contribute to small variations in plant-microbe interactions, it's likely that a combination of specific genes is necessary to achieve significant improvements in elite strains. Manipulating such a vast number of genes presents significant challenges, and alternative approaches, such as selecting improved strains through repeated reinoculation under *in vitro* conditions, should be explored (Burghardt et al. 2018).

16.3 Conceiving Novel Bioinoculants

Recognizing the crucial role of genotype-specific interactions in plant-microbe symbiosis highlights the need for customized bioinoculants. Conventional one-size-fits-all approaches may not fully exploit the potential benefits of symbiotic relationships. The quest for novel bioinoculants should involves integrating insights from comparative genomics, epigenomics, and rhizomicrobiome studies as well as developing and exploiting culture collections (Fagorzi et al. 2023). Advances in genetic engineering and synthetic biology provide the tools to engineer bioinoculants with increased specificity for particular plant genotypes (Vaccaro et al. 2022). Targeted manipulation of microbial and plant genes involved in symbiosis allows for the development of bioinoculants that can establish robust and beneficial interactions with specific crops. Experimental evolution experiments can also be performed to select strains with desirable improved features.

Particular attention should be given to non-domesticated species or non-model crops. These in fact offer the opportunity to isolate a larger microbial diversity, which may include novel PGPB effective on crops. Ideally, research priorities should be given considering plant species for which close relatives as model species exist, to allow an easier translation of methods and knowledge gained from such model species to the non-model one and, vice versa, use the microbiome from non-model species to model (and cash) crops. For instance, recently a collection of bacterial endophytes from the non-model African rice (*Oryza glaberrima*) was screened and several strains shown to be effective in improving plant growth and performing nitrogen fixation on the widely cultivated Asian rice (*Oryza sativa*) (Bianco et al. 2021).

The exploitation of non-model plant species and the integration of x-omics data, including genomics, epigenomics, transcriptomics, and metagenomics, provides a holistic understanding of the factors influencing genotype-specific interactions and pave the way for rational design of bioinoculants (Fig. 16.3). This wealth of information guides the formulation of bioinoculants, ensuring that the selected microbial consortia and genetic modifications align with the specific requirements of the target plant genotype. However, the bottleneck is in the interpretation and then use of such a wealth of information. Systems biology has emerged as an effective top-down approach for designing, predicting, and manipulating the behaviour of microbial populations. This approach can allow to rationally design novel bioinoculant strains and assemble microbial consortia, or even synthetic communities (SynCom), giving the opportunity to provide a framework to predict how the application of strains or SynComs can affect crop health and yield. One of the most popular

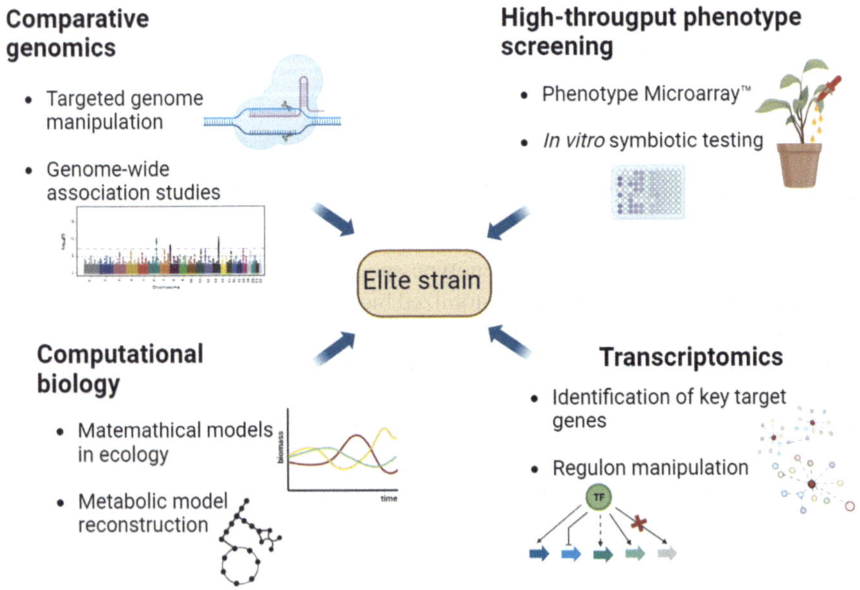

Fig. 16.3 X-omics approaches and modelling to develop élite bioinoculants

applications of a systems biology approach to plant-microbe interaction are genome-scale reconstructions (GENRE), which coupled with simulate of metabolic fluxes (e.g. with flux balance analysis, FBA), allow to predict the biomass produced by a microbial culture or by a symbiotic partnership. An *in silico* representation of the integrated metabolism of the symbiotic partnership was proposed few years ago for *Medicago truncatula-Sinorhizobium meliloti* symbiosis (diCenzo et al. 2020). This metabolic reconstruction allowed to predict the maximum level of nitrogen-fixation which can be sustained by the plant without detrimental effects (i.e. growth reduction due to excels of nutrient allocation to the symbiotic structures).

Of course, bacterial genome-wide association analysis (GWAS) is another tool as well as model of epistasis which can be applied to define higher order interactions among strains and between strains and host plant (Ludington 2022) and experimental evolution approaches seem very promising in producing novel strains with the desired improved abilities. These approaches will also have the advantages in term or regulation, since they are not genetically modified with exogenous genes. Creation of hybrid strains in species with multipartite genome structure is another alternative, which complies with current regulations on GMO, since it involves directed conjugation of megaplasmids between strains of the same species.

However, we should also consider that an ideal SynCom or a single élite strain, once inoculated in the soil, should grow, and persist in presence of the indigenous microbiome and local environmental conditions. Integrating community ecology theory with systems biology into plant microbiome research will then be needed to develop a more holistic and mechanistic synthesis in plant microbiome research (Dini-Andreote and Raaijmakers 2018).

These methods and models are valuable for translating findings from simplified and controlled plant model systems (e.g., in vitro or greenhouse studies) to non-model species, including emerging crops and field conditions. In this context, the identification of a core plant microbiome (Riva et al. 2022) and the recognition of cultivar-specific response of microbiomes are key to the development of a good predictive model and then a bioinoculant preparation, as well as a precise understanding of strain dynamics, compatibility, and persistence in the synthetic community.

Novel efforts should be devoted to overcoming the limiting factor of the culturability of strains of the microbiome, which affect their investigation and use. Strategies of culturomics, by growing microbes on plant-derived substrates, but also by growing as consortia 'per se' and not as pure cultures only, should be implemented to increase the culturable fraction of the plant-associated microbiome (Mapelli et al. 2023).

16.4 Conclusion

The progression from comparative genomics and epigenomics to understanding genotype-by-genotype interactions within the rhizomicrobiome has laid a strong foundation for designing innovative bioinoculants. The integration of omics techniques ("x-omics") has unravelled the intricacies of plant-microbe interactions, shedding light on the genetic (and in future possibly epigenetic also) factors that govern symbiotic relationships. The prospect of developing tailored bioinoculants holds great promise for sustainable agriculture, offering a targeted approach to enhance crop productivity and resilience. The impact of management practices and environmental factors on plant-microbe interactions, and consequently the effectiveness of bioinoculants, requires further attention and rigorous investigation to be integrated into omics-based modelling. Systems biology methods can aid in predicting the behaviour of bioinoculants and translating findings from model species and laboratory experiments to both model and non-model crops, contributing to food security (Vaccaro et al. 2022). As we delve into the complexities of genotype specificity, ongoing research and technological advancements will drive the development of bioinoculant formulations, fostering a more resilient and productive global agricultural system. Despite the promising advancements in bioinoculant development, challenges such as environmental impact assessments and regulatory compliance persist. Future research should prioritize addressing these challenges while continuously refining and optimizing genotype-specific bioinoculants for widespread agricultural adoption.

References

Agrahari RK, Singh P, Koyama H, Panda SK (2020) Plant-microbe interactions for sustainable agriculture in the postgenomic era. Curr Genomics 21(3):168–178

Batstone RT, O'Brien AM, Harrison TL, Frederickson ME (2020) Experimental evolution makes microbes more cooperative with their local host genotype. Science 370(6515):476–478

Batstone RT, Lindgren H, Allsup CM, Goralka LA, Riley AB, Grillo MA et al (2022) Genome-wide association studies across environmental and genetic contexts reveal complex genetic architecture of symbiotic extended phenotypes. mBio 13(6):e01823–e01822

Bellabarba A, Fagorzi C, George C, Pini F, Viti C, Checcucci A (2019) Deciphering the symbiotic plant microbiome : translating the Most recent discoveries on rhizobia for the improvement of agricultural practices in metal-contaminated and high saline lands. Agronomy 9:529

Bellabarba A, Bacci G, Decorosi F, Aun E, Azzarello E, Remm M et al (2021) Competitiveness for nodule colonization in sinorhizobium meliloti: combined in vitro-tagged strain competition and genome-wide association analysis. MSystems 6(4):10–1128

Berg G, Raaijmakers JM (2018) Saving seed microbiomes. ISME J 12(5):1167

Berninger T, González López Ó, Bejarano A, Preininger C, Sessitsch A (2018) Maintenance and assessment of cell viability in formulation of non-sporulating bacterial inoculants. Microb Biotechnol 11(2):277–301

Bianco C, Andreozzi A, Romano S, Fagorzi C, Cangioli L, Prieto P et al (2021) Endophytes from African rice (*Oryza glaberrima* L .) efficiently colonize Asian rice (*Oryza sativa* L .) stimulating the activity of its antioxidant enzymes and increasing the content of nitrogen , carbon , and chlorophyll. Microorganisms 9:1714

Burghardt LT, Guhlin J, Chun CL, Liu J, Sadowsky MJ, Stupar RM et al (2017) Transcriptomic basis of genome by genome variation in a legume-rhizobia mutualism. Mol Ecol 26(21):6122–6135

Burghardt LT, Epstein B, Guhlin J, Nelson MS, Taylor MR, Young ND et al (2018) Select and resequence reveals relative fitness of bacteria in symbiotic and free-living environments. PNAS 115(10):2425–2430

Cangioli L, Checcucci A, Mengoni A, Fagorzi C (2021) Legume tasters: symbiotic rhizobia host preference and smart inoculant formulations. Biol Communic 66(1):47–54
Cangioli L, Vaccaro F, Fini M, Mengoni A (2022a) Scent of a symbiont : the personalized genetic relationships of rhizobium-plant interaction. Int J Mol Sci 23:3358
Cangioli L, Mancini M, Napoli M, Fagorzi C, Orlandini S, Vaccaro F (2022b) Differential response of wheat rhizosphere bacterial community to plant variety and fertilization. Int J Mol Sci 23(7):3616
Cangioli L, Mancini M, Baldi A, Fagorzi C, Orlandini S, Vaccaro F et al (2022c) Effect of site and Phenological status on the potato bacterial rhizomicrobiota. Microorganisms 10(9):1743
Cao Y, Halane MK, Gassmann W, Stacey G (2017) The role of plant innate immunity in the legume-rhizobium symbiosis. Annu Rev Plant Biol 68(1):535–561
Chaudhary T, Shukla P (2018) Bioinoculant capability enhancement through metabolomics and systems biology approaches. Brief Funct Genomics 18(3):159–168
Checcucci A, Marchetti M (2020) The rhizosphere talk show: the rhizobia on stage. Front Agron 2:591494
Checcucci A, Azzarello E, Bazzicalupo M, De Carlo A, Emiliani G, Mancuso S et al (2017) Role and regulation of ACC deaminase gene in *Sinorhizobium meliloti*: is it a symbiotic, rhizospheric or endophytic gene? Front Genet 8:6
Checcucci A, DiCenzo GC, Ghini V, Bazzicalupo M, Becker A, Decorosi F et al (2018) Creation and characterization of a genomically hybrid strain in the nitrogen-fixing symbiotic bacterium *Sinorhizobium meliloti*. ACS Synth Biol [Internet] 7(10):2365–2378. https://doi.org/10.1021/acssynbio.8b00158
Checcucci A, diCenzo GC, Perrin E, Bazzicalupo M, Mengoni A (2019) Genomic diversity and evolution of rhizobia. In: Microbial diversity in the genomic era. Academic, pp 37–46
Colombi E, Hill Y, Lines R, Sullivan JT, Kohlmeier MG, Christophersen CT et al (2023) Population genomics of Australian indigenous *Mesorhizobium* reveals diverse nonsymbiotic genospecies capable of nitrogen-fixing symbioses following horizontal gene transfer. Microb Genom 9(1):1–14
de Souza RSC, Armanhi JSL, de Damasceno NB, Imperial J, Arruda P (2019) Genome sequences of a plant beneficial synthetic bacterial community reveal genetic features for successful plant colonization. Front Microbiol 10:1779
de Souza RSC, Armanhi JSL, Arruda P (2020) From microbiome to traits: designing synthetic microbial communities for improved crop resiliency. Front Plant Sci 11:1179
Deng S, Caddell DF, Xu G, Dahlen L, Washington L, Yang J et al (2021) Genome wide association study reveals plant loci controlling heritability of the rhizosphere microbiome. ISME J 15(11):3181–3194
diCenzo GC, Finan TM (2015) Genetic redundancy is prevalent within the 6.7 Mb *Sinorhizobium meliloti* genome. Mol Genet Genomics 290(4):1345–1356
diCenzo GC, MacLean AM, Milunovic B, Golding GB, Finan TM (2014) Examination of prokaryotic multipartite genome evolution through experimental genome reduction. PLoS Genet 10(10):e1004742
diCenzo GC, Checcucci A, Bazzicalupo M, Mengoni A, Viti C, Dziewit L et al (2016) Metabolic modelling reveals the specialization of secondary replicons for niche adaptation in *Sinorhizobium meliloti*. Nat Commun 7:12219
diCenzo GC, Tesi M, Pfau T, Mengoni A, Fondi M (2020) Genome-scale metabolic reconstruction of the symbiosis between a leguminous plant and a nitrogen-fixing bacterium. Nat Commun 11:2574
diCenzo GC, Cangioli L, Nicoud Q, Cheng JHT, Blow MJ, Sahpiro N et al (2022) Methylation in *Ensifer* species during free-living growth and during nitrogen-fixing symbiosis with *Medicago* spp. mSystems 7(1):e01092–e01021
Dini-Andreote F, Raaijmakers JM (2018) Embracing community ecology in plant microbiome research. Trends Plant Sci 23(6):467–469

Epstein B, Abou-Shanab RAI, Shamseldin A, Taylor MR, Guhlin J, Burghardt LT et al (2018) Genome-wide association analyses in the model rhizobium *Ensifer meliloti*. mSphere 3(5):10–1128

Escudero-Martinez C, Coulter M, Alegria Terrazas R, Foito A, Kapadia R, Pietrangelo L et al (2022) Identifying plant genes shaping microbiota composition in the barley rhizosphere. Nat Commun 13(1):1–14

Fagorzi C, Bacci G, Huang R, Cangioli L, Checcucci A, Fini M et al (2021) Nonadditive transcriptomic signatures of genotype-by-genotype interactions during the initiation of plant-rhizobium symbiosis. mSystems 6(1):10–1128

Fagorzi C, Passeri I, Cangioli L, Vaccaro F, Mengoni A (2023) When biodiversity preservation meets biotechnology: the challenge of developing synthetic microbiota for resilient sustainable crop production. J Sustain Agric Environ 2(1):5–15

Ferri L, Gori A, Biondi EG, Mengoni A, Bazzicalupo M (2010) Plasmid electroporation of *Sinorhizobium* strains: the role of the restriction gene *hsdR* in type strain Rm1021. Plasmid 63(3):128–135

French E, Kaplan I, Iyer-Pascuzzi A, Nakatsu CH, Enders L (2021) Emerging strategies for precision microbiome management in diverse agroecosystems. Nat Plants 7:256–267

Galardini M, Pini F, Bazzicalupo M, Biondi EG (2013) Replicon-dependent bacterial genome evolution: the case of *Sinorhizobium meliloti*. Genome Biol Evol 5(3):542–558. https://doi.org/10.1093/gbe/evt027

Hassani MA, Durán P, Hacquard S (2018) Microbial interactions within the plant holobiont. Microbiome 6(1):58

Jain D, Jones L, Roy S (2023) Gene editing to improve legume-rhizobia symbiosis in a changing climate. Curr Opin Plant Biol 71:102324

Jorrin B, Imperial J (2015) Population genomics analysis of legume host preference for specific rhizobial genotypes in the *Rhizobium leguminosarum* bv. *viciae* symbioses. Mol Plant-Microbe 28(3):310–318

Ke J, Wang B, Yoshikuni Y (2021) Microbiome engineering: synthetic biology of plant-associated microbiomes in sustainable agriculture. Trends Biotechnol 39(3):244–261

Klinger CR, Lau JA, Heath KD (2016) Ecological genomics of mutualism decline in nitrogen-fixing bacteria. PNAS B Biol Sci 283:1826

Kumar N, Lad G, Giuntini E, Kaye ME, Udomwong P, Shamsani NJ et al (2015) (2015) bacterial genospecies that are not ecologically coherent: population genomics of *rhizobium leguminosarum*. Open Biol 5(1):140133

Levy A, Conway JM, Dangl JL, Woyke T (2018a) Elucidating bacterial gene functions in the plant microbiome. Cell Host Microbe 24(4):475–485

Levy A, Salas Gonzalez I, Mittelviefhaus M, Clingenpeel S, Herrera Paredes S, Miao J et al (2018b) Genomic features of bacterial adaptation to plants. Nat Genet 50:138–150

Ludington WB (2022) Higher-order microbiome interactions and how to find them. Trends Microbiol 30(7):618–621

Mapelli F, Mengoni A, Riva V, Borin S (2023) Bacterial culturing is crucial to boost sustainable agriculture. Trends Microbiol 31(1):1–4

Mendes R, Garbeva P, Raaijmakers JM (2013) The rhizosphere microbiome: significance of plant beneficial, plant pathogenic, and human pathogenic microorganisms. FEMS Microb Rev 37(5):634–663

Nelson MS, Sadowsky MJ (2015) Secretion systems and signal exchange between nitrogen-fixing rhizobia and legumes. Front Plant Sci 6:491

Nicoud Q, Barrière Q, Busset N, Dendene S, Travin D, Bourge M et al (2021) Sinorhizobium meliloti functions required for resistance to antimicrobial NCR peptides and bacteroid differentiation. MBio 12(4):10–1128

Oyserman BO, Flores SS, Griffioen T, Pan X, van der Wijk E, Pronk L et al (2022) Disentangling the genetic basis of rhizosphere microbiome assembly in tomato. Nat Commun 13(1):3328

Price PA, Tanner HR, Dillon BA, Shabab M, Walker GC, Griffitts JS (2015) Rhizobial peptidase HrrP cleaves host-encoded signaling peptides and mediates symbiotic compatibility. PNAS 112(49):201417797

Raaijmakers JM, Kiers ET (2022) Rewilding plant microbiomes. Science 378(6620):599–600

Riccardi C, Passeri I, Cangioli L, Fagorzi C, Fondi M, Mengoni A (2023) Crossing bacterial genomic features and methylation patterns with MeStudio : an Epigenomic analysis tool. Int J Mol Sci 24(1):159

Riley AB, Grillo MA, Epstein B, Tiffin P, Heath KD (2023) Discordant population structure among rhizobium divided genomes and their legume hosts. Mol Ecol 32(10):2646–2659

Riva V, Mapelli F, Bagnasco A, Mengoni A, Borin S (2022) A meta-analysis approach to defining the culturable core of plant endophytic bacterial communities. Appl Environ Microbiol 88(6):e02537–e02521

Sessitsch A, Pfaffenbichler N, Mitter B (2019) Microbiome applications from lab to field: facing complexity. Trends Plant Sci 24(3):194–198

Shade A, Jacques MA, Barret M (2017) Ecological patterns of seed microbiome diversity, transmission, and assembly. Curr Opin Microbiol 37:15–22

Vaccaro F, Cangioli L, Mengoni A, Fagorzi C (2022) Synthetic plant microbiota challenges in nonmodel species. Trends Microbiol 30(10):922–924

Vaccaro F, Passeri I, Ajijah N, Bettini P, Courty PE, Dębiec-Andrzejewska K et al (2024) Genotype-by-genotype interkingdom cross-talk between symbiotic nitrogen fixing *Sinorhizobium meliloti* strains and *Trichoderma* species. Microbiol Res 285:127768

Vannier N, Mesny F, Getzke F, Chesneau G, Dethier L, Ordon J et al (2023) Genome-resolved metatranscriptomics reveals conserved root colonization determinants in a synthetic microbiota. Nat Commun 14(1):8274

Wagner MR (2021) Prioritizing host phenotype to understand microbiome heritability in plants. New Phytol 232:502–509

Wang X, Lv S, Liu T, Wei J, Qu S, Lu Y et al (2020) CRISPR/Cas9 genome editing shows the important role of AZC_2928 gene in nitrogen-fixing bacteria of plants. Funct Integr Genomics 20:657–668

Wang L, Xiao Y, Wei X, Pan J, Duanmu D (2021) Highly efficient CRISPR-mediated base editing in *Sinorhizobium meliloti*. Front Microbiol 12:686008

Concluding Section

Ayansina Segun Ayangbenro and Olubukola Oluranti Babalola

The nexus between sustainable farming practices and technological innovation is more important than ever in today's world. The microbial community beneath our feet provides potent answers to the unprecedented challenges facing global food supply, including urbanization, soil degradation and climate change. Amidst these challenges, microorganisms play a crucial role in shaping agricultural ecosystems. The various functions of these often overlooked microbial community in agriculture have been discussed in this book, and it has become clear that these organisms have the capacity to transform farming practices and help create a more sustainable future. Several important themes are revealed by combining state-of-the-art research from several areas of microbial ecology and plant-microbe interactions. The main takeaways from these investigations are compiled in this final chapter, which highlights the important roles that microbial interventions and technology advancements will play in promoting sustainable environmental practices and food production in the future.

A. S. Ayangbenro
Food Security and Safety Focus Area, Faculty of Natural and Agricultural Sciences, North-West University, Mmabatho, South Africa

School of Agricultural Sciences, Faculty of Agriculture and Natural Sciences , University of Mpumalanga, Mbombela, South Africa
e-mail: ayansina.ayangbenro@nwu.ac.za

O. O. Babalola
Food Security and Safety Focus Area, Faculty of Natural and Agricultural Sciences, North-West University, Mmabatho, South Africa

Department of Life Sciences, Imperial College London, Buckhurst Road, Ascot, Berkshire, United Kingdom
e-mail: olubukola.babalola@nwu.ac.za

O. O. Babalola, A. Segun Ayangbenro (eds.), *Microbial Allies*,
https://doi.org/10.1007/978-3-031-90530-8

Ensuring food safety is a daunting task due to the increasing interconnectivity in food supply chain networks, which are made worse by urbanization and globalization. As food trade networks develop it becomes challenging to determine the source of contamination and the source of foods. Additionally, climate change poses other threats to food security and safety by modifying the agriculture terrain, frequenting severe weather conditions and improving conditions of pathogenic microorganisms. Such changes also pose certain other questions related to the integrity of food and the frequency of contaminants or diseases in the food chain.

These difficulties have led to the development of several distinct techniques to maintain optimal yields in agricultural industry and keep food safe from contamination. Such options are closely connected with modern technologies, such as cold plasma systems, highly pressured processed food, and nanotechnology in food storage. In this respect, the lab techniques exhibit potential solutions for increasing food quality, reducing the application of chemical preservatives, and prolonging the shelf life. Modern food traceability systems are also turning to new solutions such as block chain and omics and big data, which can facilitate the proper tracking of food products from the farm to the consumer table. Consumers and food producers rely on these regulations and for them to be implemented internationally, then digital technologies play a significant role in addressing the issues of the global food trade.

We have seen throughout this book how the foundation of healthy agroecosystems is made up of complex intricate relationships between microorganisms and plants. The book highlights the important role of these often-overlooked microorganisms in maintaining the delicate balance required for sustainable agriculture. Microbial communities, which include phosphate-solubilizing and nitrogen-fixing organisms, collaborate with plants to improve soil structure, increase crop yield, and improve nutrient cycling. Beyond just providing nourishment, these beneficial interactions help plants resist abiotic challenges such as drought, and act as biocontrol agents against plant pathogens.

An untapped resource for agricultural innovation is the remarkable diversity of microorganisms found in soil and plants. Microbial communities offer a multitude of possible ecosystem services, ranging from industrial enzymes to biopesticides and biofertilizers. They shaped the ecosystem through interactions such as competition and symbiosis. To fully realize this promise, though, more funding must be allocated to fundamental studies that will clarify the intricate dynamics of microbial ecosystems and pinpoint important allies and their roles.

Microbial bioremediation with the use of microorganisms in the degradation and removal of pollutants in the environment including heavy metals, pesticides and industrial wastes is another area that was explored. This environmentally friendly approach of managing pollution not only assists in the removal of polluting elements within the agricultural soils but also promote fertility through cycling and nutrition upgrading as enhanced by microbial communities. In the use of microbial consortia, especially in the bioremediation projects, cooperation between various microbial species has been especially effective. It is therefore for this reason that the mentioned consortia are believed to offer a long-term solution to one of the major concerns of agriculture.

The diversity of microbial agents and their functions is key to inhibiting pathogens and promoting plant health. These organisms serve as biocontrol agents by reducing the virulence of pathogens. The use of microbial consortia offers synergistic effects in crop protection and plant growth promotion. Microbial consortia, comprising bacteria, fungi, or combinations of both, have demonstrated significant benefits for crop protection, plant growth promotion, and environmental health. These consortia work synergistically to improve nutrient uptake, enhance tolerance to abiotic stresses like salinity and drought, and act as biological control agents against pests and pathogens. Overall, these biocontrol agents hold significant promise in sustainable agriculture by reducing dependency on chemical pesticides, mitigating environmental harm, and addressing food security needs through enhanced crop protection and productivity.

Metagenomics and other omics such as metabolomics have revolutionised research on microbial populations and their interaction with plants. These technologies are providing possibilities of accessing relatively unknown features of plant-microbe symbioses when interfaced with bioinformatics and systems biology approaches. Using different omics as one research strategy, new chemical processes involved in these relationships can be studied, which will help promote agriculture and the health of ecosystems. While combination of omics technologies poses challenges for data integration and analysis, it opens new possibilities for radical transformation of agriculture. Researchers believe that understanding the role that microbial populations play in plant health and productivity will increase as these technologies improve. It is likely that precision agriculture and customized microbial consortia are expected to emerge more quickly as these technologies advance and become more widely available. The combination of machine learning and artificial intelligence in omics data analysis will pave way for identifying other patterns and forecast the behavior of microbial communities with higher level of precision.

Microbial interventions are of great potential in the field of agriculture but there are still many issues to be addressed. It is suggested that more attention should be paid to a number of questions; further development of large-scale production of microbial inoculants, ensuring their viability and effectiveness in different field conditions, and navigating regulatory systems. In addition, the need for tailored strategies that take into account site specific management, soil type, and plant genotype is highlighted due to the observed complexity and specificity of plant-microbe interactions. The studies described in this book focus on relationships between microbial communities, plant production, and soil quality. Subsequently, a system's approach, which will have to adopt these components, will be needed for all agricultural initiatives in the future. This could entail actions like:

- Minimizing tillage and chemical inputs to preserve soil structure and microbial habitats
- Implementing precision agriculture techniques guided by microbial indicators of soil health
- Developing crop varieties that form beneficial associations with a wide range of microbes.

As we look to the future, a number of exciting directions for study and use become apparent:

1. Designing and developing species specific and conditions specific microbial communities for attending specific crops known as 'synthetic microbial communities'.
2. Enhancing beneficial traits of microbial flora in plants through the deployment of modern bio enhancing tools such as CRISPR.
3. Dependent on microbial allies to design plants and agronomic practices that are improved for new climatic change termed as "climate-smart agriculture".
4. Investigating microbial approaches to improve the sustainability of resource use and waste management in agricultural systems.
5. To optimise farming practices, the microbial approaches are integrated with innovations in artificial intelligence, remote sensing and precision farming.

This book's investigation on plant-microbe interactions opens up a world of unrealized possibilities for sustainable agriculture. The following illustrates ways through which people can enhance food security, encourage environmental sustainability and transform farming practices in conjunction with nature's unseen partners. It will require a long term commitment across many disciplines, through research and development funding and a commitment to translating science into practical solutions for farmers worldwide to realise this potential.

Microbial allies offer a glimmer of hope as the world faces the daunting task of feeding an ever increasing population with limited resources and a changing climate. By harnessing the potential of these microbial allies, we can design resilient, productive, and sustainable agricultural systems for feeding the world for many generations to come.

Index

O. O. Babalola, A. S. Ayangbenro (eds.), *Microbial Allies*,
https://doi.org/10.1007/978-3-031-90530-8

The manufacturer's authorised representative in the EU is Springer Nature Customer Service Centre GmbH, Europaplatz 3, 69115 Heidelberg, Germany. If you have any concerns regarding our products, please contact ProductSafety@springernature.com

Printed and bound by CPI Group (UK) Ltd, Croydon, CR0 4YY

30/09/2026

02225590-0014